Optimising poultry flock health

It is widely recognised that agriculture is a significant contributor to global warming and climate change. Agriculture needs to reduce its environmental impact and adapt to current climate change whilst still feeding a growing population, i.e. become more 'climate-smart'. Burleigh Dodds Science Publishing is playing its part in achieving this by bringing together key research on making the production of the world's most important crops and livestock products more sustainable.

Based on extensive research, our publications specifically target the challenge of climate-smart agriculture. In this way we are using 'smart publishing' to help achieve climate-smart agriculture.

Burleigh Dodds Science Publishing is an independent and innovative publisher delivering high quality customer-focused agricultural science content in both print and online formats for the academic and research communities. Our aim is to build a foundation of knowledge on which researchers can build to meet the challenge of climate-smart agriculture.

For more information about Burleigh Dodds Science Publishing simply call us on +44 (0) 1223 839365, email info@bdspublishing.com or alternatively please visit our website at www.bdspublishing.com.

Related titles:

Improving gut health in poultry
Print (ISBN 978-1-78676-304-4); Online (ISBN 978-1-78676-306-8, 978-1-78676-307-5)

Achieving sustainable production of poultry meat Volume 1: Safety, quality and sustainability
Print (ISBN 978-1-78676-064-7); Online (ISBN 978-1-78676-066-1, 978-1-78676-067-8)

Achieving sustainable production of poultry meat Volume 3: Health and welfare
Print (ISBN 978-1-78676-072-2); Online (ISBN 978-1-78676-075-3, 978-1-78676-074-6)

Chapters are available individually from our online bookshop:
https://shop.bdspublishing.com

climate•SMART•publishing
IN AGRICULTURAL SCIENCE

BURLEIGH DODDS SERIES IN AGRICULTURAL SCIENCE
NUMBER 119

Optimising poultry flock health

Edited by Professor Sjaak de Wit, Royal GD and University of Utrecht, The Netherlands

burleigh dodds
SCIENCE PUBLISHING

Published by Burleigh Dodds Science Publishing Limited
82 High Street, Sawston, Cambridge CB22 3HJ, UK
www.bdspublishing.com

Burleigh Dodds Science Publishing, 1518 Walnut Street, Suite 900, Philadelphia, PA 19102-3406, USA

First published 2022 by Burleigh Dodds Science Publishing Limited

Library of Congress Control Number: 2022934575

British Library Cataloguing in Publication Data
A catalogue record for this book is available from the British Library

ISBN 978-1-78676-887-2 (print)
ISBN 978-1-78676-890-2 (PDF)
ISBN 978-1-78676-889-6 (ePub)
ISSN 2059-6936 (print)
ISSN 2059-6944 (online)

DOI 10.19103/AS.2022.0104

Typeset by Deanta Global Publishing Services, Dublin, Ireland

Contents

Series list

Title	Series number
Achieving sustainable cultivation of maize - Vol 1 From improved varieties to local applications Edited by: Dr Dave Watson, CGIAR Maize Research Program Manager, CIMMYT, Mexico	001
Achieving sustainable cultivation of maize - Vol 2 Cultivation techniques, pest and disease control Edited by: Dr Dave Watson, CGIAR Maize Research Program Manager, CIMMYT, Mexico	002
Achieving sustainable cultivation of rice - Vol 1 Breeding for higher yield and quality Edited by: Prof. Takuji Sasaki, Tokyo University of Agriculture, Japan	003
Achieving sustainable cultivation of rice - Vol 2 Cultivation, pest and disease management Edited by: Prof. Takuji Sasaki, Tokyo University of Agriculture, Japan	004
Achieving sustainable cultivation of wheat - Vol 1 Breeding, quality traits, pests and diseases Edited by: Prof. Peter Langridge, The University of Adelaide, Australia	005
Achieving sustainable cultivation of wheat - Vol 2 Cultivation techniques Edited by: Prof. Peter Langridge, The University of Adelaide, Australia	006
Achieving sustainable cultivation of tomatoes Edited by: Dr Autar Mattoo, USDA-ARS, USA and Prof. Avtar Handa, Purdue University, USA	007
Achieving sustainable production of milk - Vol 1 Milk composition, genetics and breeding Edited by: Dr Nico van Belzen, International Dairy Federation (IDF), Belgium	008
Achieving sustainable production of milk - Vol 2 Safety, quality and sustainability Edited by: Dr Nico van Belzen, International Dairy Federation (IDF), Belgium	009
Achieving sustainable production of milk - Vol 3 Dairy herd management and welfare Edited by: Prof. John Webster, University of Bristol, UK	010
Ensuring safety and quality in the production of beef - Vol 1 Safety Edited by: Prof. Gary Acuff, Texas A&M University, USA and Prof. James Dickson, Iowa State University, USA	011
Ensuring safety and quality in the production of beef - Vol 2 Quality Edited by: Prof. Michael Dikeman, Kansas State University, USA	012
Achieving sustainable production of poultry meat - Vol 1 Safety, quality and sustainability Edited by: Prof. Steven C. Ricke, University of Arkansas, USA	013
Achieving sustainable production of poultry meat - Vol 2 Breeding and nutrition Edited by: Prof. Todd Applegate, University of Georgia, USA	014

Introduction

Diseases remain a significant burden to poultry production and its future. Whilst it is widely recognised that vaccines have a major role in inducing protection, they can only be considered as part of the solution to this growing problem. Recent research efforts have focused on developing effective preventative measures that reduce the risk of disease outbreaks in flocks.

This volume reviews the wealth of recent research on the mechanisms of transmission for infectious diseases and how our understanding of this transmission can be used to improve poultry flock health. The book is split into three parts: Part 1 chapters focus on understanding infectious diseases in poultry, drawing attention to viral and bacterial disease transmission and epidemiology as well as parasite infections. Chapters in Part 2 review the ways of preventing diseases, such as improving biosecurity, using nutritional strategies to boost immune response and developments in vaccines to protect against disease. Part 3 examines optimising health at differing stages in poultry production, including chapters that focus specifically on chick health, poultry broilers, layers and breeder birds.

Part 1 Understanding infectious diseases in poultry

The first chapter of Part 1 reviews understanding the molecular biology of avian viruses and their role in poultry health. Chapter 1 first examines the two main groups of viruses, naked (or non-enveloped) and enveloped viruses and how they can survive in poultry environments. It then moves on to discuss viral capsids and viral interaction within the host cell, focusing specifically on infectious bursal disease virus VP2 protein, avian orthoreovirus capsid protein σC and fowl aviadenovirus hexon and fiber proteins. A section on viral glycoproteins is also included, which is then followed by a review of how poultry viruses can mutate via recombination and reassortment.

Chapter 2 looks at advances in understanding bacterial diseases in poultry, focusing on challenges and perspectives. The chapter first examines how identification and various pheno- and genotyping tools can be used to control and prevent bacterial diseases. It also discusses how epidemiology can help to identify the source of infection, method of transmission and host susceptibility. The chapter moves on to review disease pathogenesis and disease prevention by vaccination, as well as how biosecurity and good farm management can help to prevent the transmission of pathogens. It concludes by highlighting potential areas for future research development.

The final chapter of Part 1 focuses on advances in understanding parasite infections of poultry. Chapter 3 reviews research on understanding and controlling three of the most widespread and economically-relevant parasites of poultry and the infections they cause: *Eimeria* species (coccidiosis), *Histomonas meleagridis* (histomonosis) and *Dermanyssus gallinae* (the poultry red mite (PRM)). In each case the chapter reviews advances in areas such as: incidence and economic significance, host-pathogen interactions, population dynamics, detection, control and prevention as well as ongoing challenges in parasite management.

Part 2 Preventing diseases in poultry

Part 2 opens with a chapter that reviews improving biosecurity in poultry flocks. Chapter 4 first highlights the various ways to reduce sources of contamination, drawing specific attention to cleaning and disinfection of poultry barns, management of equipment and vehicles and water and feed hygiene. It also examines how managing insect, mite and rodent pests, wild birds and pets to help reduce contamination which is then followed by a section on manure, litter and dead bird management as well. The chapter also looks at separating healthy birds from sources of contamination through using methods such as zoning production sites and improving hatchery and regional biosecurity. Sections on biosecurity compliance and the economics of biosecurity are also provided.

Chapter 5 considers nutritional strategies to boost immune response in poultry. The chapter specifically focuses on research work from the year 2000 to present, drawing attention to immune responsiveness and disease resistance. In order to assess immune enhancement, suppression and balance, independent sections on vitamins (i.e. carotenoids and vitamin E), minerals, such as selenium and zinc and amino acids, such as arginine and threonine are presented. The chapter also includes sections on other nutrients, specifically highlighting vitamin D, manganese and copper. A section on the ingredients on immune modulation (i.e. fats, oils, wheat, rye, insect meal, algae meal, and alternatively processed oilseeds) is also included. A section on in ovo manipulation of embryo and subsequent chick immunity is presented. Studies assessing few immune parameters versus those assessing overall immune balance in the presence of challenges are discussed.

The final chapter of Part 2 examines developments in vaccines to protect poultry against diseases. Chapter 6 first introduces the three different types of vaccines currently available: inactivated, live and recombinant vaccines. The chapter then goes on to review the viral vectors used for recombinant vaccines, such as herpesviruses, Marek's disease virus-1 and -2, infectious laryngotracheitis virus, avian poxviruses, adenovirus and Newcastle disease virus. Techniques used for generating recombinant vaccines are also highlighted, which is then followed by a discussion on gene editing of herpesvirus vectors. The chapter concludes by emphasising the importance of continuing to develop vaccines for disease prevention in poultry production.

Part 3 Optimising health at differing stages in poultry production

The first chapter of Part 3 looks at incubation and chick health. The chapter begins by discussing what happens during the incubation stage and how the quality of this process can influence the health of the chick involved. The chapter then reviews day-old chick quality and the factors that can influence the chick's quality and status, focusing specifically on breeder influences such as diseases, nutritional factors, chemical intoxication and biological intoxication. It also reviews how egg quality can affect the chick's quality, drawing attention to factors such as laying house period, egg washing, egg disinfection, egg position and storage. Biological and physical risk factors during the incubation period are highlighted. A section on processing, storage and transport of the day-old chick is also included.

The next chapter focuses on optimising the health of broilers. Chapter 8 first presents the critical infectious disease challenges faced in the top five broiler-producing countries. It discusses the current and future strategies to control these diseases, strategies include biosecurity, vaccination, surveillance, diagnostics, environmental management, nutritional interventions and genetic selection. The optimisation of broiler health comes with a better understanding and monitoring of the whole production system in each location. The data presented in this chapter indicates that intestinal and respiratory diseases are the main aspects to improve worldwide.

Chapter 9 draws attention to optimising the health of poultry layers. The chapter begins by highlighting key disease issues faced by pullets, discussing the main vertically transmitted diseases, viral diseases and bacterial diseases affecting these hens. The chapter moves on to examine the key disease issues faced by laying hens, focusing specifically on the various viral and bacterial diseases that can affect them. A section on other issues affecting layers and pullets is also provided, drawing attention to the fungal infections that can occur, how feed and water sources need to be monitored closely to prevent deficiencies and excess consumption. External and internal parasites of layers are also highlighted. The chapter also reviews surveillance strategies in laying hens to ensure good bird health before concluding with an overview of why using different approaches is important to optimise layer health.

The final chapter of the book focuses on optimising the health of broiler breeds. Chapter 10 first introduces the diseases that can occur during the rearing period, drawing attention to early chick mortality and coccidiosis, various leg problems and spinal abscesses. The chapter then moves on to review the diseases that can arise during the laying period, such as *E. coli* Peritonitis Syndrome (EPS), infectious bronchitis and other respiratory infections. It also discusses fowl pox, neoplastic diseases, worm infections, histomonosis (blackhead disease), ruptured tendons and amyloidosis. Vertical transmitted diseases are also discussed, focusing on diseases such as *Mycoplasma gallisepticum*, *Salmonella* spp., chicken infectious anemia, avian encephalomyelitis (AEV), avian reovirus (ARV) and adenovirus infections. A section on vaccination programs for broiler breeders and the role of maternal immunity is also provided, which is then followed by a review of the effects of various management practices on broiler breeder health. It concludes by highlighting the effects of diet on the health of breeders.

Part 1

Understanding infectious diseases in poultry

Chapter 1

Understanding the molecular biology of avian viruses and their role in poultry health

Alejandro Banda, Mississippi State University, USA

1 Introduction

A mature, complete, and fully infectious virus is known as virion. The simplest virions consist of two basic components:

- the viral genome (single- or double-stranded RNA or DNA);
- a capsid, which is a protein coat that functions as a shell to protect the viral genome from nuclease degradation and, within some limits, from some environmental factors such as desiccation or extreme pH conditions.

Some viruses also contain an outer coat called a viral envelope. Viruses also have receptor-binding domains (RBDs) that will recognize and attach to specific receptors located on the membrane of the target cells where viruses 'deliver' their genomes. Enveloped viruses have viral glycoproteins anchored to their envelope which have critical roles in the initiation of viral infection as attachment proteins (Maclachlan et al., 2017).

Viruses are obligate intracellular parasites. Viruses lack mechanisms for energy production and protein synthesis since they do not possess cellular

http://dx.doi.org/10.19103/AS.2022.0104.01

organelles (Maclachlan et al., 2017). Viruses have been described as 'cell hijackers' because they use host cell processes to synthesize viral proteins and replicate nucleic acids. Viruses have different delivery pathways to reach host cells. Some can disseminate efficiently through direct respiratory transmission like infectious bronchitis or Newcastle disease viruses (NDVs). Some viruses such as arboviruses or poxviruses use arthropods like mosquitoes or mites, while others can infect through the oral route such as enteric viruses. Some viruses like NDV and other paramyxoviruses contain enzymes such as RNA polymerases inside the mature viral particles' proteins that are essential for the initial stages of viral replication (Maclachlan et al., 2017). In a similar way, poxviruses (fowlpox) carry inside their particles enzymes involved in genome replication (Cann, 2016). Inside the cell, the viral genome replicates and alters protein synthesis to manufacture viral proteins through interactions between cellular and viral components (Flint, 2004). Infected cells react to the presence of viruses in various ways such as apoptosis. As a response to the viral infection, innate and adaptive immune systems elicit diverse mechanisms that play major roles in the pathogenesis of viral diseases. This chapter discusses key features of viruses and their role in understanding the characteristics of common viral diseases in poultry, as well as the ways viruses evolve.

2 Enveloped and non-enveloped viruses

There are two main groups of viruses:

- naked (or non-enveloped) viruses;
- enveloped viruses.

Non-enveloped or 'naked' viruses do not have a lipid covering, and their capsid proteins directly interact with the host cell receptors to initiate the infection. Naked viruses rupture infected cells, leading to release of viral progeny (Van Der Grein et al., 2018). Examples of non-enveloped virus families affecting poultry include: *Birnaviridae* (infectious bursal disease (IBD)), *Reoviridae* (viral arthritis), *Anelloviridae* (chicken infectious anemia), *Adenoviridae* (inclusion body hepatitis), and *Picornaviridae* (avian encephalomyelitis).

Enveloped viruses are generally large particles with an outer coat, a phospholipid bilayer that originates from the cell membrane and is transformed by the virus. Examples of enveloped viral families of significance in poultry include: *Herpesviridae* (Marek disease and infectious laryngotracheitis), *Orthomyxoviridae* (avian influenza), *Paramyxoviridae* (Avian Paramyxoviruses, including Newcastle disease), and *Coronaviridae* (infectious bronchitis).

Viral envelopes mainly consist of envelope proteins (E), membrane proteins (M), and spike proteins (S). Envelope lipids are derived from the host cell, while envelope glycoproteins are encoded by the virus (Banerjee and Mukhopadhyay, 2016). The envelope acts as an anchor for different viral glycoproteins that form the 'spikes' or 'peplomeres' (Maclachlan et al., 2017). Enveloped virus particles are formed by budding through a host cell membrane, during which the particle becomes coated with a lipid bilayer derived from the cell membrane (Cann, 2016). This is important because the cell is not destroyed during the process and some of these viruses can thus induce more persistent infections.

The presence or absence of envelope will determine the stability of viruses in the environment and their resistance to chemical agents such as disinfectants. In general, enveloped viruses are more susceptible to chemical disinfectants than non-enveloped viruses. However, organic matter, especially feces or secretions, can protect these viruses and allow them to persist longer in the environment. With some minor differences, most chemical agents can be effective for infectious bronchitis, Newcastle disease, avian influenza, and infectious laryngotracheitis since these are relatively unstable in the environment. Physical factors such as heat, extreme pH, hypertonic conditions, and dryness can also inactivate them. Enveloped viruses with a low lipid content in their envelope (*Poxviridae* and *Hepadnaviridae*) are sensitive to disinfectants but more resistant than viruses with a high lipid content (*Herpesviridae, Flaviviridae, Togaviridae, Retroviridae, and Coronaviridae*), which are the most sensitive to chemical disinfectants (Tarka and Nitsch-Osuch, 2021).

On the other hand, naked viruses are very stable, and they can persist in poultry houses even when thorough cleaning and disinfection procedures were carried out. Infectious bursal disease virus (IBDV) is an example of a very hardy virus because it can survive in the environment for up to 122 days after removing infected birds from a poultry house (Benton et al., 1967). This virus can even resist cooking temperatures of over 70°C (Mandeville et al., 2000). Reoviruses can survive for at least 10 days on feathers, wood shavings, egg shells, and in feed but for at least 10 weeks in drinking water with little loss of infectivity (Savage and Jones, 2003). Non-enveloped viruses with strong hydrophilic properties (*Picornaviridae* and *Parvoviridae*) are the most resistant to chemical disinfectants in comparison with those with reduced hydrophilic properties (*Reoviridae, Rotaviridae, Adenoviridae, and Caliciviridae*) (Tarka and Nitsch-Osuch, 2021).

3 Viral capsids and viral interaction with the host cell

Viral capsids are formed as single or double protein shells and consist of only one or a few structural protein species with many copies per viral particle. These multiple protein copies must self-assemble to form a continuous three-dimensional capsid structure. These protein shells are tough and elastic. There are two kinds of capsid architectures (Cann, 2016; Maclachlan et al., 2017; Mateu, 2013):

- helical;
- icosahedral.

To form a viral capsid with helical symmetry, the protein subunits self-assemble into a helical array surrounding the nucleic acid. Such nucleocapsids may form either rigid and highly elongated rods or flexible filaments in enveloped viruses (Cann, 2016). Examples of viruses with helical symmetry are infectious bronchitis virus (IBV), avian paramyxoviruses, including NDV, and avian influenza (AI) virus.

In the icosahedral capsid, the protein subunits are arranged in the form of a hollow quasi-spherical structure enclosing the genome within, consisting of 20 triangular faces and 12 vertices. Examples of viruses with icosahedral symmetry are adenoviruses (inclusion body hepatitis, hemorrhagic enteritis of turkeys, and egg drop syndrome), herpesviruses (Marek's disease, infectious laryngotracheitis, and duck enteritis virus).

Capsid proteins are extremely important for non-enveloped viruses because they are responsible for the first interaction with the host cell and delivery of the genome in a form in which it can interact with the cell. The interaction between virus and cell is carried out through the binding of a specific virus-attachment protein to a cellular receptor molecule.

3.1 Infectious bursal disease virus VP2 protein

An example of a capsid protein is the IBDV VP2 protein. IBD is an important immunosuppressive disease in young birds, producing severe atrophy with consequent immunosuppression, an increase in mortality, and impairment of growth performance (Eterradossi and Saif, 2020). The genome of this virus codes for five proteins (VP1–VP5). VP2 is the major structural protein that builds the viral capsid and contains the antigenic domains (Lee et al., 2006). There are remarkable variations among isolates or subtypes commonly known as the hypervariable region (between amino acids 204 and 344) in VP2. This region seems to be responsible for the interaction with cellular receptors (Van Loon et al., 2002).

The VP2 protein plays a critical role in the pathogenesis of IBD. VP2 and the nonstructural protein VP5 induce programmed cell death (Huang et al., 2021; Qin and Zheng, 2017; Rodriguez-Lecompte et al., 2005; Vasconcelos and Lam, 1994). Apoptosis contributes to the depletion of lymphocytes. In addition to the rapid loss of B cells in the bursa (Fig. 1), a high level of apoptosis is found in peripheral blood lymphocytes during IBDV infection. Some level of apoptosis

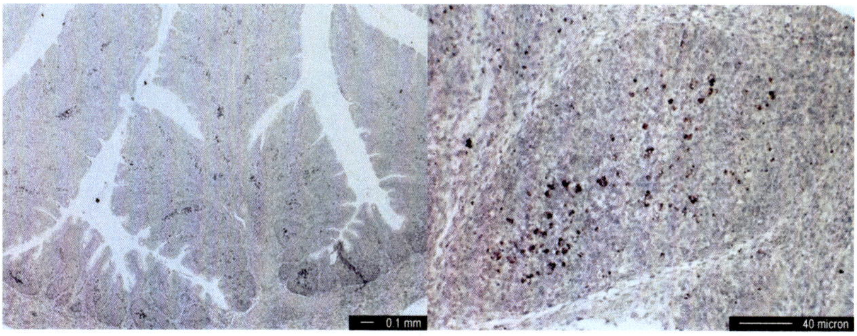

Figure 1 Bursa, commercial broilers inoculated with a Delaware E strain. Histological preparation with riboprobe in situ hybridization (ISH) targeting infectious bursal disease virus (IBDV) VP2 protein. (Left) Severe bursal atrophy with lymphoid depletion and interfollicular fibroplasia is appreciated. IBDV-infected lymphoid cells (ISH (+) in brown are located mainly in follicle centers (120× magnification). (Right) Detail of bursal follicle, immature IBDV-infected lymphoid cells, ISH (+) appear in brown in the center of the follicle.

has been observed in birds vaccinated with IBDV intermediate vaccine strains (Killian et al., 2017).

The protein VP2 has been seen as a suitable candidate for the development of recombinant vaccines. VP2 can be used to elicit a significant protective effect sufficiently comparable to that induced by an inactivated vaccine (Martinez-Torrecuadrada et al., 2003). Recombinant vaccine products using turkey herpesvirus (HVT) as a vector have been developed and have been very effective in controlling IBD even in the presence of very virulent IBDV strains (vvIBDV) (Gelb et al., 2016; Parker et al., 2014; Perozo et al., 2009; Sedeik et al., 2019). New research on vaccine development to control IBDV is based on subunit vaccines, virus-like particles, and DNA vaccines that also target VP2 to induce protection (Jackwood, 2017).

The study of IBDV VP2 has also helped in the development of efficient diagnostic tools. There are several types of IBDV, and classification or typing of strains by serological methods is difficult and lengthy and with limited discriminating power to detect minor antigenic differences between strains. Molecular methods such as reverse transcriptase-polymerase chain reaction (RT-PCR), and nucleotide sequence analysis assay have been used to identify or genotype IBDV strains. Most researchers have focused on the variable sequence region of the VP2 gene, known to encode one or more neutralizing epitopes of the virus (Jackwood, 2004; Wu et al., 2007).

3.2 Avian orthoreovirus capsid protein σC

Avian reoviruses are double-stranded, non-enveloped viruses with a segmented RNA genome. These viruses are the cause of viral arthritis/tenosynovitis in chickens and turkeys. Clinical signs of this disease include swelling of the hock joints accompanied by lesions in the gastrocnemius and digital flexor tendons causing lameness, reluctance to move (Fig. 2), and other less frequent conditions such as hydropericardium, myocarditis, enteritis, encephalitis, and bursal atrophy. Reoviruses have been also associated with enteric and respiratory disease, malabsorption, and stunting syndromes (Jones, 2000). Although not all reoviruses are pathogenic, reoviral infections may be associated with poor weight gain, increased feed conversion ratios, and rejections at the processing plant, resulting in economic losses.

Attenuated vaccines including the strain S1133 or related strains were developed in the late 1970s and 1980s and had been used in breeders and young chicks to control clinical cases of viral arthritis and tenosynovitis. However, a dramatic increase in the number of clinical cases of tenosynovitis emerged in chickens and turkeys from 2012 in the United States, France, Israel, Canada, and some countries in South America. Molecular methods identified variant reoviruses belonging to different genotypes (Sellers, 2017). Current commercial vaccines do not provide adequate protection against these new variants, requiring vaccines with homologous viruses to confer protection (Lublin et al., 2011; Troxler et al., 2013). The U.S. poultry industry is now using custom-made inactivated vaccines produced from isolates from their farms known as autogenous vaccines (Sellers, 2017).

These disease variants showed wide variations in the gene coding for the capsid protein σC (Kant et al., 2003). Avian orthoreovirus minor outer σC is the viral cell attachment protein and is the most variable protein encoded

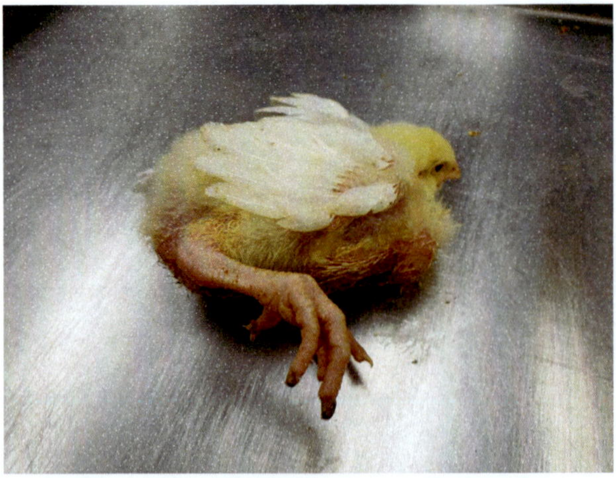

Figure 2 Viral arthritis in commercial broiler 20 days old with leg weakness, lateral recumbency, and moderate swelling of hock joints. The flock had history of lameness, leg weakness, birds with lateral recumbency, and concomitant bacterial infections. Avian reovirus was isolated from tendons after one passage in LHM cells.

by the reovirus genome. This protein contains specific epitopes which induce production of neutralizing antibodies (Ayalew et al., 2020; Benavente and Martínez-Costas, 2007). By sequencing this gene, reovirus isolates have been classified in different genotyping clusters (Kant et al., 2003; Lu et al., 2015). This gene has been targeted for viral genotyping (Kant et al., 2003; Sellers, 2017).

3.3 Fowl aviadenovirus hexon and fiber proteins

Fowl aviadenoviruses (FAdVs) are non-enveloped, double-stranded DNA viruses that display remarkable variability in their capsid proteins: hexon and fiber-1. These viruses have been associated with diseases such as inclusion body hepatitis (Fig. 3) (related to FAdV-2, -8a, -8b and -11), hepatitis-hydropericardium syndrome (associated to FAdV-4), and FAdV-induced gizzard erosion (AGE) associated to FAdV-1. Fibers and hexons are the major surface-exposed structures of the viruses and contain subgroup-specific antigenic determinants, against which neutralizing antibodies are produced (Wang et al., 2018). They are key factors in mediating infection (Wang et al., 2020). Fiber-1 provides a potential antigen for mediating a protective immune response against FAdV (Schachner et al., 2016).

As with other viruses, classical methods of serologic FAdV classification have been replaced by molecular typing of the nucleotide sequence of major immunogenic determinants, commonly based on the hexon loop-1. This gene region displays the variability required for reliable species identification with at least 12 genotypes within the currently acknowledged species (Hess, 2000).

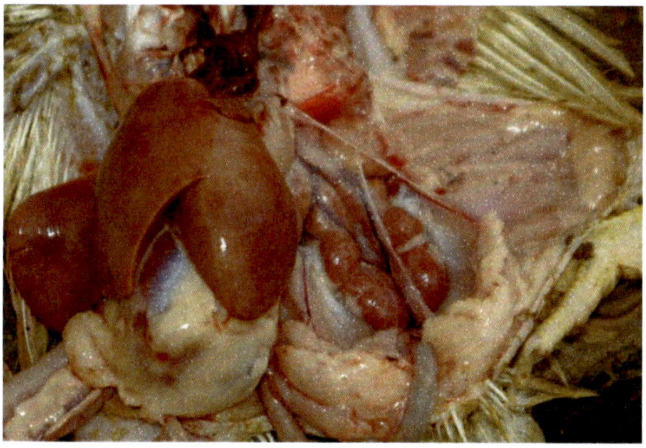

Figure 3 Inclusion body hepatitis in commercial broiler 4-week-old. Hepatomegaly with yellow to tan discoloration, with mottling and petechial hemorrhages under the capsule. *Fowl Aviadenovirus* species E, serotype 8b was isolated from the livers of these birds.

4 Viral glycoproteins

Enveloped viruses have viral glycoproteins anchored to their envelope. These form on the exterior of the virion spikes (or peplomeres) or other surface projections. They have critical roles in the initiation of viral infection as attachment proteins, facilitating the entry of the newly budded viruses into a new cell by recognizing and binding to host cell receptors. Viral glycoproteins exhibit a wide range of folds to facilitate their attachment with receptors. They may also affect membrane fusion and viral internalization. These viral glycoproteins generally have attached sugars (glycosylation) which are necessary to make the virion host compatible and for viral pathogenesis (Banerjee and Mukhopadhyay, 2016).

Glycoproteins can be recognized by the host immune system as major antigens (Cann, 2016). The infectious bronchitis S protein, AI HA protein, HN and F protein of NDV, and avian metapneumovirus F proteins are important antigens that can elicit neutralizing antibodies. These proteins have been extensively studied because of the different roles that viral glycoproteins play in viral pathogenesis and immune response. They express high variability among different viral strains, serotypes, subtypes, and genotypes. Different methods for viral genotyping are based on nucleotide or amino acid sequence analysis of these glycoproteins.

4.1 Hemagglutinin and neuraminidase of influenza viruses

Avian influenza viruses produce syndromes ranging from asymptomatic infection to respiratory disease and losses in egg production to severe, systemic disease with near 100% mortality. Gross lesions are characterized by systemic disease with necrosis and inflammation in multiple visceral organs and the nervous and cardiovascular systems, as well as in the integument (Fig. 4). AI viruses are classified in the family *Orthomyxoviridae* and genus *Influenzavirus* A (Swayne et al., 2020). There are two widely studied glycoproteins on the surface of influenza viruses as follows:

- hemagglutinin (HA);
- neuraminidase (NA).

These are the key molecules for the establishment of viral infection.

HA is the main determinant of the pathogenicity of influenza A viruses. This molecule initiates the infection by binding to cell surface receptors and by inducing membrane fusion. HA recognizes glycan structures such as sialic acid located on the cell surface. The fusion capacity of HA depends on cleavage activation by host proteases. For the virus to become infectious and continue

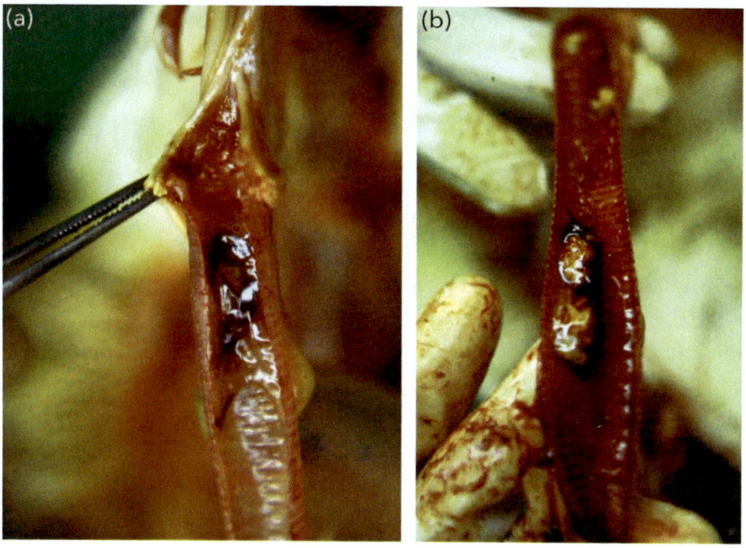

Figure 4 A and 4B highly pathogenic avian influenza Mexican H5N2 subtype. Trachea commercial broiler. Severe hemorrhagic tracheitis with mucous accumulation.

with multiple replication cycles, the cleavage of the HA protein into HA1 and HA2 subunits is essential. This cleavage occurs at a loop that protrudes from the surface of HA.

The amino acid sequence and the conformation of the loop vary with different viruses. Cleavage of HA exposes a highly conserved fusion peptide at the N-terminus of the HA2 subunit that is a prerequisite for conformational changes that trigger the fusion of the viral envelope with endosomal membranes. This releases the viral ribonucleoprotein segments into the cytoplasm and the cell nucleus where transcription and replication of the viral genome take place (Bottcher-Friebertshauser et al., 2014; Jakubcová et al., 2016).

Low virulent avian influenza (LPAI) viruses generally have only two non-consecutive basic amino acids at the carboxy-terminus of the HA1 that is only cleavable by trypsin-like proteases, mostly present in the respiratory and gastrointestinal tracts. In contrast, highly pathogenic avian influenza (HPAI) viral subtypes H5 and H7 have either multiple basic amino acids or an insertion of amino acids at the carboxy-terminal of the HA1 protein. These allow proteolytic cleavage by ubiquitous furin-like proteases present in many cells throughout the body (Senne et al., 1996). This effectively increases the cell tropism of the virus, leading to virus replication in numerous visceral organs and the nervous and cardiovascular systems, resulting in systemic disease with high mortality (Steinhauer, 1999; Swayne et al., 2020).

There are 18 known subtypes (H1–H18) of the HA protein of influenza A viruses. Of these subtypes, HA subtypes H5 and H7 are the most important for poultry because of the potential of these subtypes to become highly pathogenic. The HA subtype of the virus can be determined by serological techniques such as a hemagglutination inhibition (HI) test (Spackman, 2020). Real-time or quantitative RT-PCR is also used for more rapid detection of influenza A viral genomes and for virus subtyping (Spackman and Suarez, 2008). These techniques have been used to analyze genetic variation of influenza A viruses, especially to detect viruses of North American European and Asian H5N1 viruses (Okamatsu et al., 2016; Slomka et al., 2010; Spackman, 2020).

Neuraminidase (NA) is required to prevent aggregation at the time of budding of newly formed viral progeny from the surface of the infected cell. NA helps the virus approach the target cells by cleavage of sialic acid from respiratory tract mucins. HA also amplifies hemagglutinating activity (Shtyrya et al., 2009). The NA inhibitors (NAIs) such as oseltamivir phosphate (Tamiflu) prevent release and spread of newly formed virions from the cell surface by blocking NA anti-aggregation function (McKimm-Breschkin, 2013).

4.2 Hemagglutinin-neuraminidase and fusion proteins of avian paramyxoviruses

Newcastle disease viruses, also known as avian paramyxoviruses type 1, are members of the genus *Orthoavulavirus* in the family *Paramyxoviridae*. Over 200 avian species are susceptible. Infection with low pathogenic lentogenic viruses usually induces subclinical or mild respiratory disease with symptoms such as coughing and sneezing. Virulent viruses have three well-defined clinical forms as follows (Afonso, 2021; Miller et al., 2010; Miller and Koch, 2020):

- viscerotropic velogenic form, characterized by acute lethal infections, usually with hemorrhagic intestinal lesions (Fig. 5);
- neurotropic velogenic form, characterized by high mortality following respiratory and neurological disease, usually without lesions;
- mesogenic form with respiratory and neurological symptoms but low mortality.

Most paramyxoviruses require two distinct proteins to initiate infection. Infection is mainly mediated by a combination of the following (Harrison, 2015):

- the HA-neuraminidase (HN) protein responsible for receptor binding;
- the fusion (F) protein that drives virus-cell membrane fusion and virus entry.

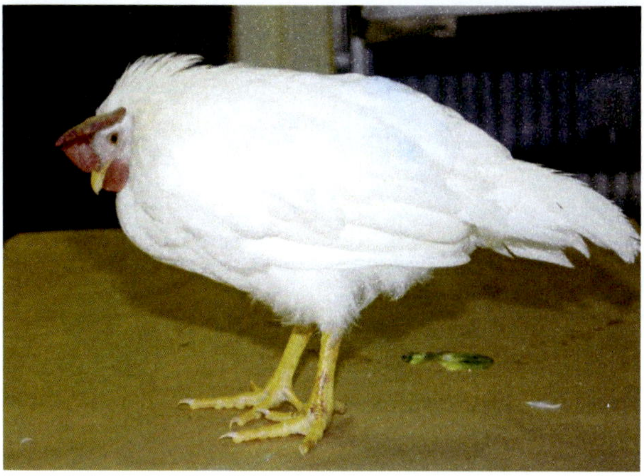

Figure 5 Virulent Newcastle disease in specific pathogen-free bird. Note ruffled cervical feathers indicating a febrile state, the bird is dehydrated with greenish diarrhea.

The paramyxovirus HN protein plays multiple roles in viral entry and egress, including binding to sialic acid receptors. It prompts the fusion (F) protein to activate membrane fusion and viral entry, cleaving sialic acid from carbohydrate chains (Yuan et al., 2011).

During infection, the F protein undergoes conformational modifications allowing the virus envelope to fuse to the host cell membranes during infection. The amino acids at the fusion protein cleavage site interact with secreted host mucosal proteases. These cleave the inactive form F0 into an active F1 and F2 active complex form, making it possible for the virions to attach, infect, and replicate in multiple tissues and organs (Afonso, 2021). The amino acid sequence at the protease cleavage site of the fusion (F) protein has been suggested as a major determinant of NDV virulence (Jardetzky and Lamb, 2014; Panda et al., 2004).

The amino acid sequence at the F protein cleavage site is different among most lentogenic, mesogenic, and velogenic NDV strains. Viruses that lack phenylalanine, and one or more of the necessary basic amino acids at the cleavage site, are cleaved by extracellular proteases which are restricted to specific tissues in the upper respiratory and enteric tracts. They result in low virulence lentogenic viruses. The F protein of all mesogenic and velogenic NDV strains has a multibasic cleavage site, cleaved by ubiquitous intracellular proteases, and which allows for systemic spread and increased virulence (Afonso, 2021; Panda et al., 2004). Real-time reverse-transcription PCR (RRT-PCR) methodologies have been developed to detect and pathotype NDV, targeted to sequences in the F protein gene that code for the cleavage site to allow detection of a diverse range of virulent and avirulent viruses (Fuller et al., 2009; Wise et al., 2004).

Avian metapneumoviruses (aMPV) possess three surface envelope glycoproteins as follows:

- the attachment glycoprotein (G);
- the fusion (F) protein;
- the small hydrophobic (SH) protein.

The G protein is a highly glycosylated protein and is known as the major viral attachment protein, contributing to viral pathogenesis and development of aMPV-specific immune response (Easton et al., 2004). The G-protein seems to have high heterogeneity even within subtypes, suggesting its role in virus evolution. Sequence analysis of the G glycoprotein has identified A, B, C, and D aMPV subgroups in Europe (Brown et al., 2014). Subtype C has been isolated in the United States, Europe (Toquin et al., 2006), and Asia (Wei et al., 2013) with no significant serological relationship with subtypes A and B (Rautenschlein, 2020). Deletion of the G gene in AMPV resulted in attenuation of the virus when it was inoculated in turkeys. The mutant virus induced less-severe clinical signs and a weaker immune response (Govindarajan et al., 2010). A recombinant LaSota NDV strain expressing the glycoprotein (G) of aMPV subgroup C induced partial protection against pathogenic aMPV-C challenge and complete protection against velogenic NDV challenge, indicating that other immunogenic proteins may be needed to induce a stronger protective immunity (Hu et al., 2011).

4.3 S1 protein of infectious bronchitis virus

Infectious bronchitis virus causes highly contagious disease in chickens characterized by upper respiratory tract and reproductive tract infections. Different IBV serotypes and genetic types of the virus have been identified. IBV is a gamma coronavirus with spike glycoproteins on its surface forming club-shaped projections (Jackwood, 2012). The IBV spike protein S is the largest structural protein emerging from the virion surface (Kant et al., 1992) (Fig. 6). It contains the RBD responsible for virus attachment to host cells (Promkuntod et al., 2014). The binding ability of this protein differs, and different attachment patterns can correlate with pathogenicity of the corresponding viruses (Wickramasinghe et al., 2011).

Determination of IBV serotypes traditionally has been done by two-way cross-virus neutralization (VN) testing in embryonated chicken eggs. Strain classification by HI tests has also been used, but most IBV strains must be treated with neuraminidase to develop hemagglutination activity (Swayne et al., 2020). Molecular assays now provide highly specific and sensitive results in a timely manner. Most RT-PCR methodologies target the hypervariable regions of the spike glycoprotein of IBV. Amplification is followed by nucleic acid sequencing to identify the type of the virus (Kingham et al., 2000; Jackwood et al., 1997;

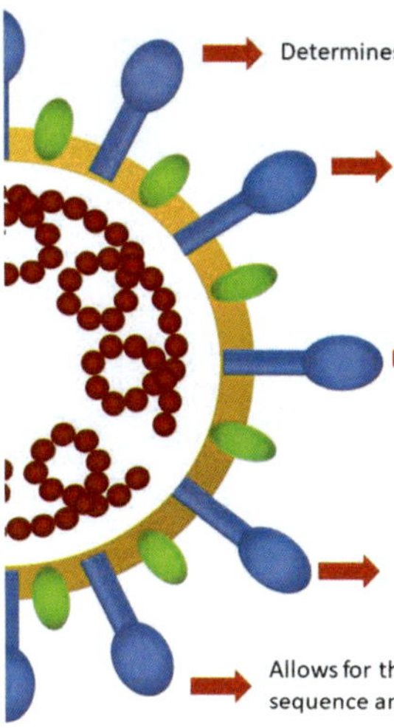

Figure 6 Activities of infectious bronchitis virus S1 protein.

Keeler et al., 1998). Type-specific molecular methods have been also developed. Cavanagh et al. (1999) used a multiplex RT-PCR for simultaneous detection of serotypes Massachusetts, D274, and 793/B in RNA extracted from mouth swabs of broilers (Cavanagh et al., 1999). A more recent approach using type-specific quantitative real-time RT-PCR assays has been used to differentiate different serotypes such as Arkansas, Massachusetts, Delaware/Georgia 98, Georgia 07, and Georgia 08 (Mo et al., 2020).

4.4 Infectious laryngotracheitis glycoproteins

Infectious laryngotracheitis (ILT) is an upper respiratory disease of chickens, pheasants, turkeys, and peafowl caused by the infectious laryngotracheitis virus (ILTV). The disease is characterized by conjunctivitis, nasal discharge, dyspnea, and lethargy. Severe symptoms include hemorrhagic tracheitis, gasping, coughing, and expectoration of bloody mucus (Garcia and Zavala, 2019).

ILTV encodes glycoprotein G (gG) conserved in most members of the subfamily *Alphaherpesvirinae*. gG functions as a soluble broad-spectrum chemokine-binding protein blocking the cell-mediated immunity of the host. A mutant gG-deficient ILTV has been shown to induce significantly milder clinical signs than wild type ILTV, though with increased inflammation in the tracheal mucosa. These findings suggest that gG plays a role in influencing inflammatory response and controlling infection (Devlin et al., 2006). Glycoprotein G deletion mutants have been considered as vaccines for ILT (Korsa et al., 2018).

Other important ILTV glycoproteins are glycoproteins B (gB) and D (gD). Glycoprotein gB is involved in virus penetration, as well as eliciting neutralizing antibodies and cell-mediated immune responses (Poulsen and Keeler, 1997). Glycoprotein gD functions as a receptor for virus binding to susceptible cells and also elicits protective immune responses (Das and Kumar, 2018). Both are also important immunogens and have been used as inserts of recombinant vaccines using fowl pox virus (FPV) (Davison et al., 2006) and HVT as vectors (Vagnozzi et al., 2012). Both recombinant vaccines can improve bird performance and reduce clinical signs of the disease, though they are not as effective as vaccination with the chicken embryo origin (CEO) vaccine (Vagnozzi et al., 2012). These recombinant vaccines thus induce partial protection against ILTV (Garcia and Zavala, 2019).

5 Mutation of poultry viruses: recombination

Three factors allow viruses to evolve: successful spreading among hosts, genetic variability, and exposure to selective pressure. Genetic diversity within a viral population, defined as a measure of all the genotypes, as well as the frequency of those genotypes in the population, is the result of selection within the cell or host, mutation, and recombination. Reassortment may also allow the spread of beneficial traits after cellular coinfection. The genetic variation of viruses is dependent on different characteristics of their replication biology. These include the fidelity of the viral genome replication, frequency of co-infection, population size of both host and virus, mode of transmission, and replication rate. All these factors influence the generation of genetic variation (Fields et al., 2001). An example of a virus that undergoes rapid evolution is IBV. Its rapid evolution has complicated the understanding of its molecular epidemiology, affected the sensitivity of diagnostic assays, limited the efficacy of vaccines, and favored episodes of immune escape, significantly complicating control of the disease (Legnardi et al., 2020).

Because of their polymerases, RNA viruses frequently introduce mutations. Mutation is the molecular synthesis or modification of a descendant gene that differs from the parental gene. This process can include deletions, insertions, or mismatches (substitutions). Viruses exhibit the broadest ranges of mutation rates among known genetic entities. The number of mutations occurring per unit of time is determined mainly by polymerase fidelity and proofreading capabilities. RNA polymerases and reverse transcriptases lack proofreading mechanisms that result in much higher mismatches in the copying event. However, coronaviruses (including IBV) have a nonstructural protein named nsp14 with exonucleolytic activity that is essential for replication proofreading activity, a property lacking in other RNA viruses that improves its fidelity replication and plays an essential role in nidoviral evolution and genome expansion (Tahir, 2021). DNA viruses that either rely on a host replication apparatus for genome duplication or that encode their own polymerase have rates of mutation significantly lower than those of RNA viruses (Fields et al., 2001). Because of their high mutation rates, RNA viruses exist as a genetically diverse population of related virus particles within a host.

The IBV S glycoprotein is involved in virus-cell membrane fusion and entry into the host cell, and it contains epitopes that induce neutralizing antibodies. Variability in the S1 gene is responsible for the multiple different genetic and antigenic types of the virus. Multiple serotypes of IBV exist that do not cross-neutralize (Jackwood and De Wit, 2020). IBV thus creates genetic diversity through rapid replication and large population sizes coupled with a high mutation rate due to a high error rate and limited proofreading capability of the viral RNA-dependent RNA polymerase (RdRP) causing insertions and deletions. Recombination increases diversity by recombining existing strains or variants and generates new IBV types with different antigenic properties or tropism. (Jackwood et al., 2012).

Recombination is the process whereby genetic material is exchanged, resulting in the formation of descendant nucleic acid composed of material from two or more parental viruses. Recombination between viruses can occur only during simultaneous co-occupation (mixed infection) of a single host cell. This introduction of foreign genetic material into a viral genome will increase the population's genetic variation and allow novel combinations of genes that can be subjected to natural selection (Fields et al., 2001).

The most widely accepted mechanism of RNA recombination is 'copy choice' recombination (Figs 7A–C). In this process, either an RdRP in most RNA viruses or a reverse transcriptase (RT) in retroviruses switches from one RNA molecule (the donor template) to another (the acceptor template) during synthesis while remaining bound to the nascent nucleic acid chain, thereby generating an RNA molecule with mixed ancestry (Simon-Loriere and Holmes, 2011).

5.1 Recombination in infectious bronchitis virus variants and turkey coronavirus

In IBV, the areas within the genome that have the highest occurrence of recombination are located in regions of the genome that code for nonstructural proteins and the structural spike glycoprotein (Thor et al., 2011). Because the S glycoprotein gene is involved in host cell attachment and contains viral neutralizing epitopes, recombination in the S glycoprotein gene can result in the emergence of new strains or serotypes of the virus as well as new viruses capable of causing disease in other host species (Jackwood et al., 2012).

One example of the emergence of a new serotype due to recombination is the emergence of Delaware 072 (DE072) strain which has a D1466-like sequence in the S1 and S2 genes. D1466 and D212 are strains of the D1466 serotype from the Netherlands (Lee and Jackwood, 2000). Two recombination events were

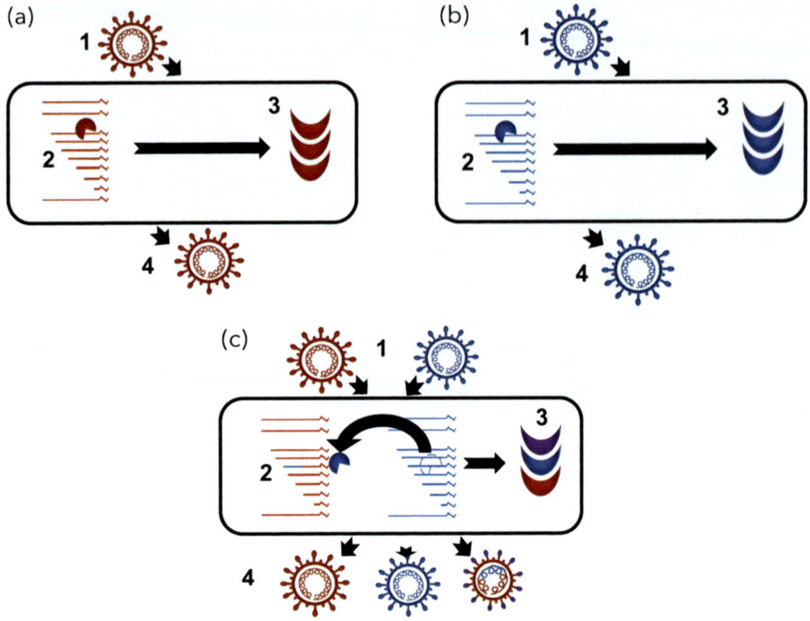

Figure 7 Recombination of infectious bronchitis virus by 'copy choice.' (a) (1) cell infection by IBV 'red serotype,' (2) genome transcription by RdRp, (3) synthesis of viral proteins, (4) viral progeny of 'red serotype' is generated; (b) (1) cell infection by IBV 'blue serotype,' (2) genome transcription by RdRp, (3) synthesis of viral proteins, (4) viral progeny of 'blue serotype' is generated; (c) (1) coinfection with both 'red' and 'blue' serotypes in the same cell, (2) both serotypes' genomes are transcribed but the RdRp from 'blue serotype' stops transcription and 'jumps' onto red genome continuing with a recombinant gene 'purple,' (3) protein synthesis of three serotypes 'red,' 'blue,' and a recombinant 'purple serotype' takes place, (4) possible outcomes of this coinfection event are the generation of viral progeny from 'red serotype,' 'blue serotype,' and a recombinant 'purple serotype.'

found between the Holland 52/Mass41-like strain and an Ark-like strain (Brooks et al., 2004). A European study reported four different recombination events involving the QX and 793B strains (Moreno et al., 2017). The emergence of Australian variants as a result of recombination has also been reported (Mardani et al., 2010). In the case of turkey coronavirus (TCoV), analysis indicated that the S gene of IBV recombined with an unknown virus, which altered its pathotype from respiratory to enteric and also modified the host specificity from chicken to turkey, resulting in the emergence of a new coronavirus (Jackwood et al., 2010).

5.2 Recombination between infectious laryngotracheitis virus vaccines and field isolates

Recombination among herpesviruses has been well documented (Loncoman et al., 2017b). This process enables genome diversification and can lead to the evolution of higher levels of viral virulence (Loncoman et al., 2017a).

Whole-genome sequencing and comparative sequence analysis have shown interspecies recombination between co-circulating Australian viruses and two European vaccines, giving rise to new virulent field strains that have become dominant. The rapid emergence of two virulent recombinants suggests that recombination between attenuated vaccines and restoration of virulence may be rare but can bring a fitness advantage, with severe consequences for poultry (Lee et al., 2012). ILTV replication and recombination are closely related and higher levels of diversity in the recombinant progeny coincide with the highest levels of viral replication (Loncoman et al., 2017a).

Recombination events between ILTV vaccines and field strains have been reported in Canada, involving two ILTV vaccine (CEO) strains as parental strains. Recombination analysis identified one ILTV isolate from Alberta as a potential parental strain for a United States origin ILTV isolate (Perez Contreras et al., 2020). The CEO vaccine viruses that circulate actively in poultry flocks in Canada added to inadequate vaccination practices favoring recombination events between vaccine and wild-type viruses. These findings represent challenges for the practical implications of ILT control using live attenuated vaccines.

5.3 The integration of reticuloendotheliosis virus into fowl pox virus

Reticuloendotheliosis virus (REV) possesses the ability to integrate proviral DNA into the genome of larger DNA viruses, such as fowl pox virus (FPV) or Marek's disease virus (MDV). Integration of REV into the FPV genome could occur during co-infection of a host with both viruses and presumably occurs by homologous recombination FPV replication (Ball, 1987). These insertions can vary from only a few nucleotides to near full-length REV provirus which is approximately 7.5 kb in length (Joshi et al., 2019). These recombination events may lead to emergence of isolates of different pathogenicity. Research indicates that FPVs with integrated REV provirus are able to induce profound but selective immunosuppressive effects on infected young chickens (Wang et al., 2006). This is an important issue in vaccine development because REV can contaminate vaccine stocks and induce congenital REV transmission among chickens (Woźniakowski et al., 2018).

6 Mutation of poultry viruses: reassortment

Reassortment is an evolutionary mechanism of segmented RNA viruses that plays an important role in virus emergence and interspecies transmission. This process is exclusive to segmented RNA viruses in which coinfection of a host cell with multiple viruses may result in the shuffling of gene segments to generate viral progeny with novel genome combinations (Vijaykrishna et al., 2015). When two or more viruses infect a single host cell, they can package each other's genome segments into a nascent virion, thereby producing hybrid progeny (Fig. 8). This reassortment generally results in segment replacement. One coinfecting virus incorporates the genome segment (or segments) of another coinfecting virus in place of its own (McDonald et al., 2016). Through reassortment, viral progeny containing genes derived from more than one parent is created, potentially conferring important fitness advantages on the progeny virus. This type of major change in the influenza A viruses is known as 'antigenic shift.'

Viruses with segmented genomes are ubiquitous in nature, and they infect a wide variety of animals. To date, 11 different segmented RNA virus families have been described in the literature (McDonald et al., 2016). Some virus members of the families *Birnaviridae* (IBD), *Orthomyxoviridae* (avian influenza), *Picobirnaviridae* (intestinal viral infections), and *Reoviridae* (viral arthritis and other syndromes) are significant in poultry health.

Genomic reassortment is central to the biology of influenza viruses and occurs frequently. Influenza viruses with epidemic and/or pandemic potential can be created in this way. As wild birds support a large diversity of AI subtypes and lineages, influenza viruses from wild birds experience frequent reassortment. Reassortment may be most frequent in the *Anseriformes* as these birds harbor the largest number of influenza subtypes. In wild bird reservoirs, reassortment is

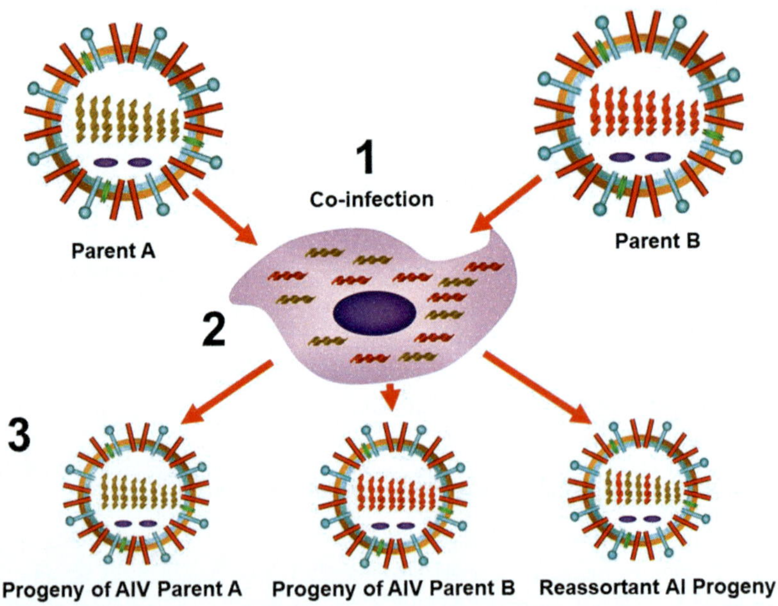

Figure 8 Ressortment of influenza viruses. (a) Coinfection by influenza viruses parent A and parent B, (b) replication and shuffling of genomic segments, (c) possible outcomes of reassortment event generation of progeny of parent A, parent B, and reassortants.

most easily observed when the viruses comprise a mosaic of Eurasian and North American origin gene segments (Wille and Holmes, 2020).

Reassortment has been pivotal to the emergence of HPAI viruses in poultry, especially the emergence of H5Nx and H7N9 viruses (Wille and Holmes, 2020). One of the most important outbreaks has been by the H5N1 influenza virus that spread directly from poultry to humans in Hong Kong in 1997. This virus denominated A/Hong Kong/156/97 H5N1 (abbreviated H5N1/97) is presumed to be a natural avian reassortant between an influenza A/Goose/Guangdong/1/96-like virus and either an H9N2 or H6N1 virus similar to those circulating in the live bird markets of Hong Kong in 1997. The H5 gene was donated by A/Goose/Guangdong/1/96-like (H5N1) virus. The N1 gene was proportioned by A/Teal/HK/W312/97 (H6N1). Finally, the internal genes were donated by ether A/Quail/HK/G1/97(H9N2)-like (H9N2) or A/Teal/HK/W312/97-like (H6N1) (Fig. 9) (Subbarao and Shaw, 2000; Webster et al., 2002). Since 2003 HPAI outbreaks caused by H5 viruses derived from the A/Goose/Guangdong/1/1996 lineage have been occurring with increasing frequency in Europe. Between 2005 and 2020, at least ten HPAI H5 incursions were identified in Europe resulting in mass mortalities among poultry and wild birds (Verhagen et al., 2021).

Another important virus generated by reassorment is influenza A H7N9 discovered in early 2013 in China that causes severe respiratory disease and mortality in humans. The data available so far indicate that this virus H7N9 was originated by a two-step reassortment process in birds and transmitted to humans through direct contact with live bird markets. This virus is a low-pathogenic avian virus but contains several molecular signatures for adaptation in mammals (Husain, 2014).

From December 2014 through to June 2015, the United States experienced an HPAI outbreak. Most cases in commercial poultry were caused by an H5N2 strain which was a reassortant with Eurasian lineage genes (including a clade 2.3.4.4 goose/Guangdong/1996 lineage hemagglutinin) and N2 genes from North American wild waterfowl LPAI viruses (Spackman et al., 2016).

6.1 Infectious bursal disease virus reassortants

Infectious bursal disease is a double-stranded RNA virus. The virus possesses a bi-segmented genome (segments A and B) within the non-enveloped icosahedral capsid. IBDV is characterized by the emergence of novel antigenic variants, known as distinct or mosaic IBDV. The occurrence of natural IBDV resultants has been reported by several groups in different parts of the world (Cui et al., 2013; Jackwood et al., 2011; Nouen et al., 2006; Pikuła et al., 2020; Soubies et al., 2017; Stoute et al., 2019; Samy et al., 2020; Wei et al., 2006).

The emergence of vvIBDV during the late 80s resulted in drastic changes to the epidemiology of IBDV with severe lesions and high mortality. Molecular

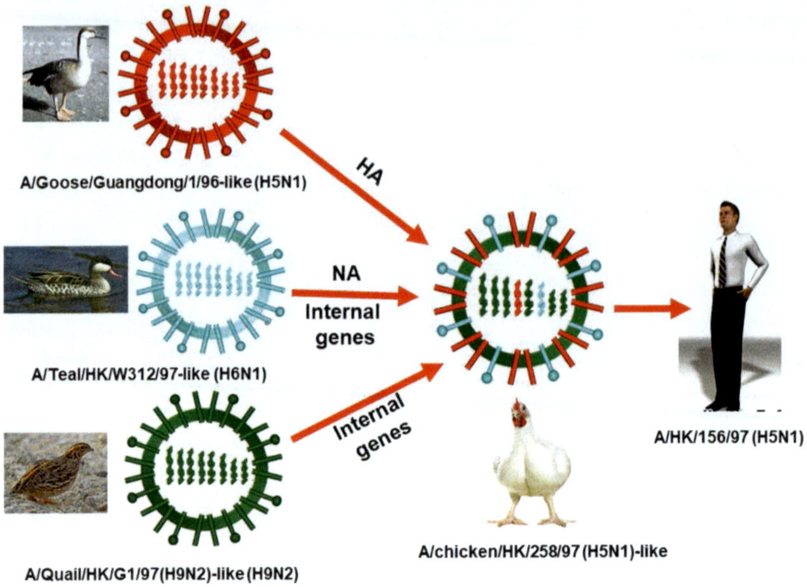

Figure 9 Origin of avian influenza virus A/Hong Kong/156/97 H5N1. This is presumed to be a natural avian reassortant between an influenza A/Goose/Guangdong/1/96-like virus and an H9N2 or H6N1 virus. The H5 gene was donated by A/Goose/Guangdong/1/96-like (H5N1) virus. The N1 gene was proportioned by A/Teal/HK/W312/97 (H6N1). Finally, the internal genes were donated by ether A/Quail/HK/G1/97(H9N2)-like (H9N2) or A/Teal/HK/W312/97-like (H6N1). The reassortant virus A/chicken/HK/258/97 (H5N1)-like underwent adaptation to infect humans as A/HK/156/97 (H5N1)-like viruses.

studies established that the emergence of these viruses was a consequence of reassortment. It is possible that the genome segment A of vvIBDV emerged some time before dissemination. However, phylogenetic analyses of segment B suggest a possible reassortment event around the mid-1980s, coinciding with the expansion of vvIBDV. It has been hypothesized that the reassortment of genome segment B initiated vvIBDV expansion in the late 1980s, possibly by enhancing the virulence of the virus synergistically with its existing genome segment A (Hon et al., 2006).

Reassortant viruses have been detected in two different locations in California. Genome segment A of these viruses was typical of vvIBDV, but genome segment B was most similar to IBDV serotype 2. These reassortants caused 20% mortality in chickens (Jackwood et al., 2011). Naturally occurring reassortant strains, with a segment A related to the vvIBDV cluster but segment B unrelated to vvIBDV, have been reported in Latin America. This strain induced lower mortality than typical vvIBDVs (Nouen et al., 2006). Novel Poland reassortant strains, containing the segment A from very virulent strains and segment B from an unidentified source, have been identified and are expanding into Finland and other European countries (Pikuła et al., 2020).

6.2 Reassortment of avian reoviruses

In addition to the accumulation of point mutations, the rapid evolution and genetic diversity of avian reoviruses are mainly acquired through genetic reassortment. The co-circulation of multiple reovirus variants from different phylogenetic lineages at the same period and same geographic location could have favored virus evolution through exchange of genetic materials between avian reoviruses of different phylogenetic lineages (Ayalew et al., 2020). Phylogeographic results indicate a widespread global intermixing of avian reovirus variants. There is co-clustering of isolates from the United States and Canada, indicating multiple cross-country introductions of variant viruses. Since Avian reoviruses are resistant to disinfection and can survive in the environment for long periods, the transfer of viruses through the importation of eggs, chickens, and processed chicken products could have taken place (Ayalew et al., 2020).

7 Conclusion

This chapter has reviewed some key features of viruses, including viral capsids and glycoproteins as well the characteristics of enveloped and non-enveloped viruses. It has also discussed the way viruses mutate via recombination and reassortment. These characteristics are fundamental to understanding viruses as well as for detection and vaccine development, as the following examples illustrate.

The protein VP2 of IBDV is the major structural protein that builds the viral capsid and is the major antigen that plays a critical role in the pathogenesis of the disease by inducing apoptosis that contributes to the depletion of lymphocytes. The study of VP2 has also helped in the development of efficient diagnostic tools. Molecular methods such as RT-PCR and nucleotide sequence analysis assays have been used to identify or genotype IBDV strains. The development of new recombinant vaccine products using HVT as vector that expresses VP2 have been developed and are now widely used. New research on vaccine development to control IBDV is based on subunit vaccines, virus-like particles, and DNA vaccines that also target VP2 to induce protection.

The viral glycoprotein HA is the main determinant of the pathogenicity of influenza A viruses as it initiates infection by binding to cell surface receptors and inducing membrane fusion. The fusion protein F of NDV allows the virus envelope to fuse to the host cell membranes to initiate viral infection. This protein is being used in developing recombinant vaccine products, and RRT-PCR methodologies targeting this gene have been developed to detect and pathotype different NDV strains.

The spike protein S of infectious bronchitis is the most significant protein for identification of virus type because it contains epitopes for neutralizing and serotype-specific antibodies. This protein contains the RBD responsible for virus attachment to host cells. This domain displays high sequence diversity among all circulating IBV strains and serotypes. Molecular assays for the detection of IBV based in this molecule are available. To date, it has not been possible to use the S protein to develop successful recombinant vaccines, but research is ongoing. The glycoprotein G (gG) of ILTV functions as a soluble broad-spectrum chemokine-binding protein blocking the cell-mediated immunity of the host. Glycoprotein G deletion mutants can be used in vaccine development, and vaccines based on these deletion mutants may be available in the future.

IBV creates genetic diversity due to a high error rate and limited proofreading capability of the viral RdRP that causes insertions and deletions. Recombination increases viral diversity by recombining existing strains or variants and creates new IBV types. These processes result in the emergence of new strains or serotypes of the virus as well as new viruses capable of causing disease in other host species. Recombination events between ILTV vaccines and field strains have been reported, enabling genome diversification and higher levels of virulence. The integration of REV into the genome of larger DNA viruses, such as FPV or MDV, can lead to isolates of different pathogenicity which can contaminate vaccine stocks and induce congenital REV transmission among chickens.

Genomic reassortment is central to the biology of influenza viruses, leading to viruses with epidemic and/or pandemic potential. The epidemic viruses such as the H5N1 influenza virus that spread directly from poultry to humans in Hong Kong in 1997, the subtype H7N9 that caused severe respiratory disease and mortality in humans in China, and the North American H5N2 subtypes that caused a severe outbreak in 2014 and 2015 were caused by reassortment events. The emergence of reassortant IBDV viruses with segments provided by vvIBDV strains has been reported in Europe, North America, and Europe.

8 Where to look for further information

8.1 Further reading

There are available several books on poultry diseases and medicine, the reader may find these sources of interest.

1. Swayne, D. (ed.) 2020. *Diseases of Poultry*, 14th Edition, Hoboken, NJ: Wiley-Blackwell.
2. Boulianne, M. (ed.) 2019. *Avian Disease Manual 7th Edition*. Jacksonville, FL: American Association of Avian Pathologists.

3 Williams, S. M. (ed.) 2016. *A Laboratory Manual for the Isolation and Characterization of Avian Pathogens*, 6th Edition. Jacksonville, FL: American Association of Avian Pathologists.
4 Abdul-Aziz, T. and Barnes, J. H. 2018. *Gross Pathology of Avian Diseases: Text and Atlas*. Jacksonville, FL: American Association of Avian Pathologists.
5 MacLachlan N. J. and Dubovi E. J. (eds) 2017. *Fenner's Veterinary Virology*, 5th Edition. San Diego CA: Academic Press.

8.2 Key journals

These indexed journals offer updated and specialized information on poultry diseases and health.

1 *Avian Diseases*. International journal published by the AAAP dedicated to publishing original basic or clinical research of the highest quality from various disciplines including microbiology, immunology, pathology and epidemiology. https://bioone.org/journals/avian-diseases.
2 *Avian Pathology*. A leading international journal for poultry disease scientists, publishing articles in the area of pathology, detection, and characterization of pathogens. https://www.tandfonline.com/toc/cavp20/current.
3 *Poultry Science*. An internationally renowned monthly journal, known as the authoritative source for a broad range of poultry information and high-caliber research.

8.3 Associations and organizations

1 American Association of Avian Pathologists (AAAP). International association whose mission is to promote scientific knowledge to enhance the health, well -being, and productivity of poultry to provide safe and abundant food for the world. https://aaap.memberclicks.net/learn-about-aaap
2 The World Veterinary Poultry Association (WVPA). A global professional association for poultry veterinarians and other avian health professionals. http://www.wvpa.net/
3 The Houghton Trust. Established with the primary objective of promoting the extension of knowledge in the field of avian pathology in all matters related to the diseases of birds, including bacteriology, epizootiology, genetics, immunology, parasitology, pathology, toxicology and virology and the dissemination of such extended knowledge to the public. http://www.houghtontrust.org.uk/
4 Western Poultry Disease Conference. https://aaap.memberclicks.net/wpdc-proceedings

8.4 Websites

1 Atlas on Avian Diseases. https://partnersah.vet.cornell.edu/avian-atlas/%23/#/.
2 The FAO picture book on avian diseases. https://www.slideshare.net/drvasuc/fao-picture-book-on-avian-diseases.
3 The Poultry Site. https://www.thepoultrysite.com/disease-guide.

9 References

Afonso, C. L. 2021. Virulence during Newcastle disease viruses cross species adaptation. *Viruses* 13(1), 1-20.

Ayalew, L. E., Ahmed, K. A., Mekuria, Z. H., Lockerbie, B., Popowich, S., Tikoo, S. K., Ojkic, D. and Gomis, S. 2020. The dynamics of molecular evolution of emerging avian reoviruses through accumulation of point mutations and genetic re-assortment. *Virus Evol.* 6(1), 1-13.

Ball, L. A. 1987. High-frequency homologous recombination in vaccinia virus DNA. *J. Virol.* 61(6), 1788-95.

Banerjee, N. and Mukhopadhyay, S. 2016. Viral glycoproteins: biological role and application in diagnosis. *VirusDisease* 27(1), 1-11.

Benavente, J. and Martínez-Costas, J. 2007. Avian reovirus: structure and biology. *Virus Res.* 123(2), 105-19.

Benton, W. J., Cover, M. S., Rosenberger, J. K. and Lake, R. S. 1967. Physicochemical properties of the infectious bursal agent (IBA). *Avian Dis.* 11(3), 438-45.

Bottcher-Friebertshauser, E., Garten, W., Matrosovich, M. and Klenk, H. D. 2014. The hemagglutinin: a determinant of pathogenicity. *Curr. Top. Microbiol. Immunol.* 385, 3-34.

Brooks, J. E., Rainer, A. C., Parr, R. L., Woolcock, P., Hoerr, F. and Collisson, E. W. 2004. Comparisons of envelope through 5B sequences of infectious bronchitis coronaviruses indicates recombination occurs in the envelope and membrane genes. *Virus Res.* 100(2), 191-8.

Brown, P. A., Lemaitre, E., Briand, F. X., Courtillon, C., Guionie, O., Allée, C., Toquin, D., Bayon-Auboyer, M. H., Jestin, V. and Eterradossi, N. 2014. Molecular comparisons of full length Metapneumovirus (MPV) genomes, including newly determined French AMPV-C and -D isolates, further supports possible subclassification within the MPV Genus. *PLoS ONE* 9(7), 1-14.

Cann, A. 2016. *Principles of Molecular Virology*, Amsterdam: Elsevier/AP, Academic Press is an imprint of Elsevier.

Cavanagh, D., Mawditt, K., Britton, P. and Naylor, C. J. 1999. Longitudinal field studies of infectious bronchitis virus and avian Pneumovirus in broilers using type-specific polymerase chain reactions. *Avian Pathol.* 28(6), 593-605.

Cui, P., Ma, S. J., Zhang, Y. G., Li, X. S., Gao, X. Y., Cui, B. A. and Chen, H. Y. 2013. Genomic sequence analysis of a new reassortant infectious bursal disease virus from commercial broiler flocks in Central China. *Arch. Virol.* 158(9), 1973-8.

Das, M. and Kumar, S. 2018. Analysis of codon usage pattern of infectious laryngotracheitis virus immunogenic glycoproteins and its biological implications. *Infect. Genet. Evol.* 62, 53-9.

Davison, S., Gingerich, E. N., Casavant, S. and Eckroade, R. J. 2006. Evaluation of the efficacy of a live fowlpox-vectored infectious laryngotracheitis/avian encephalomyelitis vaccine against ILT viral challenge. *Avian Dis.* 50(1), 50–4.

Devlin, J. M., Browning, G. F., Hartley, C. A., Kirkpatrick, N. C., Mahmoudian, A., Noormohammadi, A. H. and Gilkerson, J. R. 2006. Glycoprotein G is a virulence factor in infectious laryngotracheitis virus. *J. Gen. Virol.* 87(10), 2839–47.

Easton, A. J., Domachowske, J. B. and Rosenberg, H. F. 2004. Animal pneumoviruses: molecular genetics and pathogenesis. *Clin. Microbiol. Rev.* 17(2), 390–412.

Eterradossi, N. and Saif, Y. M. 2020. Infectious bursal disease. In: Swayne, D. E. (Ed.) *Diseases of Poultry.* 14th ed. Hoboken, NJ: Wiley-Blackwell, pp. 257–283.

Fields, B. N., Knipe, D. M., Howley, P. M. and Griffin, D. E. 2001. *Fields Virology*, Philadelphia: Lippincott Williams & Wilkins.

Flint, S. J. 2004. *Principles of Virology: Molecular Biology, Pathogenesis, and Control of Animal Viruses*, Washington, DC: ASM Press.

Fuller, C. M., Collins, M. S. and Alexander, D. J. 2009. Development of a real-time reverse-transcription PCR for the detection and simultaneous pathotyping of Newcastle disease virus isolates using a novel probe. *Arch. Virol.* 154(6), 929–37.

Garcia, M. and Zavala, G. 2019. Commercial vaccines and vaccination strategies Against infectious laryngotracheitis: what we have learned and knowledge gaps that remain. *Avian Dis.* 63(2), 325–34.

Gelb, J. Jr, Jackwood, D. J., Brannick, E. M. and Ladman, B. S. 2016. Efficacy of recombinant HVT-IBD vaccines administered to broiler chicks from a single breeder flock at 30 and 60 weeks of age. *Avian Dis.* 60(3), 603–12.

Govindarajan, D., Kim, S. H. and Samal, S. K. 2010. Contribution of the attachment G glycoprotein to pathogenicity and immunogenicity of avian metapneumovirus subgroup C. *Avian Dis.* 54(1), 59–66.

Harrison, S. C. 2015. Viral membrane fusion. *Virology* 479–480, 498–507.

Hess, M. 2000. Detection and differentiation of avian adenoviruses: a review. *Avian Pathol.* 29(3), 195–206.

Hon, C. C., Lam, T. Y., Drummond, A., Rambaut, A., Lee, Y. F., Yip, C. W., Zeng, F., Lam, P. Y., Ng, P. T. and Leung, F. C. 2006. Phylogenetic analysis reveals a correlation between the expansion of very virulent infectious bursal disease virus and reassortment of its genome segment B. *J. Virol.* 80(17), 8503–9.

Hu, H., Roth, J. P., Estevez, C. N., Zsak, L., Liu, B. and Yu, Q. 2011. Generation and evaluation of a recombinant Newcastle disease virus expressing the glycoprotein (G) of avian metapneumovirus subgroup C as a bivalent vaccine in turkeys. *Vaccine* 29(47), 8624–33.

Huang, X., Liu, W., Zhang, J., Liu, Z., Wang, M., Wang, L., Zhou, H., Jiang, Y., Cui, W., Qiao, X., Xu, Y., Li, Y. and Tang, L. 2021. Very virulent infectious bursal disease virus-induced immune injury is involved in inflammation, apoptosis, and inflammatory cytokines imbalance in the bursa of Fabricius. *Dev. Comp. Immunol.* 114, 103839 1-10.

Husain, M. 2014. Avian influenza A (H7N9) virus infection in humans: epidemiology, evolution, and pathogenesis. *Infect. Genet. Evol.* 28, 304–12.

Jackwood, D. J. 2004. Recent trends in the molecular diagnosis of infectious bursal disease viruses. *Anim. Health Res. Rev.* 5(2), 313–6.

Jackwood, D. J. 2017. Advances in vaccine research against economically important viral diseases of food animals: infectious bursal disease virus. *Vet. Microbiol.* 206, 121–5.

Jackwood, D. J., Sommer-Wagner, S. E., Crossley, B. M., Stoute, S. T., Woolcock, P. R. and Charlton, B. R. 2011. Identification and pathogenicity of a natural reassortant between a very virulent serotype 1 infectious bursal disease virus (IBDV) and a serotype 2 IBDV. *Virology* 420(2), 98–105.

Jackwood, M. W. 2012. Review of infectious bronchitis virus around the world. *Avian Dis.* 56(4), 634–41.

Jackwood, M. W., Boynton, T. O., Hilt, D. A., Mckinley, E. T., Kissinger, J. C., Paterson, A. H., Robertson, J., Lemke, C., Mccall, A. W., Williams, S. M., Jackwood, J. W. and Byrd, L. A. 2010. Emergence of a group 3 coronavirus through recombination. *Virology* 398(1), 98–108.

Jackwood, M. W. and De Wit, S. 2020. Infectious bronchitis. In: Swayne, D. E. (Ed.) *Diseases of Poultry*. 14th ed. Hoboken, NJ: Wiley-Blackwell, pp. 167–188.

Jackwood, M. W., Hall, D. and Handel, A. 2012. Molecular evolution and emergence of avian gammacoronaviruses. *Infect. Genet. Evol.* 12(6), 1305–11.

Jackwood, M. W., Yousef, N. M. and Hilt, D. A. 1997. Further development and use of a molecular serotype identification test for infectious bronchitis virus. *Avian Dis.* 41(1), 105–10.

Jakubcová, L., Hollý, J. and Varečková, E. 2016. The role of fusion activity of influenza A viruses in their biological properties. *Acta Virol.* 60(2), 121–35.

Jardetzky, T. S. and Lamb, R. A. 2014. Activation of paramyxovirus membrane fusion and virus entry. *Curr. Opin. Virol.* 5, 24–33.

Jones, R. C. 2000. Avian reovirus infections. *Rev. Sci. Tech.* 19(2), 614–25.

Joshi, L. R., Bauermann, F. V., Hain, K. S., Kutish, G. F., Armién, A. G., Lehman, C. P., Neiger, R., Afonso, C. L., Tripathy, D. N. and Diel, D. G. 2019. Detection of fowlpox virus carrying distinct genome segments of Reticuloendotheliosis virus. *Virus Res.* 260, 53–9.

Kant, A., Balk, F., Born, L., Van Roozelaar, D., Heijmans, J., Gielkens, A. and Ter Huurne, A. 2003. Classification of Dutch and German avian reoviruses by sequencing the sigma C protein. *Vet. Res.* 34(2), 203–12.

Kant, A., Koch, G., Van Roozelaar, D. J., Kusters, J. G., Poelwijk, F. A. and Van Der zeijst, B. A. 1992. Location of antigenic sites defined by neutralizing monoclonal antibodies on the S1 avian infectious bronchitis virus glycopolypeptide. *J. Gen. Virol.* 73(3), 591–6.

Keeler, C. L. Jr., Reed, K. L., Nix, W. A. and Gelb, J. Jr. 1998. Serotype identification of avian infectious bronchitis virus by RT-PCR of the peplomer (S-1) gene. *Avian Dis.* 42(2), 275–84.

Killian, M. P., Boviez, J. D., Gambarotta, M. and Lombardo, D. M. 2017. Induction of apoptosis in the bursa of Fabricius by vaccination against Gumboro disease. *Avian Pathol.* 46(5), 526–34.

Kingham, B. F., Keeler, C. L. Jr., Nix, W. A., Ladman, B. S. and Gelb, J. Jr. 2000. Identification of avian infectious bronchitis virus by direct automated cycle sequencing of the S-1 gene. *Avian Dis.* 44(2), 325–35.

Korsa, M. G., Devlin, J. M., Hartley, C. A., Browning, G. F., Coppo, M. J. C., Quinteros, J. A., Loncoman, C. A., Onasanya, A. E., Thilakarathne, D. and Diaz-Méndez, A. 2018. Determination of the minimum protective dose of a glycoprotein-G-deficient infectious laryngotracheitis virus vaccine delivered via eye-drop to week-old chickens. *PLoS ONE* 13(12), e0207611 1-12.

Lee, C. C., Ko, T. P., Chou, C. C., Yoshimura, M., Doong, S. R., Wang, M. Y. and Wang, A. H. 2006. Crystal structure of infectious bursal disease virus VP2 subviral particle at

2.6A resolution: implications in virion assembly and immunogenicity. *J. Struct. Biol.* 155(1), 74–86.

Lee, C. W. and Jackwood, M. W. 2000. Evidence of genetic diversity generated by recombination among avian coronavirus IBV. *Arch. Virol.* 145(10), 2135–48.

Lee, S. W., Markham, P. F., Coppo, M. J., Legione, A. R., Markham, J. F., Noormohammadi, A. H., Browning, G. F., Ficorilli, N., Hartley, C. A. and Devlin, J. M. 2012. Attenuated vaccines can recombine to form virulent field viruses. *Science* 337(6091), 188.

Legnardi, M., Tucciarone, C. M., Franzo, G. and Cecchinato, M. 2020. Infectious bronchitis virus evolution, diagnosis and control. *Vet. Sci.* 7(2), 1–18.

Loncoman, C. A., Hartley, C. A., Coppo, M. J. C., Vaz, P. K., Diaz-Méndez, A., Browning, G. F., García, M., Spatz, S. and Devlin, J. M. 2017a. Genetic diversity of infectious laryngotracheitis virus during in vivo coinfection parallels viral replication and arises from recombination hot spots within the genome. *Appl. Environ. Microbiol.* 83(23), 1–18.

Loncoman, C. A., Vaz, P. K., Coppo, M. J., Hartley, C. A., Morera, F. J., Browning, G. F. and Devlin, J. M. 2017b. Natural recombination in alphaherpesviruses: insights into viral evolution through full genome sequencing and sequence analysis. *Infect. Genet. Evol.* 49, 174–85.

Lu, H., Tang, Y., Dunn, P. A., Wallner-Pendleton, E. A., Lin, L. and Knoll, E. A. 2015. Isolation and molecular characterization of newly emerging avian reovirus variants and novel strains in Pennsylvania, USA, 2011–2014. *Sci. Rep.* 5, 14727 1-11.

Lublin, A., Goldenberg, D., Rosenbluth, E., Heller, E. D. and Pitcovski, J. 2011. Wide-range protection against avian reovirus conferred by vaccination with representatives of four defined genotypes. *Vaccine* 29(47), 8683–8.

Maclachlan, N. J., Dubovi, E. J., Barthold, S. W., Swayne, D. E. and Winton, J. R. 2017. *Fenner's Veterinary Virology*, Amsterdam: Elsevier/AP, Academic Press is an imprint of Elsevier.

Mandeville, W. F., Cook, F. K. and Jackwood, D. J. 2000. Heat lability of five strains of infectious bursal disease virus. *Poult. Sci.* 79(6), 838–42.

Mardani, K., Noormohammadi, A. H., Ignjatovic, J. and Browning, G. F. 2010. Naturally occurring recombination between distant strains of infectious bronchitis virus. *Arch. Virol.* 155(10), 1581–6.

Martinez-Torrecuadrada, J. L., Saubi, N., Pages-Mante, A., Caston, J. R., Espuna, E. and Casal, J. I. 2003. Structure-dependent efficacy of infectious bursal disease virus (IBDV) recombinant vaccines. *Vaccine* 21(23), 3342–50.

Mateu, M. G. 2013. Introduction: the structural basis of virus function. *Subcell. Biochem.* 68, 3–51.

McDonald, S. M., Nelson, M. I., Turner, P. E. and Patton, J. T. 2016. Reassortment in segmented RNA viruses: mechanisms and outcomes. *Nat. Rev. Microbiol.* 14(7), 448–60.

McKimm-Breschkin, J. L. 2013. Influenza neuraminidase inhibitors: antiviral action and mechanisms of resistance. *Influ. Other Respir. Viruses* 7 (Suppl. 1), 25–36.

Miller, P. J., Decanini, E. L. and Afonso, C. L. 2010. Newcastle disease: evolution of genotypes and the related diagnostic challenges. *Infect. Genet. Evol.* 10(1), 26–35.

Miller, P. J. and Koch, G. 2020. Newcastle disease. In: Swayne, D. E. (Ed.) *Diseases of Poultry*. 14th ed. Hoboken, NJ: Wiley-Blackwell, pp. 112–128.

Mo, J., Angelichio, M., Gow, L., Leathers, V. and Jackwood, M. W. 2020. Validation of specific quantitative real-time RT-PCR assay panel for Infectious Bronchitis using synthetic DNA standards and clinical specimens. *J. Virol. Methods* 276, 113773.

Moreno, A., Franzo, G., Massi, P., Tosi, G., Blanco, A., Antilles, N., Biarnes, M., Majó, N., Nofrarías, M., Dolz, R., Lelli, D., Sozzi, E., Lavazza, A. and Cecchinato, M. 2017. A novel variant of the infectious bronchitis virus resulting from recombination events in Italy and Spain. *Avian Pathol.* 46(1), 28–35.

Nouen, C. L., Rivallan, G., Toquin, D., Darlu, P., Morin, Y., Beven, V., De Boisseson, C., Cazaban, C., Comte, S., Gardin, Y. and Eterradossi, N. 2006. Very virulent infectious bursal disease virus: reduced pathogenicity in a rare natural segment-B-reassorted isolate. *J. Gen. Virol.* 87(1), 209–16.

Okamatsu, M., Hiono, T., Kida, H. and Sakoda, Y. 2016. Recent developments in the diagnosis of avian influenza. *Vet. J.* 215, 82–6.

Panda, A., Huang, Z., Elankumaran, S., Rockemann, D. D. and Samal, S. K. 2004. Role of fusion protein cleavage site in the virulence of Newcastle disease virus. *Microb. Pathog.* 36(1), 1–10.

Parker, D., De Wit, S., Houghton, H. and Prandini, F. 2014. Assessment of impact of a novel infectious bursal disease (IBD) vaccination programme in breeders on IBD humoral antibody levels through the laying period. *Vet. Rec. Open* 1(1), 1–6.

Perez Contreras, A., Van Der Meer, F., Checkley, S., Joseph, T., King, R., Ravi, M., Peters, D., Fonseca, K., Gagnon, C. A., Provost, C., Ojkic, D. and Abdul-Careem, M. F. 2020. Analysis of whole-genome sequences of infectious laryngotracheitis virus isolates from poultry flocks in Canada: evidence of recombination. *Viruses* 12(11), 1–16.

Perozo, F., Villegas, A. P., Fernandez, R., Cruz, J. and Pritchard, N. 2009. Efficacy of single dose recombinant herpesvirus of turkey infectious bursal disease virus (IBDV) vaccination against a variant IBDV strain. *Avian Dis.* 53(4), 624–8.

Pikuła, A., Śmietanka, K. and Perez, L. J. 2020. Emergence and expansion of novel pathogenic reassortant strains of infectious bursal disease virus causing acute outbreaks of the disease in Europe. *Transbound. Emerg. Dis.* 67(4), 1739–44.

Poulsen, D. J. and Keeler, C. L. Jr. 1997. Characterization of the assembly and processing of infectious laryngotracheitis virus glycoprotein B. *J. Gen. Virol.* 78(11), 2945–51.

Promkuntod, N., Van Eijndhoven, R. E., De Vrieze, G., Gröne, A. and Verheije, M. H. 2014. Mapping of the receptor-binding domain and amino acids critical for attachment in the spike protein of avian coronavirus infectious bronchitis virus. *Virology* 448, 26–32.

Qin, Y. and Zheng, S. J. 2017. Infectious bursal disease virus-host interactions: multifunctional viral proteins that perform multiple and differing jobs. *Int. J. Mol. Sci.* 18(1), 1–13.

Rautenschlein, S. 2020. Avian metapneumoviruses. In: Swayne, D. E. (Ed.) *Diseases of Poultry.* 14th ed. Hoboken, NJ: Wiley-Blackwell, pp. 135–143.

Rodriguez-Lecompte, J. C., Nino-Fong, R., Lopez, A., Frederick Markham, R. J. and Kibenge, F. S. 2005. Infectious bursal disease virus (IBDV) induces apoptosis in chicken B cells. *Comp. Immunol. Microbiol. Infect. Dis.* 28(4), 321–37.

Samy, A., Courtillon, C., Briand, F. X., Khalifa, M., Selim, A., Arafa, A. E. S., Hegazy, A., Eterradossi, N. and Soubies, S. M. 2020. Continuous circulation of an antigenically modified very virulent infectious bursal disease virus for fifteen years in Egypt. *Infect. Genet. Evol.* 78, 1–9.

Savage, C. E. and Jones, R. C. 2003. The survival of avian reoviruses on materials associated with the poultry house environment. *Avian Pathol.* 32(4), 419–25.

Schachner, A., Marek, A., Grafl, B. and Hess, M. 2016. Detailed molecular analyses of the hexon loop-1 and fibers of fowl aviadenoviruses reveal new insights into the antigenic relationship and confirm that specific genotypes are involved in field outbreaks of inclusion body hepatitis. *Vet. Microbiol.* 186, 13–20.

Sedeik, M. E., El-Shall, N. A., Awad, A. M., Abd El-Hack, M. E., Alowaimer, A. N. and Swelum, A. A. 2019. Comparative evaluation of HVT-IBD vector, immune complex, and live IBD vaccines against vvIBDV in commercial broiler chickens with high maternally derived antibodies. *Animals (Basel)* 9(3), 1–13.

Sellers, H. S. 2017. Current limitations in control of viral arthritis and tenosynovitis caused by avian reoviruses in commercial poultry. *Vet. Microbiol.* 206, 152–6.

Senne, D. A., Panigrahy, B., Kawaoka, Y., Pearson, J. E., Süss, J., Lipkind, M., Kida, H. and Webster, R. G. 1996. Survey of the hemagglutinin (HA) cleavage site sequence of H5 and H7 avian influenza viruses: amino acid sequence at the HA cleavage site as a marker of pathogenicity potential. *Avian Dis.* 40(2), 425–37.

Shtyrya, Y. A., Mochalova, L. V. and Bovin, N. V. 2009. Influenza virus neuraminidase: structure and function. *Acta Nat.* 1(2), 26–32.

Simon-Loriere, E. and Holmes, E. C. 2011. Why do RNA viruses recombine? *Nat. Rev. Microbiol.* 9(8), 617–26.

Slomka, M. J., Irvine, R. M., Pavlidis, T., Banks, J. and Brown, I. H. 2010. Role of real-time RT-PCR platform technology in the diagnosis and management of notifiable avian influenza outbreaks: experiences in Great Britain. *Avian Dis.* 54(1), 591–6.

Soubies, S. M., Courtillon, C., Briand, F. X., Queguiner-Leroux, M., Courtois, D., Amelot, M., Grousson, K., Morillon, P., Herin, J. B. and Eterradossi, N. 2017. Identification of a European interserotypic reassortant strain of infectious bursal disease virus. *Avian Pathol.* 46(1), 19–27.

Spackman, E. 2020. Avian influenza virus detection and quantitation by real-time RT-PCR. *Methods Mol. Biol.* 2123, 137–48.

Spackman, E., Pantin-Jackwood, M. J., Kapczynski, D. R., Swayne, D. E. and Suarez, D. L. 2016. H5N2 Highly Pathogenic Avian Influenza Viruses from the US 2014–2015 outbreak have an unusually long pre-clinical period in turkeys. *BMC Vet. Res.* 12(1), 260.

Spackman, E. and Suarez, D. L. 2008. Type A influenza virus detection and quantitation by real-time RT-PCR. *Methods Mol. Biol.* 436, 19–26.

Steinhauer, D. A. 1999. Role of hemagglutinin cleavage for the pathogenicity of influenza virus. *Virology* 258(1), 1–20.

Stoute, S. T., Jackwood, D. J., Crossley, B. M., Michel, L. O. and Blakey, J. R. 2019. Molecular epidemiology of endemic and very virulent infectious bursal disease virus genogroups in backyard chickens in California, 2009–2017. *J. Vet. Diagn. Invest.* 31(3), 371–7.

Subbarao, K. and Shaw, M. W. 2000. Molecular aspects of avian influenza (H5N1) viruses isolated from humans. *Rev. Med. Virol.* 10(5), 337–48.

Swayne, D. E., Suarez, D. L. and Sims, L. D. 2020. Influenza. In: Swayne, D. E. (Ed.) *Diseases of Poultry*. 14th ed. Hoboken, NJ: Wiley-Blackwell, pp. 210–256.

Tahir, M. 2021. Coronavirus genomic nsp14-ExoN, structure, role, mechanism, and potential application as a drug target. *J. Med. Virol.* 93(7), 4258–64.

Tarka, P. and Nitsch-Osuch, A. 2021. Evaluating the virucidal activity of disinfectants according to European Union standards. *Viruses* 13(4), 1–9.

Thor, S. W., Hilt, D. A., Kissinger, J. C., Paterson, A. H. and Jackwood, M. W. 2011. Recombination in avian gamma-coronavirus infectious bronchitis virus. *Viruses* 3(9), 1777–99.

Toquin, D., Guionie, O., Jestin, V., Zwingelstein, F., Allee, C. and Eterradossi, N. 2006. European and American subgroup C isolates of avian Metapneumovirus belong to different genetic lineages. *Virus Genes* 32(1), 97–103.

Troxler, S., Rigomier, P., Bilic, I., Liebhart, D., Prokofieva, I., Robineau, B. and Hess, M. 2013. Identification of a new reovirus causing substantial losses in broiler production in France, despite routine vaccination of breeders. *Vet. Rec.* 172(21), 556.

Vagnozzi, A., Zavala, G., Riblet, S. M., Mundt, A. and García, M. 2012. Protection induced by commercially available live-attenuated and recombinant viral vector vaccines against infectious laryngotracheitis virus in broiler chickens. *Avian Pathol.* 41(1), 21–31.

Van Der Grein, S. G., Defourny, K. A. Y., Slot, E. F. J. and Nolte-'t Hoen, E. N. M. 2018. Intricate relationships between naked viruses and extracellular vesicles in the crosstalk between pathogen and host. *Semin. Immunopathol.* 40(5), 491–504.

Van Loon, A. A. W. M., De Haas, N., Zeyda, I. and Mundt, E. 2002. Alteration of amino acids in VP2 of very virulent infectious bursal disease virus results in tissue culture adaptation and attenuation in chickens. *J. Gen. Virol.* 83(1), 121–9.

Vasconcelos, A. C. and Lam, K. M. 1994. Apoptosis induced by infectious bursal disease virus. *J. Gen. Virol.* 75(7), 1803–6.

Verhagen, J. H., Fouchier, R. A. M. and Lewis, N. 2021. Highly pathogenic avian influenza viruses at the wild-domestic bird interface in Europe: future directions for research and surveillance. *Viruses* 13(2), 1–34.

Vijaykrishna, D., Mukerji, R. and Smith, G. J. 2015. RNA virus reassortment: an evolutionary mechanism for host jumps and immune evasion. *PLoS Pathog.* 11(7), e1004902 1-6.

Wang, J., Meers, J., Spradbrow, P. B. and Robinson, W. F. 2006. Evaluation of immune effects of fowlpox vaccine strains and field isolates. *Vet. Microbiol.* 116(1–3), 106–19.

Wang, W., Liu, Q., Li, T., Geng, T., Chen, H., Xie, Q., Shao, H., Wan, Z., Qin, A. and Ye, J. 2020. Fiber-1, not Fiber-2, directly mediates the infection of the pathogenic Serotype 4 fowl adenovirus via its shaft and knob domains. *J. Virol.* 94(17), 1–10.

Wang, X., Tang, Q., Chu, Z., Wang, P., Luo, C., Zhang, Y., Fang, X., Qiu, L., Dang, R. and Yang, Z. 2018. Immune protection efficacy of FAdV-4 surface proteins fiber-1, fiber-2, hexon and penton base. *Virus Res.* 245, 1–6.

Webster, R. G., Guan, Y., Peiris, M., Walker, D., Krauss, S., Zhou, N. N., Govorkova, E. A., Ellis, T. M., Dyrting, K. C., Sit, T., Perez, D. R. and Shortridge, K. F. 2002. Characterization of H5N1 influenza viruses that continue to circulate in geese in southeastern China. *J. Virol.* 76(1), 118–26.

Wei, L., Zhu, S., Yan, X., Wang, J., Zhang, C., Liu, S., She, R., Hu, F., Quan, R. and Liu, J. 2013. Avian metapneumovirus subgroup C infection in chickens, China. *Emerg. Infect. Dis.* 19(7), 1092–4.

Wei, Y., Li, J., Zheng, J., Xu, H., Li, L. and Yu, L. 2006. Genetic reassortment of infectious bursal disease virus in nature. *Biochem. Biophys. Res. Commun.* 350(2), 277-87.

Wickramasinghe, I. N., De Vries, R. P., Gröne, A., De Haan, C. A. and Verheije, M. H. 2011. Binding of avian coronavirus spike proteins to host factors reflects virus tropism and pathogenicity. *J. Virol.* 85(17), 8903–12.

Wille, M. and Holmes, E. C. 2020. The ecology and evolution of influenza viruses. *Cold Spring Harb. Perspect. Med.* 10(7), 1–19.

Wise, M. G., Suarez, D. L., Seal, B. S., Pedersen, J. C., Senne, D. A., King, D. J., Kapczynski, D. R. and Spackman, E. 2004. Development of a real-time reverse-transcription PCR for detection of Newcastle disease virus RNA in clinical samples. *J. Clin. Microbiol.* 42(1), 329–38.

Woźniakowski, G., Frant, M. and Mamczur, A. 2018. Avian reticuloendotheliosis in chickens - an update on disease occurrence and clinical course. *J. Vet. Res.* 62(3), 257–60.

Wu, C. C., Rubinelli, P. and Lin, T. L. 2007. Molecular detection and differentiation of infectious bursal disease virus. *Avian Dis.* 51(2), 515–26.

Yuan, P., Swanson, K. A., Leser, G. P., Paterson, R. G., Lamb, R. A. and Jardetzky, T. S. 2011. Structure of the Newcastle disease virus hemagglutinin-neuraminidase (HN) ectodomain reveals a four-helix bundle stalk. *Proc. Natl Acad. Sci. U. S. A.* 108(36), 14920-5.

Chapter 2

Advances in understanding bacterial diseases in poultry: challenges and perspectives

Surya Paudel, Michael Hess and Claudia Hess, University of Veterinary Medicine Vienna, Austria

1 Introduction

Poultry farming holds a lion's share in global livestock production but is constantly challenged by bacterial infections due to their negative impact on animal welfare, food safety and economy. The trend of using antibiotics to constrain bacterial infections in food-producing animals is projected to increase in the future in the absence of regulatory actions (Tiseo et al., 2020). Strikingly, antimicrobial resistance has increased over time in bacterial pathogens from poultry (Nhung et al., 2017), which has already set the alarm bells ringing. Moreover, due to the fact that antimicrobial resistance is now recognized as one of the global public health threats by the World Health Organization, legislations are expected to limit the antibiotic use in poultry production, which has already come into force worldwide under the framework of antibiotic stewardship. Advances in scientific tools have contributed to a better understanding of pathogen attributes and disease pathogenesis. Based on this, substantial progress has been made in the past to improve management practices including biosecurity and surveillance programmes. Despite such improvements, vaccination against bacterial pathogens still relies largely on products developed decades ago. The introduction of autogenous vaccines in recent years reflects the increased knowledge of certain pathogens and aims to fill existing gaps in vaccination

http://dx.doi.org/10.19103/AS.2022.0104.02

programmes as the landscape of bacterial diseases has not remained stable due to changes in husbandry systems, regulatory legislations and biological complexities of pathogens themselves. The changing trend towards alternative husbandry systems with access of birds to outdoor environment can lead to widespread occurrence of varied bacteria and (re-)emergence of pathogens like *Brachyspira* species and *Campylobacter hepaticus* indicating a higher risk for infectious diseases (Zloch et al., 2018; Hess et al., 2017; Crawshaw, 2019). National surveillance and strict eradication programmes constrained the occurrence of some diseases, for instance, Pullorum disease in Europe and North America. From pathogen perspective, continuous host adaptation mechanisms might lead to phenotypic or genetic modifications and formation of novel ecological niches. Some bacteria like *Riemerella anatipestifer*, *Salmonella* Gallinarum, *Salmonella* Pullorum, *Escherichia coli*, *Mycoplasma* species, *Pasteurella multocida* or *Avibacterium paragallinarum* are known since long as causative agents of poultry diseases while others like *Gallibacterium anatis*, *Enterococcus cecorum* and *Clostridium perfringens* have a relatively short history as a pathogen. But, there are also certain bacterial species which might appear as a primary pathogen under favourable conditions as known for members of the genus *Staphylococcus*, *Streptococcus* and *Enterococcus*.

In this chapter, advances and challenges in identification and characterization/typing, epidemiology, pathogenesis, transmission and prevention strategies of bacterial pathogens in poultry medicine are discussed. Numerous research manuscripts are published on all those different subjects but only a limited number could be cited in this comprehensive book chapter due to space constraints and the overall aim to supply a more general overview valuable for readers approaching this chapter with different background.

2 Identification and typing

The success of control and prevention strategies of bacterial pathogens in the field depends on early and accurate identification from clinical cases and/or the environment. Following isolation, pheno- and genotyping tools are applied to determine the relatedness and diversity, origin and prevalence as well as transmission mapping of bacterial species (Fig. 1). Presumptive diagnosis of bacterial infections can be made based on clinical symptoms and postmortem lesions, detection of acute phase proteins or shift in lymphocyte:heterophil ratio. However, laboratory confirmations are required to correctly isolate and identify the pathogen(s) involved. Isolation of bacteria from infected organs provides an idea of the ongoing infection but results should be cautiously interpreted as some bacteria such as *E. coli*, *G. anatis*, *C. perfringens* or *Enterococcus* species are not only associated with disease conditions but are also part of the physiological microflora, already pointing towards not only heterogeneity of

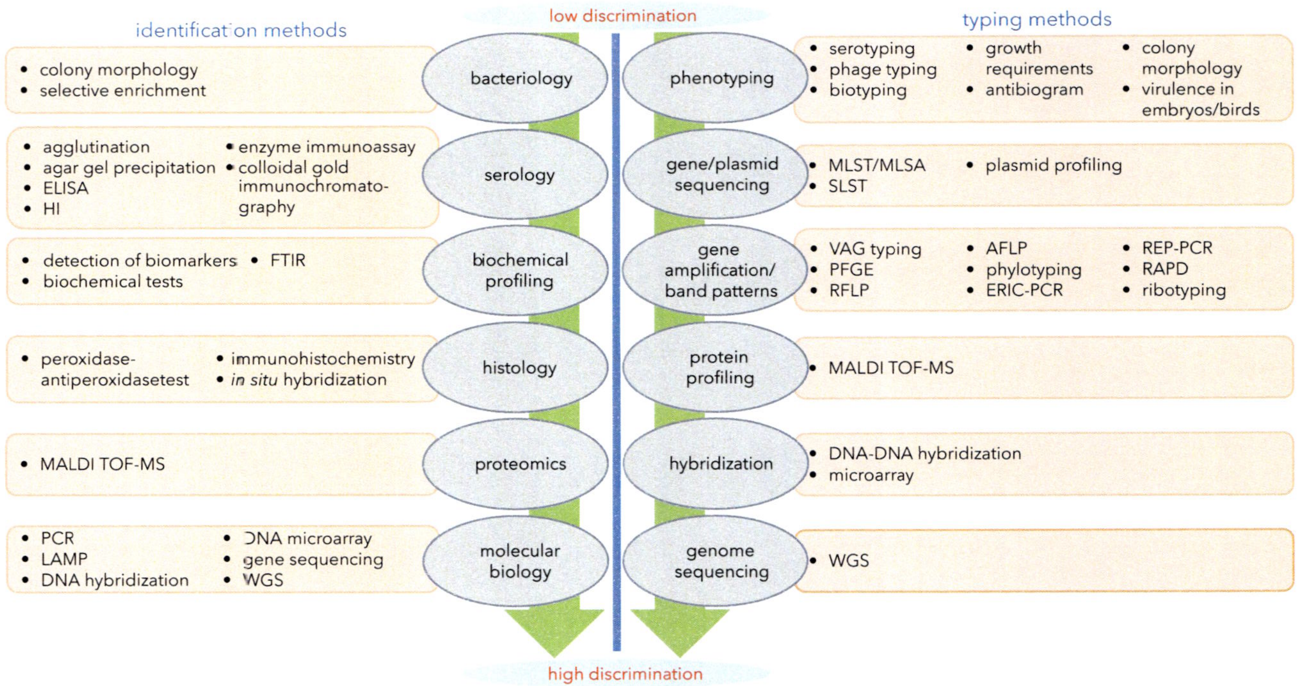

Figure 1 Commonly used identification and typing methods for bacterial pathogens in poultry diagnostic laboratories. AFLP, amplified fragment length polymorphism; ELISA, enzyme-linked immunosorbent assay; ERIC-PCR, enterobacterial repetitive intergenic consensus PCR; FTIR, Fourier transform infrared spectroscopy; HI, haemaglutination inhibition; LAMP, loop-mediated isothermal amplification; MALDI TOF-MS, matrix-assisted laser desorption ionization time of flight mass spectrometry; MLSA, multilocus sequencing analysis; MLST, multilocus sequencing typing; PCR, polymerase chain reaction; PFGE, pulsed field gel electrophoresis; RAPD, random amplified polymorphic DNA; REP-PCR, repetitive element palindromic PCR; RFLP, restriction fragment length polymorphism; SLST, single locus sequence typing; VAG, virulence-associated genes; WGS, whole-genome sequencing.

pathogens but also the host and other influences. Phenotyping relies on growth conditions, colony morphology, biochemical properties, serological reaction and functional or physiological traits (Riley, 2018). The growth behaviour of bacteria belonging to the same species might be different based on intra-species divergent nutrient requirements as seen in *A. paragallinarum* (Blackall et al., 2005). Automated systems or so-called Analytical Profile Index systems are also commercially available that produce biochemical profile indexes for bacterial identification. Susceptibility to antibiotics and determination of pathogenicity in embryos or birds categorize bacterial isolates based on their functional or physiological properties.

Phenotypic characteristics are not always consistent, thus results are often augmented with molecular tests that rely on the analysis of genomic DNA, plasmids and RNA to improve specificity of identification and to increase resolution of discrimination among bacterial isolates within a species (Riley, 2018). The use of polymerase chain reaction (PCR) techniques is quite common not only to confirm a certain pathogen but also to reveal the presence or absence of a gene or a set of genes. Based on this, PCR methods are applied to improve the knowledge on prevalence, transmission and diversity of isolates. The resolution of species differentiation varies according to the method applied. Electrophoresis based tests are used to subtype a wide variety of bacterial species. Such investigations rely on the analysis of electrophoretic band patterns generated after PCR-amplification of DNA products, commonly virulence genes and plasmid profiles. A step further in achieving more discriminative separation of bacteria belonging to the same species are methods based on the analysis of electrophoretic band patterns of whole genomic DNA (with or without the use of restriction enzymes). In this regard, methods like restriction fragment length polymorphism, amplified fragment length polymorphism genetic analysis, enterobacterial repetitive intergenic consensus PCR and pulsed field gel electrophoresis are commonly applied.

In recent years, matrix-assisted laser desorption ionization time of flight mass spectrometry (MALDI-TOF MS) was introduced for the identification and characterization of bacteria based on their protein profile spectra. The technique is very time-efficient but challenges exist as phylogenetic relationship might not harmonize with comparative genomic data and species discrimination within a genus might not be possible for certain bacteria (Alispahic et al., 2014; Lawton et al., 2018). Results of MALDI-TOF MS are highly influenced by culture conditions, the analytical approach or purity of samples, which would improve by developing robust practice guidelines for inter-laboratory reproducibility.

The promising use of 16S rRNA high throughput sequencing in metagenomics has significantly enhanced our understanding of the microbial structure, particularly of the gut microbiome (Shang et al., 2018). The advent of whole-genome sequencing has transformed genetic investigation to its ultimate standard allowing unprecedented comparison of genome-wide variations with a single nucleotide resolution. Due to easier accessibility to such technologies in recent years, the genomic traits of existing bacteria are better understood and undervalued bacteria, for instance, *Morganella morganii* or *Staphylococcus agnetis* are getting noticed as possible causes of poultry diseases (Palmieri et al., 2020; Al-Rubaye et al., 2015).

Altogether, the diagnostic toolbox in bacteriology applied on poultry pathogens has substantially expanded in recent years with access to methods having much higher accuracy and discriminative power. However, the emergence of new bacterial variants with atypical pheno- and genotypic characteristics as well as appearance of aberrant clinical symptoms might pose challenges in the future for accurate identification and typing (Table 1).

3 Epidemiology

Epidemiological investigations elucidate the source of infection, pattern of transmission, host susceptibility and determinants of disease conditions in a defined population and environment. In this regard, subtyping of bacterial isolates to species, subspecies or even to variant/clone level is required and several phenotypic, genotypic or protein-based methods are applied for

Table 1 Challenges in identification and typing of bacteria due to the occurrence of phenotypic and genotypic variants

Recent examples	Atypical characteristics	References
Salmonella species	Non-hydrogen sulphide-producing strains	Asakura et al. (2020)
Mycoplasma synoviae	Mutations leading to strains with decreased antibiotic susceptibility	Bekő et al. (2020)
Mycoplasma gallisepticum	Occurrence of new genotypes	ter Veen et al. (2021)
Escherichia coli	Emergence of mcr-1-positive colistin-resistant strains	Zhuge et al. (2019)
Gallibacterium anatis	Variabilities in antimicrobial resistance even between isolates from the same organ	Hess et al. (2020a)
Riemerella anatipestifer	Acquisition of antibiotic-resistance genes by horizontal gene transfer	Zhu et al. (2020)
Staphylococcus aureus	Methicillin-resistant strains with different resistance phenotypes and genotypic patterns	Kittler et al. (2019)
Salmonella Pullorum	Occurrence of a novel strain causing arthritis	Guo et al. (2019)
Streptococcus pasteurianus	Appearance of central nervous symptoms	Hess et al. (2020b)

characterization as described in the previous section. In addition, serological techniques mainly for monitoring the presence of antibodies by rapid plate agglutination, enzyme-linked immunosorbent assay and serotyping have become useful in epidemiological surveillance of bacterial infections in different geographical areas, as shown mainly for *Mycoplasma* and *Salmonella*. Such methods are also used to trace pathogens in order to create a link between animal production and public health.

Generally, bacteria can be introduced into a flock via vertical or horizontal transmission involving animate and inanimate sources (Table 2). Vertical transmission is of special importance as it bears a risk of spread of pathogens among flocks that are geographically unrelated with very limited epidemiological links except the origin of birds itself. The high reproduction rate of breeders underlines the substantial potential of pathogen spread due to vertical transmission. Transovarian transmission of bacteria from the breeder stock to their progenies is well known for *Salmonella* and *Mycoplasma* species (ter Veen et al., 2020; Wigley et al., 2001). With more detailed and precise discrimination methods, the importance of vertical transmission for *E. coli* was shown in recent years based on epidemiological data, highlighting the importance of biosecurity from parent flock to progenies (Christensen et al., 2021). Applying modern tools for typing as outlined above might help to (dis-) prove a link between parent flock and progenies as shown for *E. cecorum* (Jung et al., 2018). For horizontal transmission of pathogens, animated sources such as sick/carrier birds, domestic animals, wild birds, personnel, rodents, mites (e.g. *Dermanyssus gallinae)* or insects (e.g. beetles and flies) along with inanimate vectors like contaminated feed, water, litter, dust, vehicles and equipment are crucial. Poor management practices resulting in insufficient cleaning and disinfection and low-standard biosecurity measures contribute to a higher spread of bacteria within a flock, between flocks and from one farm to another.

Defining patterns of disease clusters at a certain time and place is another important aspect of epidemiological investigations. In poultry, bacterial diseases can appear as an outbreak, endemic or epidemic. An infectious disease that occurs for a particular time and place, for example, a recently reported case of pullorum disease in a young layer flock is called an outbreak (Matos et al., 2020). Infectious diseases with no major fluctuation in numbers with regard to time and place are defined as endemic. Such problems can persist or appear recurrently in farms, as known for *P. multocida* (Singh et al., 2013). Multiple appearance of a disease in different flocks/farms or geographical areas represent an epidemic as reported for avian vibrionic hepatitis (also known as spotty liver disease) caused by *C. hepaticus* in different parts of the United Kingdom, Australia and the United States (Crawshaw, 2019).

Epidemiology helps in understanding the link between pathogen spread, disease occurrence and disease determinants such as type of birds, age, sex

Table 2 Sources of infection and transmission of bacterial pathogens in poultry

Transmission	Sources	Examples	References
Vertical	Infected breeders	*Mycoplasma synoviae/gallisepticum*	ter Veen et al. (2020); Armour and Ferguson-Noel (2015)
		Salmonella Pullorum	Wigley et al. (2001)
		Escherichia coli[a]	Ronco et al. (2017)
Horizontal	Sick/carrier birds	*Pasteurella multocida*	Christensen and Bisgaard (2000)
	Domestic animals	*Pasteurella multocida*	Turni et al. (2018)
	Wild birds	*Mycoplasma gallisepticum*	Sawicka et al. (2020)
	Rodents	*Erysipelothrix rhusiopathiae*	Eriksson et al. (2014)
	Insects	*Escherichia coli*	Solà-Ginés et al. (2015)
		Salmonella Gallinarum	Cocciolo et al. (2020); Lee et al. (2020)
		Erysipelothrix rhusiopathiae	Chirico et al. (2003)
	Contaminated feed	*Clostridium perfringens*	Udhayavel et al. (2017)
	Contaminated water	*Enterococcus* species *Escherichia coli*	Di Martino et al. (2018)
	Contaminated litter	*Ornithobacterium rhinotracheale*	Lopes et al. (2002b)
	Dust	*Mycoplasma synoviae/gallisepticum*	Marois et al. (2000); Marois et al. (2002)
	Personnel		
	Vehicles/equipment		
	Ventilation	*Escherichia coli*	Li et al. (2019)

[a]Based on the similarity of genetic traits in isolates from broiler breeders and broilers.

and husbandry practices. Some bacterial pathogens are nearly exclusively restricted to certain bird species, for example, *R. anatipestifer* in waterfowl (Ruiz and Sandhu, 2020). Likewise, avian vibrionic hepatitis is primarily seen in laying hens kept in free-range systems at the peak of lay (Crawshaw, 2019). The disease 'kinky-back' caused by *E. cecorum* so far is only reported in meat-type chicken (broilers, broiler breeders) kept on deep litter starting at 4 weeks of age (Jung et al., 2018). Infections with *G. anatis* coinciding with disease are only of importance for birds in production, either hens or cockerels (Hess, 2017). Environmental conditions can also influence the outcome of a disease either by affecting the bacteria itself or by influencing its transmission and spread. An increase in temperature and CO_2 level was shown to provoke biofilm formation in *Ornithobacterium rhinotracheale* favouring persistency of resistant bacteria that can lead to re-infections or recurrent infections on a farm (De la Rosa-Ramos et al., 2015). But also, seasonal weather conditions have an influence on the airborne transmission of pathogens such as *E. coli* (Li et al., 2019). Inappropriate cleaning and disinfection measures might lead to selection of variants that are better adapted to the host. The mechanism is reflected by the occurrence of small-colony variants of *O. rhinotracheale* with increased resistance to several antibiotics (Zahra et al., 2013). For *S. aureus*, evolvement of a predominant lineage is known to originate from human transmission, which ultimately adapted to the avian host by a significant number of genetic recombination events (Haag et al., 2019). In a broad perspective, the microbiome of the host is also regarded as an environmental factor and has a complex interaction with co-residing disease causative agents. Any changes in population structure of the gut microbiome consequently have an influence on the prevalence and colonization potential of pathobionts (Kogut, 2019). Likewise, simultaneous infection of birds with multiple bacterial agents, reflecting the most common scenario in the field, can exacerbate clinical signs as demonstrated experimentally for *A. paragallinarum* and *G. anatis* (Paudel et al., 2017).

Animal health surveillance is very crucial to identify key elements in detecting and controlling disease (Bisdorff et al., 2017). Surveillance includes systematic measurement (continuous or repeated), collection, comparison and analysis of data and dissemination of gained knowledge to develop intervention strategies. The scope of such programmes can be national, continental (e.g. Animal health surveillance in Europe EU 2016/429, National Animal Health Monitoring System in the United States) or global (e.g. the FAO and the OIE). Some of the bacteria in poultry are subjected to active monitoring either because of their zoonotic importance, for example, non-typhoid *Salmonella* or due to their highly contagious nature and severe impact on the health of birds, for example, *Mycoplasma gallisepticum*/*Mycoplasma synoviae*. However, it is easy to perceive that a huge bunch of information on non-notifiable diseases is available at private companies and diagnostic laboratories without wider access for researchers.

An example of the complexity of pathogen transmission and intervention strategies is illustrated in Fig. 2 for *Salmonella*. Prevention and control of non-typhoid *Salmonella* gained more and more attention in recent years due to their public health importance. Active monitoring/surveillance coupled with good farm practices and vaccination has led to significant success in reducing *Salmonella* incidences in poultry. However, challenges still prevail due to species shift (EFSA and ECDC, 2019), multidrug resistance (Antunes et al., 2016), heterogeneity and genetic alterations of bacteria (Alba et al., 2020).

4 Pathogenesis

Disease pathogenesis is influenced by a complex triad interaction of bacterial infectivity, host susceptibility and its environment. Bacteria are well equipped with an arsenal of virulence factors that are essentially required to establish an infection (Table 3). Adhesion of bacteria in host tissues is the first step of colonization, which is usually followed by invasion of the mucosal lining such as tracheal (Pourbakhsh et al., 1997) or intestinal epithelium (Wigley, 2014), allowing bacterial cells to penetrate and spread within the host (Fig. 3). Pathogenic bacteria benefit from their host adaptation strategies such as iron acquisition, evasion of host immune cells, intracellular survivability or toxin production to reach target organs causing tissue destruction (Awad et al., 2017). Depending on the host immune response, bacterial agents, for example, *A. paragallinarum*, might reside only for a short time period (Paudel et al., 2017) or for somewhat longer in infected birds as shown for *Salmonella* Enteritidis (Berchieri et al., 2001b).

Bacterial agents differ in regards to their predilection site of infection. For instance, *M. gallisepticum*, *O. rhinotracheale* and *A. paragallinarum* are recognized respiratory pathogens, while *C. perfringens* solely affects the intestine. *E. coli* infections can be localized as seen in coliform cellulitis or leading to inflammation and necrosis of multiple systemic organs (Nolan et al., 2020). Comparing isolates derived either from cellulitis or septicemia, it was shown that the two groups of *E. coli* were similar (Jeffrey et al., 2002). This finding might lead to the hypothesis that the host and the affected tissue itself might have an influence on the course of infection resulting in different clinical and pathological pictures. Another similar example is also seen with *G. anatis*. These bacteria show the tendency to colonize both respiratory and genital tracts; however, disease consequences are due to reproductive disorders only (Paudel et al., 2014a,b). Interestingly, *G. anatis* isolates from systemic organs were more resistant to antibiotics compared to isolates from the genital tract which might be due to the niche adaptation of different genotypes in different organs (Hess et al., 2020a).

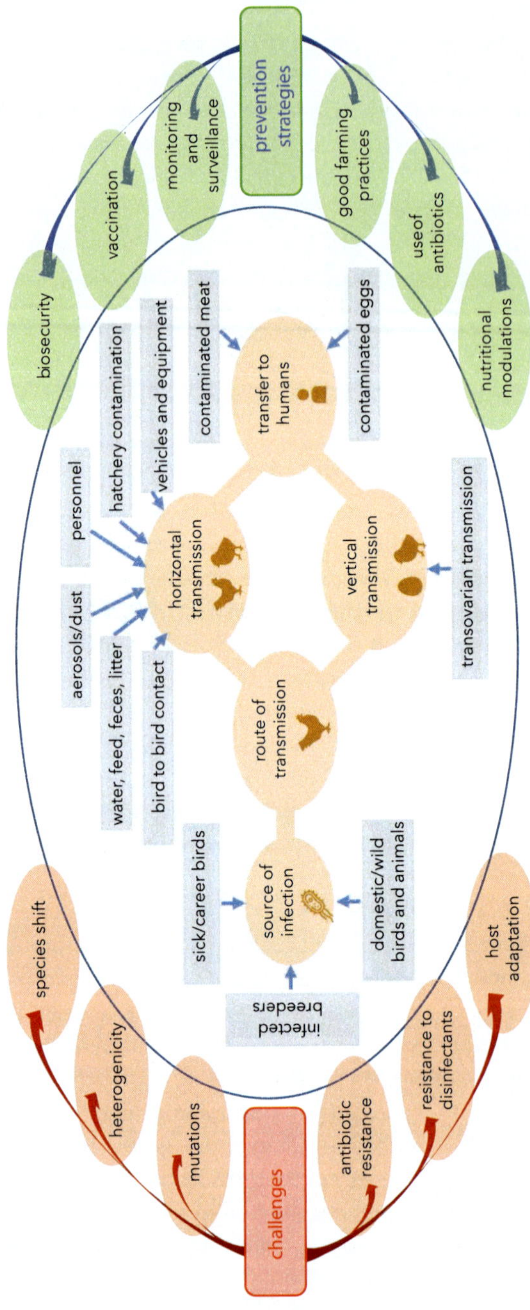

Figure 2 Existing challenges to combat *Salmonella* infections in poultry production with numerous ways of bacteria transmission including spread to humans and a wide spectrum of prevention strategies.

Table 3 Bacterial determinants of pathogenicity

Determinants	Features	Examples	References
Adherence	Flagella	*Campylobacter jejuni*	Wysok et al. (2020)
	Fimbriae	*Gallibacterium anatis*	Persson and Bojesen (2015)
	Protein A	*Staphylococcus aureus*	Gyles (2011)
Invasion	IbeB gene	*Escherichia coli*	Wang et al. (2012)
	Protein A	*Staphylococcus aureus*	Gyles (2011)
Toxin production	NetB toxin for tissue destruction	*Clostridium perfringens*	Keyburn et al. (2008)
	GtxA toxin attacking immune system	*Gallibacterium anatis*	Persson and Bojesen (2015)
Iron acquisition	Survival in the host	*Riemerella anatipestifer*	Liu et al. (2018)
Mimicry	LPS mimicry of host molecules	*Pasteurella multocida*	Harper et al. (2017)
Evading immune system	Blocking of complement system	*Riemerella anatipestifer*	Fan et al. (2017)
	Invading erythrocytes using them as hiding place	*Mycoplasma gallisepticum*	Vogl et al. (2008)
	Protein A	*Staphylococcus aureus*	Gyles (2011)

Host susceptibility towards pathogens is influenced by several factors such as genetic background (Berchieri et al., 2001a), species (Ruiz and Sandhu, 2020; Wigley, 2013), age of birds (Barrow et al., 2012) or poultry production type (Jung et al., 2018). Factors determining the pathogenicity and intra-host survivability of bacteria are elucidated by applying improved molecular techniques (Gyles, 2011). In particular, investigations on gene expression patterns are performed by nucleic acid-based methods, for example, reverse transcriptase-PCR (rt-PCR) and microarray technology. The use of whole-genome sequencing has enabled to unveil genome-wide properties of bacterial pathogens. Likewise, transcriptomic studies facilitated the understanding of

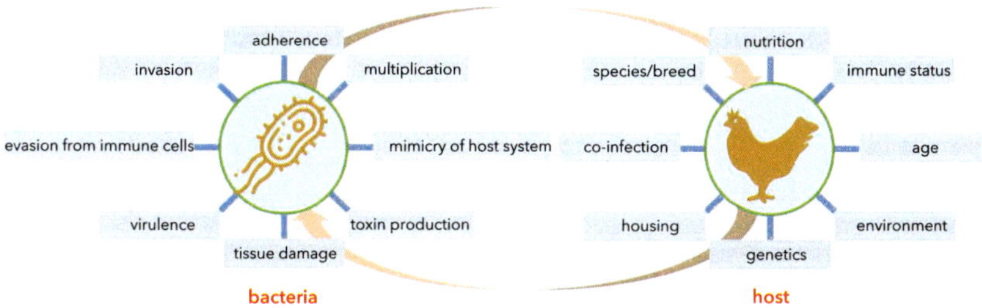

Figure 3 Bacteria and host determinants of disease outcomes.

bacterial expression profile in the host environment or the response of host tissues to infections on a wider scale (e.g. Parreira et al., 2016; Sandford et al., 2011; Li et al., 2020). In order to explore facets of host-pathogen relationship, an infection model is a fundamental tool, which still needs to be improved for

Table 4 Approaches to elucidate pathogenicity of bacterial pathogens in poultry

	Methods	Examples	References
in silico	Molecular traits (nucleic acid-based methods)	*Campylobacter jejuni*	Hu et al (2014)
		Pasteurella multocida	Boyce et al. (2004)
		Riemerella anatipestifer	Wang et al. (2019)
		Escherichia coli	Li et al. (2016)
		Mycoplasma gallisepticum	Pflaum et al. (2020)
		Mycoplasma synoviae	Cizelj et al. (2016)
		Gallibacterium anatis	Kudirkiene et al. (2014)
in vivo/ex vivo	Embryonated eggs	*Enterococcus cecorum*	Borst et al. (2014)
		Escherichia coli	Oh et al. (2012); Wooley et al. (2000)
		Riemerella anatipestifer	Seo et al. (2013)
		Gallibacterium anatis	Wang et al. (2018)
		Salmonella Gallinarum	Zhang et al. (2020)
	Tracheal organ culture	*Mycoplasma* species	Araake (1982)
			Bearson et al. (2003)
		Bordetella avium	Knab et al. (2020)
	Primary oviduct epithelial cells	*Gallibacterium anatis*	Zhang et al. (2017)
	Primary caecal epithelial cells	*Campylobacter jejuni*	Van Deun et al. (2008)
	Ligated intestinal loop model	*Clostridium perfringens*	Parent et al. (2018)
		Campylobacter jejuni	Van Deun et al. (2008)
	Ussing chamber	*Campylobacter jejuni*	Awad et al. (2020)
		Salmonella Enteritidis	Awad et al. (2012)
		Escherichia coli	Awad et al. (2014)
	Animal infection models	*Gallibacterium anatis*	Paudel et al. (2014a); Paudel et al. (2014b)
		Salmonella Pullorum	Guo et al. (2019)
		Avibacterium paragallinarum	Paudel et al. (2017)
		Riemerella anatipestifer	Tao et al. (2020)
		Mycoplasma gallisepticum	Wijesurendra et al. (2015)
		Mycoplasma synoviae	Feberwee et al. (2009)

many pathogens. Embryo lethality assays performed in embryonated chicken eggs are 'traditionally' very popular to define the virulence of bacterial strains (Table 4) but organ cultures remain more sophisticated systems to investigate the interaction of bacteria with relevant host tissues. For this purpose, tracheal organ culture, primary intestinal or oviduct epithelial organ cultures, intestinal loop model or Ussing chamber techniques are mainly used. However, none of these systems represents a complete host environment, thus the composite repertoire of the host defence mechanisms is always bypassed. In this regard, *in vivo* studies are essentially required to fully understand host reactions mimicking a natural scenario and to define pathogenicity of a microbe in the host fulfilling Koch's postulates (Hess, 2017). Overall, more standardization in animal testing and infection studies are needed to place results in a biological context and to define mechanistic disease pathway in a natural environment.

Future challenges in regard to pathogenesis research might arise due the occurrence of new variants of bacteria which can adapt to different hosts as shown lately for *A. paragallinarum* and *M. gallisepticum* (Jeong et al., 2020; Perez et al., 2020). Prevention strategies can facilitate species shift. The hypothesized outcompeting of non-typhoidal *S.* Enteritidis by *S.* Gallinarum underlines the complexity of microbial interactions and eradication programmes (Bäumler et al., 2000). In continuation, the successful elimination of *S.* Enteritidis and *S.* Typhimurium from broiler production enabled other serovars to emerge (Miller et al., 2010). Field observations like more severe clinical signs or recurrent infections in combination with loss of efficacy of vaccines (Omaleki et al., 2020), appearance of 'new' clinical signs (Matos et al., 2020), involvement of 'new' hosts (ter Veen et al., 2021), 'new' variants (e.g. via biofilm formation, De la Rosa-Ramos et al., 2015), synergistic effect of bacterial co-infections in the development of a disease (Cisneros-Tamayo et al., 2020) or the emergence of multidrug resistance strains (Zhu et al., 2020) demand for closer and deeper investigations of clinical cases.

5 Disease prevention by vaccination

Strategies for preventing diseases and minimizing the risk of pathogen spread requires consideration of the epidemiology and production specificities. Biosecurity and good farm management within the farm entities and in the whole production chain play an important role to prevent horizontally and vertically transmitted pathogens. Consequently, eradication of certain pathogens in breeding stock with intensive programmes has the highest priority to minimize the risk in progenies. In fact, eradication of *S.* Pullorum from breeders was the main focus to implement the National Poultry Improvement Plan in 1935 in the United States (http://www.poultryimprovement.org/default.cfm) and to establish the Houghton Poultry Research Station in the United Kingdom. Back in early 1900, *S.* Pullorum was demonstrated to be transmitted from hens to progeny, which was the first recognition of a vertically transmitted pathogen in avian species. The historical success story of the once very widely prevalent Pullorum disease until complete eradication in 1975 in the United States was published recently (Schat et al., 2021).

Vaccination, in addition to biosecurity, is the key element to protect birds from bacterial infections. Numerous book chapters and reviews highlighted the importance of bacterial vaccines in poultry in general (Rabie and Amin Girh, 2020; Redweik et al., 2020) or being focused on a certain disease/pathogen, for example, salmonellosis (Zhang-Barber et al., 1999; Barrow, 2007; van Immerseel et al., 2005; Wigley, 2017; Revolledo, 2018; Desin et al., 2013), mycoplasmosis (Whithear, 1996; Kleven, 2008; Ishfaq et al., 2020; Evans et al., 2005), infectious coryza (Blackall, 1995; Deshmukh, 2015), erysipelas (Opriessnig et al., 2020), necrotic enteritis (Lee et al., 2011; Mot et al., 2014), *colibacillosis* (Ghunaim et al., 2014) or *G. anatis* induced disease (Narasinakuppe Krishnegowda et al., 2020). The success of vaccination depends on numerous aspects associated with the pathogen and host, not to neglect economic constraints which are mainly driven by the way a certain vaccine has to be applied to birds. In this context, handling of individual birds for intramuscular or subcutaneous injection of killed vaccines is of major concern because of the possible impact on performance due to stress and the overall consequences of local reactions. Changing husbandry conditions to alternative housing systems also influences vaccine application; in fact, it might even make it impossible to vaccinate those birds.

Killed vaccines are more easy to produce and are safer to use as they cannot spread to other susceptible hosts or the environment and do not possess the risk to revert to virulence. To address the high variability between isolates involved in certain diseases autogenous vaccines became very popular in recent years. Such vaccines are always based upon complete bacterins and have to be applied either subcutaneously or intramuscularly. Published data about the efficacy of autogenous vaccines are very limited due to the fact that relevant testing is not obligatory. Olson et al. (1969) demonstrated already some decades ago a certain value of these vaccines to prevent the cranial form of fowl cholera in turkeys. Serotype specificity of the vaccine has important consequences to induce protection against infectious coryza (Soriano et al., 2004). As commercial products may contain different bacterins, an autogenous vaccine must not necessarily be always superior in protection (Gallardo et al., 2020). Application of an autogenous *S.* Enteritidis vaccine in pullets had no or only marginal impact on *Salmonella* prevalence in normal or stressed birds and their environment which might be due to the limited role of killed vaccines as stand-alone products in vaccination programmes (Davison et al., 1999; Nakamura et al., 1994). Therefore, non-typhoid *Salmonella* strains attract less attention in regards to autogenous vaccines. More divergent results were reported for *E. coli* demonstrating either good or nearly no protection (Li et al., 2017; Landman and van Eck, 2017). These recent studies confirm early results indicating limited heterologous protection of formalin-inactivated *E. coli* (Gross, 1957). Under practical conditions, it might well be advisable to combine a licensed live vaccine with autogenous vaccines to prevent collibacillosis in pullets (Koutsianos et al., 2020). Overall, substantial variation exists between experimental studies and the design of the vaccination/challenge model. This is even more important for pathogens like *E. coli* characterized by a huge strain variation complicating standardization (Ghunaim et al., 2014). To a certain extent, it also reflects the difficulties to reproduce clinical signs and lesions in birds under experimental conditions. Therefore, very often artificial challenge routes are used which do not mimic the situation in the field neglecting that the route of challenge can have substantial impact to assess efficacy of a vaccine, well known from fowl cholera (Maheswaran et al., 1973). This underlines the need of much more standardization in experiments to compare vaccination/challenge studies on a broader scale keeping in mind that a vaccination-challenge trial is the gold standard to gain robust safety and efficacy data and to assess the protective efficacy of a certain vaccine candidate. A vaccination-challenge trial is even relevant to test coryza vaccines as an immune response measured by haemagglutination inhibition is not a valid indicator to predict protection (García et al., 2008).

Without surprise, the portfolio of a killed vaccine varies for different pathogens. The severity of mycoplasmosis, fowl cholera, infectious coryza and erysipelas in poultry led to the development of the first inactivated vaccines, which are still in use today. Likewise, a killed vaccine was developed to protect turkeys and chickens against *O. rhinotracheale* (Van Empel and Van Den Bosch, 1998; Sprenger et al., 2000). In comparison to other bacterial infections, less research was carried out in the past on erysipelas which might be due to the fact that strains do not show much variation. Furthermore, vaccination against erysipelas is mainly applied following an outbreak in order to protect epidemiological-related flocks. Killed vaccines are more or less part of standard vaccination programmes not only in breeders but also in layers, depending on the epidemiological situation of diseases. Varying success of killed vaccines is documented in the literature to prevent non-typhoid *Salmonella* in poultry, but it is well accepted that in combination with live vaccines, they are crucial components in vaccination programmes. Killed vaccines mainly induce a humoral immune response, causing local reactions with possible consequences on welfare and growth and are not suitable for mass application. The use of an adjuvant in killed vaccines is of common practice to improve immunity and to increase efficacy of killed bacterins or subunit vaccines that need to be injected subcutaneously or intramuscularly. The risk of inducing unwanted side effects and the amyloidogenic potential of inactivated vaccines due to an inflammatory stimulation needs to be kept in mind (Rampin et al., 1989). Such a case of amyloidosis in white layer type birds is shown in Fig. 4.

Subunit vaccines reflect the achievements in characterizing bacterial components mainly surface proteins. Such vaccines were experimentally developed for numerous bacterial pathogens. Antigens might be retrieved by special extraction and purification procedures from whole bacterins, or they are recombinantly expressed in a suitable expression system, primarily *E. coli*. The formulation of killed vaccines allows the combination of different antigens, which enables extended protection to different serotypes as demonstrated for *E. coli* and *O. rhinotracheale* (Goor et al., 2017; Schuijffel et al., 2006). A hypervariable region characterized in the hemagglutinin of *A. paragallinarum* induced homologous protection in chickens (Wu et al., 2011; Noro et al., 2008). Such studies continued early success achieved with purified hemagglutinin (Iritani et al., 1980). For non-typhoid *Salmonella*, the virulence factors, *Salmonella* pathogenicity island (SPI-1 and SPI-2) were tested as subunit antigens as they are immunogenic, but the level of bacteria reduction following the challenge was rather limited which is similar to other subunit vaccines (Desin et al., 2013). The outer membrane protein H (OmpH) and outer membrane vesicles (OMVs) of *P. multocida* and *G. anatis*, respectively, were utilized to prepare subunit vaccines to induce an immune response and heterologous protection in birds (Varinrak et al., 2017; Antenucci et al., 2020).

Live vaccines possess numerous advantages, especially with regard to administration and the ability to broaden the range of defense mechanisms by triggering different arms of the immune system. The first live vaccine against fowl cholera was developed mid of the last century (Bierer and Eleazer, 1968) but certain concern was raised with regard to its spread within the environment (Hofacre and Glisson, 1986). The *S. gallinarum* 9R vaccine was also developed in the mid of last century to protect birds against fowl typhoid and is in use ever since (Wigley, 2017). The subcutaneous application is unique among live

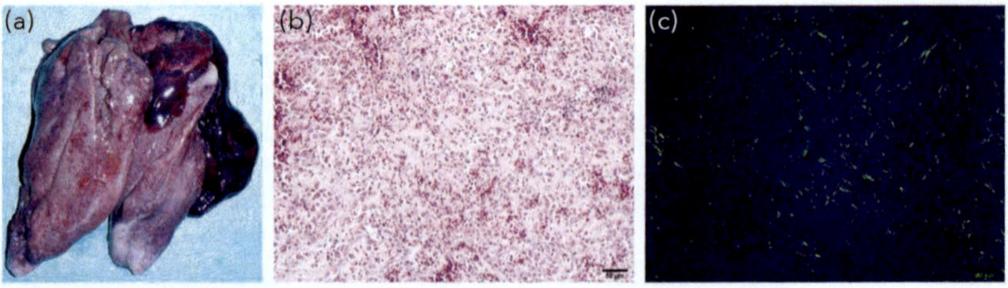

Figure 4 Clinical case of hepatic amyloidosis in 20-week-old white layer type birds. Macroscopic lesions in liver profuse and pinpoint haemorrhages with clotted blood (a); microscopic changes in liver: diffuse necrosis and severe bleedings, haematoxylin and eosin stain (b); the same sample: deposits of amyloids are present (green), Congo red staining under polarized light (c).

bacterial vaccines in poultry. Latency driven by the invasion of macrophages is a common feature of *Salmonella* spp. whereas the induction of septicemia leading to severe disease with minimal innate immune response and the poor colonization of the intestinal tract clearly differentiate typhoid from non-typhoid *Salmonella*. Non-typhoid *Salmonella* vaccination should inhibit colonization with wild-type bacteria and induce local immunity in the gut (van Immerseel et al., 2005). For this, the early oral application is needed and attenuated vaccine strains need to be differentiated from field strains which was achieved by developing metabolic drift mutants (Linde et al., 1998). Such mutants are characterized by specific nutrient requirements, antimicrobial resistance and temperature sensitivity. Additionally, the targeted mutants with reduced virulence and potential to be differentiated from field strains were reported (Desin et al., 2013). For example, deleting the pathogenicity island described above as subunit vaccine candidates in *Salmonella* spp. genomes led to the development of live vaccines for non-typhoid Salmonella (Bohez et al., 2007; Khan et al., 2003). Such strains showed reduced virulence and clearance until slaughter of broiler chickens, which are crucial requirements for live *Salmonella* vaccines. The concept was also applied in regard to prevent *S.* Pullorum infection. The reduced pathogenicity enabled the oral application of a SPI2-deleted mutant which provided efficient homologous and heterologous protection (Yin et al., 2015). The use of non-attenuated *O. rhinotracheale* in turkeys at 7 weeks of age is a very exceptional approach as birds were vaccinated at a time when they are less susceptible (Sprenger et al., 2000).

For a long time, live vaccines are considered crucial components of vaccination programmes to protect poultry against mycoplasmosis. The *M. gallisepticum* F-strain vaccine was introduced into the market several decades ago. The currently used live vaccines differ in their virulence, the F-strain being the most virulent one (Levisohn and Kleven, 2000), others are somewhat milder, such as 6/85 and ts-11 (see below) and a more recently published K-strain (Ferguson-Noel and Williams, 2015). The degree of attenuation has an influence on residual pathogenicity in a flock, especially in highly susceptible turkeys. Furthermore, this feature has also an impact on the route of vaccination, the spread within a farm and the immunity, altogether reflecting a certain balance in vaccine development. Live vaccines are able to displace wild-type *M. gallisepticum* from a holding but might circulate later if used in multi-age farms supporting the need to interrupt the infection cycle. A widely used concept to minimize the virulence of live vaccines is temperature-sensitive (ts) mutants established by chemical mutagenesis. Such bacteria possess the advantage that replication is very much restricted at bird's body temperature. Vaccines based on this concept were developed for *M. gallsepticum* (ts-11) and *M. synoviae* (MS-H), all widely used in different countries (Whithear et al., 1990; Morrow et al., 1998). The approach might also be used to improve existing live vaccines. A good example is the recovery of the MG strain ts-304 from the ts-11 vaccine strain which, due to its superior adherence to epithelial cells, offered improved efficacy despite a lower vaccine dose (Kanci Condello et al., 2020). A ts-mutant was also developed for *O. rhinotracheale* and demonstrated efficacy following vaccination of turkeys via drinking water (Lopes et al., 2002a). A similar concept to establish bacterial mutants for live bacterial vaccines is the development of auxothrophic variants in which the basic idea is a limited replication in the host tissue due to a nutritive deviation of the bacteria with consequences on the growth profile (Hoiseth and Stocker, 1981). Such *aroA* mutants were developed for *S.* Typhimurium demonstrating a reduced lethal dose and lowered excretion of virulent *S.* Typhimurium (Alderton et al., 1991). Auxotrophic *aroA* mutants of different serotypes were also developed for fowl cholera demonstrating efficacy (Scott et al., 1999). An *aroA*-deleted mutant is licensed as a live vaccine for chickens and turkeys in different countries against colibacillosis using drinking water or spray application (La Ragione et al., 2013).

Vector vaccines display a combination of protection against the vector itself and the integrated genes. Integration of 40K and *mgc* gene segments of *M. gallisepticum* in an avian poxvirus is the only licensed viral vector to be used for protecting birds against a bacterial disease. Using this vaccine, only limited efficacy data were published, which showed difficulties to withstand a very severe experimental challenge in birds (Ferguson-Noel et al., 2012). Numerous studies focused on the development of bacterial vector vaccines and the most attractive system seems to be *Salmonella* targeting the digestive tract. Such efforts led recently to the licensing of a *Salmonella*-vectored vaccine against necrotic enteritis in the United States, the first of its kind (Francis, 2021).

Overall, efforts in the past two decades focused very much on improving strategies to develop new live vaccines to be used for preventing bacterial infections in poultry. Despite substantial achievements, numerous gaps exist and only a very limited number of new products succeeded to enter the market. This reflects the high expectations on each vaccine starting from production with easy application and high safety to inducing long-lasting immunity with superior efficacy, especially cross-protection. On the contrary, autogenous vaccines are widely used in certain areas despite the lack of efficacy data, largely based on empirical evidence.

6 Future trends in research

It is easy to perceive that bacterial infections in poultry will gain much more importance in the future. The ongoing pressure to minimize the use of antibiotics with consequences on antimicrobial resistance will foster research in all areas mentioned above. Intensified monitoring will remain key to keep the high standards during production, especially in breeding stock, and to prevent vertical transmission of bacteria from hen to progeny.

It can be hypothesized that so far easy access and less controlled use of antimicrobials applied for therapeutic and prophylactic purposes were a certain hindrance to develop vaccines against various bacterial pathogens. In agreement with this, regulations on the use of vaccines in various countries, for example, import of vaccines which are not licensed, face a much tighter regulation in comparison to the use of antimicrobials, with the latter ones being more easily accessible. In consequence, it still needs substantial efforts until the view on antimicrobials will change, from an indispensable tool to improve production towards a very valuable remedy to treat diseased birds. Although a substantial reduction in the use of antimicrobials within a short period of time without major impact on production in some developed countries points towards a certain abuse before reduction, a complete ban of antimicrobials can have substantial impact on poultry health and welfare. Therefore, antimicrobial treatment remains to be a balance act deserving skilled input from veterinarians and all people involved in production.

With few exceptions, the majority of bacterial vaccines are actually produced as killed vaccines to be administered to each individual bird subcutaneously or intramuscularly. This includes licensed and autogenous vaccines. There are only very limited experimental studies focusing on the efficacy of autogenous vaccines, some of them with controversial outcomes. In contrast, live vaccines are known to trigger the whole plethora of the immune system. However, the overall portfolio of bacterial vaccines, especially live vaccines available on the market, is still very limited and needs substantial expansion. Such live vaccines should also offer the advantage of easy administration to facilitate mass vaccination. The development of bacterial vectors characterized by the integration of external genes from other important poultry pathogens might have the power to increase the acceptance in the market. But for this, more studies are needed to elucidate the genetic repertoire of bacterial pathogens and functional components of certain genes. This will help to define virulence factors, to investigate the interplay between different genes and to determine the genetic repertoire of a certain pathogen. In addition, standardized infection models should become available enabling the comparison of *in vivo* studies.

Whole-genome sequencing, proteome and metabolome data bear the option to focus on the bacteria and the host reaction in parallel thus helping to gain much more knowledge not only on the pathogen itself but also on the host reaction during all stages of infection. Despite increasing knowledge on a certain pathogen, major deficiencies in understanding the host reaction are obvious. This is also reflected by the fact that for some bacteria hardly any standardized models are established. Defining the immune reaction in the context of a certain bacterial infection is the key element to develop new vaccines and protection strategies. This should supply deeper insight in understanding the host-pathogen interaction, altogether helpful to improve the knowledge on the pathogenesis of diseases with consequences on vaccination concepts. Extending the scope towards nutrition management and housing practices and even environmental conditions should lead to a holistic approach. Altogether, this should enable new developments with high relevance for the field. Efforts to improve vaccination strategies need to include ways to improve application techniques for bacterial vaccines in order to facilitate mass vaccination without compromising safety and efficacy.

7 References

Al-Rubaye, A. A. K., Couger, M. B., Ojha, S., Pummill, J. F., Koon, J. A., Wideman, R. F. and Rhoads, D. D. (2015). Genome analysis of *Staphylococcus agnetis*, an agent of lameness in broiler chickens, *PLoS ONE* 10(11), e0143336. doi: 10.1371/journal.pone.0143336.

Alba, P., Leekitcharoenphon, P., Carfora, V., Amoruso, R., Cordaro, G., Di Matteo, P., Ianzano, A., Iurescia, M., Diaconu, E. L., Study Group, E. N., Pedersen, S. K., Guerra, B., Hendriksen, R. S., Franco, A. and Battisti, A. (2020). Molecular epidemiology of *Salmonella infantis* in Europe: insights into the success of the bacterial host and its parasitic pesi-like megaplasmid, *Microbial Genomics* 6(5). doi: 10.1099/mgen.0.000365.

Alderton, M. R., Fahey, K. J. and Coloe, P. J. (1991). Humoral responses and salmonellosis protection in chickens given a vitamin-dependent *Salmonella typhimurium* mutant, *Avian Diseases* 35(3), 435–442. doi: 10.2307/1591205.

Alispahic, M., Christensen, H., Bisgaard, M., Hess, M. and Hess, C. (2014). MALDI-TOF mass spectrometry confirms difficulties in separating species of the *Avibacterium* genus, *Avian Pathology* 43(3), 258–263. doi: 10.1080/03079457.2014.916038.

Antenucci, F., Arak, H., Gao, J., Allahgadry, T., Thøfner, I. and Bojesen, A. M. (2020). Hydrostatic filtration enables large-scale production of outer membrane vesicles that effectively protect chickens against *Gallibacterium anatis*, *Vaccines* 8(1). doi: 10.3390/vaccines8010040.

Antunes, P., Mourão, J., Campos, J. and Peixe, L. (2016). Salmonellosis: the role of poultry meat, *Clinical Microbiology and Infection* 22(2), 110–121. doi: 10.1016/j.cmi.2015.12.004.

Araake, M. (1982). Comparison of ciliostasis by Mycoplasmas in mouse and chicken tracheal organ cultures, *Microbiology and Immunology* 26(1), 1–14. doi: 10.1111/j.1348-0421.1982.tb00148.x.

Armour, N. K. and Ferguson-Noel, N. (2015). Evaluation of the egg transmission and pathogenicity of *Mycoplasma gallisepticum* isolates genotyped as ts-11, *Avian Pathology* 44(4), 296–304. doi: 10.1080/03079457.2015.1044890.

Asakura, H., Sakata, J., Yamamoto, S. and Igimi, S. (2020). Draft genome sequences of non-H 2 S-producing strains of *Salmonella* enterica serovars Infantis, Enteritidis, Berta, *Microbiology Resource Announcements* 9(30). doi: 10.1128/MRA.00335-20.

Awad, W. A., Aschenbach, J. R., Khayal, B., Hess, C. and Hess, M. (2012). Intestinal epithelial responses to Salmonella enterica serovar Enteritidis: effects on intestinal permeability and ion transport, *Poultry Science* 91(11), 2949–2957. doi: 10.3382/ps.2012-02448.

Awad, W. A., Hess, C. and Hess, M. (2017). Enteric pathogens and their toxin-induced disruption of the intestinal barrier through alteration of tight junctions in chickens, *Toxins* 9(2). doi: 10.3390/toxins9020060.

Awad, W. A., Hess, C., Khayal, B., Aschenbach, J. R. and Hess, M. (2014). In vitro exposure to *Escherichia coli* decreases ion conductance in the jejunal epithelium of broiler chickens, *PLoS ONE* 9(3), e92156. doi: 10.1371/journal.pone.0092156.

Awad, W. A., Ruhnau, D., Hess, C. and Hess, M. (2020). Campylobacter jejuni increases the paracellular permeability of broiler chickens in a dose-dependent manner, *Poultry Science* 99(11), 5407–5414. doi: 10.1016/j.psj.2020.08.014.

Barrow, P. A. (2007). *Salmonella* infections: immune and non-immune protection with vaccines, *Avian Pathology* 36(1), 1–13. doi: 10.1080/03079450601113167.

Barrow, P. A., Jones, M. A., Smith, A. L. and Wigley, P. (2012). The long view: *Salmonella* - the last forty years, *Avian Pathology* 41(5), 413–420. doi: 10.1080/03079457.2012.718071.

Bäumler, A. J., Hargis, B. M. and Tsolis, R. M. (2000). Tracing the origins of *Salmonella* outbreaks, *Science* 287(5450), 50–52. doi: 10.1126/science.287.5450.50.

Bearson, S. M. D., Collier, S. D., Bearson, B. L. and Branton, S. L. (2003). Induction of a *Mycoplasma gallisepticum* pMGA gene in the chicken tracheal ring organ culture model, *Avian Diseases* 47(3), 745–749. doi: 10.1637/6099.

Bekő, K., Kreizinger, Z., Kovács, Á. B., Sulyok, K. M., Marton, S., Bányai, K., Catania, S., Feberwee, A., Wiegel, J., Dijkman, R., Ter Veen, C., Lysnyansky, I. and Gyuranecz, M. (2020). Mutations potentially associated with decreased susceptibility to fluoroquinolones, macrolides and lincomycin in *Mycoplasma synoviae*, *Veterinary Microbiology* 248, 108818. doi: 10.1016/j.vetmic.2020.108818.

Berchieri, A., Murphy, C. K., Marston, K. and Barrow, P. A. (2001a). Observations on the persistence and vertical transmission of *Salmonella* enterica serovars Pullorum and Gallinarum in chickens: effect of bacterial and host genetic background, *Avian Pathology* 30(3), 221–231. doi: 10.1080/03079450120054631.

Berchieri, A., Wigley, P., Page, K., Murphy, C. K. and Barrow, P. A. (2001b). Further studies on vertical transmission and persistence of Salmonella enterica serovar *enteritidis* phage type 4 in chickens, *Avian Pathology* 30(4), 297–310. doi: 10.1080/03079450120066304.

Bierer, B. W. and Eleazer, T. H. (1968). Continuous use of a live vaccine in the drinking wate against fowl cholera infection in turkeys, *Poultry Science* 47(4), 1258–1260. doi: 10.3382/ps.0471258.

Bisdorff, B., Schauer, B., Taylor, N., Rodríguez-Prieto, V., Comin, A., Brouwer, A., Dórea, F., Drewe, J., Hoinville, L., Lindberg, A., Martinez Aviles, M., Martínez-López, B., Peyre, M., Pinto Ferreira, J., Rushton, J., van Schaik, G., Stärk, K. D., Staubach, C., Vicente-Rubiano, M., Witteveen, G., Pfeiffer, D. and Häsler, B. (2017). Active animal health surveillance in European Union Member States: gaps and opportunities, *Epidemiology and Infection* 145(4), 802-817. doi: 10.1017/S0950268816002697.

Blackall, P. J. (1995). Vaccines against infectious coryza, *World's Poultry Science Journal* 51(1), 17–26. doi: 10.1079/WPS19950003.

Blackall, P. J., Christensen, H., Beckenham, T., Blackall, L. L. and Bisgaard, M. (2005). Reclassification of *Pasteurella gallinarum*, *[Haemophilus] paragallinarum*, *Pasteurella avium* and *Pasteurella volantium* as *Avibacterium gallinarum* gen. nov., comb.

nov., *Avibacterium paragallinarum* comb. nov., *Avibacterium avium* comb. nov. and *Avibacterium* volantium comb. nov, *International Journal of Systematic and Evolutionary Microbiology* 55(1), 353-362. doi: 10.1099/ijs.0.63357-0.

Bohez, L., Ducatelle, R., Pasmans, F., Haesebrouck, F. and Van Immerseel, F. (2007). Long-term colonisation-inhibition studies to protect broilers against colonisation with *Salmonella enteritidis*, using Salmonella Pathogenicity Island 1 and 2 mutants, *Vaccine* 25(21), 4235–4243. doi: 10.1016/j.vaccine.2007.02.082.

Borst, L. B., Suyemoto, M. M., Keelara, S., Dunningan, S. E., Guy, J. S. and Barnes, H. J. (2014). A chicken embryo lethality assay for pathogenic *Enterococcus cecorum*, *Avian Diseases* 58(2), 244–248. doi: 10.1637/10687-101113-Reg.1.

Boyce, J. D., Wilkie, I., Harper, M., Paustian, M. L., Kapur, V. and Adler, B. (2004). Genomic-scale analysis of *Pasteurella multocida* gene expression during growth within liver tissue of chickens with fowl cholera, *Microbes and Infection* 6(3), 290–298. doi: 10.1016/j.micinf.2003.12.004.

Chirico, J., Eriksson, H., Fossum, O. and Jansson, D. (2003). The poultry red mite, *Dermanyssus gallinae*, a potential vector of *Erysipelothrix rhusiopathiae* causing erysipelas in hens, *Medical and Veterinary Entomology* 17(2), 232–234. doi: 10.1046/j.1365-2915.2003.00428.x.

Christensen, H., Bachmeier, J. and Bisgaard, M. (2021). New strategies to prevent and control avian pathogenic *Escherichia coli* (APEC), *Avian Pathology* 50(5), 370–381. doi: 10.1080/03079457.2020.1845300.

Christensen, J. P. and Bisgaard, M. (2000). Fowl cholera, *Revue Scientifique et Technique* 19(2), 626–637. doi: 10.20506/rst.19.2.1236.

Cisneros-Tamayo, M., Kempf, I., Coton, J., Michel, V., Bougeard, S., de Boisséson, C., Lucas, P., Bäyon-Auboyer, M. H., Chiron, G., Mindus, C. and Gautier-Bouchardon, A. V. (2020). Investigation on eggshell apex abnormality (EAA) syndrome in France: isolation of Mycoplasma synoviae is frequently associated with *Mycoplasma pullorum*, *BMC Veterinary Research* 16(1), 271. doi: 10.1186/s12917-020-02487-0.

Cizelj, I., Dušanić, D., Benčina, D. and Narat, M. (2016). *Mycoplasma* and host interaction: *in vitro* gene expression modulation in *Mycoplasma synoviae* and infected chicken chondrocytes, *Acta Veterinaria Hungarica* 64(1), 26–37. doi: 10.1556/004.2016.003.

Cocciolo, G., Circella, E., Pugliese, N., Lupini, C., Mescolini, G., Catelli, E., Borchert-Stuhlträger, M., Zoller, H., Thomas, E. and Camarda, A. (2020). Evidence of vector borne transmission of Salmonella enterica enterica serovar Gallinarum and fowl typhoid disease mediated by the poultry red mite, *Dermanyssus gallinae* (De Geer, 1778), *Parasites and Vectors* 13(1), 513. doi: 10.1186/s13071-020-04393-8.

Crawshaw, T. (2019). A review of the novel thermophilic *Campylobacter*, *Campylobacter hepaticus*, a pathogen of poultry, *Transboundary and Emerging Diseases* 66(4), 1481-1492. doi: 10.1111/tbed.13229.

Davison, S., Benson, C. E., Henzler, D. J. and Eckroade, R. J. (1999). Field observations with *Salmonella enteritidis* bacterins, *Avian Diseases* 43(4), 664–669. doi: 10.2307/1592735.

De la Rosa-Ramos, M. A., Rodríguez-Cruz, M., López-Villegas, E. O., Castro-Escarpulli, G. and Guerra-Infante, F. M. (2015). Conditions that induce biofilm production by *Ornithobacterium rhinotracheale*, *Avian Pathology* 44(5), 366–369. doi: 10.1080/03079457.2015.1059923.

Deshmukh, S. (2015). An update on avian infectious coryza: it's re-emerging trends on epidemiology, etiologic characterization, diagnostics, therapeutic and prophylactic advancements, *Journal of Dairy, Veterinary and Animal Research* 2(3). doi: 10.15406/jdvar.2015.02.00037.

Desin, T. S., Köster, W. and Potter, A. A. (2013). Salmonella vaccines in poultry: past, present and future, *Expert Review of Vaccines* 12(1), 87–96. doi: 10.1586/erv.12.138.

Di Martino, G., Piccirillo, A., Giacomelli, M., Comin, D., Gallina, A., Capello, K., Buniolo, F., Montesissa, C. and Bonfanti, L. (2018). Microbiological, chemical and physical quality of drinking water for commercial turkeys: a cross-sectional study, *Poultry Science* 97(8), 2880–2886. doi: 10.3382/ps/pey130.

Eriksson, H., Bagge, E., Båverud, V., Fellström, C. and Jansson, D. S. (2014). *Erysipelothrix rhusiopathiae* contamination in the poultry house environment during erysipelas outbreaks in organic laying hen flocks, *Avian Pathology* 43(3), 231–237. doi: 10.1080/03079457.2014.907485.

European Food Safety Authority and European Centre for Disease Prevention and Control (EFSA and ECDC) and ECDC (2019). The European Union one health 2018 zoonoses report, *EFSA Journal. European Food Safety Authority* 17(12), e05926. doi: 10.2903/j.efsa.2019.5926.

Evans, J. D., Leigh, S. A., Branton, S. L., Collier, S. D., Pharr, G. T. and Bearson, S. M. D. (2005). *Mycoplasma gallisepticum*: current and developing means to control the avian pathogen, *Journal of Applied Poultry Research* 14(4), 757–763. doi: 10.1093/japr/14.4.757.

Fan, M., Chen, S., Zhang, L., Bi, J., Peng, J., Huang, X., Li, X., Li, H., Zhou, Q., Jiang, S. and Li, J. (2017). *Riemerella anatipestifer* extracellular protease S blocks complement activation via the classical and lectin pathways, *Avian Pathology* 46(4), 426–433. doi: 10.1080/03079457.2017.1301648.

Feberwee, A., De Wit, J. J. and Landman, W. J. M. (2009). Induction of eggshell apex abnormalities by *Mycoplasma synoviae*: field and experimental studies, *Avian Pathology* 38(1), 77–85. doi: 10.1080/03079450802662772.

Ferguson-Noel, N., Cookson, K., Laibinis, V. A. and Kleven, S. H. (2012). The efficacy of three commercial *Mycoplasma gallisepticum* vaccines in laying hens, *Avian Diseases* 56(2), 272–275. doi: 10.1637/9952-092711-Reg.1.

Ferguson-Noel, N. M. and Williams, S. M. (2015). The efficacy of *Mycoplasma gallisepticum* K-strain live vaccine in broiler and layer chickens, *Avian Pathology* 44(2), 75–80. doi: 10.1080/03079457.2015.1005054.

Francis, M. J. (2021). Spotlight on avian pathology: the importance of recombinant vector platform technologies in poultry vaccination, *Avian Pathology* 50(2), 109–111. doi: 10.1080/03079457.2021.1875624.

Gallardo, R. A., Da Silva, A. P., Egaña-Labrin, S., Stoute, S., Kern, C., Zhou, H., Cutler, G. and Corsiglia, C. (2020). Infectious coryza: persistence, genotyping, and vaccine testing, *Avian Diseases* 64(2), 157–165. doi: 10.1637/0005-2086-64.2.157.

García, A., Romo, F., Ortiz, A. M. and Blackall, P. J. (2008). The vaccination-challenge trial: the gold standard test to evaluate the protective efficacy of infectious coryza vaccines, *Avian Pathology* 37(2), 183–186. doi: 10.1080/03079450801929941.

Ghunaim, H., Abu-Madi, M. A. and Kariyawasam, S. (2014). Advances in vaccination against avian pathogenic *Escherichia coli* respiratory disease: potentials and limitations, *Veterinary Microbiology* 172(1–2), 13–22. doi: 10.1016/j.vetmic.2014.04.019.

Goor, A. Van, Stromberg, Z. R. and Mellata, M. (2017). A recombinant multi-antigen vaccine with broad protection potential against avian pathogenic *Escherichia coli*, *PLoS ONE* 12(8), e0183929. doi: 10.1371/journal.pone.0183929.

Gross, W. B. (1957). Pathological changes of an *Escherichia coli* infection in chickens and turkeys, *American Journal of Veterinary Research* 18(69), 724–730.

Guo, R., Li, Z., Zhou, X., Huang, C., Hu, Y., Geng, S., Chen, X., Li, Q., Pan, Z. and Jiao, X. (2019). Induction of arthritis in chickens by infection with novel virulent *Salmonella* pullorum strains, *Veterinary Microbiology* 228, 165–172. doi: 10.1016/j.vetmic.2018.11.032.

Gyles, C. L. (2011). Relevance in pathogenesis research, *Veterinary Microbiology* 153(1–2), 2–12. doi: 10.1016/j.vetmic.2011.04.020.

Haag, A. F., Fitzgerald, J. R. and Penadés, J. R. (2019). *Staphylococcus aureus* in animals, *Microbiology Spectrum* 7(3). doi: 10.1128/microbiolspec.GPP3-0060-2019.

Harper, M., Wright, A., St Michael, F., Li, J., Deveson Lucas, D., Ford, M., Adler, B., Cox, A. D. and Boyce, J. D. (2017). Characterization of two novel lipopolysaccharide phosphoethanolamine transferases in *Pasteurella multocida* and their role in resistance to cathelicidin-2', *Infection and Immunity* 85(11). doi: 10.1128/IAI.00557-17.

Hess, C., Grafl, B., Bagheri, S., Kaesbohrer, A., Zloch, A. and Hess, M. (2020a). Antimicrobial resistance profiling of *Gallibacterium anatis* from layers reveals high number of multiresistant strains and substantial variability even between Isolates from the same organ, *Microbial Drug Resistance* 26(2), 169–177. doi: 10.1089/mdr.2019.0056.

Hess, C., Jandreski, C., Liebhart, D., Bilic, I. and Hess, M. (2020b). Outbreaks of Streptococcus gallolyticus subsp. *pasteurianus* in goslings characterized by central nervous symptoms, *Avian Diseases* 65(1). doi: 10.1637/aviandiseases-D-20-00101.

Hess, C., Zloch, A., Bilic, I., Hacksteiner, K., Kuchling, S. and Hess, M. (2017). High prevalence of *Brachyspira* spp. in layers kept in alternative husbandry systems associated with frequent species variations from end of rearing to slaughter, *Avian Pathology* 46(5), 481–487. doi: 10.1080/03079457.2017.1315049.

Hess, M. (2017). Commensal or pathogen–a challenge to fulfil Koch's postulates, *British Poultry Science* 58(1), 1–12. doi: 10.1080/00071668.2016.1245849.

Hofacre, C. L. and Glisson, J. R. (1986). A serotypic survey of *Pasteurella multocida* isolated from poultry, *Avian Diseases* 30(3), 632–633.

Hoiseth, S. K. and Stocker, B. A. D. (1981). Aromatic-dependent *Salmonella typhimurium* are non-virulent and effective as live vaccines, *Nature* 291(5812), 238–239. doi: 10.1038/291238a0.

Hu, Y., Huang, J. and Jiao, X. A. (2014). Screening of genes expressed in vivo during interaction between chicken and Campylobacter jejuni, *Journal of Microbiology and Biotechnology* 24(2), 217–224. doi: 10.4014/jmb.1308.08092.

Iritani, Y., Katagiri, K. and Arita, H. (1980). Purification and properties of *Haemophilus paragallinarum* hemagglutinin, *American Journal of Veterinary Research* 41(12), 2114–2118.

Ishfaq, M., Hu, W., Khan, M. Z., Ahmad, I., Guo, W. and Li, J. (2020). Current status of vaccine research, development, and challenges of vaccines for *Mycoplasma gallisepticum*, *Poultry Science* 99(9), 4195–4202. doi: 10.1016/j.psj.2020.06.014.

Jeffrey, J. S., Nolan, L. K., Tonooka, K. H., Wolfe, S., Giddings, C. W., Horne, S. M., Foley, S. L., Lynne, A. M., Ebert, J. O., Elijah, L. M., Bjorklund, G., Pfaff-McDonough, S. J., Singer, R. S. and Doetkott, C. (2002). Virulence factors of *Escherichia coli* from cellulitis or colisepticemia lesions in chickens, *Avian Diseases* 46(1), 48–52. doi: 10.1637/0005-2086(2002)046[0048:VFOECF]2.0.CO;2.

Jeong, O. M., Kang, M. S., Blackall, P. J., Jeon, B. W., Kim, J. H., Jeong, J., Lee, H. J., Kim, D. W., Kwon, Y. K. and Kim, J. H. (2020). Genotypic divergence of Avibacterium paragallinarum isolates with different growth requirements for nicotinamide adenine dinucleotide, *Avian Pathology* 49(2), 153–160. doi: 10.1080/03079457.2019.1692128.

Jung, A., Chen, L. R., Suyemoto, M. M., Barnes, H. J. and Borst, L. B. (2018). A review of *Enterococcus cecorum* infection in poultry, *Avian Diseases* 62(3), 261–271. doi: 10.1637/11825-030618-Review.1.

Kanci Condello, A., Underwood, G. J., Shil, P. K., Noormohammadi, A. H., Markham, P. F., Wawegama, N. K. and Browning, G. F. (2020). *Mycoplasma gallisepticum* strain ts-304 is a safe and effective live attenuated vaccine for use in chickens, *Veterinary Microbiology* 244, 108654. doi: 10.1016/j.vetmic.2020.108654.

Keyburn, A. L., Boyce, J. D., Vaz, P., Bannam, T. L., Ford, M. E., Parker, D., Di Rubbo, A., Rood, J. I. and Moore, R. J. (2008). NetB, a new toxin that is associated with avian necrotic enteritis caused by *Clostridium perfringens*, *PLoS Pathogens* 4(2), e26. doi: 10.1371/journal.ppat.0040026.

Khan, S. A., Stratford, R., Wu, T., Mckelvie, N., Bellaby, T., Hindle, Z., Sinha, K. A., Eltze, S., Mastroeni, P., Pickard, D., Dougan, G., Chatfield, S. N. and Brennan, F. R. (2003). *Salmonella typhi* and *S. typhimurium* derivatives harbouring deletions in aromatic biosynthesis and *Salmonella* Pathogenicity Island-2 (SPI-2) genes as vaccines and vectors, *Vaccine* 21(5–6), 538–548. doi: 10.1016/S0264-410X(02)00410-3.

Kittler, S., Seinige, D., Meemken, D., Müller, A., Wendlandt, S., Ehricht, R., Monecke, S. and Kehrenberg, C. (2019). Characteristics of methicillin-resistant *Staphylococcus aureus* from broiler farms in Germany are rather lineage- than source-specific, *Poultry Science* 98(12), 6903–6913. doi: 10.3382/ps/pez439.

Kleven, S. H. (2008). Control of avian *Mycoplasma* infections in commercial poultry, *Avian Diseases* 52(3), 367–374. doi: 10.1637/8323-041808-Review.1.

Knab, R., Petersen, H., Lin, H. J., Meixner, M., Rautenschlein, S. and Jung, A. (2020). *In vitro* characterization and genetic diversity of *Bordetella avium* field strains, *Avian Pathology* 49(1), 36–46. doi: 10.1080/03079457.2019.1660305.

Kogut, M. H. (2019). The effect of microbiome modulation on the intestinal health of poultry, *Animal Feed Science and Technology* 250, 32–40. doi: 10.1016/j.anifeedsci.2018.10.008.

Koutsianos, D., Gantelet, H., Franzo, G., Lecoupeur, M., Thibault, E., Cecchinato, M. and Koutoulis, K. C. (2020). An assessment of the level of protection against colibacillosis conferred by several autogenous and/or commercial vaccination programs in conventional pullets upon experimental challenge, *Veterinary Sciences* 7(3). doi: 10.3390/VETSCI7030080.

Kudirkiene, E., Bager, R. J., Johnson, I. J. and Bojesen, A. M. (2014). Chaperone-usher fimbriae in a diverse selection of *Gallibacterium* genomes, *BMC Genomics* 15, 1093. doi: 10.1186/1471-2164-15-1093.

La Ragione, R. M., Woodward, M. J., Kumar, M., Rodenberg, J., Fan, H., Wales, A. D. and Karaca, K. (2013). Efficacy of a live attenuated *Escherichia coli* O78:K80 vaccine in chickens and Turkeys, *Avian Diseases* 57(2), 273–279. doi: 10.1637/10326-081512-Reg.1.

Landman, W. J. M. and van Eck, J. H. H. (2017). The efficacy of inactivated *Escherichia coli* autogenous vaccines against the *E. coli* peritonitis syndrome in layers, *Avian Pathology* 46(6), 658–665. doi: 10.1080/03079457.2017.1346231.

Lawton, S. J., Weis, A. M., Byrne, B. A., Fritz, H., Taff, C. C., Townsend, A. K., Weimer, B. C., Mete, A., Wheeler, S. and Boyce, W. M. (2018). Comparative analysis of *Campylobacter* isolates from wild birds and chickens using MALDI-TOF MS, biochemical testing, and DNA sequencing, *Journal of Veterinary Diagnostic Investigation* 30(3), 354–361. doi: 10.1177/1040638718762562.

Lee, H. J., Jeong, J. Y., Jeong, O. M., Youn, S. Y., Kim, J. H., Kim, D. W., Yoon, J. U., Kwon, Y. K. and Kang, M. S. (2020). Impact of *Dermanyssus gallinae* infestation on persistent outbreaks of fowl typhoid in commercial layer chicken farms, *Poultry Science* 99(12), 6533–6541. doi: 10.1016/j.psj.2020.09.035.

Lee, K. W., Lillehoj, H. S., Jeong, W., Jeoung, H. Y. and An, D. J. (2011). Avian necrotic enteritis: experimental models, host immunity, pathogenesis, risk factors, and vaccine development, *Poultry Science* 90(7), 1381–1390. doi: 10.3382/ps.2010-01319.

Levisohn, S. and Kleven, S. H. (2000). Avian mycoplasmosis (Mycoplasma gallisepticum), *Revue Scientifique et Technique* 19(2), 425–442.

Li, L., Thøfner, I., Christensen, J. P., Ronco, T., Pedersen, K. and Olsen, R. H. (2017). Evaluation of the efficacy of an autogenous *Escherichia coli* vaccine in broiler breeders, *Avian Pathology* 46(3), 300–308. doi: 10.1080/03079457.2016.1267857.

Li, P., He, F., Wu, C., Zhao, G., Hardwidge, P. R., Li, N. and Peng, Y. (2020). Transcriptomic analysis of chicken lungs infected With avian and bovine *Pasteurella multocida* serotype A, *Frontiers in Veterinary Science* 7, 452. doi: 10.3389/fvets.2020.00452.

Li, Y., Dai, J., Zhuge, X., Wang, H., Hu, L., Ren, J., Chen, L., Li, D. and Tang, F. (2016). Iron-regulated gene ireA in avian pathogenic *Escherichia coli* participates in adhesion and stress-resistance, *BMC Veterinary Research* 12(1), 167. doi: 10.1186/s12917-016-0800-y.

Li, Z., Wang, H., Zheng, W., Li, B., Wei, Y., Zeng, J. and Lei, C. (2019). A tracing method of airborne bacteria transmission across built environments, *Building and Environment* 164, 106335. doi: 10.1016/j.buildenv.2019.106335.

Linde, K., Fthenakis, G. C. and Fichtner, A. (1998). Bacterial live vaccines with graded level of attenuation achieved by antibiotic resistance mutations: transduction experiments on the functional unit of resistance, attenuation and further accompanying markers, *Veterinary Microbiology* 62(2), 121–134. doi: 10.1016/S0378-1135(98)00201-6.

Liu, M. F., Huang, M., Shui, Y., Biville, F., Zhu, D., Wang, M., Jia, R., Chen, S., Sun, K., Zhao, X., Yang, Q., Wu, Y., Chen, X. and Cheng, A. (2018). Roles of B739_1343 in iron acquisition and pathogenesis in *Riemerella anatipestifer* CH-1 and evaluation of the RA-CH-1ΔB739_1343 mutant as an attenuated vaccine, *PLoS ONE* 13(5), e0197310. doi: 10.1371/journal.pone.0197310.

Lopes, V. C., Back, A., Shin, H. J., Halvorson, D. A. and Nagaraja, K. V. (2002a). Development, characterization, and preliminary evaluation of a temperature-sensitive mutant of *Ornithobacterium rhinotracheale* for potential use as a live vaccine in turkeys, *Avian Diseases* 46(1), 162–168. doi: 10.1637/0005-2086(2002)046[0162:DCAPEO]2.0.CO;2.

Lopes, V. C., Velayudhan, B., Halvorson, D. A. and Nagaraja, K. V. (2002b). Survival of *Ornithobacterium rhinotracheale* in sterilized poultry litter, *Avian Diseases* 46(4), 1011–1014 doi: 10.1637/0005-2086(2002)046[1011:SOORIS]2.0.CO;2.

Maheswaran, S. K., McDowell, J. R. and Pomeroy, B. S. (1973). Studies on *Pasteurella multocida*. I. Efficacy of an avirulent mutant as a live vaccine in turkeys, *Avian Diseases* 17(2), 396–405. doi: 10.2307/1589224.

Marois, C., Dufour-Gesbert, F. and Kempf, I. (2002). Polymerase chain reaction for detection of *Mycoplasma gallisepticum* in environmental samples, *Avian Pathology* 31(2), 163–168. doi: 10.1080/03079450120118658.

Marois, C., Oufour-Gesbert, F. and Kempf, I. (2000). Detection of *Mycoplasma synoviae* in poultry environment samples by culture and polymerase chain reaction, *Veterinary Microbiology* 73(4), 311–318. doi: 10.1016/S0378-1135(00)00178-4.

Matos, M., Sommer, F., Liebhart, D., Bilic, I., Hess, M. and Hess, C. (2020). An outbreak of pullorum disease in a young layer parent flock in Austria presented with central nervous system signs, *Avian Diseases* 65(1). doi: 10.1637/aviandiseases-D-20-00091.

Miller, T., Prager, R., Rabsch, W., Fehlhaber, K. and Voss, M. (2010). Epidemiological relationship between *Salmonella infantis* isolates of human and broiler origin, *Lohmann Information* 45, 27–31.

Morrow, C. J., Markham, J. F. and Whithear, K. G. (1998). Production of temperature-sensitive clones of *Mycoplasma synoviae* for evaluation as live vaccines, *Avian Diseases* 42(4), 667–670. doi: 10.2307/1592700.

Mot, D., Timbermont, L., Haesebrouck, F., Ducatelle, R. and Van Immerseel, F. (2014). Progress and problems in vaccination against necrotic enteritis in broiler chickens, *Avian Pathology* 43(4), 290–300. doi: 10.1080/03079457.2014.939942.

Nakamura, M., Nagamine, N., Takahashi, T., Suzuki, S. and Sato, S. (1994). Evaluation of the efficacy of a bacterin against *Salmonella enteritidis* infection and the effect of stress after vaccination, *Avian Diseases* 38(4), 717–724. doi: 10.2307/1592106.

Narasinakuppe Krishnegowda, D., Dhama, K., Kumar Mariappan, A., Munuswamy, P., Iqbal Yatoo, M., Tiwari, R., Karthik, K., Bhatt, P. and Reddy, M. R. (2020). Etiology, epidemiology, pathology, and advances in diagnosis, vaccine development, and treatment of *Gallibacterium anatis* infection in poultry: a review, *Veterinary Quarterly* 40(1), 16–34. doi: 10.1080/01652176.2020.1712495.

Nhung, N. T., Chansiripornchai, N. and Carrique-Mas, J. J. (2017). Antimicrobial resistance in bacterial poultry pathogens: a review, *Frontiers in Veterinary Science* 4, 126. doi: 10.3389/fvets.2017.00126.

Nolan, L. K., Vaillancourt, J.-P., Barbieri, N. L., Logue, C. M. (2020). Colibacillosis. In: Swayne, D. E. (editor in chief), *Diseases of Poultry* (14th edn.). Wiley-Blackwell Press, Iowa, 770–830.

Noro, T., Oishi, E., Kaneshige, T., Yaguchi, K., Amimoto, K. and Shimizu, M. (2008). Identification and characterization of haemagglutinin epitopes of *Avibacterium paragallinarum* serovar C, *Veterinary Microbiology* 131(3–4), 406–413. doi: 10.1016/j.vetmic.2008.04.010.

Oh, J. Y., Kang, M. S., Yoon, H., Choi, H. W., An, B. K., Shin, E. G., Kim, Y. J., Kim, M. J., Kwon, J. H. and Kwon, Y. K. (2012). The embryo lethality of *Escherichia coli* isolates and its relationship to the presence of virulence-associated genes, *Poultry Science* 91(2), 370–375. doi: 10.3382/ps.2011-01807.

Olson, L. D., McCune, E. L. and Bond, R. E. (1969). Comparison of commercial and autogenous bacterins for control of the cranial form of fowl cholera in turkeys, *Avian Diseases* 13(2), 252–260. doi: 10.2307/1588493.

Omaleki, L., Blackall, P. J., Cuddihy, T., Beatson, S. A., Forde, B. M. and Turni, C. (2020). Using genomics to understand inter-and intra-outbreak diversity of *Pasteurella multocida* isolates associated with fowl cholera in meat chickens, *Microbial Genomics* 6(3). doi: 10.1099/mgen.0.000346.

Opriessnig, T., Forde, T. and Shimoji, Y. (2020). *Erysipelothrix* spp.: past, present, and future directions in vaccine research, *Frontiers in Veterinary Science* 7, 174. doi: 10.3389/fvets.2020.00174.

Palmieri, N., Hess, C., Hess, M. and Alispahic, M. (2020). Sequencing of five poultry strains elucidates phylogenetic relationships and divergence in virulence genes in *Morganella morganii*, *BMC Genomics* 21(1), 579. doi: 10.1186/s12864-020-07001-2.

Parent, E., Burns, P., Desrochers, A. and Boulianne, M. (2018). A ligated intestinal loop model in anesthetized specific pathogen free chickens to study *Clostridium perfringens* virulence, *Journal of Visualized Experiments: JoVE* (140). doi: 10.3791/57523.

Parreira, V. R., Russell, K., Athanasiadou, S. and Prescott, J. F. (2016). Comparative transcriptome analysis by RNAseq of necrotic enteritis *Clostridium perfringens* during *in vivo* colonization and in vitro conditions, *BMC Microbiology* 16(1), 186. doi: 10.1186/s12866-016-0792-6.

Paudel, S., Hess, M. and Hess, C. (2017). Coinfection of *Avibacterium paragallinarum* and *Gallibacterium anatis* in Specific-Pathogen-Free chickens complicates clinical signs of Infectious Coryza, which can be prevented by vaccination, *Avian Diseases* 61(1), 55–63. doi: 10.1637/11481-081016-Reg.

Paudel, S., Liebhart, D., Hess, M. and Hess, C. (2014a). Pathogenesis of *Gallibacterium anatis* in a natural infection model fulfils Koch's postulates: 1. Folliculitis and drop in egg production are the predominant effects in specific pathogen free layers, *Avian Pathology* 43(5), 443–449. doi: 10.1080/03079457.2014.955782.

Paudel, S., Liebhart, D., Aurich, C., Hess, M. and Hess, C. (2014b). Pathogenesis of *Gallibacterium anatis* in a natural infection model fulfils Koch's postulates: 2. Epididymitis and decreased semen quality are the predominant effects in specific pathogen free cockerels, *Avian Pathology* 43(6), 529–534. doi: 10.1080/03079457.2014.967176.

Perez, K., Mullen, N., Canter, J. A., Ley, D. H. and May, M. (2020). Phenotypic diversity in an emerging mycoplasmal disease, *Microbial Pathogenesis* 138, 103798. doi: 10.1016/j.micpath.2019.103798.

Persson, G. and Bojesen, A. M. (2015). Bacterial determinants of importance in the virulence of *Gallibacterium anatis* in poultry, *Veterinary Research* 46, 57. doi: 10.1186/s13567-015-0206-z.

Pflaum, K., Tulman, E. R., Canter, J., Dhondt, K. V., Reinoso-Perez, M. T., Dhondt, A. A. and Geary, S. J. (2020). The influence of host tissue on *M. gallisepticum* vlhA gene expression, *Veterinary Microbiology* 251, 108891. doi: 10.1016/j.vetmic.2020.108891.

Pourbakhsh, S. A., Boulianne, M., Martineau-Doizé, B., Dozois, C. M., Desautels, C. and Fairbrother, J. M. (1997). Dynamics of *Escherichia coli* infection in experimentally inoculated chickens, *Avian Diseases* 41(1), 221–233. doi: 10.2307/1592463.

Rabie, N. S. and Amin Girh, Z. M. S. (2020). Bacterial vaccines in poultry, *Bulletin of the National Research Centre* 44(1), 15. doi: 10.1186/s42269-019-0260-1.

Rampin, T., Sironi, G. and Gallazzi, D. (1989). Episodes of amyloidosis in young hens after repeated use of antibacterial oil emulsion vaccines, *DTW. Deutsche Tierarztliche Wochenschrift* 96(4), 168–172.

Redweik, G. A. J., Jochum, J. and Mellata, M. (2020). Live bacterial prophylactics in modern poultry, *Frontiers in Veterinary Science* 7, 592312. doi: 10.3389/fvets.2020.592312.

Revolledo, L. (2018). Vaccines and vaccination against fowl typhoid and pullorum disease: an overview and approaches in developing countries, *Journal of Applied Poultry Research* 27(3), 279–291. doi: 10.3382/japr/pfx066.

Riley, L. W. (2018). Laboratory methods in molecular epidemiology: bacterial Infections, *Microbiology Spectrum* 6(6). doi: 10.1128/microbiolspec.AME-0004-2018.

Ronco, T., Stegger, M., Olsen, R. H., Sekse, C., Nordstoga, A. B., Pohjanvirta, T., Lilje, B., Lyhs, U., Andersen, P. S. and Pedersen, K. (2017). Spread of avian pathogenic *Escherichia coli* ST117 78, *BMC Genomics* 18(1), 13. doi: 10.1186/s12864-016-3415-6.

Ruiz, J. A. and Sandhu, T. S. (2020). *Riemerella anatipestifer* infection. In: David, S. (editor in chief), *Diseases of Poultry* (14th edn.). Wiley Blackwell Press, Iowa, 846–889.

Sandford, E. E., Orr, M., Balfanz, E., Bowerman, N., Li, X., Zhou, H., Johnson, T. J., Kariyawasam, S., Liu, P., Nolan, L. K. and Lamont, S. J. (2011). Spleen transcriptome response to infection with avian pathogenic *Escherichia coli* in broiler chickens, *BMC Genomics* 12, 469. doi: 10.1186/1471-2164-12-469.

Sawicka, A., Durkalec, M., Tomczyk, G. and Kursa, O. (2020). Occurrence of *Mycoplasma gallisepticum* in wild birds: a systematic review and metaanalysis, *PLoS ONE* 15(4), e0231545. doi: 10.1371/journal.pone.0231545.

Schat, K. A., Nagaraja, K. V. and Saif, Y. M. (2021). Pullorum disease: evolution of the eradication strategy, *Avian Diseases* 65(2), 227–236. doi: 10.1637/aviandiseases-D-21-00024.

Schuijffel, D. F., Van Empel, P. C., Segers, R. P., Van Putten, J. P. and Nuijten, P. J. (2006). Vaccine potential of recombinant *Ornithobacterium rhinotracheale* antigens, *Vaccine* 24(11), 1858–1867. doi: 10.1016/j.vaccine.2005.10.031.

Scott, P. C., Markham, J. F. and Whithear, K. G. (1999). Safety and efficacy of two live *Pasteurella multocida* aro-A mutant vaccines in chickens, *Avian Diseases* 43(1), 83–88. doi: 10.2307/1592765.

Seo, H. S., Cha, S. Y., Kang, M. and Jang, H. K. (2013). Chicken embryo lethality assay for determining the virulence of *Riemerella anatipestifer* isolates, *Avian Pathology* 42(4), 387–392. doi: 10.1080/03079457.2013.816654.

Shang, Y., Kumar, S., Oakley, B. and Kim, W. K. (2018). Chicken gut microbiota: importance and detection technology, *Frontiers in Veterinary Science* 5, 254. doi: 10.3389/fvets.2018.00254.

Singh, R., Blackall, P. J., Remington, B. and Turni, C. (2013). Studies on the presence and persistence of *Pasteurella multocida* serovars and genotypes in fowl cholera outbreaks, *Avian Pathology* 42(6), 581–585. doi: 10.1080/03079457.2013.854861.

Solà-Ginés, M., González-López, J. J., Cameron-Veas, K., Piedra-Carrasco, N., Cerdà-Cuéllar, M. and Migura-Garcia, L. (2015). Houseflies (*Musca domestica*) as vectors for extended-spectrum β-lactamase-producing *Escherichia coli* on Spanish broiler farms, *Applied and Environmental Microbiology* 81(11), 3604–3611. doi: 10.1128/AEM.04252-14.

Soriano, E. V., Garduño, M. L., Téllez, G., Rosas, P. F., Suárez-Güemes, F. and Blackall, P. J. (2004). Cross-protection study of the nine serovars of *Haemophilus paragallinarum* in the Kume haemagglutinin scheme, *Avian Pathology* 33(5), 506–511. doi: 10.1080/03079450400003502.

Sprenger, S. J., Halvorson, D. A., Shaw, D. P. and Nagaraja, K. V. (2000). *Ornithobacterium rhinotracheale* infection in turkeys: immunoprophylaxis studies, *Avian Diseases* 44(3), 549-555. doi: 10.2307/1593094.

Tao, Z., Zhu, C., Xu, W., Shi, Z., Zhang, S., Song, W., Liu, H. and Li, H. (2020). *Riemerella anatipestifer* infection affects intestinal barrier structure and immune reactions in the duck cecum, *Avian Pathology* 49(6), 572-580. doi: 10.1080/03079457.2020.1792414.

ter Veen, C., Dijkman, R., de Wit, J. J., Gyuranecz, M. and Feberwee, A. (2021). Decrease of *Mycoplasma gallisepticum* seroprevalence and introduction of new genotypes in Dutch commercial poultry during the years 2001-2018, *Avian Pathology* 50(1), 52-60. doi: 10.1080/03079457.2020.1832958.

ter Veen, C., de Wit, J. J. and Feberwee, A. (2020). Relative contribution of vertical, within-farm and between-farm transmission of *Mycoplasma synoviae* in layer pullet flocks, *Avian Pathology* 49(1), 56-61. doi: 10.1080/03079457.2019.1664725.

Tiseo, K., Huber, L., Gilbert, M., Robinson, T. P. and Van Boeckel, T. P. (2020). Global trends in antimicrobial use in food animals from 2017 to 2030, *Antibiotics* 9(12). doi: 10.3390/antibiotics9120918.

Turni, C., Singh, R. and Blackall, P. J. (2018). Genotypic diversity of *Pasteurella multocida* isolates from pigs and poultry in Australia, *Australian Veterinary Journal* 96(10), 390-394. doi: 10.1111/avj.12748.

Udhayavel, S., Ramasamy, G. T., Gowthaman, V., Malmarugan, S. and Senthilvel, K. (2017). Occurrence of *Clostridium perfringens* contamination in poultry feed ingredients: isolation, identification and its antibiotic sensitivity pattern, *Animal Nutrition* 3(3), 309-312. doi: 10.1016/j.aninu.2017.05.006.

Van Deun, K., Pasmans, F., Ducatelle, R., Flahou, B., Vissenberg, K., Martel, A., Van den Broeck, W., Van Immerseel, F. and Haesebrouck, F. (2008). Colonization strategy of *Campylobacter jejuni* results in persistent infection of the chicken gut, *Veterinary Microbiology* 130(3-4), 285-297. doi: 10.1016/j.vetmic.2007.11.027.

Van Empel, P. and Van Den Bosch, H. (1998). Vaccination of chickens against *Ornithobacterium rhinotracheale* infection, *Avian Diseases* 42(3), 572-578. doi: 10.2307/1592684.

van Immerseel, F., Methner, U., Rychlik, I., Nagy, B., Velge, P., Martin, G., Foster, N., Ducatelle, R. and Barrow, P. A. (2005). Vaccination and early protection against non-host-specific *Salmonella* serotypes in poultry: exploitation of innate immunity and microbial activity, *Epidemiology and Infection* 133(6), 959-978. doi: 10.1017/S0950268805004711.

Varinrak, T., Poolperm, P., Sawada, T. and Sthitmatee, N. (2017). Cross-protection conferred by immunization with an rOmpH-based intranasal fowl cholera vaccine, *Avian Pathology*. doi: 10.1080/03079457.2017.1321105.

Vogl, G., Plaickner, A., Szathmary, S., Stipkovits, L., Rosengarten, R. and Szostak, M. P. (2008). *Mycoplasma gallisepticum* invades chicken erythrocytes during infection, *Infection and Immunity* 76(1), 71-77. doi: 10.1128/IAI.00871-07.

Wang, C., Pors, S. E., Olsen, R. H. and Bojesen, A. M. (2018). Transmission and pathogenicity of *Gallibacterium anatis* and *Escherichia coli* in embryonated eggs, *Veterinary Microbiology* 217, 76-81. doi: 10.1016/j.vetmic.2018.03.005.

Wang, S., Shi, Z., Xia, Y., Li, H., Kou, Y., Bao, Y., Dai, J. and Lu, C. (2012). IbeB is involved in the invasion and pathogenicity of avian pathogenic *Escherichia coli*, *Veterinary Microbiology* 159(3-4), 411-419. doi: 10.1016/j.vetmic.2012.04.015.

Wang, Y., Yin, X., Zhou, Z., Hu, S., Li, S., Liu, M., Wang, X., Xiao, Y., Shi, D., Bi, D. and Li, Z. (2019). Cas9 regulated gene expression and pathogenicity in *Riemerella anatipestifer*, *Microbial Pathogenesis* 136, 103706. doi: 10.1016/j.micpath.2019.103706.

Whithear, K. G. (1996). Control of avian mycoplasmoses by vaccination, *Revue Scientifique et Technique* 15(4), 1527–1553. doi: 10.20506/rst.15.4.985.

Whithear, K. G., Soeripto, Harringan, K. E. and Ghiocas, E. (1990). Safety of temperature sensitive mutant *Mycoplasma gallisepticum* vaccine, *Australian Veterinary Journal* 67(5), 159–165. doi: 10.1111/j.1751-0813.1990.tb07745.x.

Wigley, P. (2013). Immunity to bacterial infection in the chicken, *Developmental and Comparative Immunology* 41(3), 413–417. doi: 10.1016/j.dci.2013.04.008.

Wigley, P. (2014). Salmonella enterica in the chicken: how it has helped our understanding of immunology in a non-biomedical model species, *Frontiers in Immunology* 5, 482. doi: 10.3389/fimmu.2014.00482.

Wigley, P. (2017). Salmonella enterica serovar Gallinarum: addressing fundamental questions in bacteriology sixty years on from the 9R vaccine, *Avian Pathology* 46(2), 119–124. doi: 10.1080/03079457.2016.1240866.

Wigley, P., Berchieri, A., Page, K. L., Smith, A. L. and Barrow, P. A. (2001). Salmonella enterica serovar Pullorum persists in splenic macrophages and in the reproductive tract during persistent, disease-free carriage in chickens, *Infection and Immunity* 69(12), 7873–7879. doi: 10.1128/IAI.69.12.7873-7879.2001.

Wijesurendra, D. S., Kanci, A., Tivendale, K. A., Bacci, B., Noormohammadi, A. H., Browning, G. F. and Markham, P. F. (2015). Development of a *Mycoplasma gallisepticum* infection model in turkeys, *Avian Pathology* 44(1), 35–42. doi: 10.1080/03079457.2014.992390.

Wooley, R. E., Gibbs, P. S., Brown, T. P. and Maurer, J. J. (2000). Chicken embryo lethality assay for determining the virulence of avian *Escherichia coli* isolates, *Avian Diseases* 44(2), 318–324. doi: 10.2307/1592546.

Wu, J. R., Wu, Y. R., Shien, J. H., Hsu, Y. M., Chen, C. F., Shieh, H. K. and Chang, P. C. (2011). Recombinant proteins containing the hypervariable region of the haemagglutinin protect chickens against challenge with *Avibacterium paragallinarum*, *Vaccine* 29(4), 660–667. doi: 10.1016/j.vaccine.2010.11.040.

Wysok, B., Wojtacka, J. and Kivistö, R. (2020). Pathogenicity of *Campylobacter* strains of poultry and human origin from Poland, *International Journal of Food Microbiology* 334, 108830. doi: 10.1016/j.ijfoodmicro.2020.108830.

Yin, J., Cheng, Z., Wang, X., Xu, L., Li, Q., Geng, S. and Jiao, X. (2015). Evaluation of the Salmonella enterica serovar Pullorum pathogenicity island 2 mutant as a candidate live attenuated oral vaccine, *Clinical and Vaccine Immunology* 22(7), 706–710. doi: 10.1128/CVI.00130-15.

Zahra, M., Ferreri, M., Alkasir, R., Yin, J., Han, B. and Su, J. (2013). Isolation and characterization of small-colony variants of *Ornithobacterium rhinotracheale*, *Journal of Clinical Microbiology* 51(10), 3228–3236. doi: 10.1128/JCM.01337-13.

Zhang, J. F., Wei, B., Cha, S. Y., Shang, K., Jang, H. K. and Kang, M. (2020). The use of embryonic chicken eggs as an alternative model to evaluate the virulence of Salmonella enterica serovar Gallinarum, *PLoS ONE* 15(9), e0238630. doi: 10.1371/journal.pone.0238630.

Zhang, X. P., Lu, C. J., Li, Y. T., Yang, X., Wang, X. W., Chang, H. T., Liu, H. Y., Chen, L., Zhao, J., Wang, C. Q. and Chang, Y. F. (2017). *In vitro* adherence and invasion of primary chicken oviduct epithelial cells by *Gallibacterium anatis*, *Veterinary Microbiology* 203, 136–142. doi: 10.1016/j.vetmic.2017.02.009.

Zhang-Barber, L., Turner, A. K. and Barrow, P. A. (1999). Vaccination for control of *Salmonella* in poultry, *Vaccine* 17(20–21), 2538–2545. doi: 10.1016/S0264-410X(99)00060-2.

Zhu, D., Yang, Z., Xu, J., Wang, M., Jia, R., Chen, S., Liu, M., Zhao, X., Yang, Q., Wu, Y., Zhang, S., Liu, Y., Zhang, L., Yu, Y., Chen, X. and Cheng, A. (2020). Pan-genome analysis of *Riemerella anatipestifer* reveals its genomic diversity and acquired antibiotic resistance associated with genomic islands, *Functional and Integrative Genomics* 20(3), 307–320. doi: 10.1007/s10142-019-00715-x.

Zhuge, X., Ji, Y., Tang, F., Sun, Y., Jiang, M., Hu, W., Wu, Y., Xue, F., Ren, J., Zhu, W. and Dai, J. (2019). Population structure and antimicrobial resistance traits of avian-origin mcr-1-positive *Escherichia coli* in Eastern China, 2015 to 2017, *Transboundary and Emerging Diseases* 66(5), 1920–1929. doi: 10.1111/tbed.13222.

Zloch, A., Kuchling, S., Hess, M. and Hess, C. (2018). Influence of alternative husbandry systems on postmortem findings and prevalence of important bacteria and parasites in layers monitored from end of rearing until slaughter, *Veterinary Record* 182(12), 350. doi: 10.1136/vr.104632.

Chapter 3

Advances in understanding parasite infections of poultry: protozoa and the red mite

Damer P. Blake, Royal Veterinary College, UK; and Dieter Liebhart, University of Veterinary Medicine Vienna, Austria

1 Introduction

A wide range of parasites can infect poultry, including protozoans, cestodes, nematodes, trematodes and arthropods. As the global chicken population undergoes dramatic expansion, production systems are increasingly moving towards drug-free and/or extensive systems in much of Europe and North America, as well as greater intensification in many tropical regions, posing new challenges to pathogen control. Parasites such as *Ascaridia galli*, *Capillaria caudinflata*, *C. obsignata*, *Heterakis gallinarum*, *Davainea proglottina* and *Raillietina cesticullus* and a range of mites and other ectoparasites are returning to significance. Others, such as the *Eimeria* species, remain challenging. Changes in legislation and husbandry systems are driving increased problems with *Histomonas meleagridis*, while acaricide resistance is exacerbating difficulties with parasites such as *Dermanyssus gallinae*. Increased parasite occurrence affects poultry performance and welfare.

In this chapter, we focus on three of the most widespread and economically relevant parasites of poultry and the infections they cause:

- *Eimeria* species: coccidiosis;
- *Histomonas meleagridis*: histomonosis; and

http://dx.doi.org/10.19103/AS.2022.0104.03

- *Dermanyssus gallinae*: the poultry red mite.

In each case, the chapter discusses:

- Incidence and economic significance;
- Host-pathogen interactions;
- Population dynamics;
- Detection;
- Control and prevention; and
- Current challenges.

The chapter concludes by looking at future challenges and opportunities in parasite control.

2 *Eimeria* species: coccidiosis

2.1 Target populations, incidence and economic relevance

All livestock and poultry can be infected by multiple *Eimeria* species (Taylor et al., 2007). Most *Eimeria* are strictly limited to a single host species, although examples such as *Eimeria innocua* can replicate successfully in domestic turkeys, grey partridge and bobwhite quail (Vrba and Pakandl, 2015). *Eimeria* that infect chickens are considered to be the most economically important, primarily due to the large number of chickens that are produced every year and their rapid population turnover. More than 72 billion chickens were produced in 2019 (FAOSTAT, 2021), and production cycles lasting just 5 to 7 weeks are common for broilers. Seven *Eimeria* species are widely recognised to infect chickens, all of which have been detected on every continent where chickens are farmed (Clark et al., 2016). *Eimeria acervulina*, *E. maxima* and *E. tenella* usually are most common in broiler chickens (Haug et al., 2008, Clark et al., 2016, Hauck et al., 2019, Kumar et al., 2014), although highly pathogenic species such as *E. necatrix* and *E. brunetti* can pose significant risks when an outbreak occurs, typically in older chickens (Sawale et al., 2018). Globally, between 2% and 80% of chicken flocks require therapeutic intervention to control coccidiosis, with between 1.5% and 7.5% of individuals expected to die during an outbreak if appropriate intervention is available (Blake et al., 2020a). Despite the significance of losses due to mortality, the cost attributed to coccidiosis is primarily associated with morbidity, where reduced weight gain is the biggest single loss (Blake et al., 2020a). The global cost incurred by *Eimeria* in chickens has recently been estimated to have exceeded UK£10.4 billion in 2016, equivalent to EU€11.9 billion or US$14.2 billion at the time of writing (February 2021).

2.2 Host-pathogen interactions

2.2.1 Clinical signs, pathology and welfare

The seven *Eimeria* species that infect chickens can cause disease with distinct, albeit overlapping, pathognomonic characteristics. Each species follows an oral-faecal life cycle involving three phases of replication: asexual (schizogony, also known as merogony) and sexual (gametogony) within the host, followed by sporulation (sporogony; Fig. 1) in the environment (Shirley et al., 2005). All seven species replicate within epithelial cells of the chicken intestine, although the precise site of infection varies from the duodenum (e.g. *E. acervulina*) to the caeca (*E. tenella*) and lower intestine (e.g. *E. brunetti*). *Eimeria brunetti*, *E. necatrix* and *E. tenella* are most pathogenic, causing a haemorrhagic form of coccidiosis in the mid (*E. necatrix*) or lower gastrointestinal tract (Reid et al., 2014). Pathology is most closely related to the asexual phase of replication (schizogony) when the large and relatively invasive schizonts rupture, resulting in deep erosions and haemorrhage in the intestinal wall. *Eimeria acervulina*, *E. maxima*, *E. mitis* and *E. praecox* tend to be less pathogenic, causing a malabsorptive form of coccidiosis associated with replication of the sexual lifecycle stages during gametogony and subsequent oocyst development,

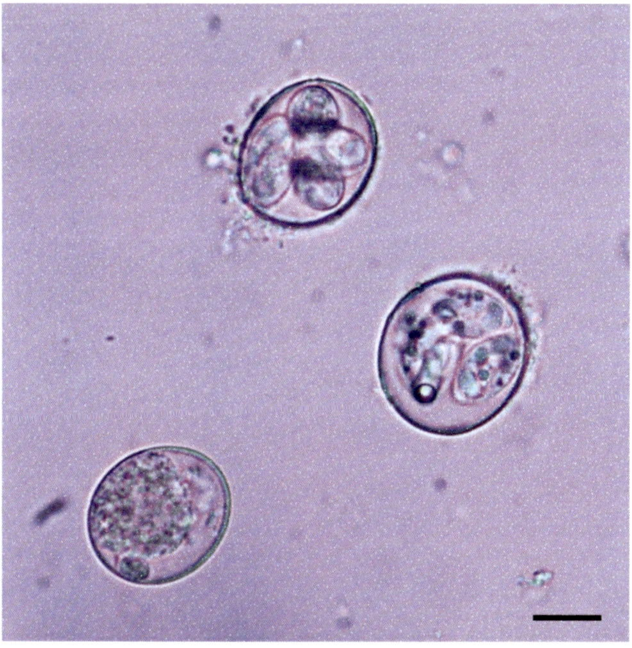

Figure 1 *Eimeria brunetti* oocysts. Bottom left = unsporulated and non-infectious. Top and right = sporulated and infectious. Scale bar = 10 μm.

although the disease can still be severe in the event of a high-level challenge. *Eimeria maxima* is the most pathogenic of the malabsorptive species, in part due to its large size. *Eimeria praecox* has been considered by some to be non-pathogenic, although evidence of pathogenic strains circulating in chicken populations has dispelled this view (Williams et al., 2009).

The clinical signs of coccidiosis, including both the location of gross pathology and the appearance of lesions, have been used to develop several lesion scoring systems to identify the infecting *Eimeria* species and describe the severity of an infection (e.g. Johnson and Reid, 1970). Less-specific signs of infection include a hunched posture, ruffled feathers, lethargy, reduced body weight gain (BWG) and increased food conversion ratio (FCR). Water consumption can also be used as a non-specific indicator of ill health.

2.2.2 Host immune responses

A protective anti-*Eimeria* immune response is primarily reliant on T lymphocytes. Treatment of chickens with cyclosporin A to prevent T lymphocyte proliferation confirmed their necessity for controlling secondary infection (Lillehoj, 1987), while the transfer of cell-mediated immunity (CMI) to *E. maxima* has also been possible (Rose and Hesketh, 1982). In studies with mice, chosen due to the availability of a more comprehensive immunological toolbox, $CD4^+$ T-cells appear to be most important in controlling primary *Eimeria* infection, supplemented by a smaller role for $CD8^+$ T-cells (Rose et al., 1992). $CD8^+$ T-cells appear to play a more significant role in secondary infections in the same study. However, several studies on chickens have suggested notable differences (Trout and Lillehoj, 1996, Cornelissen et al., 2009). A more recent study identified higher proportions of cytotoxic $CD8^+$ cells following primary infection (Wattrang et al., 2016). $CD8^+$ intraepithelial lymphocytes (IELs) are also increased following secondary *E. acervulina* infection, while genetic resistance to infection has been linked to increased $CD8^+$ IEL proportions (Lillehoj, 1994). It has been suggested that $CD8^+$ cells may function by killing infected epithelial cells (Lillehoj and Trout, 1994).

Studies of cytokines produced by T cells have indicated a key role for interferon gamma (IFN-γ) in the immune response to primary, but possibly not secondary, *Eimeria* infection. For example, blocking endogenous IFN-γ using a monoclonal antibody during *E. vermiformis* infection in mice increased susceptibility to primary but not secondary infection (Rose et al., 1989). In chickens, *E. maxima* infection has been shown to upregulate both Th1 and Th2 cytokine transcription in primary but not secondary infection (Hong et al., 2006). The contribution from other cytokines has also been assessed. For example, increased tumour necrosis factor alpha (TNF-α) is induced by primary but not secondary *E. tenella* infection (Zhang et al., 1995), although it has been suggested that this may increase pathology (Byrnes et al., 1993).

A limited role has been suggested for B lymphocytes in the natural immune response to *Eimeria* infection. Surgical bursectomy removes the ability to generate an antibody response in chickens but does not significantly reduce immune protection against secondary infection (Long and Pierce, 1963). However, it has been shown that antibodies can inhibit *Eimeria* replication under controlled circumstances, providing passive and maternal immunity against challenge and suggesting an alternative immune mechanism that is not usually induced during *Eimeria* infection (Wallach, 2010).

Considerable variation has been described in the outcome of *Eimeria* infection by individual chickens. Distinct susceptible and resistant profiles have been described in terms of performance and pathology for *E. maxima* (Boulton et al., 2018b, Hamzic et al., 2015) and *E. tenella* (Boulton et al., 2018a). Interestingly a third resistance profile, considered to be tolerant of infection as defined by good performance despite significant pathology, has been reported in commercial broiler chickens (Boulton et al., 2018a).

2.2.3 Impact on enteric dysbiosis

Beyond the direct consequences of coccidosis, *Eimeria* can also induce enteric dysbiosis. *Eimeria* infection has been linked to poor litter quality, indirectly contributing to footpad dermatitis as well as reducing overall welfare and technical performance (Abd El-Wahab et al., 2012, de Jong et al., 2014). Microbiome sequencing of enteric microbial populations has revealed notable differences in beta but not alpha diversity (i.e. variation in the levels, not presence or absence, of distinct bacterial populations), with significant variation for genera such as *Bacteroides* and *Lactobacillus* in association with *E. tenella* lesion score (Macdonald et al., 2017). Well-known interactions with specific bacteria include *Clostridium perfringens*, combining to cause necrotic enteritis (NE) (Van Immerseel et al., 2016). Less well-known interactions include increased colonisation and faecal shedding of bacterial zoonoses such as *Salmonella* Typhimurium and *Campylobacter jejuni* (Macdonald et al., 2019, Arakawa et al., 1981).

2.3 Eimeria *population dynamics*

All seven *Eimeria* species widely recognised to infect chickens have a global occurrence (Clark et al., 2016) but very little is known of their population structure or genetic diversity. It is clear from studies of antigenic diversity, using escape from strain-specific protective immunity as a phenotype, that genetic variation exists for several species including *E. acervulina* (Joyner, 1969), *E. mitis* (McDonald et al., 1985), *E. maxima* (Smith et al., 2002) and *E. tenella* (Awad et al., 2013). However, very few genetics-led studies have been undertaken. One of the most detailed studies focused on *E. tenella*, assessing variation at the apical membrane antigen 1 locus (AMA1, an anticoccidial vaccine candidate (Pastor-Fernández et al., 2020)) and a genome-wide panel of single nucleotide polymorphisms (SNPs) (Blake et al., 2015). A comparison of SNP profiles revealed notable variation between countries and regions, supporting the suggestion that variables such as climate or husbandry system shape *Eimeria* population dynamics (Blake et al., 2015, Pegg et al., 2016).

The majority of genetics-led studies for *Eimeria* have focused on single loci within the nuclear or mitochondrial genomes (e.g. the ribosomal repeat unit including internal spacer sequences and cytochrome C oxidase subunit I; Blake et al., 2020b). One such study described unexpected variation between Internal Transcribed Spacer (ITS)-2 sequences, suggesting the presence of diverse strains or cryptic species (Cantacessi et al., 2008). Recent studies including measures of oocyst morphology, pathology, genome sequencing and genetics suggest that all three, previously termed Operational Taxonomic Units (OTUs) X, Y and Z, are indeed new *Eimeria* species (Blake et al., 2021, Morgan and Godwin, 2017). The new species, tentatively been named *E. lata*, *E. nagambie* and *E. zaria*, have already been detected in parts of Africa, Asia, Australasia, North and South America, indicating a new challenge for control of coccidiosis in chickens (Clark et al., 2016, Hauck et al., 2019). The origins of these new species are not clear, although their current wide distribution suggests their appearance may represent recent recognition of pre-existing diversity rather than the emergence of genuinely new species.

2.4 Current methods of detection for Eimeria

A range of techniques and tools are available for the detection and species-specific identification of *Eimeria* infection. Routine monitoring commonly relies on microscopy to detect oocysts in faecal or litter samples (Kumar et al., 2014). Flotation using saturated saline or sucrose solutions can be used to increase sensitivity. Detection and enumeration of total eimerian oocysts are relatively straightforward, but species-specific differentiation is much more challenging and can be highly subjective (Haug et al., 2008). For example, *E. necatrix* and *E. praecox*, species defined by extremes of pathogenicity, are very difficult to differentiate by variables such as oocyst morphology alone (Long et al., 1976). Attempts to automate species identification using microscopy by systems such as COCCIMORPH offer promise (Castañón et al., 2007), although uptake has been limited. Practically, post-mortem assessment of gross pathology (lesion scoring) remains an important technique for detection and species identification. The lesion scoring system published by Johnson and Reid for five of the seven recognised species (excluding *E. mitis* and *E. praecox*) is most widely cited (Johnson and Reid, 1970). Neither of these latter species routinely results in intestinal lesions during infection.

Advances in molecular biology have improved the diagnosis of many veterinary pathogens but have proven challenging for *Eimeria*. Accessing genomic DNA for use as a template has often been limiting, requiring a laboratory for effective and reproducible extraction. Genus and species-specific detection using polymerase chain reaction (PCR) was established nearly 30 years ago (Stucki et al., 1993), with a variety of multiplex and nested options developed to improve throughput and sensitivity (Fernandez et al., 2003, Lew et al., 2003, Schwarz et al., 2009) but none have been widely adopted by industry (Fig. 2). The appearance of quantitative PCR assays specific for all recognised *Eimeria* species has had a greater impact (Vrba et al., 2010, Morgan et al., 2009), with several companies offering qPCR as a diagnostic service. A panel of loop-mediated isothermal amplification (LAMP) assays has been published for the species-specific detection of *Eimeria* that infect chickens (Barkway et al., 2011), although accessing genomic DNA as a template remains a challenge for routine application under field conditions.`

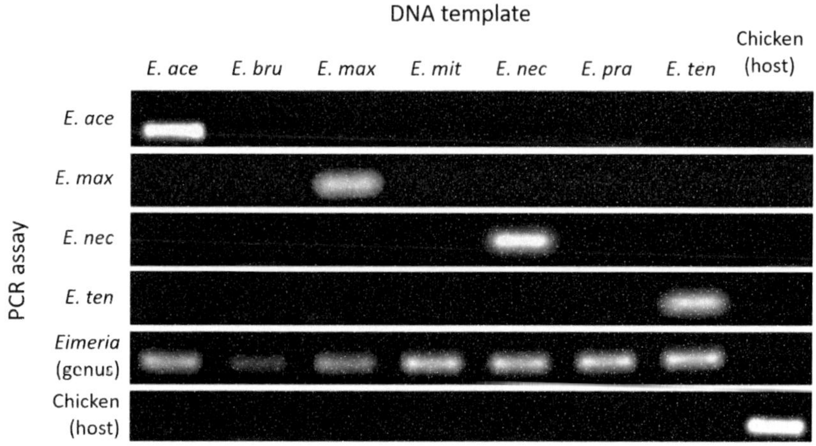

Figure 2 Example of *Eimeria* species and genus-specific PCR assays. PCR amplicons resolved for four different *Eimeria* species-specific assays targeting the internal transcribed spacer 1 (ITS1) repeat region and one *Eimeria* genus-specific assay targeting the 18S rRNA locus (all as described by Schwarz et al., 2009). DNA templates were used as indicated. The lack of cross-reactivity with the chicken host was demonstrated by the inclusion of an assay targeting the chicken glyceraldehyde-3-phosphate dehydrogenase locus (Blake et al., 2006). *E. ace* = *E. acervulina*, *E. bru* = *E. brunetti*, *E. max* = *E. maxima*, *E. mit* = *E. mitis*, *E. nec* = *E. necatrix*, *E. pra* = *E. praecox*, *E. ten* = *E. tenella*.

2.5 Current methods of control for *Eimeria*

Control of *Eimeria* relies upon good husbandry, including consideration of stocking density, ventilation rate and choice of substrate. Lower stocking densities can reduce environmental contamination with oocysts. Dry, high-quality litter can reduce oocyst sporulation, limiting infectivity (Fig. 1). However, husbandry alone is insufficient to prevent coccidiosis. Anticoccidial drugs have long dominated control of coccidiosis, including a range of synthetic or chemical anticoccidials and ionophores, which are products of fermentation (Chapman, 1997). Ionophores have been especially successful since their use permits a low level of parasite replication, even in naïve field populations, supporting the induction of a complementary protective immune response (Chapman, 1999). Importantly, ionophores are classified as antibiotics in some regions such as the USA. While ionophores are not used in human medicine and have limited direct relevance to human health, the appearance of 'no antibiotics, ever' markets for chicken products has indirectly increased the demand for alternatives (Blake et al., 2020a). Increasing reports of drug resistance and consumer concerns around drug use in livestock production have reinvigorated attempts to develop cost-effective, scalable anticoccidial vaccines. Where anticoccidial drugs remain in use, it is common to rotate between different anticoccidial drugs within and between flocks to limit and respond to selection for resistance.

The first anticoccidial vaccine was marketed in 1952 (Williams, 2002). Based upon live, unmodified *E. tenella* oocysts, the vaccine was quickly followed by other live anticoccidial vaccines including a range of different *Eimeria* species. Controlled infection using these parasite formulations induces a natural immune response and protection against subsequent challenges. Such 'wild-type' or non-attenuated vaccines are highly effective and relatively cheap to produce but risk compromising flock performance and occurrence of clinical disease if managed incorrectly (Shirley et al., 2005). The risk associated with live anticoccidial vaccines was recognised and addressed by the development of a second generation of live vaccines using attenuated parasite lines. With few exceptions, attenuation has been achieved by the selection of stable precocious lines from populations of virulent parasites. Attenuation results in shorter lifecycles and reduced replication, accompanied by lower pathogenicity while retaining immunogenicity (Shirley et al., 2005). Attenuated anticoccidial vaccines have become popular in the layer and breeder chicken sectors, but their relative cost and inherently limited production capacity have hindered application in the much larger broiler sector. However, demand for antibiotic-free poultry products is now prompting a significant shift in anticoccidial control, with ~40% of broilers sold in the USA now vaccinated using a non-attenuated product (Blake et al.,

2020a). A major selling point for live anticoccidial vaccines has been the use of drug-sensitive parasite strains, with evidence that vaccination of three or more successive flocks can significantly reduce the occurrence of drug resistance in field parasite populations (Chapman and Jeffers, 2015). The relative risk posed by these virulent vaccines can be managed using a bioshuttle approach, where vaccination of chicks on the day of hatch is followed by anticoccidial supplementation of grower and finisher diets (Kimminau and Duong, 2019). Challenges in vaccine management include ensuring effective vaccine recycling, especially for less immunogenic species such as *E. tenella* and *E. necatrix*, where multiple rounds of infection can be required to induce a robust protective immune response.

A wide range of alternatives to drugs and vaccines has been suggested to improve the control of *Eimeria*. Examples include natural herbs and botanicals or their extracts, essential oils, organic acids, immunomodulators and complex carbohydrates, probiotics and prebiotics (Khater et al., 2020). Probiotic formulations based upon *Bacillus*, *Lactobacillus* or *Saccharomyces* are becoming increasingly popular, with multiple commercial providers. *Eimeria* oocysts are relatively resistant to many common disinfectants. Historic studies indicate that ammonia gas is an effective fumigant for the control of *Eimeria* (Rawlinson et al., 1985), a finding reflected by the efficacy of ammonia-based disinfectants (Chroustová and Pinka, 1987). Formaldehyde-based disinfectants have been found to be effective during oocyst sporulation but become ineffective once sporulation is complete (Chroustová and Pinka, 1987). Phenol-based disinfectants have also been found to be effective during oocyst sporulation (Samaha et al., 2013).

2.6 Challenges and conclusions

Control of *Eimeria* and the disease coccidiosis remains a major ongoing challenge to poultry producers. Increasing public and legislative pressure to reduce the use of drugs in livestock production has prompted renewed interest in existing and novel vaccines, a range of diet-based alternatives and selective breeding of chickens for genetic resistance. It is likely that the range of drugs available to control *Eimeria* will be reduced in the future, voluntarily in some sectors and by law in others. As the industry migrates away from drugs, new challenges will emerge. Differentiation of vaccines from virulent field *Eimeria* strains is a significant gap that affects the management and application of live vaccines. The recent description of three additional *Eimeria* species that infect chickens and are capable of escaping from current vaccines indicates an immediate problem for vaccination-led control of coccidiosis, with new vaccine formulations anticipated (Blake et al., 2021). Future opportunities for genetic and genomic characterisation are becoming increasingly accessible as costs diminish and expertise is more readily available. It is likely that the coming decade will see a notable evolution in anticoccidial control for poultry.

3 *Histomonas meleagridis*: histomonosis

3.1 Target populations, incidence and economic relevance

Histomonosis is a disease of poultry with worldwide occurrence. The disease is caused by the protozoan *Histomonas meleagridis*, a flagellated parasite (Tyzzer, 1920). Gallinaceous birds can be infected with the parasite, and turkeys and chickens are the most affected hosts (Hess et al., 2015). In turkeys, histomonosis can cause severe morbidity and mortality, whereas in chickens clinical disease is less prominent (Tyzzer, 1934). Nevertheless, in both species the clinical outcome of infection can be variable, from absent clinical signs to high mortality.

Poultry production is economically affected by histomonosis as a result of retarded growth, loss in egg production and mortality. Animal welfare and economic constraints became more relevant after prophylactic and therapeutic drugs were banned for use in poultry in many countries worldwide for reasons of consumer protection (Liebhart et al., 2017). Until the twenty-first century, histomonosis could be controlled by antihistomonal drugs in Europe. However, the only applicable chemicals nitroimidazoles and nitrofurans were withdrawn in 1996 and 2003, respectively (CEC, 1995, 2002). In the USA, arsenicals were the last remaining compounds that could be used for control prior to a ban in 2015 (FDA, 2015). In many other countries, similar regulations have been applied that now preclude prophylaxis and therapy against histomonosis. Consequently, the ban on drugs effective against histomonosis has resulted in an increase in the number of cases, some of them incurring high economic losses (Hess et al., 2015, Clark and Kimminau, 2017). Severe outbreaks of the disease in turkey flocks are characterised by distinct clinical signs and pathological lesions, whereas infected chickens show less pathognomonic changes (Liebhart and Hess, 2020). However, the economic aspects of chicken production might be underestimated, as indicated in a report investigating several outbreaks in chicken flocks, especially breeder and layer flocks (Dolka et al., 2015).

In the USA, more than 100 cases of histomonosis were reported in 2016. The economic cost incurred by *H. meleagridis* worldwide is yet to be calculated but has been estimated that the economic relevance of the disease to poultry production is similar to that of coccidiosis (McDougald, 2005).

3.2 Host-pathogen interactions

3.2.1 Clinical signs and pathology

Histomonas meleagridis can cause clinical signs like apathy, ruffled feathers and drooping wings. In turkeys, diarrhoea and sulphur-coloured faeces can

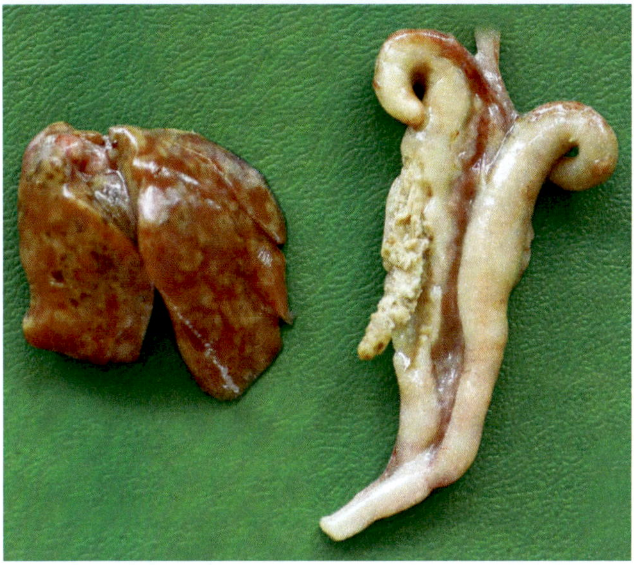

Figure 3 Lesions in liver and caeca of a turkey caused by *H. meleagridis*.

be observed as characteristic signs of the disease. In chickens, parameters like growth, weight and egg production can be the only clinical outcome.

The clinical status of birds suffering from histomonosis reflects the specific pathological changes. The caeca are the primary infected organs showing necrosis and inflammation, typically with fibrinous exudate in the lumen (Fig. 3). After tissue destruction in this part of the intestinal tract, the parasite can reach the liver via the portal vein resulting in necrosis and inflammation as indicated in Fig. 3. In chickens, lesions in the liver are less common than in turkeys, but the parasite can be distributed throughout several organs in both host species (Grabensteiner et al., 2006). However, the genotype of *H. meleagridis* has an impact on the severity of clinical signs and lesions, as outlined below.

3.2.2 Host immune responses

Variability in the clinical outcome of histomonosis in chickens and turkeys underlines differences in the immune response of each host species against *H. meleagridis*. In turkeys, which are more susceptible to histomonosis than chickens, it was shown that circulating antibodies do not have a protective effect against the disease (Clarkson, 1963, Bleyen et al., 2009b). Consequently, immune protection induced by exposure to killed histomonads was not successful (Bleyen et al., 2009b, Hess et al., 2008). These studies indicate that the systemic humoral immune response does not have a substantial impact on protection. The effect of the local humoral response against *H. meleagridis* is not elucidated but increased IgM, IgY and IgA in the caeca and other parts of the intestine in chickens have been reported (Windisch and Hess, 2010).

In contrast, the cellular immune response has been found to be crucial against histomonosis based on several studies using attenuated *H. meleagridis* to induce protective immune responses in turkeys and chickens, as outlined below (see chapter 3.6.2).

Recently, it was shown that chickens mount a faster immune response following *H. meleagridis* infection than turkeys, resulting in earlier defence mechanisms that can restrict the parasite to the caeca of infected individuals (Powell et al., 2009). In another study, flow cytometry (FCM) analyses revealed that histomonosis caused more severe changes in B cells and T-cell subsets of turkeys than in chickens which may induce immunopathogenic effects (Mitra et al., 2017). Differences in the cellular immune response of chickens and turkeys have been further investigated by determining cytokine-producing cells using *in situ* hybridisation (ISH) (Kidane et al., 2018). In this work, chickens showed a higher presence of IFN-γ producing cells in the caeca than turkeys which may influence the nature of the immune response. Studies on the mentioned immune traits against *H. meleagridis* have been summarised by Mitra and colleagues (Mitra et al., 2018).

In a recent study, it was concluded that *H. meleagridis* infection induces a type-1 differentiation of $CD4^+$ T cells but also of non-$CD4^+$ cells, in chickens based on histomonad-specific immune cells (Lagler et al., 2019). Furthermore, FCM analyses revealed significant increments of IFN-γ-producing cells within major T-cell subsets ($CD4^+$, $CD8\alpha^+$ and $CD3\varepsilon^+CD4^-CD8\alpha^-$) of the spleen and liver in infected turkeys compared to infected chickens (Lagler et al., 2021).

3.2.3 Impact on enteric dysbiosis

The growth of *H. meleagridis* is known to be highly dependent on the presence of live bacteria (Bilic and Hess, 2020). It has been shown that the parasite is unable to cause disease in gnotobiotic turkeys, requiring the presence of specific bacterial species including *Escherichia coli* (Doll and Franker, 1963, Bradley and Reid, 1966). A recent study on protection in turkeys revealed that co-cultivated bacteria like *E. coli*, *Staphylococcus aureus* and *Salmonella* Enteritidis influence the colonisation of monoxenic attenuated *H. meleagridis* (Liebhart et al., 2013a).

In the chicken host, the consequences of co-infection with *H. meleagridis* and avian pathogenic *E. coli* (APEC) have been investigated on enteric pathology, microbiota and bacterial translocation (Abdelhamid et al., 2020). It was found that such a co-infection caused caecal typhlitis and severe dysbiosis defined by a severe reduction in microbial species richness and diversity, with a relatively higher abundance of the *Escherichia* genus, *Helicobacter* and

Bacteroides revealed by 16S rRNA gene amplicon sequencing. Furthermore, lux-tagged APEC introduced into the caeca were tracked and found to be significantly increased and distributed outside the intestine in co-infected birds, indicating the role of *H. meleagridis* in supporting *E. coli* in the pathogenesis of colibacillosis in chickens.

3.3 Histomonas meleagridis *population dynamics*

3.3.1 Molecular studies on histomonads/genotypes and differences in clinical signs and pathology

In recent years, several studies focused on the molecular identification of *H. meleagridis*. Initially, the 18S rDNA sequence was determined and used as a target for the taxonomic identification of *H. meleagridis* (Gerbod et al., 2001). Taxonomically, the parasite belongs to the order Tritrichomonadida and family Dientamoebidae, showing the greatest genetic similarity to *Dientamoeba fragilis*, a parasite of humans and several other mammals. Subsequently, several studies have focused on genetic differences between isolates of *H. meleagridis* (van der Heijden et al., 2006, Hauck and Hafez, 2009, Munsch et al., 2009, Reis et al., 2009, Hauck et al., 2010, Hauck and Hafez, 2010, Gerhold et al., 2011, Lollis et al., 2011, Lynn and Beckstead, 2012).

Genetic differences in the ITS1-5.8S-ITS2 region have been detected in a clonal *H. meleagridis* line, highlighting the occurrence of sequence degeneracy between genomic copies and emphasising the requirement for appropriate interpretation of sequence analysis using this genetic region (Hauck et al., 2010). More recently, multi-locus typing using the 18S rRNA, α-actinin1 and rpb1 genes from different *H. meleagridis* isolates demonstrated the existence of two different genotypes (Bilic et al., 2014). Importantly, differences in the outcome of infection by genotypes 1 or 2 could be observed in a flock of turkeys by clinical and pathological outcomes, suggesting genotype-specific pathogenesis (Grafl et al., 2015). In contrast to genotype 1, which has been well investigated in several experimental studies (Hess et al., 2006a, Liebhart and Hess, 2009), turkeys naturally infected with genotype 2 show reduced involvement of the liver. However, infection with genotype 2 compromised growth and resulted in more than 30% mortality (Grafl et al., 2015). Experimental infection of turkeys with genotype 2 has confirmed clinical and pathological differences to genotype 1 (unpublished data).

3.3.2 Introduction to a flock and patterns of transmission

The introduction of *H. meleagridis* into a poultry flock can occur via the intermediate vector *H. gallinarum* using earthworms as a paratenic host

(Graybill and Smith, 1920, Lund et al., 1966). Following the introduction, direct transmission from bird to bird is effective and appears to play an important role (Hess et al., 2006a, Liebhart and Hess, 2009). While cyst-like stages of *H. meleagridis* have been described (Zaragatzki et al., 2010), prolonged viability of the parasite in the environment has not been reported. *In vitro* cultivated histomonads can only survive for a few hours on different materials or in media like water and faeces (Lotfi et al., 2012). However, direct transmission between individuals within a flock is rapid, supposedly below this threshold. Based on an experimental infection and the detection of *H. meleagridis* in faeces by qPCR, the basic reproduction number (R0) was estimated to be 8.4 (Landman et al., 2015). This finding might explain the rapid dissemination reported within flocks, recognising that the study detected *H. meleagridis* DNA and not infective histomonads. Furthermore, as outlined earlier for mortality and morbidity, other factors such as parasite genotype are likely to influence transmission.

3.4 *Current methods of detection for* Histomonas meleagridis

In recent years, diagnostic tools to detect *H. meleagridis* have improved in terms of sensitivity and specificity. However, older detection methods still remain widely used, depending on the specific diagnostic demand.

Histomonas meleagridis can be observed by microscopic examination, either in native samples from intestinal contents or in histological preparations (Tyzzer, 1934). The viability of histomonads in preparations from caecal content or following *in vitro* propagation is crucial, since morphology (Fig. 4) and motility

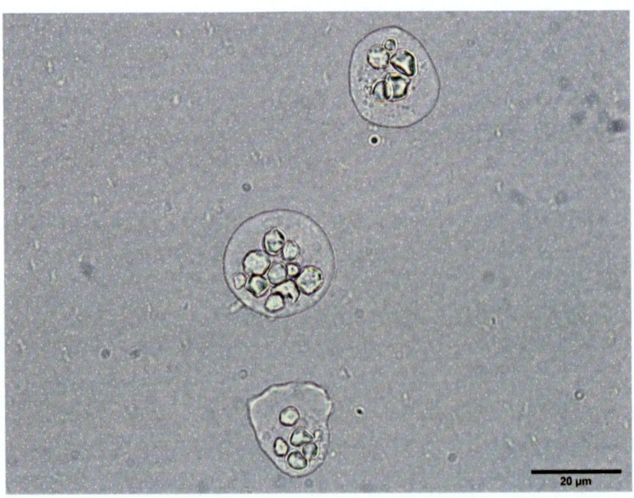

Figure 4 Cultured histomonads. Intracellular objects represent incorporated rice starch particles.

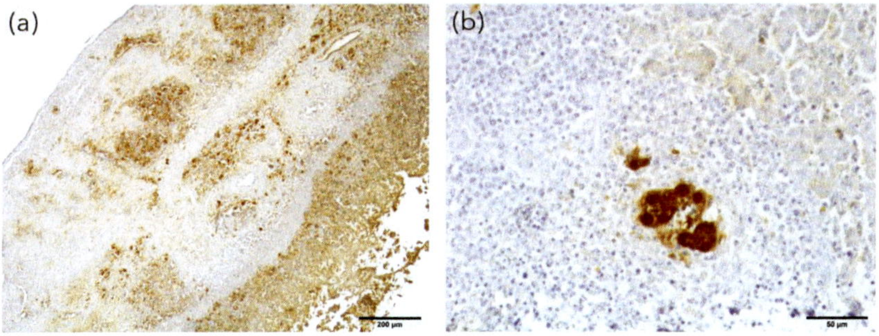

Figure 5 Immunohistochemistry for the detection and localisation of *H. meleagridis* in tissues sections. (a) Section of caecum with histomonads in all layers of the organ. (b) Liver sample showing specifically stained histomonads in the parenchyma at a higher magnification.

are characteristics of the parasite. In histological tissue samples, the flagella cannot be observed due to morphological changes in the tissue stages of the parasite. However, size, shape and the formation of a gap between the parasite and the host tissue indicate the presence of histomonads. Conventionally, Periodic Acid-Schiff (PAS) staining has been found to be most suitable for identifying *H. meleagridis* in tissue sections (Kemp and Reid, 1966). However, the occurrence of other protozoans such as trichomonads or blastocysts in host birds may impede an accurate diagnosis (Hess et al., 2006b).

Several molecular detection systems have been established in response to challenges posed by microscopy, mainly focusing on the 18S rRNA gene using conventional PCR to detect parasite DNA (Hafez et al., 2005, Huber et al., 2005, Grabensteiner and Hess, 2006, Bleyen et al., 2007). Subsequently, qPCR assays have been developed to allow the detection and quantification of *H. meleagridis* in samples (Hussain et al., 2015, Landman et al., 2015). A LAMP assay has also been published, providing high sensitivity and specificity (Xu et al., 2014). Histological examination has been improved by access to reagents for specific staining of *H. meleagridis* based on genomic sequences (*in situ* hybridisation; ISH) or antigen-antibody reactions (immunohistochemistry) in tissue sections (Fig. 5) (Liebhart et al., 2006, Singh et al., 2008). For indirect detection, a sandwich ELISA and a blocking ELISA have been set up to measure antibodies against *H. meleagridis* (Windisch and Hess, 2009, van der Heijden et al., 2010).

3.4.1 Monitoring of flocks

The introduction of *H. meleagridis* into poultry flocks and its spread from bird to bird can be monitored by direct or indirect detection systems, as described earlier. In experimental settings, necropsy and a combination of diagnostic tools including PCR, histology and re-isolation of *H. meleagridis* from cloacal swabs have been shown to give substantial results on the progression of infection (Grabensteiner et al., 2006, Hess et al., 2006a). For detailed monitoring of flocks in the field, the same methods should be applied, as described in a survey of histomonosis outbreaks in turkey flocks (Sulejmanovic et al., 2017). Additionally, the detection of specific antibodies can be used to identify infected birds by their immune response (Grafl et al., 2011, van der Heijden and Landman, 2011), although the appearance of antibodies can take at least 2 weeks in chickens and turkeys (Windisch and Hess, 2009). For example, a combination of PCR applied to faeces and dust samples with serology confirmed the infection of turkey hens and their resilience to histomonosis in barns equipped with both sexes following high mortalities in toms (Sulejmanovic et al., 2019a).

Another approach has been to monitor flocks exclusively by examination of environmental samples using PCR, as described in a recent study (Sulejmanovic et al., 2019b). Here, parasite DNA could be detected in dust samples collected from 15 of 65 investigated turkey flocks. Nine of the flocks found to be positive by PCR presented with no signs of histomonosis, indicating a high epidemiological value for histomonad detection using DNA in dust samples when negative controls are robust.

3.5 Current methods of control for Histomonas meleagridis

3.5.1 Limitations of current prophylaxis and therapy

The most effective drugs against histomonosis are nitroimidazoles, nitrofurans and arsenicals, all of which have been used for therapeutic and/or prophylactic purposes (Liebhart et al., 2017). However, as outlined earlier, these chemicals came under public and legislative pressure due to concerns around consumer health and have been banned from use in poultry production in many countries. In their absence, no effective prophylaxis or therapy is available. The antibiotic paromomycin has shown a prophylactic effect against histomonosis in turkeys (Lindquist, 1962, Bleyen et al., 2009a), but in several countries antibiotics are not licensed to be administered prophylactically. Further, the application of paromomycin after diagnosing histomonosis in commercial turkey flocks has not shown promising results when compared to untreated flocks (Sulejmanovic et al., 2017). Biosecurity and hygiene are of high importance to prevent the introduction and spread of the parasite in poultry flocks. However, the value of such measures is limited based on reports of histomonosis outbreaks in breeder birds, where high biosecurity can be presumed (Dolka et al., 2015, Aka et al., 2011).

3.6 Recent experimental approaches to control for Histomonas meleagridis

3.6.1 Plant-derived substances

The lack of available chemotherapeutics that are effective against histomonosis in poultry argued for the intensification of research into alternative substances with anti-histomonal effects (Liebhart et al., 2017). In response, several plant-derived essential oils, extracts in ethanol and water, lypholisiates, alkaloids and sesquiterpene lactones have been examined for this purpose. Several compounds have been found to reduce or suppress the propagation of *H. meleagridis in vitro*, including essential oils from cinnamon, lemon, rosemary, garlic and thyme, but confirmation *in vivo* remains to be shown (Zenner et al., 2003, Grabensteiner et al., 2007, Hauck and Hafez, 2007, van der Heijden and Landman, 2008a,b). Similarly, ethanol and water extracts from *Thymus vulgaris, Vitis vinifera, Olea europaea, Peganum harmala, Ginkgo biloba* and *Aesculus hippocastanum*, the alkaloids saponin, harmane, harmalol, harmaline and harmine, as well as artemisinin, a sesquiterpene lactone, have all shown promising results *in vitro* without equivalent results *in vivo* (Grabensteiner et al., 2007, 2008, Arshad et al., 2008, Thøfner et al., 2012).

Commercial plant-based products tested for an effect against histomonosis in turkeys include Enteroguard™ and Aromabiotic™, but neither protected against disease (van der Heijden and Landman, 2008a,b,). Application of the product Protophyt™ could not prevent clinical signs and lesions of histomonosis in infected turkeys (van der Heijden and Landman, 2008b, Hafez and Hauck, 2006). Similarly, suggestions of protection against disease using Natustat™ in turkeys kept in commercial farms are currently unconfirmed by standardised infection experiments (Duffy et al., 2005).

3.6.2 Vaccination including data on immune response following vaccination

Vaccination as a strategy to prevent histomonosis was first investigated more than 80 years ago, but attenuation of virulent *H. meleagridis* was described to be inhomogeneous, and administration routines for poultry flocks have not been developed (Tyzzer, 1934).

A prerequisite for a well-defined live vaccine was to establish a monoclonal culture of *H. meleagridis* (Hess et al., 2006b). Following prolonged culturing (295 passages *in vitro*), the clonal parasite became attenuated and could be used as a vaccine, inducing protection against severe challenges (Hess et al., 2008). Attenuated histomonads were restricted to the caecal lumen, and several *in vivo* passages did not lead to a reversion to virulence, confirming the safety of the vaccine candidate (Liebhart et al., 2011, Sulejmanovic et al.,

2016). Cross-protection against different isolates of the homologous genotype 1 has been demonstrated (Sulejmanovic et al., 2016), and a pilot study indicated protection against the heterologous genotype 2 (unpublished data). In chickens, vaccinated layers were shown to be protected against a significant reduction in egg production caused by histomonosis (Liebhart et al., 2013b). Optimisation of experimental vaccination against histomonosis could be achieved by administration of the vaccine via the oral route in day-old turkeys (Liebhart et al., 2010) and by establishing a monoxenic vaccine candidate (Ganas et al., 2012).

Besides attenuation *in vitro*, it has been shown that serial *in vivo* passage in turkeys reduces the virulence of histomonads while retaining the ability to protect turkeys from a subsequent severe challenge (Nguyen Pham et al., 2013). In contrast, the application of killed histomonads does not result in protection for challenged turkeys (Hess et al., 2008, Bleyen et al., 2009b).

Investigations to define those immune mechanisms that are relevant for protection against histomonosis have included flow cytometry, demonstrating that vaccination with attenuated histomonads induced a lower cellular immune response than virulent histomonads, causing protective immunity without an immunopathogenic effect (Mitra et al., 2017). Furthermore, the vaccination of turkeys led to increased IFN-γ producing cells in the caeca to levels comparable to naïve chickens that are innately less affected by histomonosis (Kidane et al., 2018). In studies using intracellular cytokine staining, it was found that vaccinated turkeys produce significantly more IFN-γ-producing cells in all major T-cell subsets of the spleen and liver compared to vaccinated chickens (Lagler et al., 2021). Based on these results, it can be concluded that the vaccine causes more intense systemic immune responses in turkeys, whereas in chickens protection might be driven by the local immune response.

3.7 Challenges and conclusions

The absence of effective prophylactic and therapeutic options to control histomonosis urgently requires new and improved approaches. Recent studies have focused on a wide range of plant-derived substances, but it is clear that any future product will be subject to increasingly strict regulations designed to protect consumer health. New products must be carefully selected for their safety and independence from existing or proposed products that are used in human medicine. To date, plant-derived substances have not shown substantial effects against histomonosis, but these substances may be refined and other active components may yet prove effective. Cultured *H. meleagridis* are highly suitable for efficacy tests and can be used to improve screening capacity, but it should be mandatory to confirm positive results *in vivo*.

Studies focused on vaccine development have highlighted *in vitro* attenuated histomonads as a promising new approach. Work to develop live attenuated vaccines further will require strategies for up-scaling, storage, transportation and application under field conditions. Histomonads are highly sensitive to environmental conditions, demanding innovative solutions for vaccine application to address these challenges.

4 *Dermanyssus gallinae*: the poultry red mite

4.1 Target populations, incidence and economic relevance

Dermanyssus gallinae (the poultry red mite, PRM) is an obligatory blood-feeding ectoparasite (Chauve, 1998). The PRM lifecycle includes five distinct stages: egg, larva, protonymph, deutonymph and adult (Fig. 6a) and can be completed within just 7 days under optimal conditions (i.e. temperature: 20–25°C, humidity>70%) (Koziatek and Sokół, 2015, Immediato et al., 2015, Maurer and Baumgartner, 1992). Consumption of blood is required for maturation of the protonymph, deutonymph and adult lifecycle stages, as well as the development of viable eggs. PRMs have been described from a broad host range, including horses, rodents and humans (Valiente Moro et al., 2009), but avian hosts are most common. PRMs have been reported to infest at least 28 different avian species, most notably the domestic chicken but also canaries, pigeons and doves (Roy et al., 2009b). While all chickens can be targeted by PRMs, laying and breeding stock are at the greatest risk, primarily due to their extended flock duration compared to the faster turnaround time associated with broiler chickens, providing longer opportunities for infestation and mite replication. PRMs spend

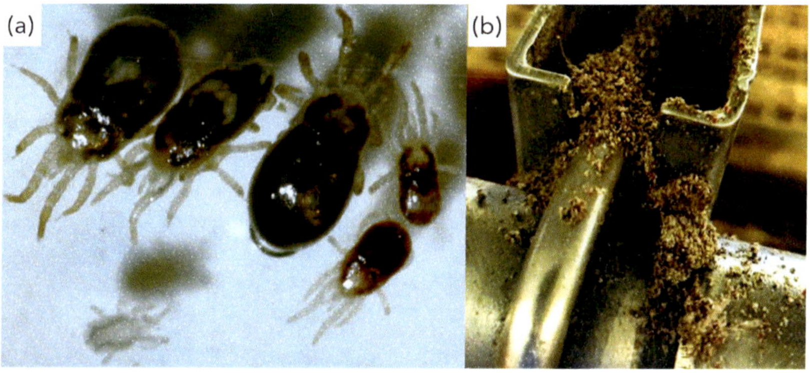

Figure 6 *Dermanyssus gallinae*, the poultry red mite (PRM). (a) Examples of a *D. gallinae* protonymph (bottom, white colouration), two deutonymphs (right) and three adults. (b) A typical hiding place for *D. gallinae* within poultry accommodation. PRMs spend most of their lives in the environment, only infesting chickens when taking a blood meal. Photographs taken by Eleanor Karp-Tatham.

the majority of their life cycle living separately from their hosts, sheltering in cracks and crevices, nests and cages (Fiddes et al., 2005) (Fig. 6B), emerging to feed when dark for approximately 30–90 minutes (Chauve, 1998).

The PRM has a global distribution although the occurrence is reported most frequently in Europe and Asia, where up to 90% of layer hen farms can be infested (Cencek, 2003, Sparagano et al., 2009, Hoglund et al., 1995, Guy et al., 2004, Marangi et al., 2012, Fiddes et al., 2005, Oh et al., 2019). Other mites such as the northern fowl mite (*Ornithonyssus sylviarum*) are commonly considered to be more important in North America. PRMs are responsible for significant economic losses from the European poultry industry, with estimates in excess of ~EU€230 million lost every year (Price et al., 2019). This cost has primarily been attributed to production losses (increased mortality, decreased egg production and quality, abbreviated laying cycle), higher feed conversion ratios and the costs of control (Sparagano et al., 2009, Wojcik et al., 2000, Sleeckx et al., 2019). Costs estimated from individual countries range from EU€3 million to EU€66.8 million for the UK, the Netherlands and Japan (Sparagano et al., 2009).

4.2 Host-pathogen interactions

4.2.1 Clinical signs, pathology and welfare

PRM infestation can impact production parameters. Direct interaction between the PRM and the host is usually restricted to feeding, when infestations in excess of 50 000 mites per hen are not uncommon (Kilpinen et al., 2005). Infestation levels of ~150 000 mites per hen have been found to result in increased restlessness, irritation, feather pecking and cannibalism, as well as anaemia and increased hen mortality (Kilpinen et al., 2005). Extended periods of infestation have also been linked with decreased body weight (Wojcik et al., 2000). Weekly mortality and laying rates, as well as egg weight, have all been shown to improve following effective anti-mite (acaricidal) treatment (Temple et al., 2020).

PRM infestation can also severely compromise hen welfare. Laying hens have been shown to change their resting and sleeping locations in response to infestation, possibly attempting to evade or reduce mite challenges (Maurer, 1993). Effective acaricidal treatment has been associated with reduced nighttime activity, including preening, head scratching and headshaking, in addition to the severity of feather peaking and aggressive behaviour during the daytime (Temple et al., 2020). Measures of comb quality, including colour and the presence of wounds, were also improved. Physiological assessments have shown that indicators of stress, such as corticosterone levels, increase in hens exposed to PRMs (Kowalski and Sokol, 2009).

4.2.2 Host immune responses

Dermanyssus gallinae demonstrate minimal host interference during feeding, incurring few significant immune responses (Harrington et al., 2010b). Humoral immune responses such as serum IgY and IgM increase with the occurrence and intensity of PRM exposure (Harrington et al., 2010b). There is some evidence for an early Th1 and pro-inflammatory cytokine response, but this is short-lived and might be down-regulated after subsequent feeding (Harrington et al., 2010a). The limited nature of the immune response induced by host-mite interaction poses a major challenge to the development of anti-PRM vaccines. Attempts have primarily focused on the development of hidden antigen vaccines, targeting mite proteins such as cathepsin-D that are not naturally exposed to the hen but can inhibit mite feeding, development or replication when targeted by antibodies (Price et al., 2019).

4.2.3 Vector capacity

It has been suggested that the PRM can serve as a vector for the transmission of several viral and bacterial pathogens (De Luna et al., 2008). PCR has been used to detect specific pathogen nucleic acids as an indication of possible transmission, recognising that detection is not evidence of viable organisms or their transmission, including Newcastle disease virus, *Mycoplasma synoviae* and *M. gallisepticum* (Huong et al., 2014, Arzey, 1990). *Mycobacterium* species DNA has also been detected in PRM eggs and unfed larvae (De Luna et al., 2008). Transmission has been demonstrated for fowlpox virus and *Pasturella multocida* (Petrov, 1975, Shirnov et al., 1972), as well as *Salmonella* Enteritidis (Valiente Moro et al., 2009). In the latter study, PRMs carrying *Salmonella* Enteritidis were found to transmit infection between chickens and to persist after cleaning and disinfection, indicating a source of transmission between individuals and flocks. Transovarial transmission was also documented, demonstrating vertical transmission between PRM generations (Valiente Moro et al., 2009).

4.3 Dermanyssus gallinae *population dynamics*

Little has been published describing population structure and dynamics for PRMs. In a series of papers, Roy and colleagues have suggested that *D. gallinae* may represent a species complex and not a single, discrete species. The complex may represent at least two morphologically indistinguishable but genetically distinct cryptic species (Roy et al., 2009a, 2010, Roy and Buronfosse, 2011). Importantly, the seemingly true (i.e. *sensu stricto*) *D. gallinae* has been detected infecting chickens and a range of other avian hosts, while the *D. gallinae* L1 lineage may be specific to pigeons. Complementary genetics-led studies have revealed distinct *D. gallinae* genotypes circulating in wild and domestic host populations in Sweden (Brännström et al., 2008). Findings that suggest distinct PRM populations in domestic and wild avians are important, since they can inform on likely sources of infestation and the dissemination of unfavourable phenotypes such as acaricide resistance. More recent studies focusing on *D. gallinae* sampled from populations in domestic chicken environments demonstrated the presence of multiple lineages in Europe (Karp-Tatham et al., 2020). Multiple genetic types were discovered, representing three haplogroups with six sub-haplogroups. Considerable variation was detected within and between countries, possibly reflecting the movement of poultry or contaminated equipment and variation in husbandry practices.

4.4 *Current methods of detection for* Dermanyssus gallinae

Adult PRMs can be visible to the human eye, but accurate enumeration for purposes such as assessment of risk or efficacy of control requires low-magnification microscopy. Importantly, PRMs are only located on their host during feeding, hiding in the environment for the majority of the time. A wide range of traps have been described to facilitate environmental sampling for PRMs, including fabric, corrugated cardboard or plastic traps that seek to create an environment that attracts mites (Kirkwood, 1965, Nordenfors and Chirico, 2001). An automated mite counting technique has been described (Mul et al., 2015), although uptake has not been high. Positioning of mite traps is key, recognising the importance of PRM feeding and aggregation behaviour (Mul et al., 2015)

4.5 *Current methods of control for* Dermanyssus gallinae

4.5.1 *Acaricides*

Control of PRMs is challenging. A wide range of organophosphates, carbamates, formamidines and pyrethroids have been used to control PRMs in the past (Abbas et al., 2014, Beugnet et al., 1997, Chauve, 1998), but public and legislative pressure have combined to limit the availability of many products. The widespread development of acaricide resistance and the scarcity of new products have added further complications (Sparagano et al., 2014, Katsavou et al., 2020). Very few products remain available and licensed for use with poultry, exceptions including the fluralaner-based Exzolt® solution (MSD Animal Health) (Temple et al., 2020). Reports of acaricide residues in poultry and poultry products for human consumption have added further pressure, with examples including carbaryl in the skin and fat of chickens (Marangi et al.,

2012) and, more recently, the scandal around fipronil residues in chicken eggs (Tu et al., 2019).

A major limitation to the use of acaricides has been the rapid emergence and dissemination of acaricide resistance (Marangi et al., 2009, 2012). The emergence of genetic (i.e. heritable) resistance to acaricides has commonly been mediated by point mutations in genes that encode proteins with key metabolic functions, contributing to the metabolism of the acaricide before it can achieve its target or enzymatic detoxification (e.g. glutathione-S-transferases and P450 monooxygenases) (Wang et al., 2020, 2021).

4.5.2 Desiccant dusts

Alternatives to chemical control include desiccant, silica or inert dusts (Steenberg and Kilpinen, 2014). It is believed that desiccant and equivalent dusts desiccate and kill PRMs and other arthropods, possibly due to cuticle abrasion and absorption of cuticular lipids (Ebeling, 1971) acting within 24 hours (Kilpinen and Steenberg, 2009). Examples of desiccant dusts include synthetic silica products and diatomaceous earth (Kilpinen and Steenberg, 2009). A comparison of a range of desiccant products revealed the importance of cation exchange, where increased capacity improved efficacy and water absorption, emphasising the importance of a dry environment (Schulz et al., 2014, Kilpinen and Steenberg, 2009). Challenges associated with the use of desiccant dusts include health and safety provisions for workers active in the area.

4.5.3 Alternatives for control of PRMs

The scale of the challenge posed by PRMs and the paucity of the controls available have prompted the development of several alternatives. The use of high heat/low humidity conditions between flocks can reduce residual PRM presence, although the approach can be costly and inappropriate for some older or extensive poultry accommodation. Predatory mites such as *Cheyletus eruditus* have been found to feed on PRMs, especially larvae (Maurer, 1993). Several predatory mite species are being developed for use in biocontrol strategies (Zriki et al., 2020). Entomopathogenic fungi have also been considered. Several fungal species have been identified with known efficacy against arthropod pests and are currently used in agriculture and forestry (de Faria and Wraight, 2007). Laboratory experiments of PRMs susceptibility to fungi such as *Beauveria bassiana* and *Metarhizium anisopliae* have suggested utility, although the process can be relatively slow (Tavassoli et al., 2011). Several vaccines are in development for use against PRMs, including vaccine candidates such as cathepsin-D (Price et al., 2019), but none are close to commercialisation (Bartley et al., 2017, Xu et al., 2020).

5 Conclusion

Control of parasites that target poultry remains a major challenge. These antigenically complex pathogens are commonly adept at evolving to escape conventional control based on husbandry and routine prophylaxis. Increasing public and legislative demands for the replacement of drugs and chemicals in livestock and poultry production is exacerbating the situation, with few or no effective products left available for control. Improved understanding of genetic diversity and population structure is beginning to support the development of novel controls, revealing previously unknown new genotypes and, in some examples, species. Attempts to modify or develop new vaccines for the three parasite groups discussed here are ongoing, offering considerable promise for the management of poultry flock health in the near future.

6 References

Abbas, R. Z., Colwell, D. D., Iqbal, Z. and Khan, A. 2014. Acaricidal drug resistance in poultry red mite (*Dermanyssus gallinae*) and approaches to its management. *World's Poultry Science Journal* 70(1), 113–124.

Abd El-Wahab, A., Visscher, C. F., Wolken, S., Reperant, J. M., Beineke, A., Beyerbach, M. and Kamphues, J. 2012. Foot-pad dermatitis and experimentally induced coccidiosis in young turkeys fed a diet without anticoccidia. *Poultry Science* 91(3), 627-635.

Abdelhamid, M. K., Quijada, N. M., Dzieciol, M., Hatfaludi, T., Bilic, I., Selberherr, E., Liebhart, D., Hess, C., Hess, M. and Paudel, S. 2020. Co-infection of chicken layers with *Histomonas meleagridis* and Avian Pathogenic *Escherichia coli* is associated with dysbiosis, cecal colonization and translocation of the bacteria from the gut lumen. *Frontiers in Microbiology* 11, 586437.

Aka, J., Hauck, R., Blankenstein, P., Balczulat, S. and Hafez, H. M. 2011. Reoccurrence of histomonosis in Turkey breeder farm. *Berliner und Münchener tierärztliche Wochenschrift* 124(1–2), 2–7.

Arakawa, A., Baba, E. and Fukata, T. 1981. *Eimeria tenella* infection enhances *Salmonella* Typhimurium infection in chickens. *Poultry Science* 60(10), 2203–2209.

Arshad, N., Zitterl-Eglseer, K., Hasnain, S. and Hess, M. 2008. Effect of *Peganum harmala* or its beta-carboline alkaloids on certain antibiotic resistant strains of bacteria and protozoa from poultry. *Phytotherapy Research : PTR* 22(11), 1533–1538.

Arzey, G. 1990. Mechanism of spread of Newcastle disease. *New South Wales Agriculture and Fisheries Bulletin* 41, 12.

Awad, A. M., El-Nahas, A. F. and Abu-Akkada, S. S. 2013. Evaluation of the protective efficacy of the anticoccidial vaccine Coccivac-B in broilers, when challenged with Egyptian field isolates of *E. tenella*. *Parasitology Research* 112(1), 113–121.

Barkway, C. P., Pocock, R. L., Vrba, V. and Blake, D. P. 2011. Loop-mediated isothermal amplification (LAMP) assays for the species-specific detection of *Eimeria* that infect chickens. *BMC Veterinary Research* 7, 67.

Bartley, K., Turnbull, F., Wright, H. W., Huntley, J. F., Palarea-Albaladejo, J., Nath, M. and Nisbet, A. J. 2017. Field evaluation of poultry red mite (*Dermanyssus gallinae*) native and recombinant prototype vaccines. *Veterinary Parasitology* 244, 25–34.

Beugnet, F., Chauve, C., Gauthey, M. and Beert, L. 1997. Resistance of the red poultry mite to pyrethroids in France. *Veterinary Record* 140(22), 577–579.

Bilic, I. and Hess, M. 2020. Interplay between *Histomonas meleagridis* and bacteria: mutualistic or predator-prey? *Trends in Parasitology* 36(3), 232–235.

Bilic, I., Jaskulska, B., Souillard, R., Liebhart, D. and Hess, M. 2014. Multi-locus typing of *Histomonas meleagridis* isolates demonstrates the existence of two different genotypes. *PLoS ONE* 9(3), e92438.

Blake, D. P., Clark, E. L., Macdonald, S. E., Thenmozhi, V., Kundu, K., Garg, R., Jatau, I. D., Ayoade, S., Kawahara, F., Moftah, A., Reid, A. J., Adebambo, A. O., Alvarez Zapata, R., Srinivasa Rao, A. S., Thangaraj, K., Banerjee, P. S., Dhinakar-Raj, G., Raman, M. and Tomley, F. M. 2015. Population, genetic, and antigenic diversity of the apicomplexan *Eimeria tenella* and their relevance to vaccine development. *Proceedings of the National Academy of Sciences of the United States of America* 112(38), E5343–E5350.

Blake, D. P., Hesketh, P., Archer, A., Shirley, M. W. and Smith, A. L. 2006. *Eimeria maxima*: The influence of host genotype on parasite reproduction as revealed by quantitative real-time PCR. *International Journal for Parasitology* 36(1), 97–105.

Blake, D. P., Knox, J., Dehaeck, B., Huntington, B., Rathinam, T., Ravipati, V., Ayoade, S., Gilbert, W., Adebambo, A. O., Jatau, I. D., Raman, M., Parker, D., Rushton, J. and Tomley, F. M. 2020a. Re-calculating the cost of coccidiosis in chickens. *Veterinary Research* 51(1), 115.

Blake, D. P., Worthing, K. and Jenkins, M. C. 2020b. Exploring *Eimeria* genomes to understand population biology: Recent progress and future opportunities. *Genes (Basel)* 11(9).

Blake, D. P., Vrba, V., Xia, D., Jatau, I. D., Spiro, S., Nolan, M. J., Underwood, G. and Tomley, F. M. 2021. Genetic and biological characterisation of three cryptic *Eimeria* operational taxonomic units that infect chickens (*Gallus gallus domesticus*). *International Journal for Parasitology* 51(8), 621–634. doi: 10.1016/j.ijpara.2020.12.004.

Bleyen, N., De Gussem, K., De Gussem, J. and Goddeeris, B. M. 2007. Specific detection of *Histomonas meleagridis* in turkeys by a PCR assay with an internal amplification control. *Veterinary Parasitology* 143(3–4), 206–213.

Bleyen, N., Gussem, K. D., Pham, A. D. N., Ons, E., Van Gerven, N. and Goddeeris, B. M. 2009a. Non-curative, but prophylactic effects of paromomycin in *Histomonas meleagridis*-infected turkeys and its effect on performance in non-infected turkeys. *Veterinary Parasitology* 165(3–4), 248–255.

Bleyen, N., Ons, E., Gussem, M. D. and Goddeeris, B. M. 2009b. Passive immunization against *Histomonas meleagridis* does not protect turkeys from an experimental infection. *Avian Pathology* 38(1), 71–76.

Boulton, K., Nolan, M. J., Wu, Z., Psifidi, A., Riggio, V., Harman, K., Bishop, S. C., Kaiser, P., Abrahamsen, M. S., Hawken, R., Watson, K. A., Tomley, F. M., Blake, D. P. and Hume, D. A. 2018a. Phenotypic and genetic variation in the response of chickens to *Eimeria tenella* induced coccidiosis. *Genetics, Selection, Evolution* 50(1), 63.

Boulton, K., Nolan, M. J., Wu, Z., Riggio, V., Matika, O., Harman, K., Hocking, P. M., Bumstead, N., Hesketh, P., Archer, A., Bishop, S. C., Kaiser, P., Tomley, F. M., Hume, D. A., Smith, A. L., Blake, D. P. and Psifidi, A. 2018b. Dissecting the Genomic Architecture of Resistance to *Eimeria maxima* Parasitism in the Chicken. *Frontiers in Genetics* 9, 528.

Bradley, R. E. and Reid, W. M. 1966. *Histomonas meleagridis* and several bacteria as agents of infectious enterohepatitis in gnotobiotic turkeys. *Experimental Parasitology* 19(1), 91–101.

Brännström, S., Morrison, D. A., Mattsson, J. G. and Chirico, J. 2008. Genetic differences in internal transcribed spacer 1 between *Dermanyssus gallinae* from wild birds and domestic chickens. *Medical and Veterinary Entomology* 22(2), 152–155.

Byrnes, S., Eaton, R. and Kogut, M. 1993. *In vitro* interleukin-1 and tumor necrosis factor-alpha production by macrophages from chickens infected with either *Eimeria maxima* or *Eimeria tenella*. *International Journal for Parasitology* 23(5), 639–645.

Cantacessi, C., Riddell, S., Morris, G. M., Doran, T., Woods, W. G., Otranto, D. and Gasser, R. B. 2008. Genetic characterization of three unique operational taxonomic units of *Eimeria* from chickens in Australia based on nuclear spacer ribosomal DNA. *Veterinary Parasitology* 152(3–4), 226–234.

Castañón, C., Fernandez, S., Fraga, J., Fontoura, L. and Gruber, A. 2007. COCCIMORPH: A real-time diagnostic tool based on automatic image recognition of protozoan parasites of the genus *Eimeria*. Proceedings of the World Association for the Advancement of Veterinary Parasitology, Gent, Belgium.

CEC 1995. Commission Regulation (EC) No 1798/95 of July 25, 1995 amending Annex IV to Council Regulation (EEC) No 2377/90 laying down a Community procedure for the establishment of maximum residue limits of veterinary medicinal products in foodstuffs of animal origin. *Official Journal*, L174 1798(95) of, 20–21.

CEC 2002. Council Regulation (EC) No 1756/2002 of 23 September 2002 amending Directive 70/524/EEC concerning additives in feedingstuffs as regards withdrawal of the authorisation of an additive and amending Commission Regulation (EC) No 2430/1999. *Official Journal* L181, 1–2.

Cencek, T. 2003. Prevalence of *Dermanyssus gallinae* in poultry farms in silesia region in Poland. *Bulletin of the Veterinary Institute in Pulawy* 47, 465–469.

Chapman, H. D. 1997. Biochemical, genetic and applied aspects of drug resistance in *Eimeria* parasites of the fowl. *Avian Pathology* 26(2), 221–244.

Chapman, H. D. 1999. Anticoccidial drugs and their effects upon the development of immunity to *Eimeria* infections in poultry. *Avian Pathology* 28(6), 521–535.

Chapman, H. D. and Jeffers, T. K. 2015. Restoration of sensitivity to salinomycin in *Eimeria* following 5 flocks of broiler chickens reared in floor-pens using drug programs and vaccination to control coccidiosis. *Poultry Science* 94(5), 943–946.

Chauve, C. 1998. The poultry red mite Dermanyssus gallinae (De Geer, 1778): Current situation and future prospects for control. *Veterinary Parasitology* 79(3), 239–245.

Chroustová, E. and Pinka, K. 1987. The efficiency of disinfectants on the oocysts of *Eimeria tenella*. *Acta Veterinaria Brno* 56(1–2), 141–149.

Clark, E. L., Macdonald, S. E., Thenmozhi, V., Kundu, K., Garg, R., Kumar, S., Ayoade, S., Fornace, K. M., Jatau, I. D., Moftah, A., Nolan, M. J., Sudhakar, N. R., Adebambo, A. O., Lawal, I. A., Alvarez Zapata, R., Awuni, J. A., Chapman, H. D., Karimuribo, E., Mugasa, C. M., Namangala, B., Rushton, J., Suo, X., Thangaraj, K., Srinivasa Rao, A. S., Tewari, A. K., Banerjee, P. S., Dhinakar Raj, G., Raman, M., Tomley, F. M. and Blake, D. P. 2016. Cryptic *Eimeria* genotypes are common across the southern but not northern hemisphere. *International Journal for Parasitology* 46(9), 537–544.

Clark, S. and Kimminau, E. 2017. Critical review: Future control of blackhead disease (histomoniasis) in poultry. *Avian Diseases* 61(3), 281–288.

Clarkson, M. J. 1963. Immunological responses to *Histomonas meleagridis* in the turkey and fowl. *Immunology* 6, 156–168.

Cornelissen, J. B., Swinkels, W. J., Boersma, W. A. and Rebel, J. M. 2009. Host response to simultaneous infections with *Eimeria acervulina*, *maxima* and *tenella*: A cumulation of single responses. *Veterinary Parasitology* 162(1–2), 58–66.

de Faria, M. Rd and Wraight, S. P. 2007. Mycoinsecticides and Mycoacaricides: A comprehensive list with worldwide coverage and international classification of formulation types. *Biological Control* 43(3), 237–256.

de Jong, I. C., Gunnink, H. and Van Harn, J. 2014. Wet litter not only induces footpad dermatitis but also reduces overall welfare, technical performance, and carcass yield in broiler chickens. *Journal of Applied Poultry Research* 23(1), 51–58.

De Luna, C. J., Arkle, S., Harrington, D., George, D. R., Guy, J. H. and Sparagano, O. A. 2008. The poultry red mite *Dermanyssus gallinae* as a potential carrier of vector-borne diseases. *Annals of the New York Academy of Sciences* 1149, 255–258.

Dolka, B., Żbikowski, A., Dolka, I. and Szeleszczuk, P. 2015. Histomonosis - An existing problem in chicken flocks in Poland. *Veterinary Research Communications* 39(3), 189–195.

Doll, J. P. and Franker, C. K. 1963. Experimental histomoniasis in gnotobiotic turkeys. I. Infection and histopathology of the bacteria-free host. *The Journal of Parasitology* 49, 411–414.

Duffy, C. F., Sims, M. D. and Power, R. F. 2005. Evaluation of dietary Natustat for control of *Histomonas meleagridis* in male turkeys on infected litter. *Avian Diseases* 49(3), 423–425.

Ebeling, W. 1971. Sorptive dusts for pest control. *Annual Review of Entomology* 16, 123–158.

FAOSTAT 2021. Food and Agriculture Organization of the United Nations FAOSTAT database [Online]. http://faostat3.fao.org/home/ [accessed 4 February 2021].

FDA 2015. Code of Federal Regulations, Title 21, Volume 6, revised as of April 1, 2015, Title 21. Food and drugs, chapter I-food and drug administration, Department of Health and Human Services, subchapter E-animal drugs, feeds, and related products, part 530 - extralabel drug use in animals, subpart E-safe levels for extralabel use of drugs in animals and drugs prohibited from extralabel use in animals. 530.41.

Fernandez, S., Pagotto, A. H., Furtado, M. M., Katsuyama, A. M., Madeira, A. M. and Gruber, A. 2003. A multiplex PCR assay for the simultaneous detection and discrimination of the seven *Eimeria* species that infect domestic fowl. *Parasitology* 127(4), 317–325.

Fiddes, M. D., Le Gresley, S., Parsons, D. G., Epe, C., Coles, G. C. and Stafford, K. A. 2005. Prevalence of the poultry red mite (*Dermanyssus gallinae*) in England. *Veterinary Record* 157(8), 233–235.

Ganas, P., Liebhart, D., Glösmann, M., Hess, C. and Hess, M. 2012. *Escherichia coli* strongly supports the growth of *Histomonas meleagridis*, in a monoxenic culture, without influence on its pathogenicity. *International Journal for Parasitology* 42(10), 893-901.

Gerbod, D., Edgcomb, V. P., Noël, C., Zenner, L., Wintjens, R., Delgado-Viscogliosi, P., Holder, M. E., Sogin, M. L. and Viscogliosi, E. 2001. Phylogenetic position of the trichomonad parasite of turkeys, *Histomonas meleagridis* (Smith) Tyzzer, inferred from small subunit rRNA sequence. *The Journal of Eukaryotic Microbiology* 48(4), 498–504.

Gerhold, R. W., Lollis, L. A., Mcdougald, L. R. and Beckstead, R. B. 2011. Partial sequence of the alpha-tubulin gene from *Histomonas meleagridis* isolates from the United States. *The Journal of Parasitology* 97(2), 354–356.

Grabensteiner, E., Arshad, N. and Hess, M. 2007. Differences in the *in vitro* susceptibility of mono-eukaryotic cultures of *Histomonas meleagridis*, *Tetratrichomonas gallinarum* and *Blastocystis* sp. to natural organic compounds. *Parasitology Research* 101(1), 193–199.

Grabensteiner, E. and Hess, M. 2006. PCR for the identification and differentiation of *Histomonas meleagridis*, *Tetratrichomonas gallinarum* and *Blastocystis* spp. *Veterinary Parasitology* 142(3–4), 223–230.

Grabensteiner, E., Liebhart, D., Arshad, N. and Hess, M. 2008. Antiprotozoal activities determined *in vitro* and *in vivo* of certain plant extracts against *Histomonas meleagridis*, *Tetratrichomonas gallinarum* and *Blastocystis* sp. *Parasitology Research* 103(6), 1257–1264.

Grabensteiner, E., Liebhart, D., Weissenböck, H. and Hess, M. 2006. Broad dissemination of *Histomonas meleagridis* determined by the detection of nucleic acid in different organs after experimental infection of turkeys and specified pathogen-free chickens using a mono-eukaryotic culture of the parasite. *Parasitology International* 55(4), 317–322.

Grafl, B., Liebhart, D., Windisch, M., Ibesich, C. and Hess, M. 2011. Seroprevalence of *Histomonas meleagridis* in pullets and laying hens determined by ELISA. *The Veterinary Record* 168(6), 160.

Grafl, B., Weise, H., Le Bris, J., Liebhart, D., Bilic, I. and Hess, M. 2015. Aberrant clinical appearance and pathomorphology noticed during an outbreak of histomonosis indicates a different pathogenesis of *Histomonas meleagridis* genotype 2. *Avian Diseases* 59(3), 452–458.

Graybill, H. W. and Smith, T. 1920. Production of fatal blackhead in turkeys by feeding embryonated eggs of *Heterakis papillosa*. *The Journal of Experimental Medicine* 31(5), 647–655.

Guy, J. H., Khajavi, M., Hlalel, M. M. and Sparagano, O. 2004. Red mite (*Dermanyssus gallinae*) prevalence in laying units in Northern England. *British Poultry Science* 45(Suppl. 1), S15–S16.

Hafez, H. M. and Hauck, R. 2006. Efficacy of a herbal product against *Histomonas meleagridis* after experimental infection of turkey poults. *Archives of Animal Nutrition* 60(5), 436–442.

Hafez, H. M., Hauck, R., Lüschow, D. and McDougald, L. 2005. Comparison of the specificity and sensitivity of PCR, nested PCR, and real-time PCR for the diagnosis of histomoniasis. *Avian Diseases* 49(3), 366–370.

Hamzic, E., Bed'hom, B., Juin, H., Hawken, R., Abrahamsen, M. S., Elsen, J. M., Servin, B., Pinard-Van Der Laan, M. H. and Demeure, O. 2015. Large-scale investigation of the parameters in response to *Eimeria maxima* challenge in broilers. *Journal of Animal Science* 93(4), 1830–1840.

Harrington, D., Robinson, K., Guy, J. and Sparagano, O. 2010a. Characterization of the immunological response to *Dermanyssus gallinae* infestation in domestic fowl. *Transboundary and Emerging Diseases* 57(1–2), 107–110.

Harrington, D. W., Robinson, K. and Sparagano, O. A. 2010b. Immune responses of the domestic fowl to *Dermanyssus gallinae* under laboratory conditions. *Parasitology Research* 106(6), 1425–1434.

Hauck, R., Balczulat, S. and Hafez, H. M. 2010. Detection of DNA of *Histomonas meleagridis* and *Tetratrichomonas gallinarum* in German poultry flocks between 2004 and 2008. *Avian Diseases* 54(3), 1021–1025.

Hauck, R., Carrisosa, M., Mccrea, B. A., Dormitorio, T. and Macklin, K. S. 2019. Evaluation of Next-Generation Amplicon Sequencing to Identify *Eimeria* spp. of Chickens. *Avian Diseases* 63(4), 577–583.

Hauck, R. and Hafez, H. M. 2007. Effect of coated plant extracts on *Histomonas meleagridis* and growth of bacteria *in vitro*. *Avian Diseases* 51(4), 880–883.

Hauck, R. and Hafez, H. M. 2009. Partial sequence of the beta-tubulin of *Histomonas meleagridis* and the activity of benzimidazoles against *H. meleagridis in vitro*. *Parasitology Research* 104(5), 1183–1189.

Hauck, R. and Hafez, H. M. 2010. Systematic position of *Histomonas meleagridis* based on four protein genes. *The Journal of Parasitology* 96(2), 396–400.

Haug, A., Gjevre, A. G., Thebo, P., Mattsson, J. G. and Kaldhusdal, M. 2008. Coccidial infections in commercial broilers: Epidemiological aspects and comparison of *Eimeria* species identification by morphometric and polymerase chain reaction techniques. *Avian Pathology* 37(2), 161–170.

Hess, M., Grabensteiner, E. and Liebhart, D. 2006a. Rapid transmission of the protozoan parasite *Histomonas meleagridis* in turkeys and specific pathogen free chickens following cloacal infection with a mono-eukaryotic culture. *Avian Pathology* 35(4), 280–285.

Hess, M., Kolbe, T., Grabensteiner, E. and Prosl, H. 2006b. Clonal cultures of *Histomonas meleagridis*, *Tetratrichomonas gallinarum* and a *Blastocystis* sp. established through micromanipulation. *Parasitology* 133(5), 547–554.

Hess, M., Liebhart, D., Bilic, I. and Ganas, P. 2015. *Histomonas meleagridis* - new insights into an old pathogen. *Veterinary Parasitology* 208(1–2), 67–76.

Hess, M., Liebhart, D., Grabensteiner, E. and Singh, A. 2008. Cloned *Histomonas meleagridis* passaged *in vitro* resulted in reduced pathogenicity and is capable of protecting turkeys from histomonosis. *Vaccine* 26(33), 4187–4193.

Hoglund, J., Nordenfors, H. and Uggla, A. 1995. Prevalence of the poultry red mite, *Dermanyssus gallinae*, in different types of production systems for egg layers in Sweden. *Poultry Science* 74(11), 1793–1798.

Hong, Y. H., Lillehoj, H. S., Lillehoj, E. P. and Lee, S. H. 2006. Changes in immune-related gene expression and intestinal lymphocyte subpopulations following *Eimeria maxima* infection of chickens. *Veterinary Immunology and Immunopathology* 114(3–4), 259–272.

Huber, K., Chauve, C. and Zenner, L. 2005. Detection of *Histomonas meleagridis* in turkeys cecal droppings by PCR amplification of the small subunit ribosomal DNA sequence. *Veterinary Parasitology* 131(3–4), 311–316.

Huong, C. T., Murano, T., Uno, Y., Usui, T. and Yamaguchi, T. 2014. Molecular detection of avian pathogens in poultry red mite (*Dermanyssus gallinae*) collected in chicken farms. *Journal of Veterinary Medical Science* 76(12), 1583–1587.

Hussain, I., Jaskulska, B., Hess, M. and Bilic, I. 2015. Detection and quantification of *Histomonas meleagridis* by real-time PCR targeting single copy genes. *Veterinary Parasitology* 212(3–4), 382–388.

Immediato, D., Camarda, A., Iatta, R., Puttilli, M. R., Ramos, R. A., Di Paola, G., Giangaspero, A., Otranto, D. and Cafarchia, C. 2015. Laboratory evaluation of a native strain of Beauveria *bassiana* for controlling *Dermanyssus gallinae* (De Geer, 1778) (Acari: Dermanyssidae). *Veterinary Parasitology* 212(3–4), 478–482.

Johnson, J. and Reid, W. M. 1970. Anticoccidial drugs: Lesion scoring techniques in battery and floor-pen experiments with chickens. *Experimental Parasitology* 28(1), 30–36.

Joyner, L. P. 1969. Immunological variation between two strains of *Eimeria Acervulina*. *Parasitology* 59(3), 725–732.

Karp-Tatham, E., Kuster, T., Angelou, A., Papadopoulos, E., Nisbet, A. J., Xia, D., Tomley, F. M. and Blake, D. P. 2020. Phylogenetic inference using cytochrome C oxidase subunit I (COI) in the poultry red mite, *Dermanyssus gallinae* in the United Kingdom relative to a European framework. *Frontiers in Veterinary Science* 7, 553.

Katsavou, E., Vlogiannitis, S., Karp-Tatham, E., Blake, D. P., Ilias, A., Strube, C., Kioulos, I., Dermauw, W., Van Leeuwen, T. and Vontas, J. 2020. Identification and geographical distribution of pyrethroid resistance mutations in the poultry red mite *Dermanyssus gallinae*. *Pest Management Science* 76(1), 125–133.

Kemp, R. L. and Reid, W. M. 1966. Staining techniques for differential diagnosis of histomoniasis and mycosis in domestic poultry. *Avian Diseases* 10(3), 357.

Khater, H. F., Ziam, H., Abbas, A., Abbas, R., Raza, M., Hussain, K., Younis, E., Radwan, I. and Selim, A. 2020. Avian coccidiosis: Recent advances in alternative control strategies and vaccine development. *Agrobiological Records* 1, 11–25.

Kidane, F. A., Mitra, T., Wernsdorf, P., Hess, M. and Liebhart, D. 2018. Allocation of interferon gamma mRNA positive cells in caecum hallmarks a protective trait against histomonosis. *Frontiers in Immunology* 9, 1164.

Kilpinen, O., Roepstorff, A., Permin, A., Norgaard-Nielsen, G., Lawson, L. G. and Simonsen, H. B. 2005. Influence of *Dermanyssus gallinae* and *Ascaridia* galli infections on behaviour and health of laying hens (Gallus *gallus domesticus*). *British Poultry Science* 46(1), 26–34.

Kilpinen, O. and Steenberg, T. 2009. Inert dusts and their effects on the poultry red mite (*Dermanyssus gallinae*). *Experimental and Applied Acarology* 48(1–2), 51–62.

Kimminau, E. A. and Duong, T. T. 2019. Longitudinal response of commercial broiler operations to bio-shuttle administration. *Journal of Applied Poultry Research* 28(4), 1389–1397.

Kirkwood, A. C. 1965. A trap perch for the control of the poultry red mite (*Dermanyssus gallinae*). *British Poultry Science* 6(1), 73–78.

Kowalski, A. and Sokol, R. 2009. Influence of *Dermanyssus gallinae* (poultry red mite) invasion on the plasma levels of corticosterone, catecholamines and proteins in layer hens. *Polish Journal of Veterinary Sciences* 12(2), 231–235.

Koziatek, S. and Sokół, R. 2015. *Dermanyssus gallinae* still poses a serious threat for the rearing of laying hens. *Polish Journal of Natural Sciences* 30, 451–463.

Kumar, S., Garg, R., Moftah, A., Clark, E. L., Macdonald, S. E., Chaudhry, A. S., Sparagano, O., Banerjee, P. S., Kundu, K., Tomley, F. M. and Blake, D. P. 2014. An optimised protocol for molecular identification of *Eimeria* from chickens. *Veterinary Parasitology* 199(1–2), 24–31.

Lagler, J., Mitra, T., Schmidt, S., Pierron, A., Vatzia, E., Stadler, M., Hammer, S. E., Mair, K. H., Grafl, B., Wernsdorf, P., Rauw, F., Lambrecht, B., Liebhart, D. and Gerner, W. 2019. Cytokine production and phenotype of *Histomonas meleagridis*-specific T cells in the chicken. *Veterinary Research* 50(1), 107.

Lagler, J., Schmidt, S., Mitra, T., Stadler, M., Patricia, W., Grafl, B., Hatfaludi, T., Hess, M., Gerner, W. and Liebhart, D. 2021. Comparative investigation of IFN-γ-producing T cells in chickens and turkeys following vaccination and infection with the extracellular parasite *Histomonas meleagridis*. *Developmental and Comparative Immunology* 116, 103949.

Landman, W. J. M., Ter Veen, C., van der Heijden, H. M. J. F. and Klinkenberg, D. 2015. Quantification of parasite shedding and horizontal transmission parameters in *Histomonas meleagridis*-infected turkeys determined by real-time quantitative PCR. *Avian Pathology* 44(5), 358–365.

Lew, A. E., Anderson, G. R., Minchin, C. M., Jeston, P. J. and Jorgensen, W. K. 2003. Inter- and intra-strain variation and PCR detection of the internal transcribed spacer 1 (ITS-1) sequences of Australian isolates of *Eimeria* species from chickens. *Veterinary Parasitology* 112(1–2), 33–50.

Liebhart, D., Ganas, P., Sulejmanovic, T. and Hess, M. 2017. Histomonosis in poultry: Previous and current strategies for prevention and therapy. *Avian Pathology* 46(1), 1–18.

Liebhart, D. and Hess, M. 2009. Oral infection of turkeys with *in vitro*-cultured *Histomonas meleagridis* results in high mortality. *Avian Pathology* 38(3), 223–227.

Liebhart, D. and Hess, M. 2020. Spotlight on histomonosis (blackhead disease): A re-emerging disease in turkeys and chickens. *Avian Pathology* 49(1), 1–4.

Liebhart, D., Sulejmanovic, T., Ganas, P., Hess, C., Ibesich, C. and Hess, M. 2013a. Impact of an experimental *Histomonas meleagridis* co-cultivated bacteria influence the colonization of attenuated *Histomonas meleagridis* following experimental vaccination of turkeys. Proceedings of the XVIIIth Congress of the World Veterinary Poultry Association, Nantnes, France, 19–23 August, p. 633.

Liebhart, D., Sulejmanovic, T., Grafl, B., Tichy, A. and Hess, M. 2013b. Vaccination against histomonosis prevents a drop in egg production in layers following challenge. *Avian Pathology* 42(1), 79–84.

Liebhart, D., Weissenböck, H. and Hess, M. 2006. In-situ hybridization for the detection and identification of *Histomonas meleagridis* in tissues. *Journal of Comparative Pathology* 135(4), 237–242.

Liebhart, D., Windisch, M. and Hess, M. 2010. Oral vaccination of 1-day-old turkeys with *in vitro* attenuated *Histomonas meleagridis* protects against histomonosis and has no negative effect on performance. *Avian Pathology* 39(5), 399–403.

Liebhart, D., Zahoor, M. A., Prokofieva, I. and Hess, M. 2011. Safety of avirulent histomonads to be used as a vaccine determined in turkeys and chickens. *Poultry Science* 90(5), 996–1003.

Lillehoj, H. S. 1987. Effects of immunosuppression on avian coccidiosis: Cyclosporin A but not hormonal bursectomy abrogates host protective immunity. *Infection and Immunity* 55(7), 1616–1621.

Lillehoj, H. S. 1994. Analysis of *Eimeria acervulina*-induced changes in the intestinal T lymphocyte subpopulations in two chicken strains showing different levels of susceptibility to coccidiosis. *Research in Veterinary Science* 56(1), 1–7.

Lillehoj, H. S. and Trout, J. M. 1994. CD8+ T cell-coccidia interactions. *Parasitology Today* 10(1), 10–14.

Lindquist, W. D. 1962. Some effects of paromomycin sulfate on blackhead in turkeys. *American Journal of Veterinary Research* 23, 1053–1056.

Lollis, L., Gerhold, R., McDougald, L. and Beckstead, R. 2011. Molecular characterization of *Histomonas meleagridis* and other parabasalids in the United States using the 5.8S, ITS-1, and ITS-2 rRNA regions. *The Journal of Parasitology* 97(4), 610–615.

Long, P. L., Millard, B. J., Joyner, L. P. and Norton, C. C. 1976. A guide to laboratory techniques used in the study and diagnosis of avian coccidiosis. *Folia Veterinaria Latina* 6(3), 201–217.

Long, P. L. and Pierce, A. E. 1963. Role of cellular factors in the mediation of immunity to avian coccidiosis (Eimeria Tenella). *Nature* 200, 426–427.

Lotfi, A. R., Abdelwhab, E. M. and Hafez, H. M. 2012. Persistence of *Histomonas meleagridis* in or on materials used in poultry houses. *Avian Diseases* 56(1), 224–226.

Lund, E. E., Wehr, E. E. and Elli, D. J. 1966. Earthworm transmission of *Heterakis* and *Histomonas* to turkeys and chickens. *Journal of Parasitology* 52(5), 899–902.

Lynn, E. C. and Beckstead, R. B. 2012. Identification of gene expression elements in *Histomonas meleagridis* using splinkerette PCR, a variation of ligated adaptor PCR. *The Journal of Parasitology* 98(1), 135–141.

Macdonald, S. E., Nolan, M. J., Harman, K., Boulton, K., Hume, D. A., Tomley, F. M., Stabler, R. A. and Blake, D. P. 2017. Effects of *Eimeria tenella* infection on chicken caecal microbiome diversity, exploring variation associated with severity of pathology. *PLoS ONE* 12(9), e0184890.

Macdonald, S. E., Van Diemen, P. M., Martineau, H., Stevens, M. P., Tomley, F. M., Stabler, R. A. and Blake, D. P. 2019. Impact of *Eimeria tenella* coinfection on *Campylobacter jejuni* colonization of the chicken. *Infection and Immunity* 87(2).

Marangi, M., Cafiero, M. A., Capelli, G., Camarda, A., Sparagano, O. A. and Giangaspero, A. 2009. Evaluation of the poultry red mite, *Dermanyssus gallinae* (Acari: Dermanyssidae) susceptibility to some acaricides in field populations from Italy. *Experimental and Applied Acarology* 48(1–2), 11–18.

Marangi, M., Morelli, V., Pati, S., Camarda, A., Cafiero, M. A. and Giangaspero, A. 2012. Acaricide residues in laying hens naturally infested by red mite *Dermanyssus gallinae*. *PLoS ONE* 7(2), e31795.

Maurer, V. 1993. *The Dynamics of Dermanyssus gallinae (Acari: Dermanyssidae) Populations Interacting with Laying Hens and the Predatory Mite Cheyletus eruditus (Acari: Cheyletidae)*. Zurich: Swiss Federal Institute of Technology Zurich.

Maurer, V. and Baumgartner, J. 1992. Temperature influence on life table statistics of the chicken mite *Dermanyssus gallinae* (Acari: *Dermanyssidae*). *Experimental and Applied Acarology* 15(1), 27–40.

McDonald, V., Shirley, M. W. and Chapman, H. D. 1985. Attenuation of *Eimeria* species: Further characterisation of two lines of *Eimeria mitis*. *Research in Veterinary Science* 39(3), 328–332.

McDougald, L. R. 2005. Blackhead disease (histomoniasis) in poultry: A critical review. *Avian Diseases* 49(4), 462–476.

Mitra, T., Gerner, W., Kidane, F. A., Wernsdorf, P., Hess, M., Saalmüller, A. and Liebhart, D. 2017. Vaccination against histomonosis limits pronounced changes of B cells and T-cell subsets in turkeys and chickens. *Vaccine* 35(33), 4184–4196.

Mitra, T., Kidane, F. A., Hess, M. and Liebhart, D. 2018. Unravelling the immunity of poultry against the extracellular protozoan parasite *Histomonas meleagridis* is a cornerstone for vaccine development: a review. *Frontiers in Immunology* 9, 2518.

Morgan, J. A., Morris, G. M., Wlodek, B. M., Byrnes, R., Jenner, M., Constantinoiu, C. C., Anderson, G. R., Lew-Tabor, A. E., Molloy, J. B., Gasser, R. B. and Jorgensen, W. K. 2009. Real-time polymerase chain reaction (PCR) assays for the specific detection and quantification of seven *Eimeria* species that cause coccidiosis in chickens. *Molecular and Cellular Probes* 23(2), 83–89.

Morgan, J. A. T. and Godwin, R. M. 2017. Mitochondrial genomes of Australian chicken *Eimeria* support the presence of ten species with low genetic diversity among strains. *Veterinary Parasitology* 243, 58–66.

Mul, M. F., Van Riel, J. W., Meerburg, B. G., Dicke, M., George, D. R. and Groot Koerkamp, P. W. 2015. Validation of an automated mite counter for *Dermanyssus gallinae* in experimental laying hen cages. *Experimental and Applied Acarology* 66(4), 589–603.

Munsch, M., Mehlhorn, H., Al-Quraishy, S., Lotfi, A. R. and Hafez, H. M. 2009. Molecular biological features of strains of *Histomonas meleagridis*. *Parasitology Research* 104(5), 1137–1140.

Nguyen Pham, A. D., Gussem, J. K. D. and Goddeeris, B. M. 2013. Intracloacally passaged low-virulent *Histomonas meleagridis* protects turkeys from histomonosis. *Veterinary Parasitology* 196(3–4), 307–313.

Nordenfors, H. and Chirico, J. 2001. Evaluation of a sampling trap for *Dermanyssus gallinae* (Acari: Dermanyssidae). *Journal of Economic Entomology* 94(6), 1617–1621.

Oh, S., Noh, G., Yi, S., DO, Y., Kim, E. and Yoo, J. 2019. Molecular epidemiological characterization of poultry red mite (*Dermanyssus gallinae*) collected from Korea. *Korean Journal of Veterinary Service* 42, 161–167.

Pastor-Fernández, I., Kim, S., Marugan-Hernandez, V., Soutter, F., Tomley, F. M. and Blake, D. P. 2020. Vaccination with transgenic *Eimeria tenella* expressing *Eimeria maxima* AMA1 and IMP1 confers partial protection against high-level *E. maxima* challenge in a broiler model of coccidiosis. *Parasites and Vectors* 13(1), 343.

Pegg, E., Doyle, K., Clark, E. L., Jatau, I. D., Tomley, F. M. and Blake, D. P. 2016. Application of a new PCR-RFLP panel suggests a restricted population structure for *Eimeria tenella* in UK and Irish chickens. *Veterinary Parasitology* 229, 60–67.

Petrov, D. 1975. Study of *Dermanyssus gallinae* as a carrier of *Pasteurella multocida*. *Veterinarno-Meditsinski Nauki* 12(5), 32–36.

Powell, F. L., Rothwell, L., Clarkson, M. J. and Kaiser, P. 2009. The turkey, compared to the chicken, fails to mount an effective early immune response to *Histomonas meleagridis* in the gut. *Parasite Immunology* 31(6), 312–327.

Price, D. R. G., Kuster, T., Oines, Ø, Oliver, E. M., Bartley, K., Nunn, F., Lima Barbero, J. F., Pritchard, J., Karp-Tatham, E., Hauge, H., Blake, D. P., Tomley, F. M. and Nisbet, A. J. 2019. Evaluation of vaccine delivery systems for inducing long-lived antibody responses to *Dermanyssus gallinae* antigen in laying hens. *Avian Pathology* 48(Suppl. 1), S60–S74.

Rawlinson, E. M., Nowell, F. and Conisbee, A. 1985. An isolator system for mice infected with *Eimeria* spp. and other parasites. *Laboratory Animals* 19(2), 92–97.

Reid, A. J., Blake, D. P., Ansari, H. R., Billington, K., Browne, H. P., Bryant, J., Dunn, M., Hung, S. S., Kawahara, F., Miranda-Saavedra, D., Malas, T. B., Mourier, T., Naghra, H., Nair, M., Otto, T. D., Rawlings, N. D., Rivailler, P., Sanchez-Flores, A., Sanders, M., Subramaniam, C., Tay, Y. L., Woo, Y., Wu, X., Barrell, B., Dear, P. H., Doerig, C., Gruber, A., Ivens, A. C., Parkinson, J., Rajandream, M. A., Shirley, M. W., Wan, K. L., Berriman, M., Tomley, F. M. and Pain, A. 2014. Genomic analysis of the causative agents of coccidiosis in domestic chickens. *Genome Research* 24(10), 1676–1685.

Reis, J. L., Beckstead, R. B., Brown, C. C. and Gerhold, R. W. 2009. *Histomonas meleagridis* and capillarid infection in a captive chukar (*Alectoris chukar*). *Avian Diseases* 53(4), 637–639.

Rose, M. E. and Hesketh, P. 1982. Immunity to coccidia in chickens: Adoptive transfer with peripheral blood lymphocytes and spleen cells. Parasite Immunology 4, 171–185.

Rose, M. E., Hesketh, P. and Wakelin, D. 1992. Immune control of murine coccidiosis: CD4+ and CD8+ T lymphocytes contribute differentially in resistance to primary and secondary infections. *Parasitology* 105(3), 349–354.

Rose, M. E., Wakelin, D. and Hesketh, P. 1989. Gamma interferon controls *Eimeria vermiformis* primary infection in BALB/c mice. *Infection and Immunity* 57(5), 1599–1603.

Roy, L. and Buronfosse, T. 2011. Using mitochondrial and nuclear sequence data for disentangling population structure in complex pest species: A case study with *Dermanyssus gallinae*. *PLoS ONE* 6(7), e22305.

Roy, L., Chauve, C. M. and Buronfosse, T. 2010. Contrasted ecological repartition of the Northern Fowl Mite *Ornithonyssus sylviarum* (Mesostigmata : Macronyssidae) and the Chicken Red Mite *Dermanyssus gallinae* (Mesostigmata : Dermanyssidae). *Acarologia* 50(2), 207–219.

Roy, L., Dowling, A. P., Chauve, C. M. and Buronfosse, T. 2009a. Delimiting species boundaries within *Dermanyssus Dugès*, 1834 (Acari: Mesostigmata) using a total evidence approach. *Molecular Phylogenetics and Evolution* 50(3), 446–470.

Roy, L., Dowling, A. P., Chauve, C. M., Lesna, I., Sabelis, M. W. and Buronfosse, T. 2009b. Molecular phylogenetic assessment of host range in five *Dermanyssus species*. *Experimental and Applied Acarology* 48(1–2), 115–142.

Samaha, H., Haggag, Y., Nossair, M. and Habib, H. 2013. Assessment efficiency of some chemical disinfectants commonly used against coccidia in poultry farms. *Alexandria Journal of Veterinary Sciences* 39, 82–90.

Sawale, G., Rambabu, D., Kommu, S., Bhandurge, M., Ramesh, G. and Lakshman, M. 2018. Outbreak of intestinal coccidiosis due to *Eimeria necatrix* in Rajasree birds: pathomorphological and electron microscopic study. *International Journal of Livestock Research* 8, 247–251.

Schulz, J., Berk, J., Suhl, J., Schrader, L., Kaufhold, S., Mewis, I., Hafez, H. M. and Ulrichs, C. 2014. Characterization, mode of action, and efficacy of twelve silica-based acaricides against poultry red mite (*Dermanyssus gallinae*) *in vitro*. *Parasitology Research* 113(9), 3167–3175.

Schwarz, R. S., Jenkins, M. C., Klopp, S. and Miska, K. B. 2009. Genomic analysis of *Eimeria* spp. populations in relation to performance levels of broiler chicken farms in Arkansas and North Carolina. *Journal of Parasitology* 95(4), 871–880.

Shirley, M. W., Smith, A. L. and Tomley, F. M. 2005. The biology of avian *Eimeria* with an emphasis on their control by vaccination. *Advances in Parasitology* 60, 285–330.

Shirnov, F., Ibragiomova, A. and Misirov, Z. 1972. he dissemination of the fowl-pox by the mite *Dermanyssus gallinae*. *Veterinarya* 4, 48–49.

Singh, A., Weissenböck, H. and Hess, M. 2008. *Histomonas meleagridis*: Immunohistochemical localization of parasitic cells in formalin-fixed, paraffin-embedded tissue sections of experimentally infected turkeys demonstrates the wide spread of the parasite in its host. *Experimental Parasitology* 118(4), 505–513.

Sleeckx, N., Van Gorp, S., Koopman, R., Kempen, I., Van Hoye, K., De Baere, K., Zoons, J. and De Herdt, P. 2019. Production losses in laying hens during infestation with the poultry red mite *Dermanyssus gallinae*. *Avian Pathology* 48(Suppl. 1), S17–S21.

Smith, A. L., Hesketh, P., Archer, A. and Shirley, M. W. 2002. Antigenic diversity in *Eimeria maxima* and the influence of host genetics and immunization schedule on cross-protective immunity. *Infection and Immunity* 70(5), 2472–2479.

Sparagano, O., Pavlicevic, A., Murano, T., Camarda, A., Sahibi, H., Kilpinen, O., Mul, M., Van Emous, R., Le Bouquin, S., Hoel, K. and Cafiero, M. A. 2009. Prevalence and key figures for the poultry red mite *Dermanyssus gallinae* infections in poultry farm systems. *Experimental and Applied Acarology* 48(1–2), 3–10.

Sparagano, O. A., George, D. R., Harrington, D. W. and Giangaspero, A. 2014. Significance and control of the poultry red mite, *Dermanyssus gallinae*. *Annual Review of Entomology* 59, 447–466.

Steenberg, T. and Kilpinen, O. 2014. Synergistic interaction between the fungus *Beauveria bassiana* and desiccant dusts applied against poultry red mites (*Dermanyssus gallinae*). *Experimental and Applied Acarology* 62(4), 511–524.

Stucki, U., Braun, R. and Roditi, I. 1993. *Eimeria tenella*: Characterization of a 5S ribosomal RNA repeat unit and its use as a species-specific probe. *Experimental Parasitology* 76(1), 68–75.

Sulejmanovic, T., Bilic, I., Hess, M. and Liebhart, D. 2016. An *in vitro* attenuated strain of *Histomonas meleagridis* provides cross-protective immunity in turkeys against heterologous virulent isolates. *Avian Pathology* 45(1), 46–53.

Sulejmanovic, T., Grafl, B., Bilic, I., Jaskulska, B. and Hess, M. 2019a. PCR and serology confirm the infection of turkey hens and their resilience to histomonosis in mixed flocks following high mortalities in toms. *Parasites and Vectors* 12(1), 228.

Sulejmanovic, T., Turblin, V., Bilic, I., Jaskulska, B. and Hess, M. 2019b. Detection of *Histomonas meleagridis* DNA in dust samples obtained from apparently healthy meat turkey flocks without effect on performance. *Avian Pathology* 48(4), 329–333.

Sulejmanovic, T., Liebhart, D., Magdefrau-Pollan, B., Sanglhuber, E. M., Wiesinger, E., Bilic, I. and Hess, M. 2017. Emergence of fatal histomonosis in meat turkey flocks in Austria from 2014 to 2016. *Wiener Tierarztliche Monatsschrift*, 277–287.

Tavassoli, M., Allymehr, M., Pourseyed, S. H., Ownag, A., Bernousi, I., Mardani, K., Ghorbanzadegan, M. and Shokrpoor, S. 2011. Field bioassay of *Metharhizium anisopliae* strains to control the poultry red mite *Dermanyssus gallinae*. *Veterinary Parasitology* 178(3–4), 374–378.

Taylor, M. A., Coop, R. L. and Wall, R. L. (Eds.) 2007. *Veterinary Parasitology*. Blackwell Publishing Ltd, Oxford.

Temple, D., Manteca, X., Escribano, D., Salas, M., Mainau, E., Zschiesche, E., Petersen, I., Dolz, R. and Thomas, E. 2020. Assessment of laying-bird welfare following acaricidal treatment of a commercial flock naturally infested with the poultry red mite (*Dermanyssus gallinae*). *PLoS ONE* 15(11), e0241608.

Thøfner, I. C. N., Liebhart, D., Hess, M., Schou, T. W., Hess, C., Ivarsen, E., Fretté, X. C., Christensen, L. P., Grevsen, K., Engberg, R. M. and Christensen, J. P. 2012. Antihistomonal effects of artemisinin and *Artemisia annua* extracts *in vitro* could not be confirmed by *in vivo* experiments in turkeys and chickens. *Avian Pathology* 41(5), 487–496.

Trout, J. M. and Lillehoj, H. S. 1996. T lymphocyte roles during *Eimeria Acervulina* and *Eimeria tenella* infections. *Veterinary Immunology and Immunopathology* 53(1–2), 163–172.

Tu, Q., Hickey, M. E., Yang, T., Gao, S., Zhang, Q., Qu, Y., Du, X., Wang, J. and He, L. 2019. A simple and rapid method for detecting the pesticide fipronil on egg shells and in liquid eggs by Raman microscopy. *Food Control* 96, 16–21.

Tyzzer, E. E. 1920. The flagellate character and reclassification of the parasite producing "blackhead" in turkeys: *Histomonas* (gen. nov.) *meleagridis* (Smith). *The Journal of Parasitology* 6(3), 124.

Tyzzer, E. E. 1934. Studies on histomoniasis, or "blackhead" infection, in the chicken and the turkey. *Proceedings of the American Academy of Arts and Sciences* 69(5), 189.

Valiente Moro, C., De Luna, C. J., Tod, A., Guy, J. H., Sparagano, O. A. and Zenner, L. 2009. The poultry red mite (*Dermanyssus gallinae*): A potential vector of pathogenic agents. *Experimental and Applied Acarology* 48(1-2), 93-104.

van der Heijden, H. M. J. F. and Landman, W. J. M. 2008a. *In vitro* effect of herbal products against *Histomonas meleagridis*. *Veterinary Parasitology* 154(1-2), 1-7.

van der Heijden, H. M. J. F. and Landman, W. J. M. 2008b. *In vivo* effect of herbal products against *Histomonas meleagridis* in turkeys. *Avian Pathology* 37(1), 45-50.

van der Heijden, H. M. J. F. and Landman, W. J. M. 2011. High seroprevalence of *Histomonas meleagridis* in Dutch layer chickens. *Avian Diseases* 55(2), 324-327.

van der Heijden, H. M. J. F., Landman, W. J. M., Greve, S. and Peek, R. 2006. Genotyping of *Histomonas meleagridis* isolates based on Internal Transcribed Spacer-1 sequences. *Avian Pathology* 35(4), 330-334.

van der Heijden, H. M. J. F., Stegeman, A. and Landman, W. J. M. 2010. Development of a blocking-ELISA for the detection of antibodies against *Histomonas meleagridis* in chickens and turkeys. *Veterinary Parasitology* 171(3-4), 216-222.

Van Immerseel, F., Lyhs, U., Pedersen, K. and Prescott, J. F. 2016. Recent breakthroughs have unveiled the many knowledge gaps in *Clostridium perfringens*-associated necrotic enteritis in chickens: The first International Conference on Necrotic Enteritis in Poultry. *Avian Pathology* 45(3), 269-270.

Vrba, V., Blake, D. P. and Poplstein, M. 2010. Quantitative real-time PCR assays for detection and quantification of all seven *Eimeria* species that infect the chicken. *Veterinary Parasitology* 174(3-4), 183-190.

Vrba, V. and Pakandl, M. 2015. Host specificity of turkey and chicken *Eimeria*: Controlled cross-transmission studies and a phylogenetic view. *Veterinary Parasitology* 208(3-4), 118-124.

Wallach, M. 2010. Role of antibody in immunity and control of chicken coccidiosis. *Trends in Parasitology* 26(8), 382-387.

Wang, C., Xu, X., Huang, Y., Yu, H., Li, H., Wan, Q., Li, H., Wang, L., Sun, Y. and Pan, B. 2021. Susceptibility of *Dermanyssus gallinae* from China to acaricides and functional analysis of glutathione S-transferases associated with beta-cypermethrin resistance. *Pesticide Biochemistry and Physiology* 171, 104724.

Wang, C., Xu, X., Huang, Y., Yu, H., Li, H., Wan, Q. and Pan, B. 2020. Transcription profiling and characterization of *Dermanyssus gallinae* cytochrome P450 genes involved in beta-cypermethrin resistance. *Veterinary Parasitology* 283, 109155.

Wattrang, E., Magnusson, S. E., Näslund, K., Thebo, P., Hagström, Å., Smith, A. L. and Lundén, A. 2016. Expression of perforin, granzyme A and Fas ligand mRNA in caecal tissues upon *Eimeria tenella* infection of naive and immune chickens. Parasite Immunology 38, 419-430.

Williams, R. B. 2002. Fifty years of anticoccidial vaccines for poultry (1952-2002). *Avian Diseases* 46(4), 775-802.

Williams, R. B., Marshall, R. N., Pages, M., Dardi, M. and Del Cacho, E. 2009. Pathogenesis of *Eimeria praecox* in chickens: Virulence of field strains compared with laboratory strains of *E. praecox* and *Eimeria Acervulina*. *Avian Pathology* 38(5), 359-366.

Windisch, M. and Hess, M. 2009. Establishing an indirect sandwich enzyme-linked-immunosorbent-assay (ELISA) for the detection of antibodies against *Histomonas meleagridis* from experimentally infected specific pathogen-free chickens and turkeys. *Veterinary Parasitology* 161(1-2), 25-30.

Windisch, M. and Hess, M. 2010. Experimental infection of chickens with *Histomonas meleagridis* confirms the presence of antibodies in different parts of the intestine. *Parasite Immunology* 32(1), 29–35.

Wojcik, A. R., Grygon-Franckiewicz, B., Zbikowska, E. and Wasielewski, L. 2000. [Invasion of *Dermanyssus gallinae* (De Geer, 1778) in poultry farms in the Torun region]. *Wiadomosci Parazytologiczne* 46(4), 511–515.

Xu, J., Qu, C. and Tao, J. 2014. Loop-mediated isothermal amplification assay for detection of *Histomonas meleagridis* infection in chickens targeting the 18S rRNA sequences. *Avian Pathology* 43(1), 62–67.

Xu, X., Wang, C., Huang, Y., Zhang, S., Yu, H., Meng, J. and Pan, B. 2020. Evaluation of the vaccine efficacy of three digestive protease antigens from Dermanyssus gallinae using an *in vivo* rearing system. *Vaccine* 38(49), 7842–7849.

Zaragatzki, E., Hess, M., Grabensteiner, E., Abdel-Ghaffar, F., Al-Rasheid, K. A. S. and Mehlhorn, H. 2010. Light and transmission electron microscopic studies on the encystation of *Histomonas meleagridis*. *Parasitology Research* 106(4), 977–983.

Zenner, L., Callait, M. P., Granier, C. and Chauve, C. 2003. *In vitro* effect of essential oils from *Cinnamomum aromaticum*, *Citrus limon* and *Allium sativum* on two intestinal flagellates of poultry, *Tetratrichomonas gallinarum* and *Histomonas meleagridis*. *Parasite* 10(2), 153–157.

Zhang, S., Lillehoj, H. S. and Ruff, M. D. 1995. *In vivo* role of tumor necrosis-like factor in *Eimeria tenella* infection. *Avian Diseases* 39(4), 859–866.

Zriki, G., Blatrix, R. and Roy, L. 2020. Predation interactions among henhouse-dwelling arthropods, with a focus on the poultry red mite *Dermanyssus gallinae*. *Pest Management Science* 76(11), 3711–3719.

Part 2

Preventing diseases in poultry

Chapter 4

Improving biosecurity in poultry flocks

Jean-Pierre Vaillancourt and Manon Racicot, Université de Montréal, Canada; and Mattias Delpont, École Nationale Vétérinaire de Toulouse, France

1 Introduction

Biosecurity is any actions or health plans designed to protect a population against infectious and transmissible agents (Toma et al., 1999). They are based on principles that have been known for centuries. Sanitation measures are detailed in *Deuteronomy*, written about 2600 years ago (Ojewole, 2011); most biosecurity measures found in modern poultry farms appear like analogies of measures prescribed to lepers during the mass of separation in the Middle Ages (Brody, 1974). Yet, we are still struggling today to get people to perform simple tasks that we know work for preventing contagious diseases on poultry

http://dx.doi.org/10.19103/AS.2022.0104.05

farms. For example, in 2011, Racicot et al. identified 44 different errors when getting in and out of poultry barns. Most errors had to do with area delimitation (clean versus contaminated areas of the entrance or anteroom of a barn), boots, and handwashing.

Biosecurity on poultry farms has been categorized in different ways (internal, external; operational, structural; bioexclusion, biomanagement, biocontainment, etc.). Ultimately, all on-farm biosecurity measures have to do with two principles: reducing sources of contamination and separating these sources from healthy flocks. In modern poultry production, a third principle linked to the latter is emerging: the need for a regional approach to biosecurity based on communication within the poultry industry and the organization of movement of people, birds, material, and equipment within any given region. Indeed, the intensification of poultry production has created an environment that may promote the spread of contagious diseases. Any poultry activity involves inherent infectious disease transmission risks, and the amplitude of these risks increases with regional poultry density (Fernandez et al., 1994).

However, no matter how we define or partition biosecurity measures, the recurring challenge is the consistent implementation of these measures by all farm personnel and visitors. Achieving high compliance with biosecurity measures is, therefore, a constant issue that needs addressing in terms of research priority.

2 Reducing sources of contamination: cleaning and disinfection of poultry barns

The decontamination of poultry barns is a multistep process. The first step is the dry removal of organic material. It consists of removing with physical force as much visible organic material as possible: litter, feed, dust, etc. The design of the building and its equipment can significantly affect cleaning (Cerf et al., 2010). For example, the floor of an anteroom with ground equipment will be more difficult to clean on a regular basis compared to one where most equipment is suspended.

The second step is the application of a detergent followed by stripping and drying: some organic materials or mineral deposits are impossible or very difficult to remove using only physical force (Step 1). Step 2 is, therefore, the step in which this material is removed using a chemical process, such as a detergent. It makes it possible to reduce the bacterial concentration of a contaminated surface by a factor of 10–1000 compared to a wash with only clean water (Course et al., 2021). Using a detergent also helps reduce the total amount of water used. Several chemicals influence the effectiveness of detergents: surfactants, wetting agents, dispersants, pH adjusters, and sequestering agents. However, two main categories are recognized: acid and alkaline detergents.

A rule of thumb is that acid detergents remove mineral deposits, while alkaline detergents remove organic waste not removed by dry cleaning (i.e. dried, greasy, or sticky material). The choice of detergent, therefore, depends on the type of material present at the start and the water used. If the cleaning water used is hard (>120 ppm), more frequent use of an acid detergent is necessary (Watkins and Venne, 2015).

When applying the detergent, there are several rules to follow. Firstly, the contact time: the use of detergent in foamy form with the help of specialized equipment is preferred because this makes it possible to optimize the contact time, which should be 10–20 min (Springthorpe, 2000). Second, the temperature of the cleaning water. If hot water is available, it should be used. Water that is too cold slows down chemical reactions. Water at 40°C is best (Böhm, 1998). Third, the flow and pressure. The flow rate and quantity of detergent are greater when applying the detergent, while the pressure is greater during stripping, which is the action following the application of the detergent. When applying the detergent, it is recommended to use a flow rate of 10–20 L/min at low pressure (300–500 psi) and to use approximately 250–500 mL of solution per square meter (Blondel et al., 2018).

If the application of the detergent takes more than 20 min, then the stripping should be done alternately with the application of the detergent. Stripping is the action of removing the rest of the organic debris that the detergent has made accessible. Using clean high-pressure water (1000–3000 psi) at a 45° angle works well for this step, as it adds mechanical force (Blondel et al., 2018). However, it is important to avoid damaging the surfaces; such pressure can damage wooden structures, creating crevices making future decontamination more difficult (Blondel et al., 2018).

Finally, it is important to let the areas dry before proceeding to the next step. Omitting this step may reduce the effectiveness of disinfection in two ways: by dilution and by inactivation on contact with organic material. The effectiveness of a disinfectant is related to its concentration when applied. The presence of water at the surface of what needs to be disinfected dilutes the disinfectant and reduces its effectiveness (Böhm, 1998).

The third step is disinfection, followed by drying. This step should only take place when all organic material has been removed and the areas are dry (Langsrud et al., 2003; Payne et al., 2005). It is important to visually monitor the area that has been cleaned before disinfection (Fig. 1). A nonpermanent marker such as a bright color aerosol hair spray or a tape can be used to identify where cleaning is not satisfactory (presence of organic material); this way, it is not necessary to completely redo the cleaning process for the entire barn. This approach may show over time areas that require special attention.

Disinfection can be carried out by vaporization (liquid or foam) and/or by fumigation/misting. Regardless of the method and product used, it is important

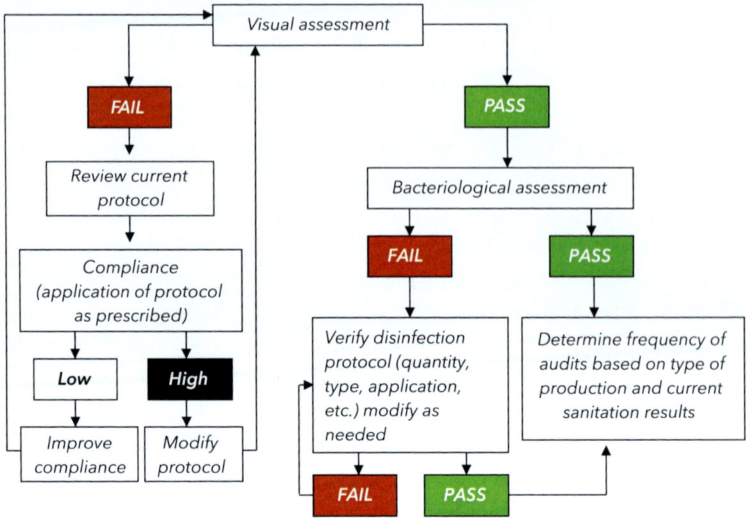

Figure 1 Pathway to assess an on-farm cleaning protocol.

to follow the manufacturer's recommendations for concentration, contact time, product mixture, etc. (Payne et al., 2005). On average, 150–300 mL/m^2 is applied (Blondel et al., 2018).

The choice of disinfectant depends on several factors: the surface to be disinfected, the pathogens targeted, the hardness and the pH of the washing water, the cost, etc. Generally, the commercial disinfectants available for livestock buildings consist of active ingredients from the following main families: halogenated derivatives, aldehydes, quaternary ammoniums, phenols, and peroxide. Drying the building after the application of the disinfectant is essential, as it prevents the multiplication of the remaining bacteria (Davies and Wray, 1995), and this desiccation period may contribute to reducing further the number of living organisms (Böhm, 1998). In some countries, heating the house (40°C) for 3–4 days has been useful (e.g. preventing the spread of *Mycoplasma gallisepticum* in North Carolina, USA). Finally, a microbiological assessment of the disinfection process may be performed as part of the monitoring (Fig. 1).

3 Reducing sources of contamination: equipment and vehicles

Vehicles and equipment may act as a mechanical vector. Dee et al. (2002) demonstrated that snow contaminated with the Porcine Reproductive and Respiratory Syndrome (PRRS) virus and applied to the undercarriage of a car can contaminate the environment for several kilometers. The inadequate management of vehicles, including where they park on a production site (too close to the barn entrance), has been associated with a greater risk of disease transmission between farms (Guinat et al., 2020). Rose et al. (2000) demonstrated an association between the surface covered by vehicles on a broiler chicken production site and the probability of isolating *Salmonella*. Snow et al. (2010) also identified a link between on-farm transportation and *Salmonella* in egg layer farms. Vehicle-sharing practices between poultry production sites are also associated with the spread of infectious laryngotracheitis and avian influenza (Nishiguchi et al., 2007; Volkova et al., 2012).

Wheel baths are assumed to be useful (Pinto and Urcelay, 2003; Zhang et al., 2013), but there is little evidence that they work. In fact, several arguments have been presented to support their lack of effectiveness. Briefly, wheels rotating at high speed at a high temperature over dry surfaces are not likely to harbor pathogens; even if they have mud on them, passing quickly in a wheel bath does not allow the disinfectant, assuming the solution is fresh, to reach the pathogens; finally, the contact time is too limited for disinfectants to reduce wheel contamination. Pathogens in vehicles are more likely to be present in the loading area and in the cab of the vehicle (Chaber and Saegerman, 2017).

The recommendations made for decontaminating poultry barns essentially apply to the decontamination of vehicles and equipment. When possible, they should be exposed to the sun, which helps the drying process as well as decontamination. Indeed, the ultraviolet radiation in sunlight works as a natural disinfectant (Kuney and Jeffrey, 2002).

4 Reducing sources of contamination: water and feed hygiene

4.1 Water hygiene

Water quality and sanitation would require its own chapter. Briefly, when using chlorine, it is best to assess the oxidation reduction potential (ORP) of water. It measures in millivolts the oxidizing potential of free chlorine residual. A strong oxidizer will effectively destroy microbes. An ORP value of 650 millivolts or greater is required. The goal should be to maintain 2–5 ppm of free chlorine in the water supply. Free chlorine is the residual available for sanitation (Watkins and Venne, 2015). For removing scale and biofilm, it is best to drop the water pH below 5 but not below 4 to prevent equipment damage. A bleach solution might be effective in removing biofilm, but it may also damage equipment. Several disinfectants are available to clean water lines, but some of the most effective and safe products for drinker systems are concentrated, stabilized hydrogen peroxides. Biofilms or any growth of bacteria, molds, and fungus in drinker systems can only be removed with cleaners that contain sanitizers (Watkins and Venne, 2015).

4.2 Feed hygiene and delivery

Basic feed mill biosecurity focuses on tight and closed silos to prevent rodent access, a continuous integrated pest management program, and sanitation procedures, mainly for silos that may have been contaminated with pathogens of importance to poultry. The main one, in terms of public health, is *Salmonella* (Jones, 2011). Li et al. (2012) reported that 8.8% of ingredients of animal origin obtained from three feed mills were contaminated with *Salmonella*, but dust samples had a higher contamination rate of 18.5%. Poultry offal meal and feather meal should be considered high-risk ingredients (Butcher and Miles, 1995). Several chemical options have been tested for the treatment of feeds to control *Salmonella*. Formaldehyde is one of the more frequently used chemicals under commercial conditions (Ricke et al., 2019). Cochrane (2016) provide detailed information on feed mill biosecurity plans.

Basic biosecurity for feed delivery starts with a check with dispatch for routing instructions to avoid passing by potentially diseased production sites or going from one to a disease-free operation. The driver of a clean feed truck must wear clean clothes and have plastic boots, hand cleaning solution, and fly spray available when entering a farm; the driver must never enter barns past the contaminated zone, in other words, must not get on the clean side of the anteroom (Jones, 2011). After all runs are completed, the driver goes directly back to the feed truck area. Trucks must be washed completely and disinfected weekly, unless the last site visited has birds infected with a disease of significance to the industry. In this case, truck decontamination must be performed after this last delivery. Breeder farms are normally visited first before meat bird operations. Delivery to diseased or quarantined farms should be done as the last load of the day (Anon., 2018).

5 Reducing sources of contamination: insect, mite and rodent pests, wild birds and pets

An often-overlooked vector, but with a significant risk of disease transmission, is the common house fly, *Musca domestica*. In 1975, more than 100 different pathogens that could be carried by flies and more than 65 diseases transmitted to humans and animals, such as salmonellosis, had already been identified. They can travel more than 30 km, but the majority of flies are confined to an area of 3 km^2. They can move quickly over long distances: 1–1.5 km in 24 h, 3–5 km in 48 h, and 6 km in 72 h (Schoof, 1959). Disease transmission can therefore occur between neighboring farms. Flies can also pose a public health problem. In 2009, researchers isolated enterococci and staphylococci in the digestive tract and on the exterior surfaces of flies caught near poultry farms. These bacteria were resistant to antibiotics of importance in human medicine (Graham et al.,

2009). A study showed that 8.2% of the flies tested within 50 m of a farm were positive for *Campylobacter jejuni* (Hald et al., 2004). A study in Japan on highly pathogenic avian influenza H5N1 showed that within a kilometer of the farm, 30% of the flies were contaminated with the same virus as the poultry; at 2 km, the percentage decreased to 10% (Sawabe et al., 2006).

Besides flies, mealworms or darkling beetles (*Alphitobius diaperinus*) are an important concern in disease control. They can transport and transmit microorganisms such as *Campylobacter*, *Salmonella*, infectious bursal disease virus, reovirus, and Eimeria. The mealworm is one of the most abundant insects in poultry farm litter, reaching levels as high as 1000 mealworms per square meter. The risk of disease dispersal can therefore be considerable when spreading litter. In addition, chickens, wild animals, and rodents feed on these insects and can thus infect themselves (Bates et al., 2004; Crippen and Sheffield, 2006). Mainly considering that darkling beetles can maintain *Salmonella* internally during pupation, the next generation of beetles can recontaminate a poultry barn and the following flocks (Roche et al., 2009). Both fly and darkling beetle populations can explode within a few weeks, even with new litter (Fig. 2). Mites (*Dermanyssus gallinae*) have also been associated with recurring site contaminations with *Salmonella Enteritidis* (Moro et al., 2009) and *Salmonella Gallinarum* (Lee et al., 2020)

Insect control requires the application of biosecurity measures for the environment, equipment, visitors, and management of organic materials. It is called integrated pest management. First, the site should always be kept free of unnecessary material since these items can harbor vermin. From the outside, a stone access to the door of the building should be favored. Indoors, rodent traps, or bait should be set in the anteroom and regularly monitored.

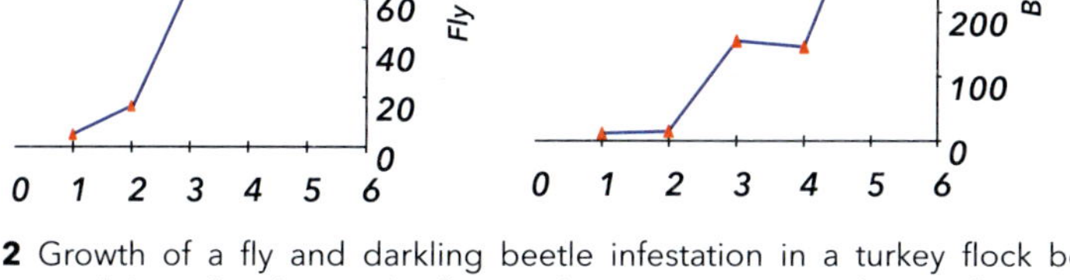

Figure 2 Growth of a fly and darkling beetle infestation in a turkey flock between placement and 6 weeks of age. The fly population was estimated using fly speck card (white card left 24 hours inside the barn and fly defecations counted); Beetle traps where located against the walls inside the barn; the number of beetles where counted after 24 hours.

Insecticides should be applied to the ground, and equipment that must be in the anteroom should be off the floor as much as possible (Villa and Velasco, 1994). It is strongly suggested to rotate between insecticides to increase efficacy and coverage and avoid the development of resistance. Since the treatment is not selective, some recommend the application of insecticides only in areas with high larval density to avoid the destruction of predatory insects, which may be beneficial (Ebeling, 1975).

In addition, feeding and watering equipment should be regularly inspected and repaired to decrease insect populations. The feed delivery system and hoses should be periodically cleaned to prevent insect nesting. To reduce the risk of introducing certain insects, it is also necessary to control employee and visitor traffic and disinfect personal effects and any equipment entering a building. These measures are particularly important when the ectoparasites can survive outside the host for a few days to several weeks. Managing organic materials, such as bird carcasses and manure, is critical. Carcasses provide a substrate and a humidity level (over 90%) favorable for the development of fly larvae, and the manure allows infestation by beetles such as mealworms and their larvae (Allen and Newell, 2005; Barrington et al., 2006).

Rodents can also be found in large numbers on farms as long as water, food, and shelter are available. Rodents are mechanical and biological vectors of several pathogens. Indeed, the same serotypes of *Pasteurella multocida* are isolated in rats and poultry (Curtis et al., 1980). Like rats, mice are involved in the transmission of *Salmonella* (Henzler and Opitz, 1992; Davies and Wray, 1995). Mice are three times more infected with *Salmonella Enteritidis* than samples taken from the environment of a poultry farm. It seems that environmental contamination decreases in the absence of mice (Henzler and Opitz, 1992). Another study supports this view: there are three times more chances that the bacteria will persist in a building when rodents are present, despite washing and disinfecting the building (odds ratio (OR) 3.1; 95% confidence interval (CI) 1.1–3.2) (Rose et al., 2000).

Control of rats and mice is only effective if the barns are rodent-proof, that is, having no openings beyond 1.3 cm (Roy and Brown, 1954). It is also important not to attract rodents to farms by picking up any feed spills, not leaving debris on the site, and regularly pruning vegetation around buildings. Vegetation provides rodents with their basic needs (water, food), protection from predators, and material for building their nests. Appropriate pest management also includes traps or traps containing rodenticides and monitoring to ensure the effectiveness of the program (Axtell, 1999; Corrigan, 2006).

Other wild animals can be reservoirs and vectors of several diseases. Raccoons, along with opossums and foxes, appear to play a role in avian influenza outbreaks. During the 2002 epidemic in Virginia, a significant association was demonstrated between their presence on a farm site and the presence of the avian influenza virus in commercial chickens (OR 1.9; 95% CI 1.0–3.4) (McQuiston et al., 2005). On the other hand, it is more often migratory birds that are involved in the initial influenza outbreaks of an epidemic. Between 1978 and 2000, turkey farms in Minnesota experienced more than 100 introductions of low-pathogenic avian influenza virus from migrating ducks (Halvorson, 2002). Moreover, free-range poultry farms, more exposed to contact with wild birds, showed a higher risk of low-pathogenic avian influenza infection in the Netherlands between 2013 and 2017 (Bouwstra et al., 2017). Several highly pathogenic avian influenza H5N8 epidemics in poultry- and duck-producing regions in Europe between 2017 and 2021 originated from wild birds. Hence, there is clear evidence that wild birds serve as reservoirs, maintaining the virus in several regions over time (Jeong et al., 2014; Verhagen et al., 2021). The mobility of wild birds is a major concern in terms of disease dispersal. They frequently contaminate the environment, vehicles, equipment, feed storage, etc., with their droppings (Davison et al., 1997). It is, therefore, important to limit the access of wild birds and to avoid contact with domestic birds (Axtell, 1999). In that regard, free-range poultry productions require increased attention to biosecurity measures related to the outdoor range (e.g. type of vegetal cover, reducing the accessibility of feed and water to wild birds) (Bestman et al., 2018; Delpont et al., 2020).

It is best to avoid having pets on a farm. Virulent forms of *Pasteurella multocida* have been isolated from the oral cavity of farm cats (Van Sambeek et al., 1995). Strains affecting cats have also been shown to infect turkeys (Curtis and Ollerhead, 1982). There is little evidence of the role of cats in the direct transmission of the bacteria, except in cases of bites (Korbel et al., 1992). However, they can transmit the infection to rats and hunted wild birds, which can subsequently infect domestic poultry. For any type of production, it is therefore inappropriate to use cats for rodent control on a farm (Corrigan, 2006). Dogs, meanwhile, have been identified as a potential carrier of the avian infectious bursal disease virus (Pagès-Manté et al., 2004). The presence of other animals on the site of a broiler farm, including pigs, cattle, sheep, and poultry other than broilers, is strongly associated with a high risk of infection with *Campylobacter* (OR 6.33; 95% CI 1.54–26.00). Cattle are also an important source of introduction of this bacterium in broiler chickens via the farmers' boots (Van De Giessen et al., 1998).

6 Reducing sources of contamination: manure, litter and dead birds

6.1 Manure and litter management

Manure and used litter, and occasionally new litter, can be very contagious material, particularly of enteric pathogens, although it is well documented that a respiratory virus like the influenza virus can persist and spread easily via this route (Duvauchelle et al., 2013; Kim et al., 2018). Kim et al. (2018) reported a reduced risk of contamination when an outside company was contracted for manure removal. However, in Ontario, in 1995, growers who required an outside company to handle used litter were on average eight times more at risk of having a flock infected with infectious laryngotracheitis in comparison to neighbors located less than 2 km from them and having birds at risk at the same moment (Vaillancourt, 1995). Manure has also been identified as a source of contamination for *Campylobacter* (Arsenault et al., 2007).

Therefore, the handling of manure, storage, decontamination (composting, incineration), and removal conditions are of critical importance.

6.2 Disposal of dead birds

In many countries, rendering is the most prevalent approach to dead bird disposal. However, it has been associated with a greater risk of disease transmission in different species, including poultry. For example, in Virginia, United States, farms using rendering instead of disposing of carcasses on site were seven times more likely to have flocks infected with avian influenza H7N2 in 2002 (McQuiston et al., 2005). The proximity of the rendering container to poultry barns may partly explain the greater risk of contamination when renderers visit the site. Another important factor is the absence of an adequate sanitation protocol for personnel visiting the site where the rendering container is located as they return to the barns.

6.3 Incineration

Incineration refers to the combustion of material to the extent that the resulting end products are heat, gaseous emissions, and residual ash. There are three types of dead animal cremation: (1) stationary cremation, (2) air curtain cremation, and (3) open-air cremation. The overriding consideration affecting the use of on-farm incineration is regulatory. In some regions or countries, regulations state that incineration equipment must contain a secondary combustion chamber to reduce particulates (i.e. 'fly ash') and other emissions, such as polluting hydrocarbons and heavy metals (Chen et al., 2004).

Fixed plant units range from business units designed for on-farm animal cremation to large cremation plants. They are normally powered by diesel, natural gas, or propane. The equipment is relatively easy to operate after a short training. Periodic observation, routine maintenance, and ash cleaning are required. Fuel consumption varies depending on incinerator design and loading rate.

Air curtain incineration involves the use of air forced mechanically through a firebox with refractory panels or a trench in the earth. Each type has distinct characteristics that may increase or limit its potential for use as a carcass disposal method. Air curtain technology was developed primarily as a means of incinerating large amounts of combustible waste resulting from land clearing or natural disasters (Ellis, 2001).

The main feature of air curtain incineration is a high-speed 'curtain' of air produced by a fan above an above-ground combustion chamber or in an earth-burning trench. The air curtain serves to contain smoke and particles in the combustion zone and provides better airflow for warmer temperatures and more complete combustion. It was used for the disposal of dead birds during the avian influenza epidemic in Virginia (United States) in 2002 (Brglez, 2003). This is a fuel-intensive process (mainly wood and diesel fuel), but its use may be justified in order to avoid moving dead animals over long distances (Ellis, 2001).

Outdoor incineration is not recommended for several reasons. The downsides include labor and fuel requirements, reliance on favorable weather conditions, potential for environmental pollution, odors, and negative public perception.

6.4 Composting

Several studies indicate that composting appears to be effective in eliminating infectious pathogens endemic to poultry. In principle, large-scale windrow composting would be effective for the disposal of a large volume of carcasses.

There is little information on the fate of prions or sporulating bacteria such as *Bacillus anthracis* when composting cadavers, which prevents it from being considered an adequate method of managing mortalities by the European Union. Deactivation of prions is difficult: it requires exposure to heat from 980 to 1100°C or to alkaline digestion. Neither of these conditions occurs in the composting process. However, there is some evidence that certain enzymes and competition from organisms may have a beneficial effect in reducing the presence of prions during composting (Bonhotal et al., 2014). In contrast, in North America, unless there is a serious suspicion of prions, which is not the case for poultry, composting is considered an excellent means of eliminating the vast majority of infectious pathogens of interest to animal industries and government authorities. In other words, the advantages of composting, including the elimination of the movement of contaminated organic material, are considered far greater than the disadvantages.

Composting is a natural biological process of decomposing organic material in a predominantly aerobic environment. During the process, bacteria, fungi, and other microorganisms break down organic material into a stable mixture (compost) while consuming oxygen and releasing heat, water, carbon dioxide (CO_2), and other gases (Keener et al., 2000). The use of compost for the routine disposal of on-farm poultry carcasses has increased in prevalence in the United States and Canada over the past 30 years (Tablante and Malone, 2006).

Four variables are considered essential for successful composting: (1) moisture content (40–60%), (2) temperature (45–60°C); (3) oxygen concentration (desirable level 10%), and (4) carbon:nitrogen (C:N) ratio (desirable range 20:1 to 30:1) (Keener et al., 2000).

The process essentially takes place in two phases - a primary thermophilic phase (temperatures up to 70°C) and a secondary mesophilic phase (usually 30–40°C) (Kalbasi et al., 2005).

Temperature and temperature maintenance are important factors in the destruction of infectious pathogens. A temperature of 54°C for 3 days, typical of carcass composting, should kill all pathogens except spores and prions (Sander et al., 2002).

To keep costs down on the farm, composting of carcasses is usually done by producing a static pile or heap that does not involve specialized mixing, crushing, turning, aeration, and screening equipment. The degradation of the carcass is initiated by natural anaerobic bacteria in the cadaver and by aerobic bacteria on the exterior surfaces. Odorous gases and liquids diffuse into drier, more aerobic plant materials, where they are ingested by microorganisms and broken down into simpler organic compounds and ultimately into CO_2 and water (Keener et al., 2000). The success of this operation relies on the careful construction of a layered heap using appropriate amounts of plant-based covering material below, between, and above the carcasses. Characteristics of effective toping materials include water-holding capacity, gas permeability or porosity (oxygen for microbial activity), biodegradability, wet strength, and sufficient carbon. These physical characteristics determine the ability of toping materials to absorb excess liquids, preventing the release of leachate and odors (King et al., 2005).

A variety of plant-based residues have been used as toping materials, including sawdust, wood chips, ground corn stalks, straw or ground hay, oat or peanut hulls, poultry or livestock bedding, dry manure, etc. (Keener et al., 2000; Glanville et al., 2006).

Turning a stack may be necessary to break up wet areas and to introduce more oxygen and moisture, if needed, to reactivate aerobic microbial activity and stimulate a secondary cycle of heat production. Once the secondary heating cycle is complete, soft tissue decomposition is usually complete and the compost is stable enough to be stored prior to field application. Based on a review of the literature, Keener et al. (2000) concluded that decomposition times are largely a function of cadaver mass, and they published weight-based prediction equations for the duration of the primary and secondary composting cycles. In practice, most of the compost is turned only once or twice. Turning accelerates the decomposition of the carcasses, but it is not essential if the carbon source used to cover the carcasses is sufficiently permeable for the diffusion of oxygen into the heap (Glanville et al., 2006).

On-farm composting is usually done in open bins or swath piles. Three-sided bins are typical, with the open end allowing access for placement, turning, and removal of compost using a tractor. Permanent structures are built with treated lumber or concrete and are usually built on a concrete platform to provide a firm work surface. Windrow composting involves the construction of long, narrow piles having a parabolic or trapezoidal cross section. Due to their shape, windrows have a large exposed surface that encourages aeration and drying (Mukhtar et al., 2004). Since the dimensions of the windrows are not constrained by the walls, their dimensions can be adapted to any size and number of carcasses, which makes them particularly useful following the mass slaughter of animals.

Although windrows do not require the construction of a structure to contain the compost, a low permeability base is recommended to avoid contamination by leachate from the underlying soil (e.g. concrete or asphalt; gravel lined with fabric plastic or geotextile; compacted soil) (Keener et al., 2000).

7 Separating healthy birds from sources of contamination: zoning production sites

Basically, zoning is about preventing the contact between susceptible birds and microbes. The barns where the birds are located must be the most restricted zone (restricted access zone or RAZ); the farm where we find the barns is a controlled access zone (CAZ) (Fig. 3). Everything else around these two zones must be considered an environment that is source of contamination (Anon., 2018).

7.1 Preserving the integrity of the controlled access zone

All farm personnel must avoid live bird markets and be in contact with any other poultry. The CAZ should be clearly marked or understood by those who must access the site. A fence is best when economically doable. The main point is that only essential vehicles, personnel, and equipment should get within the CAZ.

An integrated pest management program, including cleaning out vegetation for at least 5 m around the barns, is paramount to removing potential carriers of disease (Vaillancourt and Martinez, 2001). Standing water must be avoided on the premises. It is a breeding environment for insects, and it attracts wild birds that may be carriers of infectious pathogens such as avian influenza. Feed spills

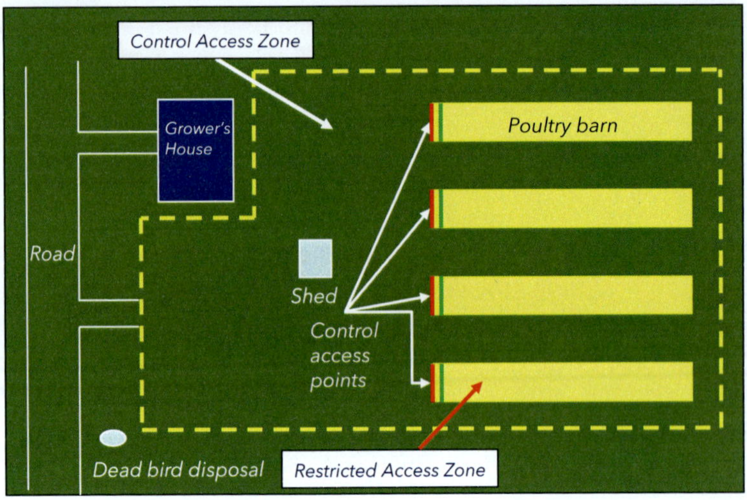

Figure 3 Control access zone (area where poultry traffic and equipment are controlled; delimitation can be virtual or physical (e.g. fence); and restricted access zone (area where birds are located). The control access points are the doors, and ideally the anteroom, providing access to the flock.

should also be avoided for the same reason. If rendering is used to dispose of dead birds, the disposal area cannot be part of the CAZ (Anon., 2018).

Finally, equipment and vehicles must be washed and disinfected, most importantly, all equipment used on more than one farm. In this case, it is best to clean and disinfect when leaving the farm and prior to use on the next farm (Anon., 2018).

7.2 Protecting the restricted access zone

The purpose is to limit access to essential visitors. When a visit is needed, it is important to make sure that visitors have taken a shower and changed clothing and footwear before visiting. Even if they have not been on another farm before, they may still have had an opportunity to get contaminated on their way to the production site. Employees and even the farm owner should consider themselves visitors when they leave the facilities and return. It is best for them to wear farm-specific clothing (laundered each day or supplied at the farm) and boots (Anon., 2018).

7.3 Anteroom: managing barn entrances

The separation between the RAZ and outside must be clearly delineated. When a shower is present, the shower itself is the line separating the two. In the absence of a shower, it is important to delineate the zones with at least a line. Anterooms separated into only two zones (external or contaminated and internal or clean) can be problematic because cross-contamination between the two zones is frequent (Figs 4 and 5). Therefore, keeping the barn entrance free of any organic matter will prevent cross-contamination of these two areas. However, it is difficult to achieve without a physical barrier between the zones. In addition, when designing an anteroom, it is recommended to have separate drainage for each area. Finally, an easier and more effective anteroom design is one with three zones (Fig. 6). Often known as the Danish entrance, the three-zone design offers an area between the contaminated and clean zones where visitors and personnel can wash their hands. This step, while transitioning between the contaminated and clean zones, is important because it ensures that hands are decontaminated before handling boots, clothing, or equipment on the clean side of the anteroom (Anon., 2018).

In an anteroom, porous surfaces (concrete, wood) are generally more prevalent than smooth surfaces (plastic, metal). For detergent application on porous surfaces, it is recommended to use a low-foaming detergent. For disinfection, the use of fumigation (0.5 μm particle) might be preferable for its ability to reach corners and penetrate cracks (Battersby et al., 2017), although research is inconclusive on this point.

The type of soil also has an impact on the sanitation of an anteroom. For example, a cracked concrete floor, a porous soil where water stagnates easily, will be much slower and more difficult to disinfect compared to a floor covered with flat epoxy. Stagnant water is a source of multiplication for microorganisms and should be given special attention.

Figure 4 Equally dirty contaminated (outside) and clean (inside) zones separated by a red line in an anteroom.

Figure 5 Examples of two-zone barn entrances. Note that a hydroalcoholic solution dispenser for hand decontamination is located at the level of the separation between the zones. The red line design is less effective for compliance than the ones with a bench.

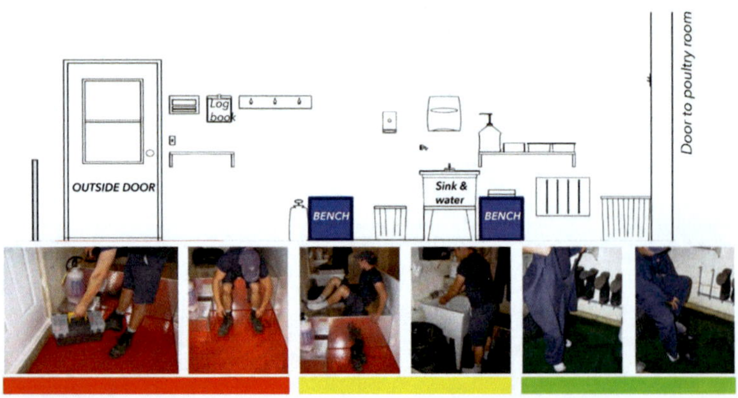

Figure 6 Layout of a three-zone barn entrance and the steps that must be followed to pass from the red zone (contaminated area) to the green zone (clean area) via the yellow zone (intermediary area for hand washing).

7.4 Footbaths

The effectiveness of a footbath is influenced by several factors. The two major factors are the presence of organic matter (on the boots or in the footbath; Fig. 7) (Amass, 1999; Amass et al., 2000; Rabie et al., 2015) and the often insufficient contact time. Other factors also influence the effectiveness of footbaths: the type and concentration of the disinfectant (Amass et al., 2000; Hauck et al.,

Figure 7 Examples of footbaths left outside and rendered ineffective due to a large amount of organic material.

2017), the location (Fig. 8), the pH and the hardness of the water, and the material of the container used (e.g. metal container). For example, if a footbath is left outside, the disinfectant may be inactivated by UV rays from the sun (e.g. iodine), diluted by rain, and affected by warm (evaporation) and cold (freezing) temperatures (Rabie et al., 2015).

Despite all these issues, some epidemiological studies have shown a protective effect of footbaths, especially in connection with contamination with *Campylobacter* spp. and with avian influenza H9N2 (Chaudhry et al., 2017). However, these results are not supported by other studies (Refrégier-Petton et al., 2001). Allen and Newell (2005) report a lower efficiency of footbaths compared to using boots dedicated to each barn.

Beyond epidemiological studies, other studies have evaluated the direct effectiveness of footbaths against infectious pathogens under different conditions (contact time, presence of organic matter, wet or dry prewash, type of disinfectant, etc.). In studies demonstrating good efficacy, Dee et al. (2004) reported that a sodium hypochlorite footbath (bleach) allowed the complete elimination of the PRRS virus with a contact time of 5 s in the disinfectant solution. It is important to note that the footbath was replaced after each use (single use), which is not done in practice. Amass et al. (2006) reported a slight reduction, insufficient to be considered significant, in bacterial contamination on lightly soiled boots when passed over a disinfectant mat. Morley et al. (2005) reported a 67–78% reduction in bacterial load after 7 min of contact time with the disinfectant solution on lightly soiled boots. Finally, Dunowska et al. (2006) reported a reduction in bacterial load of 95.4–99.8% after 10 min of contact on lightly soiled boots. The studies reporting certain effectiveness can thus

Figure 8 Examples of footbaths positioned in such a way as to prevent users to march forward as they should between two zones.

be criticized because the reduction of the microbial load is inadequate, the solution needs to be replaced after each use, or the contact time required is unrealistic in practice.

Several studies have confirmed the inadequacy of traditional footbaths as a biosecurity measure. Hauck et al. (2017) reported that quaternary ammonium and quaternary ammonium-glutaraldehyde footbaths were not able to eliminate highly pathogenic (H5N8) and low-pathogenic (H6N2) influenza. Several studies have demonstrated that the presence of organic matter prevents the effective disinfection of boots, regardless of the contact time (Amass et al., 2000; Rabie et al., 2015). In fact, the mechanical action of a brush to remove organic matter from the boots is even more effective than walking through a footbath without prewashing. It is important to note that a footbath can even be a risk factor (Amass et al., 2000) because it provides a humid environment favoring bacterial growth and sharing of pathogens between treated boots. Having this in mind, Owen and Lawlor (2012) compared phenol- and quaternary ammonium-filled footbaths to dry footbaths filled with a dry bleach powder. The average residual life of the liquid footbaths was less than 2 h and did not significantly impact the bacterial count on the boots. By contrast, the dry bleach footbath reduced the bacterial count by around 98% and had a residual life of 14 days.

7.5 Downtime

Downtime is the period when a poultry barn or an entire production site is without a flock. The concept is to remove birds in order to apply sanitation measures. Repopulation follows a few days to a few weeks later. This is called all-in/all-out (Anon., 2018). This approach has a protective effect on *Salmonella* contamination (Snow et al., 2010) and avian influenza (Kim et al., 2018). In a study on coliform cellulitis in chickens in Canada, the longer the downtime, the lower was the prevalence of cellulitis-related condemnations at slaughter (Elfadil et al., 1996). Chin et al. (2009) reported that an extended downtime of 30–91 days contributed to the regional control and eradication of infectious laryngotracheitis in California. Field experts participating in a Delphi study indicated that a 14-day downtime between meat bird flocks was the norm, with 21 days between breeder flocks (Vaillancourt and Martinez, 2001).

8 Separating healthy birds from sources of contamination: hatchery

There is more to hatchery biosecurity than just sanitation. Hatchery location, design, accessibility, workflow, and pest control all complement a stringent sanitation program to ensure good biosecurity. The temperatures and humidity required for hatching are ideal for the growth of bacteria and molds. Problems coming from the breeders can be amplified, and new ones introduced (Kim and Kim, 2010).

Embryos and newly hatched chicks can offer little disease resistance. In other words, the pressure of infection required for affecting their health and viability is much lower than for older birds. It is, therefore, essential to provide them with a clean environment during their incubation and hatching (Bennett, 2017).

The more a hatchery is located in close proximity to other poultry traffic, the more vigilant one should be about issues such as pest infestation, visitors, and ventilation-related pathogens. It is important to recognize that a lot of poultry in the area increases the risk of transmission of infectious diseases. Therefore, it is best if no poultry-related activities (feed mill, farms, live-haul routes, etc.) are conducted next to the hatchery (Bennett, 2017).

8.1 Personnel

All employees must park in a designated area. They should not be visiting other poultry facilities. On the rare occasion when this might occur, they would have to take a shower and use clean clothes before reentering the hatchery. They should use protective clothing, including boots, only used at the hatchery. If a shower in-shower out facility is available at the hatchery, it must be used consistently and never by-passed. At the very least, all employees will be expected to have taken a shower at home before coming to work, they must come to work wearing clean clothes, and they will all be required to thoroughly wash their hands when arriving at the hatchery. Therefore, a handwashing station with antibacterial soap should be readily available. Handwashing is essential before, after, and between egg and chick or poultry handling operations (e.g. setting, transfer, candling, sexing, vaccinating, packing, etc.) (Thermote, 2006; Bennett, 2017).

8.2 Hatchery workflow

The work and traffic flow should follow the same route as the hatching egg. Modern hatcheries should be designed with ventilation systems to prevent cross-contamination of the different areas of the building. The benefits of such systems will largely be negated if employees are allowed to move freely back and forth between separate areas of the building. If it is needed to 'break' the unidirectional workflow, it is important for all employees to wear clean hatchery clothing and to change it as necessary. The hatchery workflow is as follows: egg receiving area, egg holding area, egg cooler, setters, hatcher rooms, tray dumping, chick processing area, chick holding area, and chick loading area (Mauldin, 2002).

The key is to minimize contamination from one room to the next. Positive pressure rooms are important in critical areas so that contamination will not be drawn in through an open door. Doors help stop cross-contamination between rooms (Mauldin and MacKinnon, 2009).

8.3 Egg delivery

The driver cannot enter the egg holding room at the hatchery, in particular, if several breeder farms are visited. After delivery, the egg truck is washed and disinfected. Clean and disinfected trays and trolleys are returned from the hatchery to the breeder farms (Bennett, 2017).

8.4 Egg sanitation

Eggs are not laid in a sterile environment. Even before that, eggs are exposed to many microbes. The vent of the hen is a contaminated 'delivery environment'. Dirty nests, or if eggs are laid on the floor, can also contribute to egg contamination. Eggshell surfaces must be kept dry. Without humidity, most microbes will have a hard time multiplying or penetrating the egg's natural defenses. Therefore, it is important to avoid the sweating of eggs, such as by moving them from a warm to a cool environment. Bacteria inside an egg will use some of its nutrients and may affect the growth or even kill the embryo. Contaminated eggs in incubators and hatchers can also serve as a source of contamination for the other eggs and newly hatched birds.

Early embryos can also be affected by chemical vapors. So the choice of sanitizing products in the egg reception and storage areas is important (please see the following text). Eggs that have been disinfected on farm can be moved immediately to setter trays for storage. Eggs of unknown sanitation status normally should not be allowed in the hatchery. If the decision is made to allow them, they must be disinfected before setting. If dirty eggs are found (obvious presence of fecal material, litter, and feathers), they should be removed and held for breeder management to review.

Egg sanitizing (Ernst, 2004): It must be done properly; otherwise, the result may be a greater degree of egg contamination. It is important to keep sanitizing water hotter than the eggs (43–49°C). The sanitizing solution must contain a detergent sanitizer. It is best to use a washer that does not recirculate water.

Chlorine and quats (quaternary ammonium) have been excellent disinfectants. Quats offer residual activity, are safe for the operator, are relatively cheap, and are compatible with antibiotic dipping. Quats are safe for hatching eggs up to 10 000 ppm but are effective as a disinfectant at 250 ppm (large safety margin) with 10 ppm ethylenediaminetetraacetic acid. It is best to keep the solution at a pH of 8 for optimum efficacy. This can be achieved by using sodium carbonate (Rodgers et al., 2001).

Hatching eggs can also be sanitized by fumigation using 1.2 mL of formalin and 60 g of permanganate sulfate per cubic meter in a disinfection cabinet. Other products may be used for fumigation, and new technologies are also considered, such as pulsed ultraviolet light (Cassar et al., 2020) and low-energy electron beam (Steiner, 2020).

8.5 Building and equipment sanitation

Washing and disinfection of equipment are a critical component of hatchery biosecurity. A thorough cleaning of the area (setters, hatchers, floors, chick-go-rounds, vaccinators, etc.) is essential before a disinfectant can be applied. Organic material (fluff, blood, shells, and droppings) reduces the effectiveness of disinfectants. Therefore, washing is paramount.

8.6 Setters

Fog a disinfectant after eggs have been set. Multistage setters should be fogged each time new eggs are set or transferred. When setters are emptied, a thorough cleaning and disinfection are required. Like for any other areas, all debris must be removed as part of the cleaning process.

8.7 Hatchers

The hatcher is the main source of organic contamination in the hatchery (eggshells, unhatched eggs, dead chicks/poults, fluff, droppings). It is, therefore, important to use equipment handling and sanitation procedures to minimize contaminating the other areas of the hatchery.

8.8 Bird processing equipment

This equipment requires a very rigid sanitation schedule to avoid bird quality problems (Mauldin, 1983). Special attention should be paid to the vaccination equipment (syringes, vaccine containers, and hoses washed with a nonresidual disinfectant such as alcohol or chlorine).

Chick-Go-Rounds and/or chick/poultry conveyor belts and other processing equipment can contaminate day-old birds. At greater risk are chicks or poults that may have been pulled early. If their navel is incompletely closed and comes in contact with a dirty or contaminated conveyor belt, there will be many cases of infection. Therefore, it is important to clean and disinfect conveyor belts after the chicks/poults from each breeder flock are processed.

8.9 Ventilation

The hatchery ventilation system plays an important role in preventing contamination (when functional and well designed) or by being the source of the problem (e.g. aspergillosis contamination). Ideally, each room of the hatchery should have its own separate ventilation system so air does not move from one room to another. This way, the ventilation system will not introduce contaminated air from other sections of the hatchery or from outside. The system must provide clean, fresh air to the hatchery and the incubators at all times. Hatcheries should be designed to exhaust air away from intake outlets, so contaminated air is not recycled into the hatchery (Mauldin, 1983).

Maintenance of the evaporative coolers is critical. Filters must be cleaned and disinfected routinely. Disinfectants can be added to the water in the evaporative coolers, especially when starting them off in the spring.

8.10 Monitoring

Records must be kept that the biosecurity program is in place and implemented correctly and continuously. The hatchery should have a quality control program to monitor incoming eggs visually and microbiologically. A continuous monitoring program of the building and equipment must be in place to determine the microbial populations in the hatchery (Racicot et al., 2020). Samples may be collected from the tray wash area, air intakes and outlets, filters, evaporative coolers, setters, hatchers, air in chick holding and egg storage rooms, chick conveyor belts, water source to hatchery, vaccination equipment, including vaccine and diluent (Mauldin, 1983). Samples must be tested for bacteria (coliforms, *Salmonella*) and fungus.

9 Separating healthy birds from sources of contamination: regional biosecurity

The relationship between proximity to a poultry production site and the probability of contamination has been well established. When the distance between two egg layer farms increases by 1 km, the risk of colibacillosis is six times less in Belgium (Vandekerchove et al., 2004). Egg layer farms are also twice less likely to be contaminated with *Salmonella* if the nearest farm is at least 1 km away (Snow et al., 2010). The risk of infectious bursal disease (Gumboro) transmission was higher when the susceptible flock was less than 20 km away from an infected flock. Poultry traffic, wild birds, and airborne transmission might have contributed to such disease transmission in Denmark (Sanchez et al., 2005). In Australia, East et al. (2006) demonstrated that the risk of Newcastle disease infection is reduced threefold in egg layer farms if the nearest farm is at least 10 km away; fourfold for breeder farms if the nearest poultry farm is at least 1 km away, and threefold in broiler chickens if at least 500 m away. A poultry farm located less than 1 km away from an avian influenza-infected flock is about 35 times more at risk of infection compared to a farm located at least 10 km away (Boender et al., 2007). Commercial poultry farms infected with the infectious laryngotracheitis virus were 36 times more likely to be within 1.6 km from a backyard flock than disease-free farms (Johnson et al., 2005).

In Australia, veterinary authorities, in collaboration with the poultry industry, have established guidelines for the geographic distribution of poultry farms (Table 1). Although these are not yet mandated by law, they were determined and accepted collectively based on scientific evidence. Note that these are standards for regular poultry operations.

The vast majority of infectious diseases are horizontally transmitted. Cross-traffic between farms and the joint use of equipment are certainly significant

Table 1 A guide on biosecurity buffer distances[a]

Farm type	Species	Buffer (m)
New farm	Fowls/turkeys/other avian species, e.g. ratites, quail	1000
Units in large farm complex	Fowl/turkeys or other avian species	200–500
Farm complexes	Fowl/turkeys or other avian species	>2000
Breeder farms	Fowl/turkeys or other avian species	2000–5000
Duck or waterfowl farms	Duck, waterfowl	5000

[a]The buffer is measured from either the nearest barn walls for older type building or from the centroid of the mechanical ventilation system of the newer tunnel ventilation barns. Reference: http://new.dpi.vic.gov.au/notes/agg/poultry--and--other-birds/ag1155-biosecurity-guidelines-for-poultry-producers

modes of disease transmission (Guinat et al., 2020). With potential vectors such as flies and darkling beetles, known vectors of avian influenza, disease control efforts in regions of high farm concentration cannot rely solely on farm-level measures. In a competitive environment, it is tempting to limit data sharing. Although liability will always be a concern, pointing fingers has never been an effective disease control strategy, and poultry organizations sharing a region must also share the necessary information needed to prevent and contain significant contagious diseases. A regional approach to biosecurity includes establishing separate geographical working zones for service personnel. When an important disease is suspected on a farm, a self-imposed quarantine matched with communication with key personnel will go a long way to prevent a major epidemic (Vaillancourt et al., 2018).

Communication remains central to the success of regional biosecurity, especially when an outbreak occurs. In 2005, during an outbreak of infectious laryngotracheitis in California, two integrators coordinated to perform an extended downtime in a region and to set up biosecurity audits. Cooperation and communication between companies have helped eradicate the disease. The importance of communication is also highlighted in a 2001 survey of 72 North American veterinarians on the importance given to certain elements of biosecurity (Vaillancourt and Martinez, 2001).

10 Biosecurity compliance

Biosecurity compliance is defined as the behavior of a person coinciding with the recommendations of professionals in the control and prevention of infectious diseases. The measure of compliance is reported as the ratio of the number of recommendations applied to the number of recommendations prescribed; in other words, the ratio of the number of biosecurity measures applied to the number required. In practice, this is expressed as a percentage.

The constant application of biosecurity measures is essential for the success of any type of animal production. However, compliance with biosecurity measures is sporadic and variable regardless of the type of production, including pig and poultry productions (Losinger et al., 1998; Pinto and Urcelay, 2003; Boklund et al., 2004; Ribbens et al., 2008).

Similar observations are made in human medicine. Indeed, although compliance with hygienic measures among physicians is relatively high (80%) during high-risk procedures, this is not the case (30%) for low-risk procedures. In addition, it is interesting to note that compliance with medical recommendations by patients (33–54%; Haynes et al., 1979) also corresponds to what is observed in commercial poultry farms (Racicot et al., 2011).

There is a weak correlation between reported compliance and observed compliance ($r = 0.21$) for handwashing (O'Boyle et al., 2001). In veterinary medicine, few studies report compliance as observed. When it does, the results are worrying. For example, a procedure to register visitors to turkey farms in North Carolina was filmed on three farms shortly after the procedure was put in place. Compliance ranged from 7% to 49% (Vaillancourt and Carver, 1998).

A study in Quebec on compliance with biosecurity measures in the anteroom of poultry production buildings on eight farms was carried out using hidden cameras (although stakeholders on the farm were informed that they would be filmed). A total of 44 different errors were observed from 883 visits made by 102 different people. On average, four errors were recorded per visit. The maximum number of errors made by a person during a visit was 14. Twenty-seven of the 44 errors (61.4%) were related to measures relating to the respect of the zones (clean or internal versus contaminated or external), 6 for boots (13.6%), 5 for handwashing (11.4%), 3 for coveralls (6.8%), and 3 for visitation records (6.8%). The nature and frequency of errors suggest a lack of understanding of biosecurity principles (Racicot et al., 2011).

Since compliance with biosecurity is generally weak, it is essential to establish strategies to improve the implementation of biosecurity measures. A study evaluated the value of audits and visible cameras on compliance with the biosecurity measures required when entering and leaving poultry barns of 24 farms in Quebec. The evaluation was done initially during the first 2 weeks after the intervention, as well as 6 months later for another evaluation lasting 2 weeks. Nearly 2800 visits by 259 different people were recorded on video. The results showed that bimonthly audits had no impact on compliance in the medium term, 6 months after the start of the project. It should be noted, however, that the auditor was not in a position of authority. The visible cameras had an impact on the change of boots (OR = 9.6; 95% CI 1.9–48.4) and respect for areas (contaminated versus clean) during the visit (OR = 14.5; 95% CI: 1.2–175.1) for the short-term period. However, 6 months later, compliance declined and was no longer significantly different from control farms. The duration (>5 min) and the time of the visit (morning), the presence of the producer or an observer (sometimes negative effect), the design of the anteroom (in relation to the ease of application of the measures), the number of buildings (more than five buildings), the number of biosecurity measures requested, the type of boots worn (plastic boots), and being a member of the grower's family (negative effect) were significantly associated with biosecurity compliance. The same study also showed that certain personality traits were associated with compliance, as well as the number of years of experience in poultry production and the level of education of the participants. For the number of years of experience, the relationship was not linear. It is possible that early-career circumstances (e.g. avian influenza epidemic) could have a long-term impact on compliance (Racicot et al., 2012b).

Delpont et al. (2020) attempted to determine how sets of socio-psychological factors (i.e. knowledge about biosecurity and transmission of avian influenza, attitudes, personality traits, social background) affect the adoption of on-farm biosecurity practices. The study was carried out as part of 127 visits to duck farms in southwestern France. The factorial analysis of mixed data and the analysis of hierarchical grouping identified three groups of producers with different socio-psychological profiles. The first group was characterized by minimal knowledge, negative attitudes toward biosecurity, low social pressure (defined as a concern to preserve the avian influenza free status of the farm in order not to tarnish one's reputation), and a low level of conscientiousness (conscientiousness is a fundamental personality trait that reflects the tendency to be responsible, organized, hardworking, goal-oriented, and adheres to standards and rules). The second group was characterized by greater experience in poultry production, increased stress (excessive nervous tension), and social pressure. The third group was characterized by less experience in poultry production but better knowledge and positive attitudes toward biosecurity, increased self-confidence, and a focus on action. The first group had significantly lower adoption of biosecurity measures than the other two groups.

The measures implemented on farms vary greatly, depending on the animal species, type of production, region, stocking density, and possibly regional health status. For example, a Canadian survey (Young et al., 2010) of 642 broiler breeder owners indicates that 13% of respondents do not allow access to visitors. Of the rest, 36% require them to wash their hands and 50% require additional protective clothing to be worn before entering a barn. The main reasons for not requiring visitors to wash their hands were the lack of perceived need (20%), the time required by this measure (20%), and the lack of facilities (18%). Almost a third of farm owners also did not require the use of coveralls because of the time required to put them on, and 20% found it sufficient to wear only boots.

The known risk of disease transmission could favor the better implementation of measures (Dorea et al., 2010). The implementation of biosecurity measures also depends on the individuals being questioned. Indeed, producers and technical staff responsible for health monitoring do not always agree on what should be done and even what is done on a given farm. For example, in a study carried out in Canada, a weak to slight agreement was observed when these technicians were asked about the restrictions required by the farm owner to have access to the farm and their responses were compared to those of these owners (Nespeca et al., 1997). It is therefore important to provide each employee with written plans for the required biosecurity measures and to ensure a continuous training program (England, 2002).

10.1 Obstacles to compliance

Several reasons are given to explain the lack of compliance with biosecurity measures. Lack of knowledge or understanding of measurements is the main reason (Lotz, 1997; Barcelo and Marco, 1998; Sanderson et al., 2000). Lack of communication, time, incentives (positive and negative) to follow the rules, lack of audit programs, apathy or denial of potential risks, and economic constraints are also considered to be important factors (Vaillancourt and Carver, 1998). Communication is particularly necessary during epidemics. In 2005, during an outbreak of infectious laryngotracheitis in chickens in California, USA, two integrated companies coordinated to carry out an extended downtime and implement extensive audits. Cooperation and communication between companies have helped eradicate the virus (Chin et al., 2009). Gunn et al. (2008) also underline the importance of better collaboration between producers, veterinarians, and technical staff working on farms. To ensure good communication, the messenger is essential, but the content of the message is decisive. Several private and public organizations produce training materials for producers, but their content varies widely. This lack of harmony between training programs, and the resulting confusion, likely contribute to the lack of application of biosecurity (Jardine and Hrudey, 1997; Moore et al., 2008).

For a given disease, the perceived risk of infection influences the implementation of biosecurity measures. Unfortunately, the perception of risk differs greatly between producers, making consistency in the application of biosecurity measures difficult. The perception of risk is strongly linked to the understanding of the principles of biosecurity and the regional incidence of diseases. Perception of risk is also influenced by the intensity of traffic on the farm. Thus, producers of small farms or family farms often consider themselves less at risk compared to producers of intensive commercial livestock farms (Larsen, 2009).

In addition to the perception of risk, other individual factors may influence the application of recommendations. These are the attitudes, personality traits of the individual, experience with the disease, level of education and personal beliefs about animal health, incidence, prevention, and disease control (Buckalew and Sallis, 1986; Strömberg et al., 1999; Delabbio et al., 2005).

10.2 Intervention strategies

Haynes et al. (1979) describe different strategies to improve adherence to medical recommendations. Reminders of recommendations increase compliance from 24% to 70%. Thus, proper instructions with clear information and repeated feedback to stakeholders should improve compliance.

When it comes to handwashing, studies show that increasing accessibility to washing stations, training programs, and frequent feedback significantly increase compliance. This can go from less than 20% to more than 70% after the establishment of a training program and a monitoring system with frequent feedback (Tibballs, 1996; Colombo et al., 2002). However, compliance declines over time. This deterioration appears to be multifactorial and probably associated with the lack of a continuing education program (Conly et al., 1989).

10.3 Motivation of farm personnel

Motivation is another important element. Compliance with industrial safety rules is intimately linked to the motivation of employees to comply (Vroom, 1994). Clear goal statements and performance feedback have an optimal motivational effect when these two elements are combined. To increase motivation toward desired behavior such as compliance with biosecurity measures, the ability to achieve goals and a sense of self-efficacy are important (Bandura and Cervone, 1983). Self-efficacy is defined as 'the ability to implement the behavior essential to obtain a given result' (Pervin and John, 2001). According to Bandura and Cervone (1983), 'Our ability to deal with a situation and control the outcome is the key element that actually influences behavior. It is by manipulating the feeling of self-efficacy (e.g. by means of feedback announcing to the person that, compared to the performance obtained by others, their own performance is very good), rather than by giving feedback regarding a risky behavior, that a change in behavior is likely (Pervin and John, 2001).

Millman et al. (2017) have clearly demonstrated the perverse effect of a contradiction between what is required and what can be achieved. They studied the behavior of poultry catchers who had to follow several biosecurity procedures when the time allowed to do so was not sufficient. This situation led catchers to openly question the protocols and the need to respect biosecurity measures. Their noncompliance with the requirements was then seen as resourcefulness in the face of unrealistic conditions. Timmermans and Berg (1997) also stress the importance of 'local universality'; that is, standardized practices, such as biosecurity measures, can only be universal (achievable in different places and times) if they can be adapted locally.

10.4 Training program

A training program should include an evaluation of the participants before the training (theoretical and practical parts) and at the end. It is necessary that the trainer be credible and have the confidence of the participants so that the message is delivered well. Participants should also leave the session with a copy of the training material. A website accessible to participants will include materials, as well as program updates, upcoming training schedules, and links to other sites providing information on biosecurity. Conly et al. (1989) showed that training programs are not sufficient to maintain long-term compliance. Feedback and reinforcements (incentives) are needed. Reminders in various forms are also useful, such as messages in agricultural journals, mailings, contests and awards at annual meetings (Bradley, 2007).

10.5 Audits

Three audits over a 6-month period had no impact on biosecurity compliance in eight poultry farms in Canada (Racicot et al., 2012a). So it is not always necessarily impactful. In order to be valid, the audit process must be relevant, objective, quantifiable, repeatable, and capable of suggesting changes to be made. Standards must be clearly identified. These standards must reconcile current practices with protocols already defined, evaluated, and published. The data must collectively be the subject of regular reports showing the points assessed, the improvements or shortcomings identified, the corrective measures and their results. Finally, audits must be subject to independent evaluation to improve the process (Shaw and Costain, 1989; Smith, 1990). When these conditions are met, on-farm biosecurity audits have been shown to be very effective in aquaculture in New Zealand (Georgiades et al., 2016). In Belgium, an audit system, Biocheck.UGent, has been developed for swine and poultry and has made it possible to establish a link between the level of biosecurity and production performances (Gelaude et al., 2014; Rodrigues da Costa et al., 2019).

11 The economics of biosecurity

There is a paucity of scientific articles directly investigating the economic impact of biosecurity. An economic assessment of biosecurity in broiler breeders in 1987 indicated that a benefit-cost ratio of at least 3 should be expected for a farm considered at a 30% risk of being infected by an agent causing a severe disease (Gifford et al., 1987). Morris (1995) states that the primary purpose of any business should be to maximize return on investment over the long term. This is an important concept because a comprehensive biosecurity program does not necessarily offer a quick pay back. Yet, it is, or should be, an essential component of a farm owner's long-term strategy for success.

Biosecurity investments involve a mix of fixed and variable costs. Knowledge of costs would help inform cost-sharing programs related to animal disease mitigation efforts (Pudenz et al., 2019). In British Columbia, shortly after the H7N3 outbreak in 2004, these two types of costs were reported to represent about 4.5% of the cost of production for duck farms (JP Vaillancourt, pers. Comm.). However, in determining the economic impact of biosecurity measures, one must also consider the benefits in terms of reduced losses associated with infectious diseases, including the reduction in the use of antibiotics (Rojo-Gimeno et al., 2016). In pig production, a negative correlation between enzootic pneumonia, pleurisy, acute pleuropneumonia, and internal biosecurity underscores the importance of good biosecurity in reducing health problems (Pandolfi et al., 2018). In South Africa, financial modeling over a 3-year period of a farrow to finish pig farm of 122 sows estimated that, in the absence of disease, the implementation of biosecurity resulted in a 9.70% reduction in the total annual profit. In contrast, the study found that the implementation of biosecurity and its effective monitoring would prevent losses due to Africa Swine Fever with an impressive benefit-cost ratio of 29 (Fasina et al., 2012).

The perceptions of pastoralists on the cost of biosecurity are important for the adoption of these measures. If the perceived costs are excessive, producers may prefer not to apply the measures. This is in accordance with Casal et al. (2007), who found that perceptions about biosecurity measures and their use were interrelated. Other studies such as Fraser et al. (2010), Valeeva et al. (2011), and Toma et al. (2013) also found that the costs and benefits of biosecurity contribute to the adoption of biosecurity measures (Niemi et al., 2016). But the likelihood of adopting biosecurity increases inelastically as perceived costs decrease. This means that the relative change (percent) in biosecurity adoption is smaller than the relative change in assumed cost. On the other hand, if the grower receives information about a low-cost biosecurity technology that is available, it increases the rate of biosecurity adoption.

12 Conclusion and future trends

We essentially know the risk factors associated with infectious disease transmission in poultry. Biosecurity measures have been developed over the years. Two main challenges require research for on-farm measures: (a) finding ways to make specific measures easier, faster, and cheaper to perform; (b) how to increase compliance. Any progress in point (a) will favor point (b). But this second point has been associated with many factors, namely, lack of knowledge; economic constraints; lack of training, communication, incentives, time; difficulty in applying the requested measures; lack of audits; lack of consistency in the information available; beliefs, attitudes, perceptions, education, experience, and personality traits of farm workers. It is essential to consider several of these factors at the same time if one wants to have a significant impact on biosecurity compliance.

Finally, research is needed regarding regional biosecurity measures, such as the management of poultry traffic, zone raising (regional downtime), and zoning.

13 Where to look for further information

Good introductions to the subject for non-specialists:

- Dewulf, J. and Van Immerseel, F. (Eds). (2019). *Biosecurity in Animal Production and Veterinary Medicine*. CABI. 523 pp.
- Owen, R. L. (2011). *A Practical Guide for Managing Risk in Poultry Production* (No. V480 OWEp). 276 pp.
- Racicot, M., Venne D., Durivage A. and Vaillancourt J.-P. (2011). Description of 44 biosecurity errors while entering and exiting poultry barns based on video surveillance in Quebec, Canada. *Prev. Vet. Med.* 100, 193–199.
- Conan, A., Goutard, F. L., Sorn, S. and Vong, S. (2012). Biosecurity measures for backyard poultry in developing countries: a systematic review. *BMC Veterinary Research* 8(1), 1–10.
- Scott, A. B., Singh, M., Groves, P., Hernandez-Jover, M., Barnes, B., Glass, K., Moloney, B., Black, A. and Toribio, J. A. (2018). Biosecurity practices on Australian commercial layer and meat chicken farms: Performance and perceptions of farmers. *PLoS ONE* 13(4), e0195582.

Key research centers readers can investigate, for example, for possible collaboration as well as to keep up with research trends:

1. Prof. Dr. Jeroen Dewulf: https://biocheck.ugent.be/en:
   Veterinary Epidemiology Unit, Department of Reproduction, Obstetrics and Herd Health.

Faculty of Veterinary Medicine; Ghent University, Belgium: https://www.ugent.be/di/vvb/en/research/research-epidemiology.

2. Prof. Jean-Pierre Vaillancourt: Department of Clinical Sciences; Faculty of Veterinary Medicine, Université de Montréal, Canada. Jean-pierre.vaillancourt@umontreal.ca.
3. Jean-Luc Guérin. Poultry Biosecurity Chair, École Nationale Vétérinaire de Toulouse; jl.guerin@envt.fr: http://www.envt.fr/content/recherche-publication-de-la-chaire-de-biosécurité-aviaire.
4. Alberto Oscar Allepuz Palau: Alberto.Allepuz@uab.cat: Centre de Recerca en Sanitat Animal IRTA-CReSA; University of Barcelona, Spain.
5. Armin R. W. Elbers; armin.elbers@wur.nl: Department of Epidemiology, Bioinformatics and Animal Models, Wageningen Bioveterinary Research, The Netherlands.

14 References

Allen, V. and Newell, D. (2005). *Evidence for the Effectiveness of Biosecurity to Exclude Campylobacter from Poultry Flocks*. Commissioned Project MS0004. Food Standards Agency Report. Available at: https://pdfs.semanticscholar.org/4236/1dd1fb80d51742c54629ee46cb315ef936a3.pdf?_ga=2.63452588.287620099.1573885102-1033270699.1521297455 (accessed 22 March 2021).

Amass, S. F. (1999). Biosecurity considerations for pork production units, *Swine Health and Production* 7(5), 12.

Amass, S. F., Vyverberg, B. D., Ragland, D., Dowell, C. A., Anderson, C. D., Stover, J. H. and Beaudry, D. J. (2000). Evaluating the efficacy of boot baths in biosecurity protocols, *Journal of Swine Health and Production* 8(4), 169–173.

Amass, S. F., Arighi, M., Kinyon, J. M., Hoffman, L. J., Schneider, J. L. and Draper, D. K. (2006). Effectiveness of using a mat filled with a peroxygen disinfectant to minimize shoe sole contamination in a veterinary hospital, *Journal of the American Veterinary Medical Association* 228(9), 1391–1396.

Anon. (2018). National on-farm avian biosecurity standards. Available at: https://inspection.canada.ca/animal-health/terrestrial-animals/biosecurity/standards-and-principles/national-avian-on-farm-biosecurity-standard/eng/1528732756921/1528732872665?chap=0 (accessed 18 April 2022).

Arsenault, J., Letellier, A., Quessy, S., Normand, V. and Boulianne, M. (2007). Prevalence and risk factors for *Salmonella* spp. and *Campylobacter* spp. caecal colonization in broiler chicken and turkey flocks slaughtered in Quebec, Canada, *Preventive Veterinary Medicine* 81(4), 250–264.

Axtell, R. C. (1999). Poultry integrated pest management: status and future, *Integrated Pest Management Reviews* 4(1), 53–73.

Bandura, A. and Cervone, D. (1983). Self-evaluative and self-efficacy mechanisms governing the motivational effects of goal systems, *Journal of Personality and Social Psychology* 45(5), 1017–1028. doi: 10.1037/0022-3514.45.5.1017.

Barcelo, M. and Marco, E. (1998). On farm biosecurity. *Proceedings of the 15th Intl Pig. Veterinary Society Congress*. Nottingham University Press, Birmingham, England.

Barrington, G. M., Allen, A. J., Parish, S. M. and Tibary, A. (2006). Biosecurity and biocontainment in alpaca operations, *Small Ruminant Research* 61(2), 217–225. doi: 10.1016/j.smallrumres.2005.07.011.

Bates, C., Hiett, K. L. and Stern, N. J. (2004). Relationship of campylobacter isolated from poultry and from darkling beetles in New Zealand, *Avian Diseases* 48(1), 138–147.

Battersby, T., Walsh, D., Whyte, P. and Bolton, D. (2017). Evaluating and improving terminal hygiene practices on broiler farms to prevent Campylobacter cross-contamination between flocks, *Food Microbiology* 64, 1–6. doi: 10.1016/j.fm.2016.11.018.

Bennett, B. (2017). The importance of biosecurity in the modern day hatchery, *International Hatchery Practice* 31, 21–23.

Bestman, M., de Jong, W., Wagenaar, J. and Weerts, T. (2018). Presence of avian influenza risk birds in and around poultry free-range areas in relation to range vegetation and openness of surrounding landscape, *Agroforestry Systems* 92(4), 1001–1008. doi: 10.1007/s10457-017-0117-2.

Blondel, V., Huard, G., Vaillancourt, J. P. and Racicot, M. (2018). Base du nettoyage et de la désinfection dans les exploitations agricoles. Available at: https://www.agrireseau.net/documents/98011/bases-du-nettoyage-et-de-la-desinfection-dans-les-exploitations-agricoles (accessed 7 October 2021).

Boender, G. J., Meester, R., Gies, E. and De Jong, M. C. (2007). The local threshold for geographical spread of infectious diseases between farms, *Preventive Veterinary Medicine* 82(1–2), 90–101. doi: 10.1016/j.prevetmed.2007.05.016.

Böhm, R. (1998). Disinfection and hygiene in the veterinary field and disinfection of animal houses and transport vehicles, *International Biodeterioration and Biodegradation* 41(3–4), 217–224.

Boklund, A., Alban, L., Mortensen, S. and Houe, H. (2004). Biosecurity in 116 Danish fattening swineherds: descriptive results and factor analysis, *Preventive Veterinary Medicine* 66(1–4), 49–62. doi: 10.1016/j.prevetmed.2004.08.004.

Bonhotal, J., Schwarz, M. and Rynk, R. (2014). Composting animal mortalities. Cornell Waste Management Institute, 1–23. Available at: https://hdl.handle.net/1813/37369.

Bouwstra, R., Gonzales, J. L., de Wit, S., Stahl, J., Fouchier, R. A. M. and Elbers, A. R. W. (2017). Risk for low pathogenicity avian influenza virus on poultry farms, the Netherlands, 2007–2013, *Emerging Infectious Diseases* 23(9), 1510–1516. doi: 10.3201/eid2309.170276.

Bradley, F. A. (2007). Biosecurity: educational programs, *Journal of Applied Poultry Research* 16(1), 77–81. doi: 10.1093/japr/16.1.77.

Brglez, B. (2003). *Disposal of Poultry Carcasses in Catastrophic Avian Influenza Outbreaks*. doi: 10.17615/vxd0-mv88.

Brody, S. N. (1974). The disease of the soul: leprosy in medieval literature (1st ed.). Ithaca: Cornell University Press.

Buckalew, L. W. and Sallis, R. E. (1986). Patient compliance and medication perception, *Journal of Clinical Psychology* 42(1), 49–53. doi: 10.1002/1097-4679(198601)42:1<49::AID-JCLP2270420107>3.0.CO;2-F.

Butcher, G. D. and Miles, R. D. (1995). *Minimizing Microbial Contamination in Feed Mills Producing Poultry Feed*. Veterinary Medicine-Large Animal Clinical Sciences Department, Florida Cooperative Extension Service, Institute of Food and

Agricultural Sciences and University of Florida. Available at: http://www.nutritime.com.br/arquivos_internos/artigos/Artigo78_VM054002.pdf.

Casal, J., De Manuel, A., Mateu, E. and Martín, M. (2007). Biosecurity measures on swine farms in Spain: perceptions by farmers and their relationship to current on-farm measures, *Preventive Veterinary Medicine* 82(1–2), 138–150. doi: 10.1016/j.prevetmed.2007.05.012.

Cassar, J. R., Bright, L. M., Patterson, P. H., Mills, E. W. and Demirci, A. (2020). The efficacy of pulsed ultraviolet light processing for table and hatching eggs, *Poultry Science* 100(3), 100923.

Cerf, O., Carpentier, B. and Sanders, P. (2010). Tests for determining in-use concentrations of antibiotics and disinfectants are based on entirely different concepts: "Resistance" has different meanings, *International Journal of Food Microbiology* 136(3), 247–254.

Chaber, A. L. and Saegerman, C. (2017). Biosecurity measures applied in the United Arab Emirates – a comparative study Between livestock and wildlife sectors, *Transboundary and Emerging Diseases* 64(4), 1184–1190. doi: 10.1111/tbed.12488.

Chaudhry, M., Ahmad, M., Rashid, H. B., Sultan, B., Chaudhry, H. R., Riaz, A. and Shaheen, M. S. (2017). Prospective study of avian influenza H9 infection in commercial poultry farms of Punjab Province and Islamabad Capital Territory, Pakistan, *Tropical Animal Health and Production* 49(1), 213-220. doi: 10.1007/s11250-016-1159-6. PMID: 27761776; PMCID: PMC7088531.

Chen, S. J., Hung, M. C., Huang, K. L. and Hwang, W. I. (2004). Emission of heavy metals from animal carcass incinerators in Taiwan, *Chemosphere* 55(9), 1197–1205.

Chin, R. P., García, M., Corsiglia, C., Riblet, S., Crespo, R., Shivaprasad, H. L., Rodríguez-Avila, A., Woolcock, P. R. and França, M. (2009). Intervention strategies for laryngotracheitis: impact of extended downtime and enhanced biosecurity auditing, *Avian Diseases* 53(4), 574–577. doi: 10.1637/8873-041309-Reg.1.

Cochrane, R. A. (2016). Feed mill biosecurity plans: a systematic approach to prevent biological pathogens in swine feed, *Journal of Swine Health and Production* 24(3), 154–164.

Colombo, C., Giger, H., Grote, J., Deplazes, C., Pletscher, W., Lüthi, R. and Ruef, C. (2002). Impact of teaching interventions on nurse compliance with hand disinfection, *Journal of Hospital Infection* 51(1), 69–72. doi: 10.1053/jhin.2002.1198.

Conly, J. M., Hill, S., Ross, J., Lertzman, J. and Louie, T. J. (1989). Handwashing practices in an intensive care unit: the effects of an educational program and its relationship to infection rates, *American Journal of Infection Control* 17(6), 330–339. doi: 10.1016/0196-6553(89)90002-3.

Corrigan, R. M. (2006). Overview of rodent control for commercial pork operation. Available at: https://porkgateway.org/resource/an-overview-of-rodent-control-for-commercial-pork-production-operations/.

Course, C. E., Boerlin, P., Slavic, D., Vaillancourt, J. P. and Guerin, M. T. (2021). Factors associated with *Salmonella enterica* and *Escherichia coli* during downtime in commercial broiler chicken barns in Ontario, *Poultry Science* 100(5), 101065.

Crippen, T. L. and Sheffield, C. (2006). External surface disinfection of the lesser mealworm (Coleoptera: Tenebrionidae), *Journal of Medical Entomology* 43(5), 916–923.

Curtis, P. E., Ollerhead, G. E. and Ellis, C. E. (1980). *Pasteurella multocida* infection of poultry farm rats, *Veterinary Record* 107(14), 326–327.

Curtis, P. E. and Ollerhead, G. E. (1982). *Pasteurella multocida* infection of cats on poultry farms, *Veterinary Record* 110(1), 13–14.

Davies, R. H. and Wray, C. (1995). Observations on disinfection regimens used on *Salmonella enteritidis* infected poultry units, *Poultry Science* 74(4), 638–647.

Davison, S. A., Dunn, P. A., Henzler, D. J., Knabel, S. J., Patterson, P. H. and Schwartz, J. H. (1997). *Preharvest HACCP in the Table Egg Industry*. PA State: The Pennsylvania State University, 1-36.

Dee, S., Deen, J., Rossow, K., Wiese, C., Otake, S., Joo, H. S. and Pijoan, C. (2002). Mechanical transmission of porcine reproductive and respiratory syndrome virus throughout a coordinated sequence of events during cold weather, *Canadian Journal of Veterinary Research* 66(4), 232–239.

Dee, S., Deen, J. and Pijoan, C. (2004). Evaluation of 4 intervention strategies to prevent the mechanical transmission of porcine reproductive and respiratory syndrome virus, *Canadian Journal of Veterinary Research* 68(1), 19–26.

Delabbio, J. L., Johnson, G. R., Murphy, B. R., Hallerman, E., Woart, A. and McMullin, S. L. (2005). Fish disease and biosecurity: attitudes, beliefs, and perceptions of managers and owners of commercial finfish recirculating facilities in the United States and Canada, *Journal of Aquatic Animal Health* 17(2), 153–159. doi: 10.1577/H04-005.1.

Delpont, M., Racicot, M., Durivage, A., Fornili, L., Guerin, J. L., Vaillancourt, J. P. and Paul, M. C. (2020). Determinants of biosecurity practices in French duck farms after a H5N8 Highly Pathogenic Avian Influenza epidemic: the effect of farmer knowledge, attitudes and personality traits, *Transboundary and Emerging Diseases* 68(1), 51–61.

Dorea, F. C., Berghaus, R., Hofacre, C. and Cole, D. J. (2010). Survey of biosecurity protocols and practices adopted by growers on commercial poultry farms in Georgia, USA, *Avian Diseases* 54(3), 1007–1015.

Dunowska, M., Morley, P. S., Patterson, G., Hyatt, D. R. and Van Metre, D. C. (2006). Evaluation of the efficacy of a peroxygen disinfectant-filled footmat for reduction of bacterial load on footwear in a large animal hospital setting, *Journal of the American Veterinary Medical Association* 228(12), 1935–1939.

Duvauchelle, A., Huneau-Salaün, A., Balaine, L., Rose, N. and Michel, V. (2013). Risk factors for the introduction of avian influenza virus in breeder duckflocksduring the first 24weeks of laying, *Avian Pathology* 42(5), 447–456.doi: 10.1080/03079457.2013.823145.

East, I., Kite, V., Daniels, P. and Garner, G. (2006). A cross-sectional survey of Australian chicken farms to identify risk factors associated with seropositivity to Newcastle-disease virus, *Preventive Veterinary Medicine* 77(3-4), 199–214. doi: 10.1016/j.prevetmed.2006.07.004.

Ebeling, W. (1975). *Urban Entomology*. Los Angeles: University of California.

Elfadil, A. A., Vaillancourt, J. P. and Meek, A. H. (1996). Management risk factors associated with cellulitis in broiler chickens in Southern Ontario: a retrospective study, *Avian Diseases* 40(3), 699–706.

Ellis, D. B. (2001). Carcass disposal issues in recent disasters, accepted methods, and suggested plan to mitigate future events. Available at: https://digital.library.txstate.edu/handle/10877/3502 (accessed 18 April 2022).

England, J. J. (2002). Biosecurity: safeguarding your veterinarian:client:patient relationship, *Veterinary Clinics of North America. Food Animal Practice* 18(3), 373-8, v. doi: 10.1016/S0749-0720(02)00033-6.

Ernst, A. R. (2004). *Hatching Egg Sanitation: The Key Step in Successful Storage and Production*. University of California, Division of Agriculture and National Resources: Publication 8120.

Fasina, F. O., Lazarus, D. D., Spencer, B. T., Makinde, A. A. and Bastos, A. D. S. (2012). Cost implications of African swine fever in smallholder farrow-to-finish units: economic benefits of disease prevention through biosecurity, *Transboundary and Emerging Diseases* 59, 244–255.

Fernandez, D., et al. (1994). Farm location as a determinant to production performance in turkeys. Annual Meeting of the American Association of Avian Pathologists, Annual Convention AVMA, San Francisco.

Fraser, R. W., Williams, N. T., Powell, L. F. and Cook, A. J. C. (2010). Reducing campylobacter and salmonella infection: two studies of the economic cost and attitude to adoption of on-farm biosecurity measures, *Zoonoses and Public Health* 57(7–8), e109–e115.

Gelaude, P., Schlepers, M., Verlinden, M., Laanen, M. and Dewulf, J. (2014). Biocheck. UGent: a quantitative tool to measure biosecurity at broiler farms and the relationship with technical performances and antimicrobial use, *Poultry Science* 93(11), 2740–2751.

Georgiades, E., Fraser, R. and Jones, B. (2016). Options to strengthen on-farm biosecurity management for commercial and non-commercial aquaculture. Aquaculture Unit. Technical Paper No: 2016/47.

Gifford, D. H., Shane, S. M., Hugh-Jones, M. and Weigler, B. J. (1987). Evaluation of biosecurity in broiler breeders, *Avian Diseases* 31(2), 339–344. doi: 10.2307/1590882.

Glanville, T. D., Ahn, H. K., Richard, T. L., Harmon, J. D., Reynolds, D. L. and Akinc, S. (2006). Environmental impacts of emergency livestock mortality composting–leachate release and soil contamination. 2006 ASAE Annual Meeting (p. 1). American Society of Agricultural and Biological Engineers.

Graham, J. P., Price, L. B., Evans, S. L., Graczyk, T. K. and Silbergeld, E. K. (2009). Antibiotic resistant enterococci and staphylococci isolated from flies collected near confined poultry feeding operations, *Science of the Total Environment* 407(8), 2701–2710.

Guinat, C., Comin, A., Kratzer, G., Durand, B., Delesalle, L., Delpont, M., Guérin, J. L. and Paul, M. C. (2020). Biosecurity risk factors for highly pathogenic avian influenza (H5N8) virus infection in duck farms, France, *Transboundary and Emerging Diseases* 67(6), 2961–2970. doi: 10.1111/tbed.13672.

Gunn, G. J., Heffernan, C., Hall, M., McLeod, A. and Hovi, M. (2008). Measuring and comparing constraints to improved biosecurity amongst GB farmers, veterinarians and the auxiliary industries, *Preventive Veterinary Medicine* 84(3–4), 310–323. doi: 10.1016/j.prevetmed.2007.12.003.

Hald, B., Skovgard, H., Bang, D. D., Pedersen, K., Dybdahl, J., Jespersen, J. B. and Madsen, M. (2004). Flies and campylobacter infection of broiler flocks, *Emerging Infectious Diseases* 10(8), 1490–1492.

Halvorson, D. A. (2002). The control of H5 or H7 mildly pathogenic avian influenza: a role for inactivated vaccine. *Avian Pathology* 31(1):5–12. doi: 10.1080/03079450120106570. PMID: 12430550.

Hauck, R., Crossley, B., Rejmanek, D., Zhou, H. and Gallardo, R. A. (2017). Persistence of highly pathogenic and low pathogenic avian influenza viruses in footbaths and poultry manure. *Avian Diseases* 61(1), 64–69. doi: 10.1637/11495-091916-Reg.

Haynes, R. B., Taylor, D. W. and Sackett, D. L. (Eds) (1979). *Compliance in Health Care*. Baltimore: Johns Hopkins University Press.

Henzler, D. J. and Opitz, H. M. (1992). The role of mice in the epizootiology of *Salmonella enteritidis* infection on chicken layer farms, *Avian Diseases* 36(3), 625–631.

Jardine, C. G. and Hrudey, S. E. (1997). Mixed messages in risk communication, *Risk Analysis* 17(4), 489–498. doi: 10.1111/j.1539-6924.1997.tb00889.x.

Jeong, J., Kang, H. M., Lee, E. K., Song, B. M., Kwon, Y. K., Kim, H. R., Choi, K. S., Kim, J. Y., Lee, H. J., Moon, O. K., Jeong, W., Choi, J., Baek, J. H., Joo, Y. S., Park, Y. H., Lee, H. S. and Lee, Y. J. (2014). Highly pathogenic avian influenza virus (H5N8) in domestic poultry and its relationship with migratory birds in South Korea during 2014, *Veterinary Microbiology* 173(3–4), 249–257.

Johnson, Y. J., Gedamu, N., Colby, M. M., Myint, M. S., Steele, S. E., Salem, M. and Tablante, N. L. (2005). Wind-borne transmission of infectious laryngotracheitis between commercial poultry operations, *International Journal of Poultry Science* 4(5), 263-267. doi: 10.3923/ijps.2005.263.267.

Jones, F. T. (2011). A review of practical Salmonella control measures in animal feed, *Journal of Applied Poultry Research* 20(1), 102–113.

Pinto, C. J. and Urcelay, V. S. (2003). Biosecurity practices on intensive pig production systems in Chile, *Preventive Veterinary Medicine* 59(3), 139–145. doi: 10.1016/S0167-5877(03)00074-6.

Kalbasi, A., Mukhtar, S., Hawkins, S. E. and Auvermann, B. W. (2005). Carcass composting for management of farm mortalities: a review, *Compost Science and Utilization* 13(3), 180–193.

Keener, H. M., Elwell, D. L. and Monnin, M. J. (2000). Procedures and equations for sizing of structures and windrows for composting animal mortalities, *Applied Engineering in Agriculture* 16(6), 681–692.

Kim, J. H. and Kim, K. S. (2010). Hatchery hygiene evaluation by microbiological examination of hatchery samples, *Poultry Science* 89(7), 1389–1398.

Kim, W. H., An, J. U., Kim, J., Moon, O. K., Bae, S. H., Bender, J. B. and Cho, S. (2018). Risk factors associated with highly pathogenic avian influenza subtype H5N8 outbreaks on broiler duck farms in South Korea, *Transboundary and Emerging Diseases* 65(5), 1329–1338.

King, M. A., Seekins, B., Hutchinson, M. and MacDonald, G. (2005). Observations of static pile composting of large animal carcasses using different media. Symposium on Composting Animal Mortalities and Slaughterhouse Residuals in South Portland, ME.

Korbel, R., Gerlach, H., Bisgaard, M. and Hafez, H. M. (1992). Further investigations on *Pasteurella multocida* infections in feral birds injured by cats, *Zentralblatt Fur Veterinarmedizin. Reihe B. Journal of Veterinary Medicine. Series B* 39(1), 10–18.

Kuney, D. R. and Jeffrey, J. S. (2002). Cleaning and disinfecting poultry facilities. In: Bell, D. D., Weaver, W. D. (Eds) *Commercial Chicken Meat and Egg Production*, 557–564. Boston: Springer. doi: 10.1007/978-1-4615-0811-3_29.

Langsrud, S., Møretrø, T. and Sundheim, G. (2003). Characterization of *Serratia marcescens* surviving in disinfecting footbaths, *Journal of Applied Microbiology* 95(1), 186–195. doi: 10.1046/j.1365-2672.2003.01968.x.

Larsen, A. F. (2009). Semi-subsistence producers and biosecurity in the Slovenian alps, *Sociologia Ruralis* 49(4), 330–343. doi: 10.1111/j.1467-9523.2009.00481.x.

Lee, H. J., Jeong, J. Y., Jeong, O. M., Youn, S. Y., Kim, J. H., Kim, D. W., Yoon, J. U., Kwon, Y. K. and Kang, M. S. (2020). Impact of Dermanyssus gallinae infestation on persistent outbreaks of fowl typhoid in commercial layer chicken farms, *Poultry Science* 99(12), 6533–6541.

Li, X., Bethune, L. A., Jia, Y., Lovell, R. A., Proescholdt, T. A., Benz, S. A., Schell, T. C., Kaplan, G. and McChesney, D. G. (2012). Surveillance of *Salmonella* prevalence in animal feeds and characterization of the *Salmonella* isolates by serotyping and antimicrobial susceptibility, *Foodborne Pathogens and Disease* 9(8), 692–698. doi: 10.1089/fpd.2011.1083.

Losinger, W. C., Bush, E. J., Hill, G. W., Smith, M. A., Garber, L. P., Rodriguez, J. M. and Kane, G. (1998). Design and implementation of the United States National Animal Health Monitoring System 1995 National Swine Study, *Preventive Veterinary Medicine* 34(2–3), 147–159. doi: 10.1016/s0167-5877(97)00076-7.

Lotz, J. M. (1997). Special topic review: viruses, biosecurity and specific pathogen-free stocks in shrimp aquaculture, *World Journal of Microbiology and Biotechnology* 13, 9.

Mauldin, J. M. (1983). *Hatchery and Breeder Flock Sanitation Guide*. Bulletin-Cooperative Extension Service. Available at: agris.fao.org.

Mauldin, J. M. (2002). Hatchery planning, design, and construction. In: Bell, D. D., Weaver, W. D. and North, M. O. (Eds) *Commercial Chicken Meat and Egg Production*, 661–683. Boston: Springer.

Mauldin, J. M. and MacKinnon, I. R. (2009). Hatchery ventilation and environmental control, *Avian Biology Research* 2(1–2), 87–91.

McQuiston, J. H., Garber, L. P., Porter-Spalding, B. A., Hahn, J. W., Pierson, F. W., Wainwright, S. H., Senne, D. A., Brignole, T. J., Akey, B. L. and Holt, T. J. (2005). Evaluation of risk factors for the spread of low pathogenicity H7N2 avian influenza virus among commercial poultry farms, *Journal of the American Veterinary Medical Association* 226(5), 767–772.

Millman, C., Christley, R., Rigby, D., Dennis, D., O'Brien, S. J. and Williams, N. (2017). "Catch 22": biosecurity awareness, interpretation and practice amongst poultry catchers, *Preventive Veterinary Medicine* 141, 22–32.

Moore, D. A., Merryman, M. L., Hartman, M. L. and Klingborg, D. J. (2008). Comparison of published recommendations regarding biosecurity practices for various production animal species and classes, *Journal of the American Veterinary Medical Association* 233(2), 249–256. doi: 10.2460/javma.233.2.249.

Morley, P. S., Morris, S. N., Hyatt, D. R.and Van Metre, D. C. (2005). Evaluation of the efficacy of disinfectant footbaths as used in veterinary hospitals, *Journal of the American Veterinary Medical Association* 226(12), 2053–2058.

Moro, C. V., De Luna, C. J., Tod, A., Guy, J. H., Sparagano, O. A. and Zenner, L. (2009). The poultry red mite (Dermanyssus gallinae): a potential vector of pathogenic agents. In: Sparagano, O. A. (Ed.) *Control of Poultry Mites (Dermanyssus)*, 93–104. Dordrecht: Springer.

Morris, M. P. (1995). Economic considerations in prevention and control of poultry disease. In: *Biosecurity in the Poultry Industry*, Shane, S. M., Halvorson, D., Hill, D., Villegas, P. and Wages, D. (Eds), 4–16. Kennett Square: American Association of Avian Pathologists.

Mukhtar, S., Kalbasi, A. and Ahmed, A. (2004). *Composting: A Comprehensive Review*. Kansas State University, National Agricultural Biosecurity Center. Available at: https://krex.k-state.edu/dspace/bitstream/handle/2097/662/Chapter3.pdf?sequence=16&isAllowed=y.

Nespeca, R., Vaillancourt, J. P. and Morrow, W. E. (1997). Validation of a poultry biosecurity survey, *Preventive Veterinary Medicine* 31(1–2), 73–86.

Niemi, J. K., Sahlström, L., Kyyrö, J., Lyytikäinen, T. and Sinisalo, A. (2016). Farm characteristics and perceptions regarding costs contribute to the adoption of biosecurity in Finnish pig and cattle farms, *Review of Agricultural, Food and Environmental Studies* 97(4), 215–223. doi: 10.1007/s41130-016-0022-5.

Nishiguchi, A., Kobayashi, S., Yamamoto, T., Ouchi, Y., Sugizaki, T. and Tsutsui, T. (2007). Risk factors for the introduction of avian influenza virus into commercial layer chicken farms During the outbreaks caused by a low-pathogenic H5N2 virus in Japan in 2005, *Zoonoses and Public Health* 54(9–10), 337–343. doi: 10.1111/j.1863-2378.2007.01074.x.

O'Boyle, C. A., Henly, S. J. and Larson, E. (2001). Understanding adherence to hand hygiene recommendations: the theory of planned behavior, *American Journal of Infection Control* 29(6), 352–360. doi: 10.1067/mic.2001.18405.

Ojewole, A. O. (2011). Biblical mandates for sustainable sanitation, *Millennium Development Goals (MdGS) as Instrument for Development in Africa*, 598–613. Available at: https://d1wqtxts1xzle7.cloudfront.net/49544347/How_green_are_hotels_in_Accra_Environmen20161012-27178-u5ze10-with-cover-page-v2.pdf?Expires=1650324230&Signature=JFBCUcC8IbV5l4WMUpjOQBc-xpAsgUu0WZKcq-ZnnZ7pq2Ua~ccfwUSN-6YyQkCAoTIxr1h0qks8EVWkpaA3m315SoPjCoFUEGFka9-3QYp9wC7rRDRSAX7a~g9EgMH253tYuJW5Z67hMcLH5mNQwdLwmCFQmwAaLzzvk4c-xh5V6I-1Y576Klzv8SkCS0giTlCCqRfmb-Tv5hmh9JoKz9K0C7qRmHu0XlmMf31j6NecmUm-Xhh0PwelFFxkoGL~BzwkZ5TQ~Upyb7RefKAOjze65X2XdCCuri-14XVRvBK~n-lTgh3FoX-73onLAx0QfXr7ACdxvSgotDmDxYvmBQ__&Key-Pair-Id=APKAJLOHF5GGSLRBV4ZA#page=609.

Owen, R. L. and Lawlor, J. (2012). A novel approach to foot dipping. Available at: https://fr.slideserve.com/bernad/a-novel-approach-to-foot-dipping.

Pagès-Manté, A., Torrents, D., Maldonado, J. and Saubi, N. (2004). Dogs as potential carriers of infectious bursal disease virus, *Avian Pathology* 33(2), 205–209.

Pandolfi, F., Edwards, S. A., Maes, D. and Kyriazakis, I. (2018). Connecting different data sources to assess the interconnections between biosecurity, health, welfare, and performance in commercial pig farms in Great Britain, *Frontiers in Veterinary Science* 5, 41. doi: 10.3389/fvets.2018.00041.

Payne, J. B., Kroger, E. C. and Watkins, S. E. (2005). Evaluation of disinfectant efficacy when applied to the floor of poultry grow-out facilities, *Journal of Applied Poultry Research* 14(2), 322–329.

Pervin, L. A. and John, O. P. (2001). *Personality: Theory and Research* (8th edn.). New York: John Wiley & Sons Inc.

Pudenz, C. C., Schulz, L. L. and Tonsor, G. T. (2019). Adoption of secure pork supply plan biosecurity by US Swine producers, *Frontiers in Veterinary Science* 6, 146.

Rabie, A. J., McLaren, I. M., Breslin, M. F., Sayers, R. and Davies, R. H. (2015). Assessment of anti-Salmonella activity of boot dip samples, *Avian Pathology* 44(2), 129–134.

Racicot, M., Venne, D., Durivage, A. and Vaillancourt, J. P. (2011). Description of 44 biosecurity errors while entering and exiting poultry barns based on video surveillance in Quebec, Canada, *Preventive Veterinary Medicine* 100(3–4), 193–199. doi: 10.1016/j.prevetmed.2011.04.011.

Racicot, M., Venne, D., Durivage, A. and Vaillancourt, J. P. (2012a). Evaluation of strategies to enhance biosecurity compliance on poultry farms in Québec: effect of audits and cameras, *Preventive Veterinary Medicine* 103(2–3), 208–218. doi: 10.1016/j.prevetmed.2011.08.004.

Racicot, M., Venne, D., Durivage, A. and Vaillancourt, J. P. (2012b). Evaluation of the relationship between personality traits, experience, education and biosecurity compliance on poultry farms in Québec, Canada, *Preventive Veterinary Medicine* 103(2–3), 201–207. doi: 10.1016/j.prevetmed.2011.08.011.

Racicot, M., Comeau, G., Tremblay, A., Quessy, S., Cereno, T., Charron-Langlois, M., Venne, D., Hébert, G., Vaillancourt, J. P., Fravalo, P., Ouckama, R., Mitevski, D., Guerin, M. T., Agunos, A., DeWinter, L., Catford, A., Mackay, A. and Gaucher, M. L. (2020). Identification and selection of food safety-related risk factors to be included in the Canadian Food Inspection Agency's Establishment-based Risk Assessment model for Hatcheries, *Zoonoses and Public Health* 67(1), 14–24.

Refrégier-Petton, J., Rose, N., Denis, M. and Salvat, G. (2001). Risk factors for Campylobacter spp. contamination in French broiler-chicken flocks at the end of the rearing period, *Preventive Veterinary Medicine* 50(1–2), 89–100. doi: 10.1016/s0167-5877(01)00220-3.

Ribbens, S., Dewulf, J., Koenen, F., Mintiens, K., De Sadeleer, L., de Kruif, A. and Maes, D. (2008). A survey on biosecurity and management practices in Belgian pig herds, *Preventive Veterinary Medicine* 83(3–4), 228–241. doi: 10.1016/j.prevetmed.2007.07.009.

Ricke, S. C., Richardson, K. and Dittoe, D. K. (2019). Formaldehydes in feed and their potential interaction with the poultry gastrointestinal tract microbial community-a review, *Frontiers in Veterinary Science* 6, 188. doi: 10.3389/fvets.2019.00188.

Roche, A. J., Cox, N. A., Richardson, L. J., Buhr, R. J., Cason, J. A., Fairchild, B. D. and Hinkle, N. C. (2009). Transmission of *Salmonella* to broilers by contaminated larval and adult lesser mealworms, *Alphitobius diaperinus* (Coleoptera: Tenebrionidae), *Poultry Science* 88(1), 44–48.

Rodgers, J. D., McCullagh, J. J., McNamee, P. T., Smyth, J. A. and Ball, H. J. (2001). An investigation into the efficacy of hatchery disinfectants against strains of Staphylococcus aureus associated with the poultry industry, *Veterinary Microbiology* 82(2), 131–140.

Rodrigues da Costa, M., Gasa, J., Calderón Díaz, J. A., Postma, M., Dewulf, J., McCutcheon, G. and Manzanilla, E. G. (2019). Using the biocheck. UGent™ scoring tool in Irish farrow-to-finish pig farms: assessing biosecurity and its relation to productive performance, *Porcine Health Management* 5, 4.

Rojo-Gimeno, C., Postma, M., Dewulf, J., Hogeveen, H., Lauwers, L. and Wauters, E. (2016). Farm-economic analysis of reducing antimicrobial use whilst adopting improved management strategies on farrow-to-finish pig farms, *Preventive Veterinary Medicine* 129, 74–87. doi: 10.1016/j.prevetmed.2016.05.001.

Rose, N., Beaudeau, F., Drouin, P., Toux, J. Y., Rose, V. and Colin, P. (2000). Risk factors for Salmonella persistence after cleansing and disinfection in French broiler-chicken houses, *Preventive Veterinary Medicine* 44(1–2), 9–20. doi: 10.1016/S0167-5877(00)00100-8.

Roy, D. N. and Brown, A. W. A. (1954). *Entomology : (Medical & Veterinary) Including Insecticides & Insect & Rat Control*. Calcutta: Excelsior Press.

Sambeek, F. V., McMurray, B. L. and Page, R. K. (1995). Incidence of *Pasteurella multocida* in poultry house cats used for rodent control programs, *Avian Diseases* 39(1), 145–146.

Sanchez, J., Stryhn, H., Flensburg, M., Ersbøll, A. K. and Dohoo, I. (2005). Temporal and spatial analysis of the 1999 outbreak of acute clinical infectious bursal disease in broiler flocks in Denmark, *Preventive Veterinary Medicine* 71(3-4), 209-223. doi: 10.1016/j.prevetmed.2005.07.006.

Sander, J. E., Warbington, M. C. and Myers, L. M. (2002). Selected methods of animal carcass disposal, *Journal of the American Veterinary Medical Association* 220(7), 1003-1005.

Sanderson, M. W., Dargatz, D. A. and Garry, F. B. (2000). Biosecurity practices of beef cow-calf producers, *Journal of the American Veterinary Medical Association* 217(2), 185-189. doi: 10.2460/javma.2000.217.185.

Sawabe, K., Hoshino, K., Isawa, H., Sasaki, T., Hayashi, T., Tsuda, Y., Kurahashi, H., Tanabayashi, K., Hotta, A., Saito, T., Yamada, A. and Kobayashi, M. (2006). Detection and isolation of highly pathogenic H5N1 avian influenza A viruses from blow flies collected in the vicinity of an infected poultry farm in Kyoto, Japan, 2004, *American Journal of Tropical Medicine and Hygiene* 75(2), 327-332.

Schoof, H. F. (1959). How far do flies fly and what effect does flight pattern have on their control, *Pest Control* 27(4), 16-24.

Shaw, C. D. and Costain, D. W. (1989). Guidelines for medical audit: seven principles, *BMJ* 299(6697), 498-499.

Smith, T. (1990). Medical audit, *BMJ* 300(6717), 65-65.

Snow, L. C., Davies, R. H., Christiansen, K. H., Carrique-Mas, J. J., Cook, A. J. and Evans, S. J. (2010). Investigation of risk factors for Salmonella on commercial egg-laying farms in Great Britain, 2004-2005, *The Veterinary Record* 166(19), 579-586. doi: 10.1136/vr.b4801.

Springthorpe, S. (2000). La désinfection des surfaces et de l'équipement, *Journal of Cancer Dent. Assoc* 66, 558-560.

Steiner, J. J. (2020). *Disinfection of Hatching Eggs Using Low-Energy Electron Beam* (Doctoral dissertation, University of Zurich).

Strömberg, A., Broström, A., Dahlström, U. and Fridlund, B. (1999). Factors influencing patient compliance with therapeutic regimens in chronic heart failure: a critical incident technique analysis, *Heart and Lung* 28(5), 334-341. doi: 10.1053/hl.1999.v28.a99538.

Tablante, N. L. and Malone, G. W. (2006). Controlling avian influenza through in-house composting of depopulated flocks: sharing Delmarva's experience. Proceedings of the 2006 National Symposium on Carcass Disposal. Available at: https://www.animalmortmgmt.org/wp-content/uploads/2014/04/Controlling-AI-through-in-House-Composting-of-De-Populated-F.pdf.

Thermote, L. (2006). Effective hygiene within the hatchery, *International Hatchery Practice* 20(5), 18-21.

Tibballs, J. (1996). Teaching hospital medical staff to handwash, *The Medical Journal of Australia* 164(7), 395-398.

Timmermans, S. and Berg, M. (1997). Standardization in action: achieving local universality through medical protocols, *Social Studies of Science* 27(2), 273-305.

Toma, B., Vaillancourt, J.-P., Dufour, B., Eliot, M., Moutou, F., Marsh, W., Benet, J.-J., Sanaa, M. and Michel, P. (1999). *Dictionary of Veterinary Epidemiology*. Wiley.

Toma, L., Stott, A. W., Heffernan, C., Ringrose, S. and Gunn, G. J. (2013). Determinants of biosecurity behaviour of British cattle and sheep farmers-a behavioural economics analysis, *Preventive Veterinary Medicine* 108(4), 321-333.

Vaillancourt J-P. (1995). Infectious laryngo-tracheitis in broilers: a case-control study, *Annual Meeting of the American Association of Avian Pathologists*, July 11, 1995, Pittsburgh, Pennsylvania

Vaillancourt, J.-P. and Martinez, A. (2001). Relative importance of biosecurity measures: a delphi study. Annual Meeting of the American Association of Avian Pathologists, 138th Annual Convention AVMA, Boston.

Vaillancourt, J.-P. and Carver, D. K. (1998). Biosecurity: perception is not reality, *Poultry Digest* 57(6), 28–36.

Vaillancourt, J.-P., Delpont, M., Racicot, M., Paul, M. and Guérin, J.-L. (2018). Une perspective régionale de la biosécurité. Le nouveau praticien vétérinaire; 10/n°40156.

Valeeva, N. I., van Asseldonk, M. A. and Backus, G. B. (2011). Perceived risk and strategy efficacy as motivators of risk management strategy adoption to prevent animal diseases in pig farming, *Preventive Veterinary Medicine* 102(4), 284–295.

Van De Giessen, A. W., Tilburg, J. J. H. C., Ritmeester, W. S. and Van Der Plas, J. (1998). Reduction of campylobacter infections in broiler Flocks by application of hygiene measures, *Epidemiology and Infection* 121(1), 57–66.

Vandekerchove, D., De Herdt, P., Laevens, H. and Pasmans, F. (2004). Colibacillosis in caged layer hens: characteristics of the disease and the aetiological agent, *Avian Pathology* 33(2), 117–125. doi: 10.1080/03079450310001642149.

Verhagen, J. H., Fouchier, R. A. M. and Lewis, N. (2021). Highly pathogenic avian influenza viruses at the wild–domestic bird interface in Europe: future directions for research and surveillance, *Viruses* 13(2), 212.

Villa, B. and Velasco, A. (1994). Integrated pest management of the rat Rattus norvegicus in poultry farms, *Veterinaria México* 25(3), 247–249.

Volkova, V., Thornton, D., Hubbard, S. A., Magee, D., Cummings, T., Luna, L., Watson, J. and Wills, R. (2012). Factors associated with introduction of infectious laryngotracheitis virus on broiler farms during a localized outbreak, *Avian Diseases* 56(3), 521–528. doi: 10.1637/10046-122111-Reg.1.

Vroom, V. H. (1994). *Work and Motivation*. San Francisco, CA: John Wiley & Sons.

Watkins, S. and Venne, D. (2015). Water quality. In: *Manuel de pathologie aviaire*, Brugère-Picoux, J., Vaillancourt, J.-P., Bouzouaia, M., Shivaprasad, H. L. and Venne, D. (Eds), 560–569. Paris: AFAS.

Young, I., Rajić, A., Letellier, A., Cox, B., Leslie, M., Sanei, B. and McEwen, S. A. (2010). Knowledge and attitudes toward food safety and use of good production practices among Canadian broiler chicken producers, *Journal of Food Protection* 73(7), 1278–1287. doi: 10.4315/0362-028x-73.7.1278.

Zhang, Y. H., Li, C. S., Liu, C. C. and Chen, K. Z. (2013). Prevention of losses for hog farmers in China: insurance, on-farm biosecurity practices, and vaccination, *Research in Veterinary Science* 95(2), 819–824.

Chapter 5

Nutritional strategies to boost immune response in poultry

M. T. Kidd, Center of Excellence for Poultry Science, University of Arkansas, USA; and P. F. Surai, Vitagene and Health Research Center, UK, Moscow State Academy of Veterinary Medicine and Biotechnology named after K.I. Skryabin, Russia, Trakia University, Bulgaria, Szent Istvan University, Hungary and Saint-Petersburg State Academy of Veterinary Medicine, Russia

1 Introduction

Commercial live poultry-production systems are geared to optimize high-quality broilers to market. Losses in poultry livability result in lesser quantities reaching the market, and health-related issues result in poor quality or uniformity of the broilers marketed. The broilers' immune system functionality affects both of these live-production issues. Moreover, both the ingredients offered in the matrix of formulation and the nutritional profiles set in the nutrient matrix of the formulation are known to impact the immune system functionality of poultry. In addition, a hen with a robust immune system will pass these traits on to progeny chicks.

Nutritional modulation of immunity in poultry has been reviewed (Kidd, 2004; Klasing, 2007; Kogut, 2009; Korver, 2012). The review of Kidd (2004) was

http://dx.doi.org/10.19103/AS.2022.0104.08

a sequel to a Poultry Science Association Annual Meeting 2003 symposia and outlined specific considerations on nutritional modulation of immune function in order for poultry nutritionists and researchers to work closer together on immune system strategies, resulting in improved health of poultry in practice. The review by Klasing (2007) was a sequel to the 2007 Gordon Memorial Lecture and outlined the need to understand the evolution underling nutrition by immunity and host pathogens in poultry with specific reference to current consumer demands. It was demonstrated that diet fortification of nutrients known to improve immune system functions during host--pathogen interactions is not always beneficial (Klasing, 2007). Kogut's review (2009) was a sequel to the 2008 Annual Poultry Science Associations Informal Nutrition Symposium and provided insights for nutritionists and immunologists to collaborate on strategies that boost the innate immune system of poultry against pathogens in practice. The review of Korver (2012) was a component of a special journal issue entitled 'Nutrition and Pathology of Non-Ruminants' and outlines pitfalls in the current literature on the nutrition and immunity of poultry. It was concluded that a more precise understanding of avian immunity will be required to take advantage of improving the health of poultry in practice via nutrition (Korver, 2012).

Specific focus in this chapter is given to the effect of vitamins (i.e. carotenoids and vitamin E), minerals (i.e. selenium and zinc), and amino acids (i.e. arginine and threonine) on the immunity of poultry. It was with purpose that two key mediators in the former class of nutrients were chosen with a specific focus. In the sequel thereof, however, brief attention is given to other dietary nutrients and ingredients on the immunity of poultry. Citations on the former have been confined to the period 2000–present.

2 Carotenoids

Carotenoid compounds encompass natural pigments consisting of more than 1000 molecules (Yabuzaki, 2017). In nature, carotenoids provide a variety of bright colors in senescent leaves, flowers (e.g. narcissus, marigold), fruits (e.g. pineapple, citrus fruit, paprika), vegetables (e.g. carrots, tomatoes), insects (e.g. ladybird), bird plumage (e.g. flamingo, cock of rock, ibis, canary), and marine animals (e.g. crustaceans, salmon), as reviewed by Surai and Kochish (2020). These pigments are responsible for different colors - from light yellow to dark red - and, when complexed with proteins, can also produce green and blue colorations (Ong and Tee, 1992). It is well known that carotenes are precursors of vitamin A (e.g. beta-carotene), and this conversion has been well studied; however, less than 10% of known carotenoids can be converted to vitamin A (Surai et al., 2001a,b). Further, since commercial premixes are supplemented with vitamin A at dosages much higher than the potential of dietary pro-vitamin A carotenoids, the effect of dietary beta-carotene on the immunity of commercially growing birds seems to be minimal. Clearly defined roles for non-provitamin carotenoids require more research, but it is believed that carotenoids participate in (Surai and Kochish, 2020):

- antioxidant protection;
- promotion of cell differentiation;
- regulation of intracellular communication via gap junctions;
- Nrf2 upregulation;
- NF-κB downregulation; and
- enhancement of immune function.

The aforementioned functions of carotenoids are thought to be responsible for the various health-promoting properties these pigments provide. However, the physiological requirement for carotenoids in avian species, if any, is not yet established and more work in this field is needed to understand how carotenoids fulfil any essential nutritional function in poultry. However, high concentrations of carotenoids in egg yolks of several wild avian species (Surai et al., 2001c) and their effective transfer to the developing embryo (Surai et al.,1996) with maximum concentration at the time of hatching (Surai, 2002; Surai et al., 2016), in addition to protective effects against oxidative stress (Surai, 2012a,b), suggest possible roles of carotenoids in avian embryonic development. Therefore, poultry breeding diet fortification of carotenoids and transfer into the egg yolk could be an important strategy to deal with oxidative stress related to the hatching process (Surai and Kochish, 2020). Though in wild birds, carotenoid-based plumage coloration reflects the health status (i.e. high immunocompetence), resulting in a criterion for females choosing males for progeny health status (Blount et al., 2003). Applying this concept to commercial poultry production warrants future research. Furthermore, the main difference between corn-based (i.e. North and South American) and wheat-based (i.e. European) diets was shown to be related to the lutein and zeaxanthin content of diets (Surai and Sparks, 2001). It is not clear at present, however, if high-carotenoid American-type diets have any advantages over low-carotenoid wheat-based European-type diets in terms of immunocompetence of poultry. The immunomodulatory properties of carotenoids were previously reviewed (Moller et al., 2000; Surai, 2002), and an updated assessment of avian immunity is presented herein (Table 1).

The carotenoid most tested experimentally in relation to avian immunity was lutein, which is known to be the main carotenoid of poultry diets (Surai and Sparks, 2001). Canthaxanthin (Johnson-Dahl et al., 2017) and astaxanthin (Takimoto et al., 2007) were also shown to have immunomodulating properties, while lycopene was shown (Olson et al., 2008) to be ineffective.

Table 1 Carotenoids and immunity

Nutritional effector	Response	Reference
Cobb broiler chicks, d1 to d22 of age, BD or BD + LTN at 50 mg/kg. At d20 birds were injected with LPS.	LTN decreased plasmatic NO oxide levels, ameliorated LPS-associated upregulation of pro-inflammatory cytokines (IL-1β and IL-12) in the liver, decreased splenic but increased hepatic TLR-4 mRNA.	Moraes et al. (2016)
Day-old chicks, BD or BD + LTN at 25 or 50 mg/kg feed for 23 days.	During LPS stimulation, LTN (25 and 50 mg/kg) showed anti-inflammatory response by increasing the expression of PPAR and RXR isomer mRNAs.	Selvaraj et al. (2010)
1-d-old Arbor Acres broilers, BD or BD + LTN at 200 mg/kg for 42 days. Vaccination against ND and AI on d 10. At 16, 18, and 20 d of age, chicks were injected with LPS (250 mg/kg).	At d30, the ND titers were increased in the LPS-induced LTN-supplemented chickens. No effect on AI titers or B- and T-lymphocyte proliferation.	Rajput et al. (2013)
Four-week-old layer chicks (Hy-Line), BD or BD + LTN (50 mg/kg) for 4 weeks. Chickens were challenged with LPS.	LTN increased by 52.9% plasma IgA concentration in comparison to the vitamin E-fed birds. No effect on females. No effect on immune-related gene expression (IL-1β, IL-12, TLR4).	Wils-Plotz and Klasing (2017)
19-week-old Leghorn birds (layers), BD or BD + LTN (125 or 250 mg/kg) for 20 weeks.	Secondary Ab response to IB vaccination was improved by LTN (125 mg/kg). No effect of 250 mg/kg. No effect of LTN (both doses) on cutaneous hypersensitivity to PHA-P.	Bédécarrats and Leeson (2006)
One-day-old turkey chicks, BD or BD + LTN (25 or 50 mg/kg of feed) for 50 days.	In the LPS injected groups, LTN50 decreased the IL-1β mRNA amount and increased IL-10 mRNA content in the spleen compared with the BD group. No effect of LTN25.	Shanmuga-sundaram and Selvaraj (2011)
Breeding hens at 34 weeks of age, BD or BD + XTH, 40% of LTN and 60% of zeaxanthin) at 20 or 40 mg/kg for 35 days.	XTH (both doses) decreased proinflammatory IL-1β mRNA in the liver and jejunum, IL-6 mRNA in the liver, IFN-γ mRNA in the jejunum and LITAF mRNA in the liver compared to the BD group.	Gao et al. (2012)
Progeny chicks hatched from hens fed on BD or BD + 40 mg/kg XTH were fed on BD or BD + 40 mg/kg XTH. The liver, duodenum, jejunum and ileum were collected at 0, 7, 14 and 21 d after hatching.	In ovo XTH decreased proinflammatory cytokine expression (IL-1β, IL-6, IFN-γ and LITAF) in the liver, duodenum, jejunum and ileum and increased anti-inflammatory cytokine expression (IL-4 and IL-10) in the liver, jejunum and ileum mainly at 0–7 d after hatching. Dietary XTH modulated proinflammatory cytokines (IL-1β, IL-6 and IFN-γ) in the liver, duodenum, jejunum and ileum and anti-inflammatory cytokine (IL-10) in the liver and jejunum mainly from 2 weeks onwards.	Gao et al. (2012)

(*Continued*)

Table 1 (*Continued*)

Nutritional effector	Response	Reference
1-day-old Arbor Acres broiler chicks, BD or BD + marigold flower extract (source of LTN) at 0, 100 150 and 200 (MG200) mg/kg of feed for 42 d. Birds were vaccinated against ND and AIV at 10 days of age and re-vaccinated against ND via drinking water at 21 days of age.	Ab titers against ND were increased at days 20 and 30 by MG200; Ab titers against AIV were increased at day 30 by MG150 and MG200.	Rajput et al. (2012)
Broiler breeders, BD or BD + 6 or 12 mg/kg Cxn	Maternal Cxn (6 mg/kg) increased progeny chick *E. coli* bactericidal capacity and d1 oxidative burst; Cxn at12 mg/kg was not effective; phagocytosis was unaffected by Cxn.	Johnson-Dahl et al. (2017)
One-week-old chicks, BD or BD + Ax (Ax, 100 mg/kg diet) for 14 d.	Plasma IgG concentration was increased by Ax feeding; Ax enhanced splenocyte proliferation stimulated by both concanavalin A and pokeweed mitogen.	Takimoto et al. (2007)
Hyline W36 laying hens (23 or 39 wk-old). BD or BD + Lycopene at 65, 257, 420, 650 420, or 840 mg/kg diet.	Immune responses (inflammatory immune response after LPS stimulation, cutaneous basophil hypersensitivity after stimulation with PHA, primary and secondary antibody response to vaccination) were induced but were not affected by dietary lycopene	Olson et al. (2008)

Ab = antibody
AI = avian influenza
AIV = Avian Influenza virus
Ax = Astaxanthin
BD = basal diet
Cxn = canthaxanthin
IB = infectious bronchitis
IFN = interferon
Ig = immunoglobulin
IL = interleukin
LITAF = lipopolysaccharide-induced TNF-α factor
LPS = lipopolysaccharide
LTN = lutein
MG = marigold flower extract
ND = Newcastle disease
NO = nitric oxide
PHA = phytohemagglutinin
PPAR = peroxisome proliferators activated receptors
RXR = retinoic acid X receptor
TNF = Tumor Necrosis Factor alpha
XTH = xanthophylls

Marigold flower extract (Rajput et al., 2012) and xanthophyll mixture (Gao et al., 2012) were also studied. Carotenoids were tested at 25–250 mg/kg, and the active concentrations were shown to be in a range of 25–125 mg/kg, as 250 mg/kg was not effective (Bédécarrats and Leeson, 2006). Most research related to the immune-modulating properties of carotenoids was conducted with broiler chicks; however, immunomodulating properties of lutein in turkey poults were studied (Shanmugasundaram and Selvaraj, 2011). Furthermore, the effects of a xanthophyll mixture containing 40% of lutein and 60% of zeaxanthin (Gao et al., 2012) and canthaxanthin (Johnson-Dahl et al., 2017) in the maternal diet on immunity indexes in progeny chicks were also reported. Carotenoids were shown to improve humoral immunity, indicated by improved primary antibody production, including immunoglobulin A (Wils-Plotz and Klasing, 2017) and immunoglobulin G (e.g. analogous to avian IgY) (Takimoto et al., 2007) for Newcastle Disease (Rajput et al., 2012; 2013), avian influenza virus (Rajput et al., 2012), as well as secondary antibody response to infectious bronchitis (Bédécarrats and Leeson, 2006). Furthermore, carotenoids were shown to decrease the expression of proinflammatory cytokines and increase the anti-inflammatory cytokine expression (Moraes et al., 2016; Shanmugasundaram and Selvaraj, 2011; Gao et al., 2012). Carotenoids showed an anti-inflammatory response by increasing the expression of peroxisome proliferator-activated receptors and retinoic acid X receptor isomer mRNAs (Selvaraj et al., 2010). Furthermore, canthaxanthin in the breeder diet was shown to increase *E. coli* bactericidal capacity and day-one oxidative burst in progeny chicks (Johnson-Dahl et al., 2017), and astaxanthin improved splenocyte proliferation stimulated by both concanavalin A and pokeweed mitogen (Takimoto et al., 2007). It should be noted that the immunomodulatory actions of carotenoids are condition-dependent. Indeed, in many studies, only a few immune indexes were improved, while others were not affected, including no effect on avian influenza titers or B and T lymphocyte proliferation (lutein at 200 mg/kg, Rajput et al., 2013). Additional work is provided whereby carotenoid diet fortification was without effect: immune-related gene (IL-1β, IL-12, TLR4) expression (lutein, 50 mg/kg; Wils-Plotz and Klasing, 2017), cutaneous basophil hypersensitivity to phytohemagglutinin-P mitogen (lutein, 125 or 250 mg/kg, Bédécarrats and Leeson, 2006), and phagocytosis (canthaxanthin, Johnson-Dahl et al., 2017).

Based on the aforementioned data (Table 1), it could be suggested that the main molecular mechanisms of the immunomodulating properties of carotenoids are related to their regulatory roles in the expression of Nrf2, NF-κB,and pro-/anti-inflammatory cytokines (Surai and Kochish, 2020); upregulation of vitagenes (Surai, 2020); and redox balance maintenance (Surai et al., 2021). Because these classes of compounds are fat-soluble, levels above those considered adequate could produce certain toxicities, but the economic scrutiny of premixes should prevent this in practices, as opposed to humans (i.e. canthaxanthin damage to the retina). In future, the potential anti-inflammatory properties of carotenoids would be of great importance, since sub-chronic and chronic inflammation are associated with various poultry disorders, including gut health problems. Hence, the effect of carotenoids and polyphenolics on gut health deserves more attention.

3 Vitamin E

Vitamin E is the main biological antioxidant that plays an important role in animal/poultry physiology. In general, feed-derived vitamin E is not taken into account when vitamin E supplementation is considered (Surai, 2002), and commercial vitamin E supplementation varies from 20 to 60 mg/kg for broilers and up to 100 mg/kg for broiler breeders. Vitamin E is easily transferred from the feed to the egg yolk and developing embryo and is the major element of the antioxidant defense network of the developing embryo (Surai et al., 1996, 2016, 2019a). Since commercial poultry production is associated with a range of environmental, technological (Surai and Fisinin, 2016a), nutritional, and internal/biological (Surai and Fisinin, 2016b; Surai et al., 2019b) stresses, the protective effect of vitamin E in practice is paramount (Surai et al., 2019a). Interestingly, there is a range of avian species in the wild where vitamin E concentration in egg yolks is much higher than that in eggs from commercial breeders supplemented with 100 mg/kg of vitamin E (Surai, 2002; Surai et al., 2001). The biological significance of high vitamin E levels in the egg yolks from wild birds requires further investigations. Early publications related to the immunomodulating properties of vitamin E have been characterized previously (Surai, 2002), and in this chapter, an update on the immunomodulating properties of vitamin E is presented (Table 2).

Most studies related to the immunomodulating properties of vitamin E were conducted with broiler birds, and only a few trials were related to broiler breeders or shell egg layers. In broilers, immune response (e.g. primarily antibody responsiveness) is improved at vitamin E doses up to 50 mg/kg (Leshchinsky and Klasing, 2001). In many other broiler studies, vitamin E supplementation at 50 mg/kg was not tested, and it is not known if it would be as effective as 100 mg/kg (Singh et al., 2006; Niu et al., 2009b; Rostami et al., 2018), 110 mg/kg (Konjufca et al., 2004), 125 mg/kg (Habibian et al., 2014), 250 mg/kg (Muir et al., 2002) or 300 mg/kg (Boa-Amponsem et al., 2000), since intermediate treatment doses between 0 and 100 mg/kg were absent. Similar problems with establishing an optimal immunomodulating dose of vitamin E have been observed with layers. Immune response to increased vitamin E supplementation is condition-dependent and could differ substantially

Table 2 Vitamin E and immunity

Nutritional effector	Response	Reference
	Broilers	
Broilers (BD containing 10.2 mg vit.E/kg. BD + vit.E (VE, 10, 17.5, 25, 37.5, 50, 100, and 200 mg/kg)	BD→VE25, a dose-dependent increase in Ab production in response to attenuated IBV, no further increase at higher VE levels; Ab to SRBC - highest at VE50; VE25 and VE50 suppressed mitogenic responses for ConA and VE25 suppressed response for PHA; heterophil was lowest at VE50, 6 h after LPS injection; cutaneous basophil hypersensitivity to PHA, and lipopolysaccharide induction of acute-phase proteins and Ab production in response to *Brucella abortus* were not affected by VE.	Leshchinsky and Klasing (2001)
Broilers: BD, BD + VE (100 or 200 mg/kg)	VE100 vs BD: antibody titers against SRBC↑ at the 24th day No effect of VE200; no effect of VE100 or VE200 on immune response after vaccination for avian influenza, Newcastle disease or infectious bronchitis disease.	Rostami et al. (2018)
Broilers, VE: 25, 250, 2500 or 5000 mg/kg	VE5000 vs other VE doses: total IgA in the intestine↑; VE250 vs other VE doses: serum IgA↑ at d21 and d 42; VE250 and VE5000 vs other VE doses: anti-T. toxoid IgA Ab titers in the intestine and in the serum at day 42↑, following immunization with T. toxoid; VE250 vs other VE doses: percentage of T-helper cells in peripheral blood on day 26↑; other dosses were not effective.	Muir et al. (2002)
Broilers, VE: 50 or 300 mg/kg	VE300 vs VE50: relative weight of bursa↑; no effect on spleen; Proinflammatory eicosanoids in blood: LTB4↓; TBX2↓ No effect on swelling index.	Konieczka et al. (2017)
Broilers, VE: 10 or 300 mg/kg	VE300 vs VE10 H/L index↑; Cutaneous basophil hypersensitivity response↓; primary SRBC Ab responses at day 6: line C↓; lines A and B - no effect; day 20 - no effect; Secondary SRBC Ab responses at day 6: Line C↓; Lines A and B - no effect; day 20 - no effect.	Boa-Amponsem et al. (2000)
Broilers, VE: 16, 110 or 220 mg/kg	Phagocytic function of MΦ at wk 3, 5, or 7 was not affected by VE. VE110 and 220 vs VE16: increased % of phagocytically active MΦ and the number of SRBC phagocytosed per MΦ in 3-wk-old birds.	Konjufca et al. (2004)
Broilers, VE: 10 or 50 mg/kg, LPS challenge	VE50 vs VE10: decreased expression of nuclear NF-κB p65 and increased levels of I-κBα in the liver simultaneously with reduced plasma protein levels of both proinflammatory cytokines (IFN-γ, IL-1β, IL-2, and IL-6) and anti-inflammatory cytokines (IL-4 and IL-10).	Zhang et al. (2010)
Broilers, VE: 14.11-14.91 mg/kg and 4.38–4.63 mg/kg for 21 or 42 days	VE14.11–14.91 vs VE4.38–4.63 mg/kg: liver NF-κB p65 levels↓, liver IκB-α levels↑ - inhibition of inflammation.	Zhan et al. (2020)

(Continued)

Table 2 (*Continued*)

Nutritional effector	Response	Reference
Broilers, d-α-Tocopherol succinate: 10, 30, or 50 mg/ kg	VE50 vs VE30 or VE10: splenic T- and B-lymphocyte proliferation↑.	Zhang et al. (2009)
Broilers, BD, BD + VE: 33, 65, 100, broilers	VE65 vs VE33 vs BD: decreased linearly expression of pro-inflammatory (IFN-γ, IL-1β, IL-6) and anti-inflammatory (IL4, IL-10, TGF-β4) cytokine genes in the jejunum. There were no differences in effects between VE65 and VE100.	Pitargue et al. (2019)
Broilers, BD, BD + VE: 125, 250 mg/ kg, broilers, thermoneutral (TN) or heat stress (HS)	VE125 orVE250 vs BD primary (IgM, IgG, total Ab) and secondary (IgM, IgG, total Ab responses↑ in both TN and HS broilers. No difference between VE125 and VE250.	Habibian et al. (2014)
Broilers, BD, BD + VE: 100, 200 mg/kg	VE 200 or VE100 vs BD primary (IgM, IgG, total Ab) and secondary (IgM, IgG, total Ab responses↑; phagocytic ability of MΦ for opsonized and unopsonized SRBC↑; No difference between VE200 and VE100, except total Ab in primary response. No effect on lymphoid organ weight.	Niu et al. (2009b)
Broilers, BD, BD + VE: 100, 200 mg/kg	VE100 vs VE0: Ab titers and total Ig↑; bursa as % of body weight↑; VE200 vsVE100 vs BD: HI titers↑; no effect on spleen and thymus weight; no effect on circulatory immune complexes.	Singh et al. (2006)
	Laying hens	
Layers, 32 weeks old, BD, BD + VE 20 mg/kg	VE30 vs BD: IgA↑(44 and 46 w), IgG↑(44 w), IgM↑(44 and 46 w); No effect after salmonella challenge.	Liu et al. (2019)
Layers, BD, BD + VE: 40, 100 mg/kg	VE40 or 100 IU/kg vs BD H/L value↓, IL-1↑ in plasma at 12 h after the LPS; No effect on Ab in laying hens. In offspring chicks at 7th week of age - Ab titers against AIV and ND↑.	Zhao et al. (2011)
layers under tropical summer conditions, VE: 25, 125, 250 g/ kg	VE125 or VE250 vs VE25: SRBC1 titer↑; ND titer↑; lymphocyte proliferation ratio↑; no difference between VE125 and VE250; no effect on cutaneous basophilic hypersensitivity.	Panda et al. (2008)
4-week-old layer chicks, BD containing 8 mg/ kg, BD + VE: 90 mg/kg	VE98 vs BD: Hepatic IL4↓; No effect on hepatic IL12.	Wils-Plotz and Klasing (2017)

(*Continued*)

Table 2 (*Continued*)

Nutritional effector	Response	Reference
	Roosters	
45-week-old breeder roosters, challenged with dexamethasone, BD or BD + VE: 200 mg/kg for 35 days	VE200 vs BD: Ab titers against ND virus↑, no effects on IgA, IgG, and IgM.	Min et al. (2018)

AB = antibody
AIV = avian influenza virus
BD = basal diet
Con A = concanavalin A
d = day
H/L ratio = Heterophil/lymphocyte ratio
HI = haemagglutination inhibition
HS = heat stress
IBV = infectious bronchitis virus
Ig = immunoglobulin
I-κBα = inhibitor of nuclear factor kappa B
LBT4 = Leukotriene
MΦ = macrophages
ND = Newcastle disease
PHA = phytohemagglutinin
SRBC = Sheep Red Blood Cells
T. toxoid = tetanus toxoid
TBX = thromboxane
TGF-β4 = Transforming growth factor beta
TN = thermoneutral
VE = vitamin E

depending on the genetics of the birds (Boa-Amponsem et al., 2000). In many cases, improvements in immune responses were seen in comparison to non-supplemented birds, a situation that is irrelevant to the poultry industry. In most cases, there was no further improvement in immune responses at doses higher than 100 mg/kg. Because commercial broiler feed supplementation with vitamin E range from 55 mg/kg to 80 mg/kg globally, it seems likely that current recommendations on vitamin E supplementation meet immunocompetence needs. The same is true for broiler breeder diets, where recommended vitamin E supplementation is 100 mg/kg. Since broiler production is associated with a range of stresses, a vitamin E requirement of growing birds in stress conditions needs further research. Further, with current vitamin E levels fed in rearing shell egg-laying hens (Hy-Line) of 25 mg/kg and shell egg-laying hens of 20 mg/kg, it could well be that those doses need to be increased to maintain robust immunocompetence. On the other hand, vitamin E supplementation for the shell egg layer breeder birds of 75 mg/kg could be close to supporting optimal immunocompetence.

It seems likely that vitamin E affects both innate and adaptive immunity with a specific emphasis on humoral immunity. Hence, vitamin E supplementation is associated with increased antibody production, including total intestine IgA, serum IgA, anti-T. toxoid IgA, antibody titers in the intestine and in the serum (Muir et al., 2002), primary (IgM, IgG, total antibody titers) and secondary (IgM, IgG, total Ab) responses in both thermoneutral and heat-stressed broilers (Habibian et al., 2014), total immunoglobulins (Singh et al., 2006), antibody production in response to attenuated infectious bronchitis virus (Leshchinsky and Klasing, 2001), antibody titers against Newcastle Disease virus (Min et al., 2018; Zhao et al., 2011; Panda et al., 2008), antibody titers against avian influenza virus (Zhao et al., 2011), and antibody production to sheep red blood cells (Leshchinsky and Klasing, 2001; Rostami et al., 2018; Panda et al., 2008). Vitamin E was also shown to improve splenic T- and B-lymphocyte proliferation (Zhang et al., 2009), increase the percentage of T-helper cells in peripheral blood (Muir et al., 2002), increase the percentage of mature phagocytic macrophages, increase the number of sheep red blood cells phagocytosed per macrophage (Konjufca et al., 2004), and decrease proinflammatory eicosanoids (e.g. Leukotriene and thromboxane) in the blood (Konieczka et al., 2017). Importantly, vitamin E was shown to decrease the expression of nuclear NF-κB p65 and increase the levels of inhibitor of nuclear factor kappa B (I-κBα) in the liver simultaneously with reduced plasma protein levels of both proinflammatory cytokines (IL, interleukin) (interferon-γ, IL-1β, IL-2, and IL-6) and anti-inflammatory cytokines (IL-4 and IL-10; Zhang et al., 2010; Zhan et al., 2020; Pitargue et al., 2019). Interestingly, vitamin E (98 mg/kg) was shown to decrease hepatic IL-4, while not affecting hepatic IL-12 (Wils-Plotz and Klasing, 2017). The immunomodulating properties of vitamin E are condition-dependent in many cases, which is similar to carotenoids. For example, vitamin E at 50 mg/kg suppressed mitogenic responses for concanavalin A and vitamin E at 25 mg/kg suppressed the response for phytohemagglutinin in broilers (Leshchinsky and Klasing, 2001), which could be considered a suppressed innate immune response in the chicken. In most studies reviewed herein, vitamin E effects were dose-dependent, and it affected only some of the immunity indexes, while others were not affected. For example, vitamin E did not affect cutaneous basophil hypersensitivity to phytohemagglutinin, and lipopolysaccharide induction of acute-phase proteins and antibody production in response to *Brucella abortus* were not affected by vitamin E as well (Leshchinsky and Klasing, 2001). Similarly, cutaneous basophil hypersensitivity swelling index (Konieczka et al., 2017) or antibody production post avian influenza, Newcastle Disease virus, or infectious bronchitis disease vaccinations (Rostami et al., 2018) were not affected by vitamin E. Therefore, it is difficult to assess the result of increased vitamin E supplementation in terms of chicken immunocompetence and, most likely, commercially supplemented levels of vitamin E are adequate to maintain optimal immunity. However, under various stress conditions, including heat stress and mycotoxin stress, increased vitamin E supplementation could be beneficial (e.g. 150 to 250 mg/kg of diet). However, since commercial stresses differ from one poultry production complex to another, it is difficult to conclude on an exact level of vitamin E for optimal immunocompetence.

4 Selenium

In terms of the immunomodulating properties of trace elements, selenium has many special considerations. On the one hand, most commercial diets for poultry and farm animals are Se-deficient (0.1 to 0.3 mg/kg of feed) and require Se supplementation (Surai, 2006, 2018). On the other hand, Se is an essential part of at least 25 selenoproteins deeming Se as the 'chief-executive' of the antioxidant system responsible for antioxidant system regulation and optimal activity maintenance (Surai and Kochish, 2019). It is known that Se deficiency is immune-suppressive, which has recently been reviewed (Surai, 2018). It should be mentioned that Se excess is toxic, and therefore the usage of high Se doses is restricted by its toxicological and legal limits. The U.S. maximal Se supplementation is set at 0.3 mg/kg, while in Europe the total Se level in the feed should not exceed 0.5 mg/kg. Taking into account Se levels in commercial poultry diets in Europe (0.1–0.2 mg/kg), the legal limit of Se supplementation is similar in the U.S. and Europe. However, recently, in the EU, a legal limit of organic Se supplementation was set at 0.2 mg/kg (Surai, 2018).

Recent developments in relation to the immunomodulating properties of Se are related to the organic Se usage in poultry diets. Indeed, the organic Se concept is based on the understanding that the natural form of Se in major feed ingredients (e.g. corn, wheat, barley, soya, etc.) is Se methionine (SeMet), while the supplemental form of Se widely used in poultry diets is sodium selenite (SS). There is a great difference in Se assimilation, accumulation, and action between SS and organic forms of Se (Surai et al., 2018). In addition, Se ingredient price varies, with SS priced consistently lower than organic Se. However, in stressful commercial conditions, the advantages of organic forms of Se (SeMet and its derivatives) are related to the possibility of building Se reserves in muscles, which can be used in stress conditions when Se requirement is increased and feed consumption is decreased. Therefore, there are a range of organic Se commercial products, including Se-enriched yeast (SeY), SeMet, zinc (Zn)-SeMet, and OH-SeMet, on the market. The advantages and disadvantages of each form of organic Se have been recently reviewed (Surai et al., 2018). In general, organic Se is believed to be an optimal form of Se under conditions of precise nutrition application in commercial poultry nutrition under various stress conditions (Surai, 2018; Surai and Kochish, 2019).

Table 3 Selenium and immunity

Nutritional effector	Response	Reference
One-day-old male broilers (Ross × Ross 308), sodium selenite (SS), Se-enriched yeast (SeY) and DL-selenomethionine (SeMet) at 0.1 or 0.4 mg/kg Se for 42 days.	Total anti-SRBC and IgG titers and hypersensitivity were enhanced by increasing Se doses and using organic sources of Se rather than SS.	Bakhshalinejad et al. (2018)
One-day-old male broiler chicks, BD or BD + Se (0.2 and 0.4 mg/kg as SeMet) for 6 weeks. SRBC were used as a test antigen to quantify specific antibody responses on day 32.	Total Ab for primary response increased with increasing dietary Se (0.4>0.2>BD). IgM and IgG: 0.4>BD; 0.2 = 0.4; 0.2 = BD. In secondary Ab response: total Ab, IgM and IgG: 0.4 = 0.2 > BD. Phagocytic activity of macrophages: 0.4 = 0.2 > BD.	Niu et al. (2009a)
Broiler chickens raised under either thermoneutral (TN, 23.9°C constant) or heat stress (HS, 23.9 to 37°C cycling) conditions, BD or BD + Se (0.5 and 1 mg/kg as SeMet). SRBC injection on day 32 followed by evaluation of serum for Ab titers in primary and secondary responses 7 days after injection.	Se had no effect on the primary Ab responses with the exception of total Ab titers, which were increased in TN broilers fed 1 mg/kg Se compared to BD. In the secondary response, Se at 1 mg/kg increased Ab titers in HS broilers, no effect on the Ab titers of broilers in TN conditions.	Habibian et al. (2014)
One-day-old female broiler chicks (Cobb, 500), BD of BD + 0.3 mg/kg as SS. Vaccination against IB and ND on day 7, and against IBD on day 14.	At day 42, IgA and IgG levels increased in SS group compared to BD group. No effect on IgM. No effect on IgA, IgG and IgM at day 21. SS also affected cytokines gene expression in spleen at d 42: TNF-α↑, IL-2↑, IL-10↑, IFN-γ↓.	Dalia et al. (2018)
22-day-old Arbor Acres male chicks reared under heat-stressed condition, BD, containing 0.027 mg Se/kg, or BD + SS or BD + SeY at 0.15 or 0.3 mg/kg for 20 days.	In heat-stressed chicks Se increased serum Ab titers against AI (Re-4 strain) and AI (Re-5 strain) vaccine compared with the BD chicks. No difference between Se sources and doses. No Se effect on serum Ab titer against ND and on blood lymphocyte proliferation.	Liao et al. (2012)
Ross 308 broilers (7 days old), BD containing 0.04 mg Se/kg; control, BD + 0.05, 0.15 or 0.25 mg/kg Se as SS or SeMet for 42 days.	Se improved immune organ index and serum immunoglobulins (IgA, IgG, and IgM). Se-Met increased thymus index, the amounts of IgA, IgG, IgM in comparison to SS-supplemented birds. Serum IgA and IgM levels of broilers fed 0.15 mg Se/kg were significantly higher than those of broilers fed 0.05 or 0.25 mg Se/kg.	Wang et al. (2016)

(*Continued*)

Table 3 (*Continued*)

Nutritional effector	Response	Reference
One-day-old chicks were fed from hatching on a BD diet or BD + Se (0.25, 0.50, and 1.00 mg/kg as SS). Chickens were orally infected with Eimeria maxima at 14 d of age and with Clostridium perfringens 4 days later. Blood samples were collected at day 6 post-infection. The spleen and jejunum tissues were collected at 2 days post-infection with *C. perfringens*.	Se (0.25 mg/kg) reduced NE-induced gut lesions compared with infected birds given BD diet. Serum Ab against the NetB toxin in the chicks fed with 0.25 and 0.50 mg/kg Se were significantly higher than the BD group. The transcripts for IL-1β, IL-6, IL-8 were increased in the intestine and spleen of Se-supplemented birds. No difference between 0.25 and 0.5 mg/kg, 1 mg/kg was not effective.	Xu et al. (2015)
One-day-old SPF chickens, BD or BD + Se (0.30 mg/kg and 0.15 mg/kg of feed) as SeY or SS for 2 weeks followed by low pathogenicity avian influenza virus challenge.	The cloacal shedding of virus was significantly decreased in all the Se supplemented groups in comparison to the BD group. The oropharyngeal shedding of virus in chickens fed with SeY was significantly lower than that in the chickens that fed on either SS or BD.	Shojadoost et al. (2019)
One-day-old SPF chickens, BD or BD + Se (0.30 mg/kg or 0.15 mg/kg of feed) as SeY or SS. At d14 days chickens were challenged with avian influenza H9N2 strain.	SeY (0.15 mg/kg) and SS (0.30 mg/kg) increased transcription of IFN-α, IFN-β and IFN-γ genes in the cecal tonsils and spleens of chickens at post virus challenge time points compared to BD.	Shojadoost et al. (2019)
One-day-old SPF chickens, two Se doses (0.30 mg/kg or 0.15 mg/kg of feed) as SeY or SS. All birds (except negative control) were vaccinated against low pathogenicity avian influenza virus intramuscularly at days 14 and 21 post-hatch.	SeY at 0.3 mg/ kg elicited higher Ab titers than the birds received 0.15 or 0.3 mg/kg SS at day 16 post-vaccination; no changes were evident for any Se treatment when compared to the vaccinated BD group at any time point. Chickens receiving SeY (0.15) showed lower cloacal virus titers at day 3, 5 and 7 d post-challenge, SeY (0.3) significantly reduced virus titers only at day 5 post-challenge, compared to the challenged BD group. SS (0.15) reduced cloacal virus titers at day 5 and 7 post-challenge, SS (0.3) did not affect them. SeY and SS reduced oropharyngeal virus titers at day 3 post-challenge.	Shojadoost et al. (2020)
Ross 308 broiler chickens, BD or BD + SeMet at 0.5, 0.8, and 1.2 mg/kg as SeMet. Newcastle and influenza inocula were injected into the 8- and 18-day-old chickens. The first and second blood samples were collected on days 10 and 20 after inoculation.	SeMet improved Ab titers against ND at first sampling. No difference between doses used. At the second sampling only 1.2 mg/kg SeMet improved Ab. Titers against ND. SeMet also improved Ab titers against Influenza at first time (at 1.2 mg/kg) and second time (all doses). No effect of SeMet on immune responses of broiler chickens after SRBC injection.	Shabani et al. (2019)

(*Continued*)

Table 3 (*Continued*)

Nutritional effector	Response	Reference
47-week-old Hy-Line W-36 laying hens. BD, containing 0.04 mg/kg Se or BD + Se (0.2 or 0.4 mg/kg as Zn-SeMet) for 70 days. Vaccination against NDV at d 38 of trial period, sera samples were collected for Ab assay at day 10 postvaccine inoculation. On days 54 and 62 birds were injected with SRBC and at day 7 after each inoculation primary and secondary antibody responses to SRBC were measured.	Zn-SeMet supplementation increased Ab response to NDV at day 10 postvaccine inoculation. No difference between 0.2 and 0.4 ppm Se. Se (at 0.4 ppm) improved SRBC titers during the secondary response. No effect on primary response.	Laika and Jahanian (2015)
One-day-old quail chicks, BD containing 0.2 mg Se/kg and BD + Se (0.5 or 1.0 mg Se/kg as SS). On d28, antibody responses to inoculated SRBC were determined.	Se supplementation increased Ab titers. No difference between SS0.5 and SS1.0. During week 4, the response to intradermally injected PHA increased in the Se-supplemented birds. SS0.5 was more effective than SS1.0.	Biswas et al. (2006)
One-day-old geese, BD or BD + Se (0.10, 0.30, 0.50 mg/kg as SeY for 63 days.	SeY enhanced the organ and cellular immunity of geese. Thymus index: $BD < 0.1 = 0.3 = 0.5$Se; Spleen Index: $BD < 0.3 = 0.5$Se; Bursa Index: $BD < 0.3 = 0.5$Se. Lymphocyte transformation rate: $BD < 0.1 < 0.3 = 0.5$ Se. Se did not affect the humoral immunity: no effect on titer of goose paramyxovirus antibody.	Baowei et al. (2011)

AB = antibody
AI = avian influenza
BD = basal diet
BD = basal diet
C. perfringens = *Clostridium perfringens*
HS = heat stress
HS = heat stress
IB = infectious bronchitis
IBD = infectious bursal disease
Ig = Immunoglobulin
Ig = immunoglobulin
ND = Newcastle Disease
ND = Newcastle disease
OH-SeMet = hydroxy-selenomethionine
SeCys = selenocysteine
SeMet = selenomethionine
SeY = Se-enriched yeast
SRBC = Sheep Red Blood Cells
SS = sodium selenite
TN = thermoneutral
TN = thermoneutral
TNF = Tumor Necrosis Factor alpha

Most studies related to the immunomodulating properties of Se were conducted with growing chickens and, to a lesser extent, laying hens, growing quail, and geese (Table 3). Selenium doses used in the aforementioned studies varied from 0.1 up to 1.2 mg/kg. It should be mentioned that the validity of testing Se in excess of legal limits is questionable. Hence, high Se doses can cause toxicity. Further, increased Se supplementation beyond its requirement does not increase Se-protein expression and does not further improve antioxidant defenses (Surai, 2018), rendering any positive effects on the immunity of high doses difficult to explain. For example, Se studies have shown there was no difference between 0.5 and 1.0 mg/kg SS (Biswas et al., 2006), between 0.2 and 0.4 mg/kg SeMet (Niu et al., 2009a), between 0.5, 0.8, and 1.2 SeMet (Shabani et al., 2019), between 0.15 and 0.3 mg/kg SeY (Liao et al., 2012), between 0.3 and 0.5 SeY (Baowei et al., 2011), and between 0.2 and 0.4 mg/kg ZnSeMet (Laika and Jahanian, 2015). In the former references, cases exist where treatments were implemented with birds fed on diets without Se supplementation, a situation which is not relevant to commercial poultry production. When SS was used as the sole Se source, positive effects on immunity were observed with 0.3 mg/kg (Dalia et al., 2018), 0.25 mg/kg (Xu et al., 2015), or 0.5 mg/kg (Biswas et al., 2006). However, in a study conducted by Biswas et al. (2006), the lowest dose tested was 0.5 mg/kg, and it is not known if lower doses (e.g. 0.25–0.30 mg/kg) would have been as affective. In another study, increasing SS supplementation from 0.25 mg/kg to 0.5 mg/kg did further increase immunity, but adding 1.0 mg/kg SS reduced it (Xu et al., 2015). When a direct comparison between SS and organic sources of Se was made, advantages to the following organic Se were observed: SeMet (Wang et al., 2016); SeY (Shojadoost et al., 2020); and both SeY and SeMet (Bakhshalinejad et al., 2018). However, Liao et al. (2012) reported no difference between SS and SeY. Interestingly, it was also reported that a lower dietary SeY level (0.15 mg/kg) was as effective as a higher dietary SS (0.30 mg/kg) level in increasing the transcription of interferon-α, β, and γ genes in the cecal tonsils and spleens of chickens at post virus challenge (Shojadoost et al., 2019). Recently, a range of papers related to nano-Se effects on poultry have been published. However, it is not known how metallic selenium in the nano-form can be converted into selenite/selenate, thus becoming an effective source of Se for Se cysteine (Cys) synthesis (Surai et al., 2017; Surai and Kochish, 2021). Thus, an understanding of the fundamental process usage of nano-Se in poultry production, including its possible immunomodulating properties, is needed before commercial use can be expected.

Therefore, the immunomodulating properties of Se are proven; however, the outcome of dietary supplementation in terms of immunity enhancement would depend on the diet Se level and form (organic Se vs. SS) and level of challenges/stresses. In most cases, only a few indexes of immunity have been used, and it is not clear what would be the health result in terms of disease resistance due to the various doses and forms of Se used. Furthermore, similar to results obtained with other nutrients (e.g. carotenoids and vitamin E), only limited indexes of immunity were improved while others were not affected. Since all arms of immunity work together in providing an effective protection against pathogens/damages, more research in this exciting area is required. In addition, further work is warranted to assess damage to immune receptors under various stress conditions as a driving mechanism of immune disbalance/suppression (Surai, 2002, 2018, 2020). This mechanism could be an important tool to help explain the beneficial/detrimental effects of various nutrients on immunocompetence under various commercially relevant stress conditions.

5 Zinc

Zinc is an essential trace element for poultry, which is required for growth, development, and immune system functioning. It is essential for cellular metabolism in terms of structure and enzyme functionality. Hence, the cellular Zn-metalloenzymes are vast and are represented in all six enzyme classes (i.e. isomerase, lyase, ligase, hydrolase, and oxido-reductase; Vallee and Auld, 1990). The former catalytic functioning of metalloenzymes via Zn approximates 2 000 enzymes (Andreini and Bertini, 2012). Some of the discoveries by Prasad in the 1960s in human patients (e.g. growth depression, skin abnormalities, and poor immunity; Prasad et al., 1963) are relevant in poultry today. For example, skin scratches and cellulitis can result in processing plant downgrades in broilers. Downs et al. (2000) demonstrated that the combined effects of vitamin E and Zn reduced cellulitis severity. Further, poor litter quality in poultry houses can cause paws (i.e. feet) of birds to develop hyperkeratosis, followed by epidermis lesions, resulting in footpad dermatitis. A study by Manangi et al. (2012) demonstrated that the combination of dietary Zn, manganese, and copper in a chelated form improved footpad health and litter condition when compared to birds fed at higher levels of Zn, manganese, and copper in inorganic forms. The original work on dietary Zn bioavailability in broilers was assessed with the Zn-Met complex because field reports in the 1980s indicated that it was improving hoof conditions in cattle. Indeed, the ZnMet complex has been correlated to improved immunity based on its improved bioavailability (Chien et al., 2006). A second-generation Zn-amino acid complex (e.g. a feed ingredient whereby Zn is complexed to a non-specific amino acid) has been shown to be more bioavailable in broilers when tested against the inorganic standard $ZnSO_4$ (Star et al., 2012). Further, inorganic Zn salt ingredients that are known to have poor bioavailability can be improved as nanosize ZnO has been shown to increase intestinal Zn absorption and physiological Zn functions in comparison to ZnO (Tsai et al., 2016).

All poultry diets contain an ingredient mineral premix. As such, it would appear that Zn deficiencies do not occur. As global poultry formulators continue to create diets devoid of animal protein meals, phytate sequestration of Zn must be closely monitored (Walk et al., 2016). It is important to remind the reader that Zn salts, complexes, and chelates vary tremendously in terms of bioavailability. A discussion on dietary Zn in poultry for immune functionality cannot take place without discussing the Zn source. Furthermore, diet additions of more available forms of Zn can not only improve health but reduce Zn excretion by the bird into the litter (Table 4).

A recent meta-analysis on Zn and immunity in poultry was conducted using 41 peer-reviewed manuscripts (Hidayat et al., 2020). This work assessed dietary Zn levels from 0 to 200 mg/kg as the primary predicted independent variable, in addition to Zn source defined as control, inorganic, or organic. It was determined that as broilers are fed increasing Zn levels, regardless of source, the tissue Zn increases quadratically, and stress (e.g. as measured by the heterophil to lymphocyte ratio) decreases linearly. This section of the chapter focuses on recent work on Zn in immunity and health of avian species (Table 4).

Feng et al. (2010) fed broilers a dose–response of Zn (0 to 120 mg Zn/kg) using Zn glycine (Gly) with a 120 mg/kg Zn control from $ZnSO_4$. Immunological differences between diets containing 120 mg/kg Zn did not occur, and ZnGly doses between 60 mg/kg and 90 mg/kg optimized lymphoid organs and immunoglobulins, with no effect on the complement (Feng et al., 2010). Gene expression and cytokine evaluation of broilers fed 100 mg Zn/kg from $ZnSO_4$ or ZnGly from diets with and without 500 FTU phytase were assessed from placement to 42 days of age (Jarosz et al., 2017a). It was established that organic Zn, but not inorganic Zn, allowed for a Th1 to Th2 balance from gut-associated lymphoid tissue and immunoglobulin expression (Jarosz et al., 2017a). In a similar experiment with identical treatments, Jarosz et al. (2017b) again showed heightened efficacy of ZnGly in helping to improve stability between the Th1 and Th2 response, but 100 mg/kg of $ZnSO_4$ exerted a negative effect on inflammation. In addition, the same treatments were assessed on acute-phase proteins and birds fed $ZnSO_4$ exhibited local inflammatory responses, whereas birds fed ZnGly had an improved regulation of the intensity of the immune response as measured by alpha-1-acid glycoprotein and transferrin (Jarosz et al., 2019).

A study was conducted with laying pullets or hens to assess diet Zn (Zn amino acid complex (AA)) or drinking water Zn from the same Zn source on vaccine stress (Nelson et al., 2020). Both Leghorn pullets and hens had improved stress as measured by the heterophil to lymphocyte ratio by all Zn treatments, but treatments providing Zn in drinking water prior to vaccinations improved adaptive immunity, in addition to reduced stress (Nelson et al., 2020).

Table 4 Zinc and immunity

Nutritional effector	Response	Reference
	Broilers	
Broilers fed diets from 0 to 120 mg ZnGly/kg of diet with a 120 mg $ZnSO_4$/kg control	Optimal lymphoid organ responses occurred between 60 and 90 mg/kg Zn (0–21 d), and only thymus was improved (60 mg Zn/kg) 22–42 d. Ig levels were improved with 60 mg/kg Zn with 120 mg ZnSO4/kg being intermediate.	Feng et al. (2010)
Broiler fed 100 mg Zn/kg from $ZnSO_4$ or ZnGly and without or with 500 FTU phytase	ZnGly improved immunity, the balance between the Th1 and Th2 response, and Ig production.	Jarosz et al. (2017a)
Broiler fed 100 mg Zn/kg from $ZnSO_4$ or ZnGly and without or with 500 FTU phytase	ZnGly improved balance between the Th1 and Th2 response, and $ZnSO_4$ was without effect, except a negative effect on inflammation.	Jarosz et al. (2017b)
Broiler fed 100 mg Zn/kg from $ZnSO_4$ or ZnGly and without or with 500 FTU phytase	Broilers fed ZnGly had improved regulation of the immune response from acute phase protein measurements, whereas $ZnSO_4$ increased local inflammation.	Jarosz et al. (2019)
Half the broilers with challenged with *Eimeria tenella* and all were fed 0, 20, 40, and 60 mg Zn/kg from methionine hydroxyl analog-Zn chelate	An interaction occurred whereby increasing Zn from 0 to 20 mg/kg improved GPx in unchallenged birds, but challenged birds had elevated GPx and Zn responses were less clear. Secretory IGA was improved with 20 mg/kg Zn 7 or 14 d post challenge, independent of *Eimeria*.	Bun et al. (2011)
Broilers were fed 40, 120, and 200 mg Zn/kg from $ZnSO_4$ and challenged with *Eimeria tenella*	Broilers fed 120 and 200 mg/kg Zn had higher white blood cell counts, and serum nitrite and nitrate.	Sajadifar et al. (2013)
Broiler were fed 0 mg/kg Zn (30 mg/kg from basal), 60 mg/kg Zn from Zn SO_4, or a diet with a 1:1 blend (30 mg/kg Zn from $ZnSO_4$ and 30 mg Zn/kg from ZnAA). A coccidiosis vaccination stress was implemented.	Little effect of Zn source or level was noted.	Troche et al. (2015)
Broiler fed 0 mg/kg Zn, 90 mg Zn/kg from $ZnSO_4$ or Zn proteinate with and without *Clostridium perfringens* challenge.	Birds fed Zn proteinate had lower IL-8 and INF-γ during challenge, suggesting a reduced inflammatory response.	Bortoluzzi et al. (2019)

(*Continued*)

Table 4 (*Continued*)

Nutritional effector	Response	Reference
Broilers reared under 24°C or 24 to 35°C and fed 34, 68, or 181 mg/kg total dietary Zn	Independent of temperature, 181 mg Zn/kg improved macrophage function. Birds reared in 24°C had improved titers when fed 181 mg Zn/kg.	Bartlett and Smith, (2003)
Broilers fed Zn (0, 100, 200, and 400 mg/kg) in the form of ZnMet and vitamin E (0, 150, and 300 mg/kg)[1]	Total SRBC and NDV titers were improved at 300 mg/kg vitamin E and 400 mg/kg Zn. Antioxidant status, as measured by malondialdehyde decrease, was improved in a dose dependent manner with vitamin E and Zn.	Akbari Moghaddam Kakhki et al. (2018)
	Laying hens	
White Leghorn pullets or hens received stress, and dietary ZnAA or Zinpro LQ in drinking water at a rate of 200g/L stock solution	All Zn administered treatments improved stress as measured by the heterophil to lymphocyte ratio, but Zn provided in the drinking water of pullets and hen prior to stress not only improved stress, but also adaptive immunity.	Nelson et al. (2020)

AA = amino acid
d = day
Gly = glycine
GPx = glutathione peroxidase in plasma
Ig = immunoglobulin
IL = interleukin
$INF_{-\gamma}$ = interferon gamma
NDV = Newcastle Disease Virus
SRBC = sheep red blood cells
Vitamin E = alpha-tocopheryl acetate
Zn = zinc
ZnMet = zinc methionine

To improve the health of poultry via nutrition, independent laboratories must assess numerous immune parameters and birds in different environments. Similarly, feed ingredients known to heighten immunity should be tested simultaneously when experimental designs allow for such comparisons. For example, Bartlett and Smith (2003) evaluated Zn from an unsupplemented test diet (34 mg/kg) or diets supplemented to 68 mg Zn/kg and 181 mg Zn/kg of diet and fed to broilers reared in two temperature conditions. Although heat stress conditions decreased performance and immunity, macrophage effector functions were improved in birds fed the highest Zn level, regardless of the temperature (Bartlett and Smith, 2003). Akbari Moghaddam Kakhki et al. (2018) fed Ross 308 male broilers combinations of diets differing in Zn (i.e. ZnMet) or vitamin E (i.e. alpha-tocopheryl acetate) and assessed their immunity. Interactive effects occurred with Zn and vitamin E, improving antioxidant status and adaptive immunity (Akbari Moghaddam Kakhki et al., 2018).

As mentioned previously, assessing dietary treatment regimes in poultry with health limitations, mimicking field scenarios, is paramount in finalizing the nutrition by immune interaction and level/source needed. The later scientific contributions assess Zn needs and responses in birds with either coccidiosis or necrotic enteritis. Bun et al. (2011) evaluated immunity in broilers challenged with *Eimeria tenella* and fed Zn titration from 0 to 60 mg Zn/kg in 20 mg/kg increments. The zinc source in the former work was methionine hydroxyl analog-Zn chelate. Feeding both nonchallenged and challenged birds increased Zn-optimized antioxidant status and secretory IgA (Bun et al., 2011). Sajadifar et al. (2013) fed broilers 40, 120, and 200 mg Zn/kg from $ZnSO_4$ and challenged all birds with *Eimeria tenella*. Both high-dietary Zn treatments resulted in significant improvements in the cutaneous basophil hypersensitivity test, white blood cells, and serum nitrites and nitrates (Sajadifar et al., 2013). Troche fed broilers 0 mg/kg Zn (30 mg/kg from ingredients), 60 mg/kg Zn from Zn SO_4, or a diet with a 1:1 blend (30 mg/kg Zn from $ZnSO_4$ and 30 mg Zn/kg from ZnAA) during a coccidiosis vaccination stress and demonstrated few Zn level or source improvements in immunity. From the same laboratory as the former work, Bortoluzzi et al. (2019) implemented a 3 × 2 factorial design in broilers fed 0 mg/kg Zn, 90 mg Zn/kg from $ZnSO_4$ or Zn proteinate, with and without *Clostridium perfringens* challenge to induce necrotic enteritis. An interaction offered in jejunal IL-8 and INF-$_\gamma$ during challenge whereby the challenge upregulated expressions, but not in birds fed the Zn proteinate, suggesting a reduced inflammatory response (Bortoluzzi et al., 2019).

6 Arginine

Because avian species lack a functional urea cycle, the lack of Arg *de novo* synthesis from urea formation renders Arg essential (Scott et al., 1982). In a review on immunity, Kidd (2004) presented original work on Arg relative to Lys (0.9, 1.1, and 1.3) from 1 to 42 days of age, and at day 28, the birds were given a challenge of *Eimeria acervulina*, *Eimeria tenella*, and *Eimeria maxima* from a commercial field isolate. Broilers fed the highest Arg:Lys ratio had reduced mortality. The increased need for Arg during a coccidial infection may be indicative of Arg production of nitric oxide to break the oocyst replication cycle in the intestine (Rochell et al., 2017). This section on Arg is dedicated to work on delineating its role in immunity and disease mitigation (Table 5).

Kidd et al. (2001) conducted three experiments to assess Arg need for immunity in chicks reared in battery brooders. Arg variations were 10% below and up to 20% above (up to 1.5% of diet, total) the NRC (1994) recommended level. Arg effects on immunity did not occur, and it was concluded that

Table 5 Arginine and immunity

Nutritional effector	Response	Reference
Arg:Lys of 0.9, 1.1, and 1.3 from 1 to 42 days of age with *E. acervuline*, *tenella*, and *maxima* challenge at d 28.	Broiler fed 1.3 Arg:Lys ratio had reduced d 42 mortality compared to the other two dietary treatments.	Kidd (2004)
Three experiments with Arg (fed in ranges of 90% to 120% NRC) in chicks were conducted in batteries. In one experiment chicks were challenged with ND and IB vaccine.	Arg did not impact innate or adaptive immunity, and it was concluded that more should be done on Arg in disease conditions.	Kidd et al. (2001)
Broiler were fed dietary total Arg of 1.20% and 1.35% (Exp 1) and 1.10%, 1.30%, and 1.50% (Exp 2)	1.35% Arg in Exp 1 showed some positive immune effects, but 1.50% Arg in Exp 2 resulted more blood monocytes and increased mRNA markers for Arg use in developing immune cells.	D'Amato and Humphrey (2010)
Broilers were fed Arg ranges from 80% to 120% of NRC and two CP levels (19.0% or 22.4%).	Feeding birds 80% of NRC Arg depressed relative spleen and thymus, in addition to CBH and titer responses. It was also shown that the Arg need may be lower in birds few low CP diets.	Jahanian (2009)
Broiler were challenged with HPSV and one group of birds were supplemented with 2.0% Arg on top of the BD.	Based on innate, adaptive, and HPSV detection, it appears that higher levels of Arg have antiviral properties.	Munir et al. (2009)
Broilers were assessed after being fed Mn and Arg (1.12% or 1.20% of diet).	Arg did not impact lymphoid organs or macrophage phagocytic activity, but some T cell functions were improved with marginally increased Arg.	Burin Junior et al. (2019)
Broiler were fed the BD, a test diet with 1.87% Arg, or a mixture of these diets. The other factor was C. per. challenge which occurred in half the birds.	In challenged birds L-Arg supplementation increased expression of IFN-γ, IL-10 and NOD1 mRNA expression, and thus alleviated intestinal insult from challenge.	Zhang et al. (2017)
Treatments tested in broilers were: control, challenged with C. per., or challenged with C. per. and fed BD plus 0.3% L-Arg.	In addition to repeating the former effects on gut health (Zhang et al., 2017), it was shown that birds fed 0.3% L-Arg had gut microbiota that mimicked unchallenged broilers.	Zhang et al. (2018)

(Continued)

Table 5 (*Continued*)

Nutritional effector	Response	Reference
Broilers were fed 1.05, 1.42, or 1.90 with and without LPS injection in Exp 1. In Exp 2, broilers were fed a dose of Arg: 0.99%, 1.39%, 1.76%, 2.13%, and 2.53%.	In Exp 1, Arg supplementation decreased $CD11^+$, $CD14^+$ and B cells in the spleen. Also, Arg supplementation decreased IL-1b, TLR4 and PPARg mRNA in the spleen and IL-1b, IL-10, TLR4 and NF-kB mRNA in the caecal tonsils. In Exp 2, Arg supplementation quadratically suppressed (1.76%) blood and splenic B cells.	Tan et al. (2014a)
Broilers were fed a 3 (1.11%, 1.33%, and 2.02% dietary Arg) by 2 (with and without coccidial challenge) factorial design.	Dietary Arg decreased intestinal mucosal damage through suppression of the Toll-like receptor 4 pathway and activation of the mTOR complex 1 pathway.	Tan et al. (2014b)
Broilers were fed a 5 x 2 factorial design of Arg (Starter:.99%, 1.39%, 1.76%, 2.13%, and 2.53% of diet; Grower: 0.95%, 1.35%, 1.71%, 2.01%, and 2.36% of diet) with and without an IBD challenge.	As dietary Arg increased, challenged birds had increased $CD3^+$ T cells, and $CD4^+$ and $CD8^+$ T cells, serum IL-6, and IBD titers.	Tan et al. (2015)

Arg = arginine
BD = basal diet
C. per. = *Clostridia perfringens*
CP = crude protein
Exp = experiment
d = day
E = *Eimeria*
IB = infectious bronchitis virus
IBD = infectious bursal disease virus
Ig = immunoglobulin
IL = interleukin
IFN-γ = interferon γ
H = heterophil
HPSV = hydropericardium syndrome virus
L = lymphocyte
LPS = lipopolysaccharide
Lys = lysine
K = potassium
ND = Newcastle Disease virus
NOD1 = nucleotide-binding oligomerization domain 1
NF-kB = Nuclear Factor kappa-light-chain-enhancer of activated B cells
PPAR γ = peroxisome proliferator-activated receptor γ
TLR4 = Toll-like receptor 4
Thr = threonine
SRBC = sheep red blood cells

Arg research needs to be conducted in practical environments with wider variations in diet Arg levels. A dose-response of Arg at two crude protein (CP) levels (19.0 and 22.4%) was fed to broilers, and it was shown that the 80% NRC Arg treatment suppressed both innate and adaptive immune responses, indicating that in the former work, Arg was not low enough to initiate a response (Jahanian, 2009). D'Amato and Humphrey (2010) fed similar levels of Arg to chicks as in the previous studies but provided one treatment with 'excess' Arg at 1.5% of the diet. The excess Arg treatment resulted in a higher percentage of monocytes but also increased mRNA markers for Arg use in developing immune cells (D'Amato and Humphrey, 2010). Further, although Arg has been shown to optimize macrophage effector function via arginase and nitric oxide synthase pathways (Tsai et al., 2002), Burin Junior et al. (2019) found that macrophage phagocytosis activity was not affected with Arg levels fed near industry practice. However, excess Arg (2.0% supplemental Arg) improved innate and adaptive immunity of broilers challenged with hydropericardium syndrome virus. The work on Arg needs assessed herein indicates that more work is needed to qualify the level of Arg needed to optimize immunity in commercial conditions.

Professor Yuming Guo and colleagues assessed bird recovery from *Clostridium perfringens* when fed dietary Arg. In their first work (Zhang et al., 2017), the Arg level fed was 1.87%, and it resulted in improved gastrointestinal function and immune function. Subsequent work was carried out to assess if dietary Arg (0.3% L-Arg added to a control diet) could change gut microbiota post challenge (Zhang et al., 2018). The birds fed increased dietary Arg had improved intestinal health and immunity, mirroring the previous work, but it was also shown that birds fed dietary Arg post challenge had gut microbiota composition mimicking unchallenged birds, an interesting finding on Arg's ability to maintain gut homeostasis.

Tan et al. (2014a,b, 2015) conducted a series of experiments to assess Arg during inflammation. Using a lipopolysaccharide model (Tan et al., 2014a), it was demonstrated that Arg suppressed Toll-like receptor 4 and $CD14^+$ cells, thereby reducing the overexpression of pro-inflammatory cytokines. A factorial design of 1.11%, 1.33%, and 2.02% dietary Arg with and without coccidial challenge in broilers demonstrated that Arg lessens intestinal mucosal damage through suppression of the Toll-like receptor 4 pathway and activation of the mTOR complex 1 pathway (Tan et al., 2014b). Lastly, as dietary Arg increased in a linear manner, immunosuppression effects were noted in broilers challenged with infectious bursal disease virus, which resulted in lessoning of the infection as facilitated via T cell modulation (Tan et al., 2015). In addition to Arg on the health status of poultry, threonine has also received considerable attention, and a recent typescript concerned with intestinal health has reviewed them together (Bortoluzzi et al., 2018).

7 Threonine

In the 1990s, Thr became available as a commercially produced amino acid (AA) (L-Thr) for monogastric diet formulation. Wide variation in Thr needs occurred in poultry from requirement studies. Although it was known that Thr had a significant role in maintenance in poultry, it was not until Bertolo et al. (1998) determined that its need for piglet total parenteral nutrition is 45% of oral ad libitum intake needs that researchers realized the large gastrointestinal need for Thr. The former work, and the poultry industry's position to reduce feed antibiotics, places Thr not only as the third most limiting AA in diet formulation but a critical AA for gastrointestinal functionality and thus poultry efficiency. Recent Thr studies on immunity and gut health are presented in Table 6.

As Thr is typically limiting after Lys in diet formulation, its feed-grade inclusion in the diet results in a reduced dietary inclusion of oilseed meals (e.g. usually soybean meal), which results in a decrease in dietary potassium. Zarrin-Kavyani et al. (2018) fed Ross broilers Thr at 100%, 110%, and 120% of breeder guidelines and potassium at two levels (0.85% and 0.94%) from 0 to 42 days of age. This work focused on innate immunity and found that the heterophil to lymphocyte ratio increased as Thr increased from 100% to 110% of recommendation in a low potassium diet, but the increased heterophil effect was diminished in birds fed high potassium (Zarrin-Kavyani et al., 2018). Sigolo et al. (2017) conducted a similar design using four levels of Thr (100%, 110%, 120%, and 130% of Ross recommendations) and two protein levels. This work focused on adaptive immunity and determined that interactions did not occur; however, increasing dietary Thr quadratically improved titers to Newcastle Disease virus (*viscerotropic velogenic* strain by drinking water at days 1, 7, and 20) and sheep red blood cells, but not avian influenza (H9N2 strain by oral-spray at day 1) (Sigolo et al., 2017).

Chen et al. (2017) fed Arbor Acre Plus chicks 0.79%, 0.89%, or 1.09% total Thr and evaluated immune and intestinal health parameters. Chicks fed 0.89% Thr had improved goblet cell numbers, but villus height to crypt depth was increased at each dietary Thr level increase. Jejunal IgG, IgM, and secretory IgA were improved in chicks fed 0.89% Thr, whereas the highest level of Thr upregulated micin-2 mRNA expression (Chen et al., 2017). Zhang et al. (2017) fed chicks either deficient dietary Thr (0.49% total) or adequate Thr (0.90% total) and excised ileal for culture assays. It was shown that Ileal cultures from chicks fed adequate Thr had increased mRNA levels of mucin-2, IL-8, and IgA after LPS exposure, but cultures from chicks fed Thr deficient diets did not. Ji et al. (2019a) fed variations in dietary Thr (85%, 100%, 125%, and 150% of NRC) from 0 to 21 and 22 to 42 days of age. Increasing Thr quadratically improved secretory IgA and mucin-2 mRNA expressions, demonstrating its impact on gut barrier function (Ji et al. (2019a). In a study using identical

Table 6 Threonine and immunity

Nutritional effector	Response	Reference
Boilers were fed Ross recommendations for Thr (100%, 110%, and 120%) with two K levels (0.85% and 0.94 % of diet) from 0 to 42 d.	A H increase from feeding 110% Thr with low K resulted in an increased H:L ratio, and this negative effect was diminished in high K diets.	Zarrin-Kavyani et al. (2018)
Broilers were fed Ross recommendations of 100% to 130% in 10% increments and two CP levels.	Innate immunity was not assessed, but Thr quadratically improved titers to NDV and SRBC, independent of CP.	Sigolo et al. (2017)
Broiler chicks were fed 0.79%, 0.89%, or 1.09% total Thr.	Chicks fed 0.89% Thr had improved intestinal function and jejunal Ig expression, whereas chicks fed 1.09% Thr had increased mucin-2 mRNA expression.	Chen et al. (2017)
Ileal culture from broiler chicks fed either deficient (0.49%) or adequate (0.90%) total dietary Thr.	Ileal cultures from Thr adequate chicks exposed to LPS had increased mRNA levels of mucin-2, IL-8, and IgA, whereas cultures from Thr deficient chicks did not.	Zhang et al. (2017)
Broiler were fed averaged Thr/Lys ratios of 60, 63, 69, 75, and 81 to 21 d of age.	Increasing starter diet Thr levels quadratically improved heterophil, villus, and goblet cell functions.	Debnath et al. (2019)
Broiler were fed Thr in relation to NRC recommendations as: 85%, 100%, 125%, and 150% from 0 to 42 d of age.	Intestinal barrier function was improved as noted by quadratic Thr increases for secretory IgA and mucin-2 mRNA expression.	Ji et al. (2019a)
Broiler were fed Thr in relation to NRC recommendations as: 85%, 100%, 125%, and 150% from 0 to 42 d of age.	The former responses (Ji et al., 2019a) were confirmed as quantified by upregulation of gene expression.	Ji et al. (2019b)
Broilers were fed corn-canola or -soybean meal diets from 0 to 35 d with Thr variations of NRC (100%, 110%, or 120%).	Increasing Thr from 100% to 110% NRC increased infectious bursal disease titers, but not Newcastle disease titers. Thr fed to broilers above NRC also increased villus height and crypt depth.	Ahmed et al. (2020)
Broiler received 0.49% vs 0.84% total of dietary Thr or pectin, with and without *Eimeria maxima* challenge.	Challenged birds had improved immunity and overall performance as Thr increased.	Wils-Plotz et al. (2013)

(Continued)

Table 6 (*Continued*)

Nutritional effector	Response	Reference
Broiler chicks were fed 0.49% or 0.90% Thr from 1 to 21 d and received feed withdrawal and coccidial vaccine.	The mechanism behind Thr deficiency impairing broiler response from feed withdrawal and coccidial vaccine was impaired gut function, impaired gut barrier function, and lymphocyte cytokine profiles.	Zhang et al. (2017)

CP = crude protein
d = day
Ig = immunoglobulin
H = heterophil
L = lymphocyte
LPS = lipopolysaccharide
Lys = lysine
K = potassium
NDV = Newcastle Disease virus
Thr = threonine
SRBC = sheep red blood cells

treatments, Ji et al. (2019b) assessed transcription sequencing and found the chicks fed 125% NRC recommended Thr needs had upregulated genes, indicating improved antioxidant capacity, immunity, and viral resistance. Debnath et al. (2019) fed varying Thr/Lys ratios across starter diets to 21 days of age (60, 63, 69, 75, and 81) in broilers and noted quadratically improved neutrophil phagocytic index, in addition to improved quadratic responses for numerous villus indexes and goblet cell functions. Further, Ahmed et al. (2020) fed corn-canola or corn-soybean meal diets to broilers from 0 to 35 d with Thr variations of NRC (100%, 110%, or 120%). Similar to previous references, improvements in villus height and crypt depth were noted in broilers fed Thr in excess of NRC levels. Further, adaptive immunity improvements were also noted with higher Thr in titer responses to infectious bursal disease virus (Ahmed et al., 2020).

Wils-Plotz et al. (2013) fed Thr from deficient to adequacy (0.49% vs 0.84% total of diet) and challenged birds with *Eimeria maxima*. Pectin was also compared to Thr, and although it provided some health improvements, adding Thr to adequacy improved intestinal immunity and restored overall performance (Wils-Plotz et al., 2013). In a similar study whereby Zhang et al. (2017) evaluated Thr adequacy or deficiency in broilers exposed to feed withdrawal or coccidial infection, it was noted that the poor performance from the Thr deficiency was caused by a loss of gut functionality and lymphocyte cytokine cascades.

8 Additional nutritional factors

8.1 Breeder and in-ovo feeding for chick health

The viability of the chick post-hatch is dependent on nutrients assembled in the fertile egg by the hen or hatchery fortification via *in ovo* feeding. Unlike mammalians, the early growth and health of poultry species are contingent upon a very small window of time for nutrient assimilation to support life and immune functionality. It has recently been shown that chicks from young hens (e.g. prior to 30 weeks) or mature hens (e.g. past 60 weeks) produce chicks with lowered innate immune systems as opposed to chicks from mid-cycle hens (Saunders-Blades and Korver, 2015). Furthermore, research has shown that variations in hen diet and *in ovo* feeding can change the immune system in the post-hatch chick (Kidd, 2003). For example, chicks from broiler breeders fed organic minerals have been shown to have improved immunity (Zn, Kidd et al., 1992) and heart functionality (Zn and Mn, Virden et al., 2004). Regarding *in ovo* feeding, Tako et al. (2004) injected a solution containing organic Zn into fertile egg amnions at incubation day 17, resulting in improved intestinal function in chicks post-hatch. Zhu et al. (2019) injected fertile chicken eggs with 0 or 3 mg vitamin C at day 15 of incubation and evaluated chick immunity post-hatch. Chicks that received 3 mg of vitamin C *in ovo* had improved antioxidant capacity and reduced expression of splenic pro-inflammatory cytokines (Zhu et al., 2019). Research regarding hen nutrition and *in ovo* feeding to optimize post-hatch chick immunity and survival warrants continued attention. Furthermore, recent attention has been given to conjugating known *in ovo* immunomodulators (i.e. Arg) with inorganic and organic synthesized silver nanoparticles to mitigate embryo toxicity (Subramaniyan et al., 2020).

8.2 Other vitamins

In addition to the carotenoids and vitamin E, the bulk of other recent work on vitamins and immunity in poultry is concerned with vitamin D. Moreover, much of the work on vitamin D has been concerned with source, specifically, sources of cholecalciferol (D_3) and 1,25-dihydoxycholecalciferol (25-OH-D_3) in diets being expressed as IU/kg and μg/kg, respectively. Fritts et al. (2004) fed broiler diets varying in vitamin D (125, 250, 500, 1000, 2000, and 4000 IU D_3/kg and 3.1, 6.3, 12.5, 25.0, 50.0, and 100.0 μg 25-OH-D_3/kg) and assessed cutaneous basophil hypersensitivity, macrophage function, and blood parameters. Immune response differences to vitamin D level or source did not occur, and it was concluded that the commercial premix range (2000 to 4000 IU) of vitamin D is sufficient to support normal immune function (Fritts et al., 2004). Although the former work did not assess adaptive immunity, Chou et al. (2009) fed 25-OH-D_3 to broilers and challenged them to *Salmonella* Typhimurium and noted improved titers to *S.* Typhimurium in birds fed 25-OH-D_3. In addition, numerous intestinal morphology criteria were assessed and 25-OH-D_3 improved intestinal development and function (Chou et al., 2009), and similar to the work of Fritts et al. (2004), no improvements were found in broiler performance. However, other work on broiler fed 69 μg 25-OH-D_3 /kg of feed has noted improved innate immunity (Gómez-Verduzco et al., 2013; Vazquez et al., 2018). Further, Rodriguez-Lecompte et al. (2016) fed D_3 and 25-OH-D3 to broilers with adequate or low calcium and phosphorus levels and measured numerous cytokines using Quantitative Reverse Transcription PCR analyses. It was demonstrated that both D_3 and 25-OH-D_3 improved immunity in Ross 308 broilers, especially when fed lower Ca and P levels, as noted by a shift from the Th1 to the Th2 immune response (Rodriguez-Lecompte et al., 2016). Although the crystalline form of cholecalciferol has a rich history as a poultry supplement feed additive, continued work on vitamin D source on immunity is warranted due to the complex metabolism of cholecalciferol to become 25-hydroxycholecalciferol in the liver and 1,25-dihydroxycholecalciferol in the kidney. Hence, highly available and bioactive forms of vitamin D may heighten overall immunity in birds based on absorption, retention, and biopotency.

8.3 Other minerals

Although Zn and Se have been discussed, it must be pointed out that much work has been conducted with multiple trace elements as dietary treatments. Thus, additional citations of Zn and Se are discussed in this section in research evaluating several trace elements. Echeverry et al. (2016) fed Ross broilers diets from 1 to 42 days of age with partially (50%) replaced trace inorganic minerals with organic trace minerals (i.e. manganese (Mn), Zn, iron (Fe), copper (Cu), iodine, and Se), in addition to a treatment with inorganic trace minerals and bacitracin methylene disalicylate. The performance of broilers was not improved with organic minerals, which resulted in low overall mineral diet levels. However, broilers fed organic trace minerals had improved intestinal health and antioxidant status (Echeverry et al., 2016). Similarly, Yang et al. (2011) found that variations in inorganic levels of Cu, Fe, Zn, and Mn did not affect broiler performance but positively impacted meat color and water-holding capacity.

The role of Mn in the avian immune response has been studied due to its support of the antioxidant immune system through Mn, providing a catalytic metal, in addition to Zn and Cu, for the activation of superoxide dismutase. Ghosh et al. (2016) fed unsexed broiler chicks diets containing Mn (0, 50, 75, and 100 mg/kg) and phytase (0 and 500 units/kg) from placement to 42 days of age. Phytase and Mn did not interact, but it was concluded that higher levels of Mn were favorable based on the enhancement of some immune parameters and reduced leg abnormalities and carcass fat (Ghosh et al., 2016). With specific reference to phytase, Khodambashi Emami et al. (2013) found that the concurrent supplementation of phytase and an organic acid can improve immunity in broilers when fed diets marginal in phosphorus. Diet Mn source (oxide and nanoparticle) was assessed in feed with three levels (10, 50, and 100 mg/kg) fed to growing turkeys (Jankowski et al., 2019). Lower Mn levels resulted in more favorable immune responses, and it was concluded that turkey premix Mn levels to support immunity warrant further attention (Jankowski et al., 2019). Burin Junior et al. (2019) fed Cobb 500 broilers a 2 × 2 factorial arrangement of Mn (inorganic and organic) and Arg (1.12 and 1.20 Arg/Lys), while adding a challenging factor of *S.* Enteritidis vaccine. Immunoglobulin IgM production to *S.* Enteritidis vaccine increased with increasing the Arg/Lys ratio with inorganic Mn, but a decreased with organic Mn, pointing to the importance of better understanding mineral source effects (Burin Junior et al., 2019).

Mineral source effects of chromium (Cr) have been fed to broilers to assess immunity (Bahrami et al., 2012; Lu et al., 2019). In order to evaluate Cr needs in broilers exposed to stress, Bahrami et al. (2012) fed 0, 0.8, or 1.2 mg Cr/kg of feed in two forms (Cl and Met) to birds housed in an environment with high (33 ± 3C) ambient temperature. Both Cr sources improved some immune endpoints, but organic Cr improved more immune function measurements (e.g. titers to Newcastle Disease and serum IgG) than inorganic Cr (Bahrami et al., 2012). Research from Lu et al. (2019) compared Cr chloride to Cr picolinate at four levels (0.4, 0.8, 1.6, and 3.2 mg/kg of diet) in a 2 × 4 factorial dietary design in broilers vaccinated to avian influenza virus from 1 to 42 days of age. Although both Cr sources had positive effects on broiler immunity, the effects differed by Cr source (Lu et al., 2019).

An understanding of mineral toxicity and resultant immune response effects in poultry is required. Liu et al. (2018) fed Hy-Line hens arsenic (Ar) as Ar_2O_3 (30 mg/kg) and Cu as $CuSO_4$ (300 mg/kg) to evaluate immunity and heat shock protein function. Dietary Ar and/or Cu impaired the hen's thymus function resulting in inflammatory responses and elevated heat shock protein levels (Liu et al., 2018).

8.4 Fats and oils

Feed fats and oils differ in terms of gastrointestinal absorption and available energy for poultry and. thus, immune system functionality in poultry. Moreover, fat and oil sources differ in terms of oxidation, which dictates differing antioxidant needs in the feed for poultry. For example, Mohiti-Asli and Ghanaatparast-Rashti (2018) assessed immunity in broilers fed soybean or canola oil and vitamins C and E and determined that vitamin E decreased lipid oxidation in the thigh and breast muscles of broilers, but birds fed canola oil did not need a dietary antioxidant.

Al-Khalaifah and Al-Nasser (2020) tested oil sources on innate immunity on Cobb 500 broilers and determined that oils rich in n-3 fatty acids improved immunity as measured by a cutaneous basophil hypersensitivity test. Cinnamon bark oil, clove bud oil, and ajwain seed oil were supplemented to diets fed to broilers in order to test the effectiveness of replacing bacitracin methylene disalicylate (Chowdhury et al., 2018). In addition to decreasing serum cholesterol, it was shown that cinnamon bark oil may be more advantageous than feed antibiotics as measured by intestinal health, antioxidant status, and adaptive and innate immunity (Chowdhury et al., 2018).

8.5 Wheat and rye

Soluble non-starch polysaccharides in the cereals wheat and rye are known to increase intestinal viscosity and decrease nutrient absorption when fed to broilers. However, it has also been shown that broiler diets high in wheat and rye negatively impact immunity by increasing gut T lymphocyte infusion, increasing goblet cell size, and increasing mucosal cell apoptosis (Teirlynck et al., 2009). Furthermore, van Krimpen et al. (2017) found similar trends in rye's ability to negatively affect immunity in broilers, even in cases where 5% to 10% of rye diet inclusion was without effect on broiler performance.

8.6 Insect meal

Insect meals are being assessed in poultry diets for the energy and digestible amino acid values. Lee et al. (2018) evaluated the effect of Black Soldier fly larva meal on the health of broilers exposed to *S.* Gallinarum. The insect meal (1% to 3% of diet) improved splenic function, spleen lymphocyte proliferation, and the frequency of CD4+T lymphocytes. Furthermore, birds fed the insect meal were able to clear the *S.* Gallinarum bacteria at a heightened rate (Lee et al., 2018). Another study was conducted to assess smaller amounts (0.2% to 0.3% of diet) of full-fat insect meal (*Tenebrio molitor* and *Zophobas morio* larvae) on immunity of Ross 308 broilers and resulted in a lower expression of IgM and IgY.

8.7 Algae meal

The increased production of microalgae has resulted in it being tested as a high protein feed ingredient. In addition, the immune stimulating properties of microalgae are promising. Kang et al. (2013) fed *Chlorella vulgaris*, a unicellular green microalga, to broilers in an experiment designed to assess its efficacy in comparison to feed-based antibiotics. Broilers fed the microalgae had improved plasma IgA and intestinal microflora as measured by an increased gut *Lactobacillus* population (Kang et al., 2013). Recent research by one of the author's collaborators, Professor S. Dridi, molecular biology laboratory at the University of Arkansas, has been concerned with bird health as affected by dietary CP reduction and algae (*Spirulina platensis*) inclusion. Reducing dietary CP by 30 g/kg increased circulating pro-inflammatory markers in broilers (Mullenix et al., 2021). However, the inclusion of 10% Spirulina Platensis, in lieu of soybean meal, lowered systemic inflammation and bacterial translation compared to broilers fed a similar protein level without algae (Mullenix et al., 2021).

8.8 Fermented and enzymatically processed oilseeds

Fermentation of oilseeds can be used to reduce the anti-nutritional factors and increase the nutritional value, as opposed to traditional oilseed processing. Wang et al. (2011) produced enzymolytic soybean meal via fermentation and protease hydrolysis to replace conventionally produced soybean meal in increments of 50, 100, or 150 g/kg for dietary treatments for Cobb 500 boilers. It was concluded that future research should be carried out to evaluate the health of broilers fed fermented soybean meal because the birds fed the fermented soya had increased whole blood relative levels of T lymphocytes (i.e. CD4 and CD8) and immunoglobulins (i.e. IgA, IgG, and IgM) (Wang et al., 2011).

Wang et al. (2017) created fermented cottonseed meal with *Bacillus subtilis* ST-141 and *Saccharomycetes* N5 to replace conventionally processed cottonseed meal or to partially replace conventionally proceed soybean meal in diets fed to Arbor Acre Plus broilers from hatch to 42 days of age. It was demonstrated that birds fed 89 g/kg fermented cottonseed meal had improved antioxidant capacity and adaptive immunity, in addition to improved *Lactobacilli* and total anaerobic bacteria amounts in cecal samples (Wang et al., 2017).

In a study to assess broiler immunity from diet replacement of soybean meal with peas, the treatments assessed consisted of raw, fermented, or enzymatically proceeded peas (Röhe et al., 2017). It was determined that all pea-based diets, which resulted in feed formulation levels of corn and soybean being reduced, impacted mucosal immunity of broilers (Röhe et al., 2017). Further work is needed to assess gut function of broilers fed pea-based diets to determine if mucosal responses are beneficial or harmful.

9 Conclusion

Indeed, there are numerous nutrients and ingredients that improve the immune system function of poultry. One gap that exists is correlating both innate and adaptive immune modulators to improved poultry performance in practice.

Field tests whereby nutritional modulators of immunity are tested in birds in pens added in a commercial facility would add practical validity to many laboratory/small scale research reports, which could further be heightened pending field report dissemination in the peer review press. Moreover, nutrition by immune interactions depends upon the level of stress exposure. On the one hand, most stresses are immunosuppressive. On the other hand, stresses can compromise nutrient assimilation and change nutrient requirements. As such, when birds are selected for improved growth rate and diet optimization takes place, some immune parameters should also be considered. The problem, however, remains: which parameters are predictive of future health and what degree of stress to impose.

10 Where to look for further information

Both authors have a keen interest in improving the immune responsiveness of poultry, and as such, have been involved in much research looking into poultry health. In poultry, health-related benefits as achieved via nutrition will continue to be of high importance. In addition to poultry production-based scientific journals highlighting poultry health, two books are listed below:

- Surai, P. F. (2018). Selenium in Poultry Nutrition and Health. Wageningen Academic Publishers, The Netherlands.
- Surai, P. F. (2020). Vitagenes in Avian Biology and Poultry Health. Wageningen Academic Publishers, The Netherlands.

11 References

Ahmed, I., Qaisrani, S. N., Azam, F., Pasha, T. N., Bibi, F., Naveed, S. and Murtaza, S. (2020). Interactive effects of threonine levels and protein source on growth performance and carcass traits, gut morphology, ileal digestibility of protein and amino acids, and immunity in broilers, *Poult. Sci.* 99(1):280–289.

Akbari Moghaddam Kakhki, R., Bakhshalinejad, R. and Zoidis, E. (2018). Interactive effects of α-tocopheryl acetate and zinc supplementation on the antioxidant and immune systems of broilers, *Br. Poult. Sci.* 59(6):679–688.

Al-Khalaifah, H. and Al-Nasser, A. (2020). Dietary supplementation with various fat oils affect phytohemagglutinin skin test in broiler chickens, *Front. Immunol.* 11:1735.

Andreini, C. and Bertini, I. (2012). A bioinformatics view of zinc enzymes, *J. Inorg. Biochem.* 111:150–156.

Bahrami, A., Moeini, M. M., Ghazi, S. H. and Targhibi, M. R. (2012). The effect of different levels of organic and inorganic chromium supplementation on immune function of broiler chicken under heat-stress conditions, *J. Appl. Poult. Res.* 21(2):209–215.

Bakhshalinejad, R., Akbari Moghaddam Kakhki, R. and Zoidis, E. (2018). Effects of different dietary sources and levels of selenium supplements on growth performance, antioxidant status and immune parameters in Ross 308 broiler chickens, *Br. Poult. Sci.* 59(1):81–91.

Baowei, W., Guoqing, H., Qiaoli, W. and Bin, Y. (2011). Effects of yeast selenium supplementation on the growth performance, meat quality, immunity, and antioxidant capacity of goose, *J. Anim. Physiol. Anim. Nutr. (Berl)* 95(4):440–448.

Bartlett, J. R. and Smith, M. O. (2003). Effects of different levels of zinc on the performance and immunocompetence of broilers under heat stress, *Poult. Sci.* 82(10):1580–1588.

Bédécarrats, G. Y. and Leeson, S. (2006). Dietary lutein influences immune response in laying hens, *J. Appl. Poult. Res.* 15(2):183–189.

Benzertiha, A., Kierończyk, B., Kołodziejski, P., Pruszyńska-Oszmałek, E., Rawski, M., Józefiak, D. and Józefiak, A. (2020). Tenebrio molitor and Zophobas morio full-fat meals as functional feed additives affect broiler chickens' growth performance and immune system traits, *Poult. Sci.* 99(1):196–206.

Bertolo, R. F. P., Chen, C. Z. L., Law, G., Pencharz, P. B. and Ball, R. O. (1998). Threonine requirement of neonatal piglets receiving total parenteral nutrition is considerably lower than that of piglets receiving an identical diet intragastrically, *J. Nutr.* 128(10):1752–1759.

Biswas, A., Mohan, J. and Sastry, K. V. (2006). Effect of higher levels of dietary selenium on production performance and immune responses in growing Japanese quail, *Br. Poult. Sci.* 47(4):511–515.

Blount, J. D., Metcalfe, N. B., Birkhead, T. R. and Surai, P. F. (2003). Carotenoid modulation of immune function and sexual attractiveness in zebra finches, *Science* 300(5616):125-127.

Boa-Amponsem, K., Price, S. E., Picard, M., Geraert, P. A. and Siegel, P. B. (2000). Vitamin E and immune responses of broiler pureline chickens, *Poult. Sci.* 79(4):466–470.

Bortoluzzi, C., Rochell, S. J. and Applegate, T. J. (2018). Threonine, arginine, and glutamine: influences on intestinal physiology, immunology, and microbiology in broilers, *Poult. Sci.* 97(3):937–945.

Bortoluzzi, C., Vieira, B. S., Lumpkins, B., Mathis, G. F., King, W. D., Graugnard, D., Dawson, K. A. and Applegate, T. J. (2019). Can dietary zinc diminish the impact of necrotic enteritis on growth performance of broiler chickens by modulating the intestinal immune-system and microbiota?, *Poult. Sci.* 98(8):3181–3193.

Bun, S. D., Guo, Y. M., Guo, F. C., Ji, F. J. and Cao, H. (2011). Influence of organic zinc supplementation on the antioxidant status and immune responses of broilers challenged with *Eimeria tenella*, *Poult. Sci.* 90(6):1220–1226.

Burin Junior, A. M., Fernandes, N. L. M., Snak, A., Fireman, A., Horn, D. and Fernandes, J. I. M. (2019). Arginine and manganese supplementation on the immune competence of broilers immune stimulated with vaccine against *Salmonella* enteritidis, *Poult. Sci.* 98(5):2160–2168.

Chen, Y. P., Cheng, Y. F., Li, X. H., Yang, W. L., Wen, C., Zhuang, S. and Zhou, Y. M. (2017). Effects of threonine supplementation on the growth performance, immunity, oxidative status, intestinal integrity, and barrier function of broilers at the early age, *Poult. Sci.* 96(2):405–413.

Chien, X. X., Zafra-Stone, S., Bagchi, M. and Bagchi, D. (2006). Bioavailability, antioxidant and immune-enhancing properties of zinc methionine, *BioFactors* 27(1–4):231–244.

Chou, S. H., Chung, T. K. and Yu, B. (2009). Effects of supplemental 25-hydroxycholecalciferol on growth performance, small intestinal morphology, and immune response of broiler chickens, *Poult. Sci.* 88(11):2333-2341.

Chowdhury, S., Mandal, G. P., Patra, A. K., Kumar, P., Samanta, I., Pradhan, S. and Samanta, A. K. (2018). Different essential oils in diets of broiler chickens: 2. Gut microbes and morphology, immune response, and some blood profile and antioxidant enzymes, *Anim. Feed Sci. Technol.* 236:39–47.

Coleman, J. E. (1992). Zinc proteins: enzymes, storage proteins, transcription factors, and replication proteins, *Annu. Rev. Biochem.* 61:897–946.

D'Amato, J. L. and Humphrey, B. D. (2010). Dietary arginine levels alter markers of arginine utilization in peripheral blood mononuclear cells and thymocytes in young broiler chicks, *Poult. Sci.* 89(5):938–947.

Dalia, A. M., Loh, T. C., Sazili, A. Q., Jahromi, M. F. and Samsudin, A. A. (2018). Effects of vitamin E, inorganic selenium, bacterial organic selenium, and their combinations on immunity response in broiler chickens, *BMC Vet. Res.* 14(1):249.

Debnath, B. C., Biswas, P. and Roy, B. (2019). The effects of supplemental threonine on performance, carcass characteristics, immune response and gut health of broilers in subtropics during pre-starter and starter period, *J. Anim. Physiol. Anim. Nutr. (Berl)* 103(1):29–40.

Downs, K. M., Hess, J. B., Macklin, K. S. and Norton, R. A. (2000). Dietary zinc complexes and vitamin E for reducing cellulitis incidence in broilers, *J. Appl. Poult. Res.* 9(3):319–323.

Echeverry, H., Yitbarek, A., Munyaka, P., Alizadeh, M., Cleaver, A., Camelo-Jaimes, G., Wang, P. O. K. and Rodriguez-Lecompte, J. C. (2016). Organic trace mineral supplementation enhances local and systemic innate immune responses and modulates oxidative stress in broiler chickens, *Poult. Sci.* 95(3):518–527.

Feng, J., Ma, W. Q., Niu, H. H., Wu, X. M., Wang, Y. and Feng, J. (2010). Effects of zinc glycine chelate on growth, hematological, and immunological characteristics in broilers, *Biol. Trace Elem. Res.* 133(2):203–211.

Fritts, C. A., Erf, G. F., Bersi, T. K. and Waldroup, P. W. (2004). Effect of source and level of vitamin D on immune function in growing broilers, *J. Appl. Poult. Res.* 13(2):263–273.

Gao, Y. Y., Xie, Q. M., Jin, L., Sun, B. L., Ji, J., Chen, F., Ma, J. Y. and Bi, Y. Z. (2012). Supplementation of xanthophylls decreased proinflammatory and increased anti-inflammatory cytokines in hens and chicks, *Br. J. Nutr.* 108(10):1746–1755.

Ghosh, A., Mandal, G. P., Roy, A. and Patra, A. K. (2016). Effects of supplementation of manganese with or without phytase on growth performance, carcass traits, muscle and tibia composition, and immunity in broiler chickens, *Livest. Sci.* 191:80–85.

Gómez-Verduzco, G., Morales-López, R. and Avila-Gozàlez, E. (2013). Use of 25-hydroxycholecalciferol in diets of broiler chickens: effects on Growth Performance, Immunity and Bone Calcification, *J. Poult. Sci.* 50(1):60–64.

Habibian, M., Ghazi, S., Moeini, M. M. and Abdolmohammadi, A. (2014). Effects of dietary selenium and vitamin E on immune response and biological blood parameters of broilers reared under thermoneutral or heat stress conditions, *Int. J. Biometeorol.* 58(5):741–752.

Hidayat, C., Jayanegara, A. and Wina, E. (2020). Effect of zinc on the immune response and production performance of broilers: a meta-analysis, *Asian-Australas J. Anim. Sci.* 33(3):465–479.

Jahanian, R. (2009). Immunological responses as affected by dietary protein and arginine concentrations in starting broiler chicks, *Poult. Sci.* 88:1818–1824.

Jankowski, J., Ognik, K., Stępniowska, A., Zduńczyk, Z. and Kozłowski, K. (2019). The effect of the source and dose of manganese on the performance, digestibility and distribution of selected minerals, redox, and immune status of turkeys, *Poult. Sci.* 98(3):1379–1389.

Jarosz, Ł., Marek, A., Grądzki, Z., Kwiecien, M., Zylinska, B. and Kaczmarek, B. (2017a). Effect of feed supplementation with zinc glycine chelate and zinc sulfate on cytokine and immunoglobulin gene expression profiles in chicken intestinal tissue, *Poult. Sci.* 96(12):4224–4235.

Jarosz, Ł., Marek, A., Grądzki, Z., Kwiecień, M. and Kalinowski, M. (2017b). The effect of feed supplementation with zinc chelate and zinc sulphate on selected humoral and cell-mediated immune parameters and cytokine concentration in broiler chickens, *Res. Vet. Sci.* 112:59–65.

Jarosz, Ł., Marek, A., Grądzki, Z., Laskowska, E. and Kwiecien, M. (2019). Effect of zinc sulfate and zinc glycine chelate on concentrations of acute phase proteins in chicken serum and liver tissue, *Biol. Trace El. Res.*, 187:258–272.

Ji, S., Qi, X., Ma, S., Liu, X., Liu, S. and Min, Y. (2019a). A deficient or an excess of dietary threonine level affects intestinal mucosal integrity and barrier function in broiler chickens, *J. Anim. Physiol. Anim. Nutr. (Berl)* 103(6):1792–1799.

Ji, S., Qi, X., Ma, S., Liu, X. and Min, Y. (2019b). Effects of dietary threonine levels on intestinal immunity and antioxidant capacity based on cecal metabolites and transcription sequencing of broiler, *Animals (Basel)* 9(10):739; doi:10.3390/ani9100739.

Johnson-Dahl, M. L., Zuidhof, M. J. and Korver, D. R. (2017). The effect of maternal canthaxanthin supplementation and hen age on breeder performance, early chick traits, and indices of innate immune function, *Poult. Sci.* 96(3):634–646.

Kang, H. K., Salim, H. M., Akter, N., Kim, D. W., Kim, J. H., Bang, H. T., Kim, M. J., Na, J. C., Hwangbo, J., Choi, H. C. and Suh, O. S. (2013). Effect of various forms of dietary Chlorella supplementation on growth performance, immune characteristics, and intestinal microflora population of broiler chickens, *J. Appl. Poult. Res.* 22(1):100–108.

Khodambashi Emami, N., Zafari Naeini, S. and Ruiz-Feria, C. A. (2013). Growth performance, digestibility, immune response and intestinal morphology of male broilers fed phosphorus deficient diets supplemented with microbial phytase and organic acids, *Livest. Sci.* 157(2–3):506–513.

Kidd, M. T. (2003). A treatise on chicken dam nutrition that impacts progeny, *Worlds Poult. Sci. J.* 59(4):475–494.

Kidd, M. T. (2004). Nutritional modulation of immune function in broilers, *Poult. Sci.* 83(4):650–657.

Kidd, M. T., Anthony, N. B. and Lee, S. R. (1992). Progeny performance when dams and chicks are fed supplemental zinc, *Poult. Sci.* 71(7):1201–1206.

Kidd, M. T., Peebles, E. D., Whitmarsh, S. K., Yeatman, J. B. and Wideman Jr., R. F. (2001). Growth and immunity of broiler chicks as affected by dietary arginine, *Poult. Sci.* 80(11):1535–1542.

Klasing, K. C. (2007). Nutrition and the immune system, *Br. Poult. Sci.* 48(5):525–537.

Kogut, M. H. (2009). Impact of nutrition on the innate immune response to infection in poultry, *J. Appl. Poult. Res.* 18(1):111–124.

Konieczka, P., Barszcz, M., Chmielewska, N., Cieślak, M., Szlis, M. and Smulikowska, S. (2017). Interactive effects of dietary lipids and vitamin E level on performance, blood

eicosanoids, and response to mitogen stimulation in broiler chickens of different ages, *Poult. Sci.* 96(2):359–369.
Konjufca, V. K., Bottje, W. G., Bersi, T. K. and Erf, G. F. (2004). Influence of dietary vitamin E on phagocytic functions of macrophages in broilers, *Poult. Sci.* 83(9):1530–1534.
Korver, D. R. (2012). Implications of changing immune function through nutrition in Poultry, *Anim. Feed Sci. Technol.* 173(1-2):54–64.
Laika, M. and Jahanian, R. (2015). Dietary supplementation of organic selenium could improve performance, antibody response, and yolk oxidative stability in laying hens fed on diets containing oxidized fat, *Biol. Trace Elem. Res.* 165(2):195–205.
Lee, J., Kim, Y. M., Park, Y. K., Yang, Y. C., Jung, B. G. and Lee, B. J. (2018). Black soldier fly (Hermetia illucens) larvae enhances immune activities and increases survivability of broiler chicks against experimental infection of Salmonella gallinarum, *J. Vet. Med. Sci.* 80(5):736–740.
Leshchinsky, T. V. and Klasing, K. C. (2001). Relationship between the level of dietary vitamin E and the immune response of broiler chickens, *Poult. Sci.* 80(11):1590–1599.
Liao, X., Lu, L., Li, S., Liu, S., Zhang, L., Wang, G., Li, A. and Luo, X. (2012). Effects of selenium source and level on growth performance, tissue selenium concentrations, antioxidation, and immune functions of heat-stressed broilers, *Biol. Trace Elem. Res.* 150(1-3):158–165.
Liu, J., Zhao, H., Wang, Y., Shao, Y., Zhang, L. and Xing, M. (2018). Impacts of simultaneous exposure to arsenic (III) and copper (II) on inflammatory response, immune homeostasis, and heat shock response in chicken thymus, *Int. Immunopharmacol.* 64:60–68.
Liu, Y. J., Zhao, L. H., Mosenthin, R., Zhang, J. Y., Ji, C. and Ma, Q. G. (2019). Protective Effect of vitamin E on laying performance, antioxidant capacity, and immunity in laying hens challenged with Salmonella enteritidis, *Poult. Sci.* 98(11):5847–5854.
Lu, L., Zhao, L. L., Dong, S. Y., Liao, X. D., Dong, X. Y., Zhang, L. Y. and Luo, X. G. (2019). Dietary supplementation of organic or inorganic chromium modulates the immune responses of broilers vaccinated with Avian influenza virus vaccine, *Animal* 13(5):983–991.
Manangi, M. K., Vazquez-Añon, M., Richards, J. D., Carter, S., Buresh, R. E. and Christensen, K. D. (2012). Impact of feeding lower levels of chelated trace minerals versus industry levels of inorganic trace minerals on broiler performance, yield, footpad health, and litter mineral concentration, *J. Appl. Poult. Res.* 21(4):881–890.
Min, Y. N., Niu, Z. Y., Sun, T. T., Wang, Z. P., Jiao, P. X., Zi, B. B., Chen, P. P., Tian, D. L. and Liu, F. Z. (2018). Vitamin E and vitamin C supplementation improves antioxidant status and immune function in oxidative-stressed breeder roosters by up-regulating expression of GSH-Px gene, *Poult. Sci.* 97(4):1238–1244.
Mohiti-Asli, M. and Ghanaatparast-Rashti, M. (2018). Feeding antioxidant vitamin and vegetable oils to broilers: vitamin E reduced negative effect of soybean oil on immune response and meat lipid oxidation, *Anim. Prod. Sci.* 58(10):1829–1836.
Moller, A. P., Biard, C., Blount, J., Houston, D. C., Ninni, P., Saino, N. and Surai, P. F. (2000). Carotenoid dependent signals: indicators of foraging efficiency, immunocompetence or detoxification ability?, *Poult. Avian Biol. Rev.* 11:137–159.
Moraes, M. L., Ribeiro, A. M., Santin, E. and Klasing, K. C. (2016). Effects of conjugated linoleic acid and lutein on the growth performance and immune response of broiler chickens, *Poult. Sci.* 95(2):237–246.

Muir, W. I., Husband, A. J. and Bryden, W. L. (2002). Dietary supplementation with vitamin E modulates avian intestinal immunity, *Br. J. Nutr.* 87(6):579–585.

Mullenix, G. J., Greene, E. S., Emami, N. K., Tellez-Isaias, G., Bottje, W. G., Erf, G. F., Kidd, M. T. and Dridi, S. (2021). Spirulina platensis inclusion reverses circulating pro-inflammatory (chemo)cytokine profiles in broilers fed low-protein diets, *Front. Vet. Sci.* 8:640968.

Munir, K., Muneer, M. A., Masaoud, E., Tiwari, A., Mahmud, A., Chaudhry, R. M. and Rashid, A. (2009). Dietary arginine stimulates humoral and cell-mediated immunity in chickens vaccinated and challenged against hydropericardium syndrome virus, *Poult. Sci.* 88(8):1629–1638.

Nelson, J. R., Sobotik, E. B., Rebollo, M. A. and Archer, G. S. (2020). Supplementation of a water-soluble zinc-AA complex to reduce stress in broilers, pullets, and layers1, *J. Anim. Sci.* 98(1); doi:10.1093/jas/skz361.

Niu, Z., Liu, F., Yan, Q. and Li, L. (2009a). Effects of different levels of selenium on growth performance and immunocompetence of broilers under heat stress, *Arch. Anim. Nutr.* 63(1):56-65.

Niu, Z. Y., Liu, F. Z., Yan, Q. L. and Li, W. C. (2009b). Effects of different levels of vitamin E on growth performance and immune responses of broilers under heat stress, *Poult. Sci.* 88(10):2101–2107.

NRC (1994). *Nutrient Requirements of Poultry* (9th edn.), The National Academies Press, Washington, DC.

Olson, J. B., Ward, N. E. and Koutsos, E. A. (2008). Lycopene incorporation into egg yolk and effects on laying hen immune function, *Poult. Sci.* 87(12):2573–2580.

Ong, A. S. H. and Tee, E. S. (1992). [14] Natural sources of carotenoids from plants and oils. In: *Methods in Enzymology 'Carotenoids: Part A. Chemistry, Separation, Quantitation and Antioxidation'*, Packer, L. (Ed.), Academic Press, San Diego, CA. Vol. 213, pp.142–167.

Panda, A. K., Ramarao, S. V., Raju, M. V. and Chatterjee, R. N. (2008). Effect of dietary supplementation with vitamins E and C on production performance, immune responses and antioxidant status of White Leghorn layers under tropical summer conditions, *Br. Poult. Sci.* 49(5):592–599.

Pitargue, F. M., Kim, J. H., Goo, D., Delos Reyes, J. B. and Kil, D. Y. (2019). Effect of vitamin E sources and inclusion levels in diets on growth performance, meat quality, alpha-tocopherol retention, and intestinal inflammatory cytokine expression in broiler chickens, *Poult. Sci.* 98(10):4584–4594.

Prasad, A. S., Miale, A., Jr., Farid, Z., Sandstead, H. H. and Schulert, A. R. (1963). Zinc metabolism in patients with the syndrome of iron deficiency anemia, hepatosplenomegaly, dwarfism, and hypognadism, *J. Lab. Clin. Med.* 61:537–549.

Rajput, N., Naeem, M., Ali, S., Rui, Y. and Tian, W. (2012). Effect of dietary supplementation of marigold pigment on immunity, skin and meat color, and growth performance of broiler chickens, *Rev. Bras. Cienc Avic* 14(4):291–295.

Rajput, N., Naeem, M., Ali, S., Zhang, J. F., Zhang, L. and Wang, T. (2013). The effect of dietary supplementation with the natural carotenoids curcumin and lutein on broiler pigmentation and immunity, *Poult. Sci.* 92(5):1177–1185.

Rochell, S. J., Helmbrecht, A., Parsons, C. M. and Dilger, R. N. (2017). Interactive effects of dietary arginine and Eimeria Acervulina infection on broiler growth performance and metabolism, *Poult. Sci.* 96(3):659–666.

Rodriguez-Lecompte, J. C., Yitbarek, A., Cuperus, T., Echeverry, H. and van Dijk, A. (2016). The immunomodulatory effect of vitamin D in chickens is dose-dependent and influenced by calcium and phosphorus levels, *Poult. Sci.* 95(11):2547–2556.

Röhe, I., Göbel, T. W., Goodarzi Boroojeni, F. and Zentek, J. (2017). Effect of feeding soybean meal and differently processed peas on the gut mucosal immune system of broilers, Poult. Sci. 96:2064–2073.

Rostami, H., Seidavi, A., Dadashbeiki, M., Asadpour, Y., Simões, J., Shah, A. A., Laudadio, V., Losacco, C., Perillo, A. and Tufarelli, V. (2018). Supplementing dietary rosemary (Rosmarinus officinalis L.) powder and vitamin E in broiler chickens: evaluation of humoral immune response, lymphoid organs, and blood proteins, *Environ. Sci. Pollut. Res. Int.* 25(9):8836–8842.

Sajadifar, S., Miranzadeh, H. and Moazeni, M. (2013). Effect of zinc on humoral and cell-mediated immunity of broilers vaccinated against coccidiosis, *Iran, J. Parasitol.* 8(3):474–480.

Saunders-Blades, J. L. and Korver, D. R. (2015). Effect of hen age and maternal vitamin D source on performance, hatchability, bone mineral density, and progeny in vitro early innate immune function, *Poult. Sci.* 94(6):1233–1246.

Scott, M. L., Nesheim, M. C. and Young, R. J. (1982). *Nutrition of the Chicken*, 3rd edn. Scott, M. L., and Associates (Eds), Ithaca, NY.

Selvaraj, R. K., Shanmugasundaram, R. and Klasing, K. C. (2010). Effects of dietary lutein and PUFA on PPAR and RXR isomer expression in chickens during an inflammatory response, *Comp. Biochem. Physiol. A Mol. Integr. Physiol.* 157(3):198–203.

Shabani, R., Fakhraei, J., Yarahmadi, H. M. and Seidavi, A. (2019). Effect of different sources of selenium on performance and characteristics of immune system of broiler chickens, *R. Bras. Zootec.* 48:e20180256.

Shanmugasundaram, R. and Selvaraj, R. K. (2011). Lutein supplementation alters inflammatory cytokine production and antioxidant status in F-line turkeys, *Poult. Sci.* 90(5):971–976.

Shojadoost, B., Kulkarni, R. R., Yitbarek, A., Laursen, A., Taha-Abdelaziz, K., Negash Alkie, T., Barjesteh, N., Quinteiro-Filho, W. M., Smith, T. K. and Sharif, S. (2019). Dietary selenium supplementation enhances antiviral immunity in chickens challenged with low pathogenic avian influenza virus subtype H9N2', *Vet. Immunol. Immunopathol.* 207:62–68.

Shojadoost, B., Taha-Abdelaziz, K., Alkie, T. N., Bekele-Yitbarek, A., Barjesteh, N., Laursen, A., Smith, T. K., Shojadoost, J. and Sharif, S. (2020). Supplemental dietary selenium enhances immune responses conferred by a vaccine against low pathogenicity avian influenza virus, *Vet. Immunol. Immunopathol.* 227:110089; doi:10.1016/j.vetimm.2020.110089.

Sigolo, S., Zohrabi, Z., Gallo, A., Seidavi, A. and Prandini, A. (2017). Effect of a low crude protein diet supplemented with different levels of threonine on growth performance, carcass traits, blood parameters, and immune responses of growing broilers, *Poult. Sci.* 96(8):2751–2760.

Singh, H., Sodhi, S. and Kaur, R. (2006). Effects of dietary supplements of selenium, vitamin E or combinations of the two on antibody responses of broilers, *Br. Poult. Sci.* 47(6):714–719.

Star, L., van der Klis, J. D., Rapp, C. and Ward, T. L. (2012). Bioavailability of organic and inorganic zinc sources in male broilers, *Poult. Sci.* 91(12):3115–3120.

Subramaniyan, S. A., Kang, D., Siddiqui, S. H., Park, J., Tian, W., Park, B. and Shim, K. (2020). Effects of in ovo supplementation with nanonutrition (L-arginine conjugated with Ag NPs) on muscle growth, immune response and heat shock proteins at different chicken embryonic development stages, *Animals (Basel)* 10(4):564; doi:10.3390/ani10040564.

Surai, P. F. (2002). *Natural Antioxidants in Avian Nutrition and Reproduction*, Nottingham University Press, Nottingham, UK.

Surai, P. F. (2006). *Selenium in Nutrition and Health*, Nottingham University Press, Nottingham, UK.

Surai, P. F. (2012a). The antioxidant properties of canthaxanthin and its potential effects in the poultry eggs and on embryonic development of the chick. Part 1, *Worlds Poult. Sci. J.* 68(3):465–476.

Surai, P. F. (2012b). The antioxidant properties of canthaxanthin and its potential effects in the poultry eggs and on embryonic development of the chick. Part 2, *Worlds Poult. Sci. J.* 68(4):717–726.

Surai, P. F. (2018). *Selenium in Poultry Nutrition and Health*, Wageningen Academic Publishers, Wageningen, The Netherlands.

Surai, P. F. (2020). *Vitagene in Avian Biology and Poultry Health*, Wageningen Academic Publishers, Wageningen, The Netherlands.

Surai, P. F. and Fisinin, V. I. (2016a). Vitagenes in poultry production. Part 1. Technological and environmental stresses, *Worlds Poult. Sci. J.* 72(4):721–734.

Surai, P. F. and Fisinin, V. I. (2016b). Vitagenes in poultry production. Part 2. Nutritional and internal stresses, *Worlds Poult. Sci. J.* 72(4):761–772.

Surai, P. F. and Kochish, I. I. (2019). Nutritional modulation of the antioxidant capacities in poultry: the case of selenium, *Poult. Sci.* 98(10):4231–4239.

Surai, P. F. and Kochish, I. I. (2020). Carotenoids in aviculture. In: *Pigments from Microalgae Handbook*, Jacob-Lopes, E., Queiroz, M. I. and Zepka, L. Q. (Eds), Springer Nature, Switzerland. pp.515–540.

Surai, P. F. and Kochish, I. I. (2021). Food for thought: nano-selenium in poultry nutrition and health, *Anim. Health Res. Rev.* 21(2):2021.

Surai, P. F. and Sparks, N. H. C. (2001). Comparative evaluation of the effect of two maternal diets on fatty acids, vitamin E and carotenoids in the chicken embryo, *Br. Poult. Sci.* 42(2):252–259.

Surai, P. F., Bortolotti, G. R., Fidgett, A. L., Blount, J. D. and Speake, B. K. (2001). Effects of piscivory on the fatty acid profiles and antioxidants of avian yolk: studies on eggs of the gannet, skua, pelican and cormorant, *J. Zool. Lond.* 255(3):305–312

Surai, P. F., Fisinin, V. I. and Karadas, F. (2016). Antioxidant systems in chick embryo development. Part 1. Vitamin E, carotenoids and selenium, *Anim. Nutr.* 2(1):1–11.

Surai, P. F., Kochish, I. I., Fisinin, V. I. and Velichko, O. A. (2018). Selenium in poultry nutrition: from sodium selenite to organic Se sources, *J. Poult. Sci.* 55(2):79–93.

Surai, P. F., Kochish, I. I. and Kidd, M. T. (2021). Redox homeostasis in poultry: regulatory roles of NF-κB, *Antioxidants (Basel)* 10(2):186.

Surai, P. F., Kochish, I. I. and Velichko, O. A. (2017). Nano-Se assimilation and action in poultry and other monogastric animals: is gut microbiota an answer?, *Nanoscale Res. Lett.* 12(1):612.

Surai, P. F., Kochish, I. I., Romanov, M. N. and Griffin, D. K. (2019a). Nutritional modulation of the antioxidant capacities in poultry: the case of vitamin E, *Poult. Sci.* 98(9):4030–4041.

Surai, P. F., Kochish, I. I., Fisinin, V. I. and Kidd, M. T. (2019b). Antioxidant defence systems and oxidative stress in poultry biology: an update, *Antioxidants (Basel)* 8(7):235; doi:10.3390/antiox8070235.

Surai, P. F., Speake, B. K. and Sparks, N. H. C. (2001a). Carotenoids in avian nutrition and embryonic development. 1. Absorption, availability and levels in plasma and egg yolk, *J. Poult. Sci.* 38(1):1–27.

Surai, P. F., Speake, B. K. and Sparks, N. H. C. (2001b). Carotenoids in avian nutrition and embryonic development. 2. Antioxidant properties and discrimination in embryonic tissues, *J. Poult. Sci.* 38(2):117–145.

Surai, P. F., Noble, R. C. and Speake, B. K. (1996). Tissue-specific differences in antioxidant distribution and susceptibility to lipid peroxidation during development of the chick embryo, *Biochim. Biophys. Acta* 1304(1):1–10.

Takimoto, T., Takahashi, K. and Akiba, Y. (2007). Effect of dietary supplementation of astaxanthin by Phaffia rhodozyma on lipid peroxidation, drug metabolism and some immunological variables in male broiler chicks fed on diets with or without oxidised fat, *Br. Poult. Sci.* 48(1):90–97.

Tako, E., Ferket, P. R. and Uni, Z. (2004). Zinc-methionine enhances intestinal development and functionality in the late term embryos and chicks, *Poult. Sci.* 83(Suppl. 1):267.

Tan, J., Liu, S., Guo, Y., Applegate, T. J. and Eicher, S. D. (2014a). Dietary L-arginine supplementation attenuates lipopolysaccharide-induced inflammatory response in broiler chickens, *Br. J. Nutr.* 111(8):1394–1404.

Tan, J., Applegate, T. J., Liu, S., Guo, Y. and Eicher, S. D. (2014b). Supplemental dietary L-arginine attenuates intestinal mucosal disruption during a coccidial vaccine challenge in broiler chickens, *Br. J. Nutr.* 112(7):1098–1109.

Tan, J. Z., Guo, Y. M., Applegate, T. J., Du, E. C. and Zhao, X. (2015). Dietary L-arginine modulates immunosuppression in broilers inoculated with an intermediate strain of infectious bursa disease virus, *J. Sci. Food Agric.* 95(1):126–135.

Teirlynck, E., Bjerrum, L., Eeckhaut, V., Huygebaert, G., Pasmans, F., Haesebrouck, F., Dewulf, J., Ducatelle, R. and van Immerseel, F. (2009). The cereal type in feed influences gut wall morphology and intestinal immune cell infiltration in broiler chickens, *Br. J. Nutr.* 102(10):1453–1461.

Troche, C., Eicher, S. D. and Applegate, T. J. (2015). The influence of dietary zinc source and coccidial vaccine exposure on intracellular zinc homeostasis and immune status in broiler chickens, *Br. J. Nutr.* 114(2):202–212.

Tsai, H. J., Shang, H. F., Yeh, C. L. and Yeh, S. L. (2002). Effects of arginine supplementation on antioxidant enzyme activity and macrophage response in burned mice, *Burns* 28(3):258–263.

Tsai, Y. H., Mao, S. Y., Li, M. Z., Huang, J. T. and Lien, T. F. (2016). Effects of nanosize zinc oxide on zinc retention, eggshell quality, immune response and serum parameters of aged laying hens, *Anim. Feed Sci. Technol.* 213:99–107.

Vallee, B. L. and Auld, D. S. (1990). Zinc coordination, function, and structure of zinc enzymes and other proteins, *Biochemistry* 29(24):5647–5659.

van Krimpen, M. M., Torki, M. and Schokker, D. (2017). Effects of rye inclusion in grower diets on immune competence-related parameters and performance in broilers, *Poult. Sci.* 96(9):3324–3337.

Vazquez, J. R., Gómez, G. V., López, C. C., Cortés, A. C., Díaz, A. C., Fernández, S. R. T., Rosales, E. M. and Avila, A. G. (2018). Effects of 25-hydroxycholecalciferol with two

D3 vitamin levels on production and immunity parameters in broiler chickens, *J. Anim. Physiol. Anim. Nutr. (Berl)* 102(1):e493–e497.

Virden, W. S., Yeatman, J. B., Barber, S. J., Willeford, K. O., Ward, T. L., Fakler, T. M., Wideman, Jr., R. F. and Kidd, M. T. (2004). Immune system and cardiac functions of progeny chicks from dams fed diets differing in zinc and manganese level and source, *Poult. Sci.* 83(3):344–351.

Walk, C. L., Kuhn, I., Stein, H. H., Kidd, M. T. and Rodehutscord, M. (2016). *Phytate Destruction: Consequences for Precision Animal Nutrition*, Wageningen Academic Publishers, The Netherlands.

Wang, J. P., Liu, N., Song, M. Y., Qin, C. L. and Ma, C. S. (2011). Effect of enzymolytic soybean meal on growth performance, nutrient digestibility and immune function of growing broilers, *Anim. Feed Sci. Technol.* 169(3–4):224–229.

Wang, Y., Wang, H. and Zhan, X. (2016). Effects of different dl-selenomethionine and sodium selenite levels on growth performance, immune functions and serum thyroid hormones concentrations in broilers, *J. Anim. Physiol. Anim. Nutr. (Berl)* 100(3):431–439.

Wang, Y., Deng, Q., Song, D., Wang, W., Zhou, H., Wang, L. and Li, A. (2017). Effects of fermented cottonseed meal on growth performance, serum biochemical parameters, immune functions, antioxidative abilities, and cecal microflora in broilers, *Food Ag. Immunol.* 28(4):725–738.

Wils-Plotz, E. L. and Klasing, K. C. (2017). Effects of immunomodulatory nutrients on growth performance and immune-related gene expression in layer chicks challenged with lipopolysaccharide, *Poult. Sci.* 96(3):548–555.

Wils-Plotz, E. L., Jenkins, M. C. and Dilger, R. N. (2013). Modulation of the intestinal environment, innate immune response, and barrier function by dietary threonine and purified fiber during a coccidiosis challenge in broiler chicks, *Poult. Sci.* 92(3):735–745.

Xu, S. Z., Lee, S. H., Lillehoj, H. S. and Bravo, D. (2015). Dietary sodium selenite affects host intestinal and systemic immune response and disease susceptibility to necrotic enteritis in commercial broilers, *Br. Poult. Sci.* 56(1):103–112.

Yabuzaki, J. (2017). Carotenoids database: structures, chemical fingerprints and distribution among organisms, *Database (Oxford)* 2017(1); doi:10.1093/database/bax004.

Yang, X. J., Sun, X. X., Li, C. Y., Wu, X. H. and Yao, J. H. (2011). Effects of copper, iron, zinc, and manganese supplementation in a corn and soybean meal diet on the growth performance, meat quality, and immune responses of broiler chickens, *J. Appl. Poult. Res.* 20(3):263–271.

Zarrin-Kavyani, Sh., Khatibjoo, A., Fattahnia, F. and Taherpour, K. (2018). Effect of threonine and potassium carbonate supplementation on performance, immune response and bone parameters of broiler chickens, *J. Appl. Anim. Res.* 46(1):1329–1335.

Zhan, T., Han, Y., Tang, C., Zhao, Q., Sun, D., Li, Y., Jia, X., Zhou, L. and Zhang, J. (2020). Metabolism and biological activity of α-tocopherol derived from vitamin E-enriched transgenic maize in broilers, *J. Sci. Food Agric.* 100(11):4319–4328.

Zhang, B., Lv, Z., Li, H., Guo, S., Liu, D. and Guo, Y. (2017a). Dietary L-arginine inhibits intestinal Clostridium perfringens colonisation and attenuates intestinal mucosal injury in broiler chickens, *Br. J. Nutr.* 118(5):321–332.

Zhang, Q., Chen, X., Eicher, S. D., Ajuwon, K. M. and Applegate, T. J. (2017b). Effect of threonine on secretory immune system using a chicken intestinal *ex vivo* model with lipopolysaccharide challenge, *Poult. Sci.* 96(9):3043–3051.

Zhang, Q., Chen, X., Eicher, S. D., Ajuwon, K. M. and Applegate, T. J. (2017c). Effect of threonine deficiency on intestinal integrity and immune response to feed withdrawal combined with coccidial vaccine challenge in broiler chicks, *Br. J. Nutr.* 116:2330-2043.

Zhang, B., Lv, Z., Li, Z., Wang, W., Li, G. and Guo, Y. (2018). Dietary L-arginine supplementation alleviates the intestinal injury and modulates the gut microbiota in broiler chickens challenged by *Clostridium perfringens*, *Front. Microbiol.* 9:1716; doi:10.3389/fmicb.2018.01716.

Zhang, X., Zhong, X., Zhou, Y., Wang, G., Du, H. and Wang, T. (2010). Dietary RRR-α-tocopherol succinate attenuates lipopolysaccharide-induced inflammatory cytokines secretion in broiler chicks, *Br. J. Nutr.* 104(12):1796-1805.

Zhang, X. H., Zhong, X., Zhou, Y. M., Du, H. M. and Wang, T. (2009). Effect of RRR-alpha-tocopherol succinate on the growth and immunity in broilers, *Poult. Sci.* 88(5):959-966.

Zhao, G. P., Han, M. J., Zheng, M. Q., Zhao, J. P., Chen, J. L. and Wen, J. (2011). Effects of dietary vitamin E on immunological stress of layers and their offspring, *J. Anim. Physiol. Anim. Nutr. (Berl)* 95(3):343-350.

Zhu, Y. F., Li, S. Z., Sun, Q. Z. and Yang, X. J. (2019). Effect of in ovo feeding of vitamin C on antioxidation and immune function of broiler chickens, *Animal* 13(9):1927-1933.

Chapter 6

Developments in vaccines to protect poultry against diseases

Yongxiu Yao and Venugopal Nair, Viral Oncogenesis Group - Pirbright Institute, UK

1 Introduction

Poultry is one of the fastest growing sectors of livestock farming contributing immensely to global food security and alleviation of poverty. Poultry production is expected to grow by 24% in each of the next decades, and it is estimated that poultry will account for the largest proportion of livestock output. In order to meet the increased demand and cost-effectiveness, a large part of the poultry industry has transitioned into intensive production systems rearing large numbers of birds in closed confinement. While this has certainly helped to maximise efficiency and productivity, it has also significantly increased the risk and impact of many infectious diseases, as a direct result of the potential for rapid spread and emergence of new and mutant pathogens. Among the different infectious agents that threaten poultry health, viral infections have the most impact causing significant effects on poultry health, welfare and productivity (Brun, 2016). Thus, sustainable poultry production will only be feasible if losses due to infectious diseases, particularly viral diseases, are successfully controlled through improved biosecurity measures to reduce pathogen exposure, and through innovations in disease diagnosis, prevention and control through enhanced immune responses by vaccination.

Vaccination, one of the great success stories in human and veterinary medicine in the fight against infectious diseases, is widely used for protection

http://dx.doi.org/10.19103/AS.2022.0104.09

of poultry against multiple viral pathogens (Schijns et al., 2014; Stewart-Brown, 2015). Effective vaccination will significantly reduce the risk of birds developing clinical disease and of associated economic losses when exposed to the multitude of pathogens in the poultry house environment. Some of the vaccines are also valuable to protect the next generation through the maternally derived antibodies passed to the progeny through the eggs. While the effectiveness of the vaccine is very much dependent on the vaccine itself, factors such as the maintenance of the appropriate cold chain during storage and transport and correct application of the vaccines are also important.

2 Types of poultry vaccines

Vaccines used to protect against avian diseases can be broadly grouped into (a) inactivated (killed) vaccines, (b) live (attenuated) vaccines and (c) recombinant vaccines (Collet and Smith, 2020).

2.1 Inactivated vaccines

Inactivated vaccines, generated by inactivating the infectivity of pathogens, are safe and can trigger immune responses. However, immune responses induced by these vaccines are weaker, mainly restricted to humoral responses, needing previous priming to induce good protection. Moreover, inactivated vaccines are usually given as repeated injections into individual birds, in many instances with adjuvants, and hence not ideal for widespread applications in intensive production systems. Nevertheless, inactivated vaccines are useful to vaccinate layer and breeder birds prior to the laying period to boost immune responses for increasing the level of protection during the laying period and transferring maternal antibodies to the chicks through the eggs.

2.2 Live vaccines

Live vaccines, consisting of attenuated strains of the pathogen, are widely used by the poultry industry mainly because of its use in mass vaccination. Through their ability to replicate *in vivo*, live vaccines can induce both innate and adaptive immune responses mediated through B- and T-lymphocyte populations in the vaccinated host. Attenuation of live poultry vaccines is achieved usually by serial passage in tissue culture, eggs or embryo-derived tissues to a level that maintains the immune response of the pathogen, but reduces the ability to cause disease or immunosuppressive effects. As live vaccines, these have to be stored and transported in refrigerated conditions or in freeze-dried form. Additionally, some of the live cell-associated vaccines, such as those against Marek's disease (MD), have to be stored in liquid nitrogen to maintain cell viability that is important for vaccine efficacy. Shelf life of these vaccines varies depending on the type of vaccines, and it is very important for strict adherence to the expiration date and the procedures for vaccine preparation and administration. Immune response to live vaccines is influenced by the type of the vaccine strain, degree of attenuation, delivery route, site of replication, as well as the age and immune status of the birds.

2.3 Recombinant vaccines

While conventional live and killed vaccines have been largely successful in preventing losses from most poultry diseases, these may have limitations including poor and short duration of immunity, interference from maternal antibodies and the potential for reversion to virulence of some live vaccine strains. More challengingly, constant and rapid evolution of pathogens in many instances resulted in the breakdown of vaccine-induced immunity (Devaux, 2012). Hence, there is an urgent need for innovative vaccines that are faster to develop than the conventional attenuation or inactivation methods (Liang et al., 2016). Advances in molecular biology and the identification of major protective antigens of different pathogens have contributed to the development of new-generation vaccines, broadly termed recombinant vaccines. These vaccines consist of vectors that deliver the gene(s) coding for the protective subunit antigen of other infectious agents against which immunity is desired. Majority of the recombinant vaccines rolled out against important infectious diseases of poultry are based on viral vectors (Baron et al., 2018), although some vaccines based on bacteria as well as DNA vector-based vaccines have also been successful under experimental conditions. In addition, subunit vaccines comprising of antigens such as NetB, alpha-toxin and metallopeptidase; outer membrane protein F or superoxide dismutase B have been demonstrated to be effective against diseases caused by pathogens *Clostridium perfringens*, *Salmonella enteritidis* and *Campylobacter jejuni,* respectively. This chapter describes the major recombinant viral vector-based vaccines against poultry pathogens. This will also include the most frequently used methodologies in the generation of recombinant viral vectors and summarise the state-of-the-art technologies related to the development of viral-vectored vaccines to find innovative solutions for the continuing challenges in the control of poultry diseases.

3 Viral vectors used for recombinant vaccines

Advances in recombinant DNA technology have made it possible to design new, innovative, genetically engineered viral-vectored vaccines with improved safety profiles and greater protective efficacy. Many viruses have been used as recombinant vectors to deliver subunit immunogenic antigens to provide effective protective immunity against major avian pathogens. The number of vectored vaccines licensed for protecting poultry has increased significantly over time. While most of the initial efforts in the development of viral vector vaccines were focused on DNA viruses such as herpesviruses, poxviruses and adenoviruses (Yokoyama et al., 1997), advances in reverse genetics technologies have also enabled genetic engineering of RNA viruses such as Newcastle disease virus (NDV) as efficient viral vectors (Stobart and Moore, 2014). Today, a number of poultry vaccine viruses such as herpesvirus of turkey (HVT), fowl pox virus (FPV), adenovirus, infectious laryngotracheitis virus (ILTV), MD virus (MDV) and NDV have been engineered as vaccine vectors for combatting diseases caused by multiple pathogens of economic importance to the poultry industry. Some of the commercial recombinant viral-vectored vaccines currently used to protect poultry are listed in Table 1.

3.1 Herpesviruses

3.1.1 Herpesvirus of turkey

HVT (Meleagridis herpesvirus 1), a non-pathogenic alphaherpesvirus of the genus Mardivirus, was originally isolated from turkeys (Kawamura et al., 1969; Witter et al., 1970). HVT has a significant antigenic and genetic relationship with the highly pathogenic MDV that induces MD, characterised by rapid-onset T-cell lymphomas and lymphoid infiltration into the peripheral nerves of poultry (Calnek, 2001; Gimeno et al., 2016). Antigenic relatedness of HVT with MDV was exploited in vaccination tactics, and HVT vaccination has been extensively used in the last five decades to induce durable and protective immunity against MD (Gimeno et al., 2016). Despite the evolution of virulence of MDV strains and the introduction of new generations of MD vaccines to protect birds from increasingly virulent MDV strains, HVT-based vaccines are still being used widely (Bublot, 2004), particularly in combination with other strains to exploit the synergistic protective effects (Witter and Lee, 1984). HVT has also emerged as the preferred vector of choice for the expression of heterologous antigens against a number of avian diseases such as ND, avian influenza (AI), infectious bursal disease (IBD), ILT, lymphoid leukosis and coccidiosis (Bublot, 2004; Cronenberg et al., 1999; Heckert et al., 1996; Li et al., 2011; Reddy et al., 1996; Tsukamoto et al., 2002). In addition, the ability to distinguish between infected and vaccinated animals is a common advantage of all recombinant-vectored vaccines, and the attractive features of using HVT as vector include the ability to accommodate the insertion of several multiple foreign genes, its suitability and safety for both *in ovo* and subcutaneous hatchery administration, its effectiveness even in the presence of maternal antibodies and its capability for

Table 1 List of some of the commercially available recombinant viral-vectored vaccines

Generation	Viral vector	Commercial name	Virus strain	Expressed antigen	Route of administration	Manufacturer
First-generation vaccine	FPV	TROVAC®-NDV	G IV Texas strain	NDV HN, F	s/c	Boehringer Ingelheim
		Vectormune®FP ND	G I D26 strain	NDV HN	s/c, w/w	CEVA AH
Second-generation vaccine	HVT	Vectormune®HVT ND	G I D26 strain	NDV F	s/c, *in ovo*	CEVA AH
		Innovax® ND	G II Clone 30	NDV F	s/c, *in ovo*	Merck AH
		Newxxitek® HVT+ND	G IV Texas strain	NDV F	s/c, *in ovo*	Boehringer Ingelheim
		Procerta® HVT-ND	NDV strain	NDV F	s/c, *in ovo*	Zoetis Inc.
		Farmune HVT-NWG12	NDV strain	NDV F	s/c, *in ovo*	Farvet
		Vaxxitek® HVT+IBD	Faragher 52/70	IBDV VP2	s/c, *in ovo*	Boehringer Ingelheim
		Vectorvac HVT-IBDV	IBDV strain	IBDV VP2	s/c, *in ovo*	Farvet
Third-generation vaccine	HVT	Innovax® ND-ILT	G II Clone 30/ ILTV 83-2	NDV F, ILTV gI-gD	s/c, *in ovo*	Merck AH
		Innovax® ND-IBD	G II Clone 30/ ILTV 83-2	NDV F, IBDV VP2	s/c, *in ovo*	Merck AH
		Vaxxitek® HVT+IBD+ND	G VII 1.1 Faragher 52/70	NDV F, IBDV VP2	s/c, *in ovo*	Boehringer Ingelheim
		Ultifend®IBD ND	G I D26 strain STC strain	NDV F, IBDV VP2	s/c, *in ovo*	CEVA AH
		Farmune HVT-IBDV-LT	NDV + ILTV	NDV F, ILTV Ag	s/c, *in ovo*	Farvet
LTR insertion vaccine	MDV-1	Prevexxion-RN	Chimeric Md5 & CVI988	REV LTR	s/c, *in ovo*	Boehringer Ingelheim

s/c: subcutaneous. w/w: wing web.

providing life-long immunity (Tang et al., 2020). However, disadvantages such as interference between individual HVT vaccines, the need for liquid nitrogen storage and transportation and the lack of vaccine administration to individual birds during mass vaccination still remain as some of the limitations of HVT-based vaccines.

3.1.2 Marek's disease virus-1 and -2

MDV-1 (Gallid herpesvirus (GaHV)-2) and MDV-2 (GaHV-3) are the other members of the genus Mardivirus (which include the strains CVI988 (also called Rispens) and SB-1, respectively) used as highly effective vaccines against MD (Calnek et al., 1983; Witter, 2001). CVI988 is considered the most effective gold standard vaccine strain against MD, as it induces strongest protection against highly virulent MDV pathotypes (Witter, 2001). As a highly effective vaccine strain against diverse MDV pathotypes, it could also be an ideal vector for the construction of a recombinant vaccine against other avian diseases. CVI988 expressing NDV F protein has been shown to be effective against both MD and ND in commercial chickens with maternal antibodies (Sonoda et al., 2000). Antigens such as Infectious Bursal Disease Virus VP2 (IBDV-VP2) (Tsukamoto et al., 1999) and Infectious Bronchitis Virus S1 (IBV-S1) (Zhang et al., 2012) have also been expressed in CVI988. Some of the other MDV-1 strains have also been used as viral vectors. For example, MDV-1 vaccine strain 814 has been shown as an effective recombinant viral vector for expression of IBDV-VP2 (Li et al., 2016a,b, 2017) and avian leukosis virus subgroup J (ALV-J) envelope glycoprotein (Liu et al., 2016). In addition, vaccine strain 814 has also been used as a vector delivering clustered regularly interspaced short palindromic repeats (CRISPR)/Cas9 components for direct targeting of ALV-J (Liu et al., 2020) and reticuloendotheliosis virus (REV) (Li et al., 2020) infections *in vitro* and *in vivo*. Compared to HVT and MDV-1 vaccine strains, MDV-2 has not been extensively explored as a vector for the development of recombinant vaccines. However, recombinant SB-1 expressing IBDV-VP2 induced neutralising antibody responses at levels comparable to those induced by commercial HVT-based VAXXITEK HVT+IBD vaccine. Moreover, the birds were protected against clinical disease after challenge with the very virulent UK661 IBDV isolate, further demonstrating its value as a recombinant herpesviral vector for developing multivalent vaccines against avian diseases (Sadigh et al., 2018).

3.1.3 Infectious laryngotracheitis virus

ILTV (GaHV-1), a member of the genus Iltovirus, is the causative agent of ILT, a highly infectious upper respiratory disease of fowl associated with weight loss, drop in egg production, increased mortality and predisposition to secondary bacterial infections (Garcia and Spatz, 2020). Control of ILT is achieved through vaccination using live-attenuated vaccines, deployed individually as eye drops or mass application through coarse spray or drinking water for replication in the upper respiratory tract (Fuchs et al., 2007). ILTV has the advantages of accommodating heterologous genes and the potential to induce cellular and humoral arms of immunity. In recognition of these advantages, different gene-deletion versions of ILTV have been experimentally deployed as a vector platform for the development of recombinant vaccines against multiple avian viruses including NDV (Atasoy et al., 2019; Shao et al., 2018) and influenza H5 strains (Pavlova et al., 2009a,b).

3.2 Avian poxviruses

Avian poxviruses are important pathogens of poultry associated with cutaneous, diphtheritic or systemic infections in different avian species. These are grouped within the genus Avipoxvirus of the Family *Poxviridae*, with FPV as the prototypic member (Afonso et al., 2000). The use of live-attenuated FPV as effective vaccines in chickens was reported as early as the 1920s (Skinner et al., 2005). With the advent of recombinant DNA techniques in the 1980s, FPV was extensively used in the construction of recombinant vaccine vectors against many poultry diseases (Weli and Tryland, 2011). FPV vector has several advantages, including large capacity for foreign gene expression, ability to elicit efficient immune responses, strict host range, and no risk of spreading virus to the environment, and offers the ability to distinguish between infected and vaccinated animals, but has the disadvantage that the vaccine has to be administered to the birds individually (Skinner et al., 2005). FPV has been successfully used to express protective antigens from several avian viruses, including MDV (Omar et al., 1998; Vagnozzi et al., 2012), NDV (Boursnell et al., 1990; Taylor et al., 1996), Avian Influenza Virus (AIV) (Swayne et al., 2000) and ILTV (Chen et al., 2020; Davison et al., 2006; Johnson et al., 2010; Tong et al., 2001). Canary pox virus, another Avipoxvirus, is also used as recombinant viral vector (Weli and Tryland, 2011).

3.3 Adenovirus

Adenoviruses have most of the qualities that are required for a successful vaccine delivery tool (Ferreira et al., 2005; Fougeroux and Holst, 2017). Adenovirus vectors replicate rapidly and efficiently to high titres *in vitro*, are stable, maintain their infectivity well and are easy to prepare and purify. The genome can accommodate ~2 kb of foreign DNA without significant effects on stability or infectivity, and foreign genes are expressed at high levels under the control of an appropriate promoter. There are simple and well-established techniques for constructing recombinant adenoviruses (Altaras et al., 2005; Ross and Parks, 2009) and for growing these to high titre (Kallel and Kamen, 2015; Silva et al., 2014). Most importantly, they elicit strong humoral and cell-mediated antigen-specific responses (Ertl, 2016; Yang et al., 2003) and can be delivered both by injection and orally (Babiuk and Tikoo, 2000). Recombinant adenovirus vaccines have been developed and tested successfully in mammals, including vaccines against human immunodeficiency virus, rabies virus in man [20,21], and for the newly emerged coronavirus, severe acute respiratory syndrome-coronavirus 2 (Folegatti et al., 2020; Logunov et al., 2020; Zhu et al., 2020). Vectors based on either human adenovirus or fowl adenovirus have been proven effective against pathogenic viruses in poultry including IBV (Johnson et al., 2003; Zeshan et al., 2010), AIV (Boyd et al., 2013; Gao et al., 2006; Steitz et al., 2010; Toro et al., 2007), IBDV (Francois et al., 2004) and MDV (Baigent et al., 2016).

3.4 Newcastle disease virus

Virulent NDV is the causative agent of ND, a devastating disease with huge economic losses prevalent in poultry worldwide, particularly in Asia, the Middle East and Africa (Hines and Miller, 2012). For several decades, the global poultry industry has used live-attenuated or killed vaccines to control and prevent ND. Conventional ND vaccines are highly immunogenic and induce protective immunity in poultry. Importantly, live-attenuated ND vaccines can be administrated in drinking water, eye drops and sprays, making it possible for large-scale application in different production systems. NDV vaccines can be combined with other poultry vaccines and used as polyvalent vaccines to control economically important poultry diseases. The production of NDV vaccines is highly cost-effective since they grow to very high titers in embryonated specific pathogen-free chicken eggs and in cell culture. All these characteristics make NDV attractive viral vectors for the development of polyvalent vaccines against pathogens that devastate the poultry industry (Bello et al., 2020; Kim and Samal, 2019). NDV-vectored AIV vaccines which express the HA protein of AIV are suitable for mass vaccination of poultry on a commercial scale and have a proven safety profile (Ge et al., 2007; Park et al., 2006). To date, NDV-vectored AIV vaccines have been administered nationwide to poultry in China (Kortekaas et al., 2010) and Mexico (Suarez and Pantin-Jackwood, 2017). Other bivalent NDV-vectored vaccines, which have been developed to prevent diseases of economic importance to the poultry industry, include IBD (Huang et al., 2004), IB (Toro et al., 2014), ILT (Kanabagatte Basavarajappa et al., 2014; Zhao et al., 2014) and avian metapneumovirus (aMPV) diseases (Hu et al., 2011).

4 Techniques used for generating recombinant vaccines

Recombinant vaccines developed over the last several years can be broadly placed into different generations (Table 1). The first and second generations of vaccines comprised viral vectors expressing single insert antigens, while the third generation recombinant vaccines contained dual inserts capable of inducing simultaneous protection against two diseases, in addition to the vector-induced protection. More recently, a new generation of long terminal repeat LTR insertion chimeric viral vector vaccines have also been introduced. Generation of recombinant vaccines, regardless of the viral vector used, will require genetic manipulation of the viral genomes using one of the following approaches.

4.1 Homologous recombination and marker rescue

Homologous recombination, where homologous regions of the DNA undergo genetic recombination, has been widely used in different biological systems including genetic manipulation of viral genomes. In early research on herpesviruses, this approach was widely used for the generation of mutant viruses together with marker rescue, where a selectable marker was introduced into the viral genome to assist in the rescue and purification of the mutants (Bublot et al., 1999). While this approach was useful, it was very laborious and time-consuming, particularly in cell-associated avian herpesviruses such as HVT.

4.2 Recombination of overlapping cosmid/fosmid clones containing viral genomes

Cosmids are plasmids that contain bacteriophage packaging sequences to enable insertion of the cloned DNA into λ phage heads and have the ability to accommodate up to 45 kb of foreign DNA. Cosmid vectors have been widely used for a number of years for cloning and analysing genomic DNA. Fosmids are similar to cosmids but are based on the bacterial F-plasmid. Fosmids can hold DNA inserts of up to 40 kb in size. In the search for efficient and faster methods of generating recombinant viruses, cosmid clones became attractive, since it was demonstrated that series of cosmid clones containing overlapping regions to cover the full-length viral genome transfected into cells can result in successful rescue of pure infectious clones of viruses (Liao et al., 2021; van Zijl et al., 1988). Fosmid clones have also been used similarly for the generation of recombinant avian herpesvirus vaccines (Li et al., 2016a).

4.3 Recombineering of bacterial artificial chromosome clones of viral genomes

A bacterial artificial chromosome (BAC) is an engineered DNA molecule used to clone DNA sequences in bacterial cells (e.g. *Escherichia coli*). DNA sequences ranging from 100 kb to 300 kb can be inserted into BACs. The ability to clone and maintain the full-length genomes of large DNA viruses as BAC in bacteria (Kúdelová et al., 2013) was a significant advance that allowed rapid engineering of the viral genomes (Tischer et al., 2006). This approach has since been widely used not only for the manipulation of the viral genome to identify pathogenic and immunogenic determinants but also for the generation of recombinant vaccines in multiple avian viral vectors (Liao et al., 2021).

4.4 Clustered regularly interspaced short palindromic repeats-Cas9-based gene editing to develop recombinant viral vaccines

Recent innovations in targeted gene editing using CRISPR-Cas9, an ancient antiviral defence system in bacteria using an RNA-guided endonuclease (Tang et al., 2018), have opened up enormous opportunities for manipulation of viral genomes for identifying pathogenic determinants, as well as in finding solutions through development of innovative recombinant viral vaccines to protect poultry health. Compared with traditional technologies for modification of viral vectors, the novel CRISPR/Cas9 approach is more efficient, convenient and fast. Its high efficiency, specificity, versatility and low cost have provided great inspirations to the field of vaccinology to solve some of the bottlenecks in the development of recombinant viral vaccines.

5 Gene editing of herpesvirus vectors

Since the first report of CRISPR-Cas9-mediated modification on Herpes simplex virus and adenovirus vectors (Bi et al., 2014), a variety of CRISPR-Cas9 strategies have been used as basic tools for precise editing of other large DNA viruses that can be further optimised for the development of recombinant vaccines. Herpesvirus-based vectors have been widely used in the development of variety of vaccines and therapeutics due to their inherent advantages, including limited host range, large genome size capable of accommodating multiple inserts, stable attenuation, ability for inducing strong cell-mediated and humoral immune responses, stable and robust expression of foreign genes, as well as good safety records. For these reasons, a number of human and animal herpesviruses have been successfully developed as recombinant vectors (Baron et al., 2018). We have recently developed a simple and rapid CRISPR-Cas9-based strategy for constructing recombinant HVT vector delivering the VP2 gene of IBDV through a non-homologous end joining (NHEJ)-dependent repair pathway, the primary pathway for the repair of DNA double-strand breaks (DSBs) throughout the cell cycle and a Cre-LoxP system to eliminate the reporter protein (Tang et al., 2018, 2019). This strategy, used for insertion of genes encoding foreign antigens into the HVT genome, can lead to the rapid construction of recombinant vaccines for efficacy studies. Using this approach, expression cassettes of ILTV glycoproteins gD and gI and AIV H9-HA have been inserted sequentially into the distinct locations of the recombinant HVT-VP2 viral genome, generating a triple insert recombinant vaccine candidate (Tang et al., 2020). This approach offers the prospect of rapidly developing multivalent HVT-vectored vaccines capable of inducing simultaneous protection against multiple avian diseases, particularly to overcome the interference between recombinant HVTs expressing different antigens. Other herpesvirus vaccine vectors such as duck enteritis virus (Chang et al., 2018; Zou et al., 2017) and ILTV (Atasoy et al., 2019) have also been developed via NHEJ-CRISPR-Cas9 and Cre-Lox Systems. While NHEJ repair pathways have been used for most herpesvirus recombinant vaccine development using the CRISPR/Cas9 system, an error-free homology-directed repair approach has also been successfully utilised to generate HVT-AIV recombinant that could be identified by an easy and quick erythrocyte adsorption assay which speeds up the whole screening process significantly (Chang et al., 2019). In addition to the generation of a recombinant vector expressing the foreign antigens, recombinant MDV vector carrying all the CRISPR-Cas9 components using an attenuated MDV vaccine strain 814 for *in vivo* editing of REV and ALV-J has also been reported (Li et al., 2020; Liu et al., 2020). These studies demonstrate the potential for efficient delivery of Cas9 and multiple gRNAs into chickens, for targeted editing and reduction of REV and ALV-J viral load, paving the way for preventing clinical disease and REV-associated symptoms. Thus, CRISPR-Cas9-based gene editing has emerged as an effective, fast and simple tool for the generation of viral-vectored vaccine candidates.

5.1 Generation of recombinant herpesvirus of turkey-vectored vaccines

Among live replicating vaccine vectors used in poultry, HVT is the most widely used vector due to its suitability and safety for *in ovo* and subcutaneous administration at the hatchery, its effectiveness even in the presence of maternal antibodies, and its potential for induction of strong life-long immunity due to the long periods of viral persistence and replication in vaccinated hosts. Here we describe the detailed methodology for the generation of recombinant HVT-vectored vaccine using NHEJ-CRISPR-Cas9 combined with Cre-LoxP system for insertion of foreign gene (Tang et al., 2018). The detailed procedures of donor template construction and generation of recombinant HVT expressing IBDV VP2 are shown in Fig. 1.

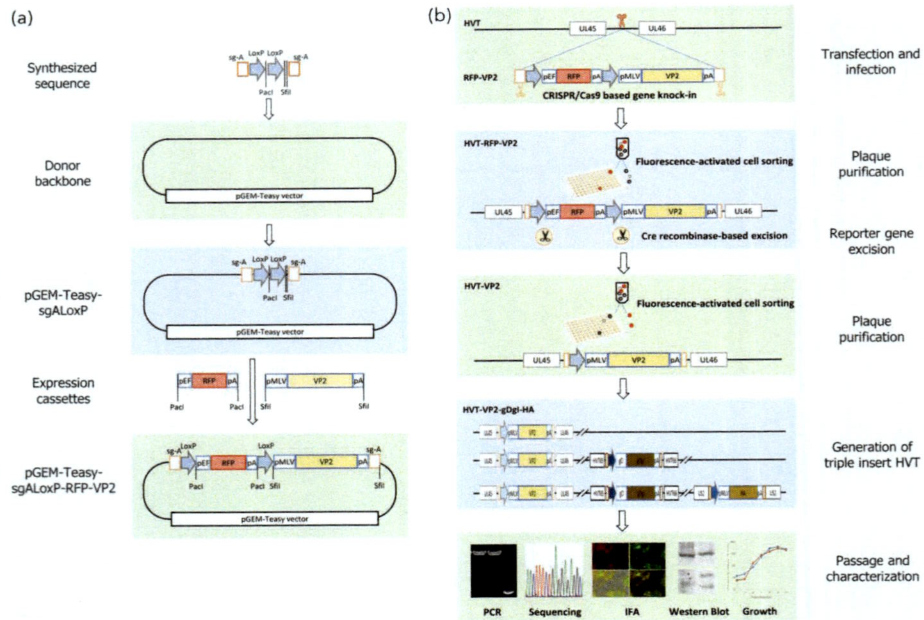

Figure 1 Strategy for the generation of a recombinant HVT-vectored vaccine. (a) Schematic representation of the cloning strategy for donor plasmid construction. The elements including two Cas9 target sites (sgA) at both ends for releasing insert, a PacI site flanked with two LoxP sequences for the RFP reporter cassette cloning and excision and two SfiI sites for the cloning of the VP2 expression cassette were first cloned into pGEM-Teasy vector (any cloning vector can be used to construct donor plasmid) which is followed by sequential cloning of RFP and VP2 expression cassettes. (b) Overview of a two-step gene knock-in strategy. The insert fragment of the RFP and the VP2 expression cassettes is released by Cas9/sgA cleavage and inserted into the HVT genome at UL45/46 loci via NHEJ-CRISPR/Cas9. The RFP-positive recombinant virus is then sorted and purified by single-cell fluorescence-activated cell sorting (FACS). Subsequently, the RFP reporter gene is excised by Cre recombinase and the recombinant virus is purified and characterized. The same procedure can be followed for the insertion of ILTV gDgI and AIV HA into the HVT-VP2 to generate triple insert recombinant HVT-VP2-gDgI-HA. The resulting recombinant can be characterised by PCR sequencing, immunofluorescence assay, Western blot and growth property as described in the main text.

5.1.1 Selection of the insertion sites

Selection of the target insertion site is one of the important steps for the construction of recombinant viruses. The primary consideration to choose a target region for insertion of a foreign gene is to minimise the effect on virus replication. Generally, the intergenic regions and the genes that are non-essential for virus replication are the common and preferred target sites for insertion of foreign genes. A number of insertion sites have been tested and shown to be suitable for insertion of foreign genes in HVT. For instance, the intergenic region between UL45 and UL46 has been used for insertion of the NDV F gene (Esaki et al., 2013a), ILTV glycoprotein B (gB) gene (Esaki et al., 2013b) and IBDV VP2 gene (Tang et al., 2018, 2019, 2020; Tsukamoto et al., 2002). Other intergenic regions UL3-UL4, UL22-23 and US10-SORF3 were also evaluated for the insertion of IBDV-VP2 (Andoh et al., 2017). HVT065-066 region was used for insertion of VP2 gene in commercial recombinant Vaxxitek HVT-IBD vaccine (Tang et al., 2018) and for the insertion of ILTV gD-gI genes into HVT (Tang et al., 2020). Gao et al. demonstrated that US2 insertion site for AIV HA gene provided more effective protection than recombinant HVT delivering the AIV HA gene at the US10 locus (Gao et al., 2011). HVT US2 locus, used for the insertion of BAC sequence in the recombinant HVT-BAC, also was a useful insertion site that conferred protection against virulent MDV infection (Baigent et al., 2006). Similarly, NDV F gene inserted in the HVT US10 locus was also capable of inducing dual protection against ND and MD (Morgan et al., 1992; Sakaguchi et al., 1998).

5.1.2 Construction of Cas9/gRNA expression plasmid and donor template

Generation of recombinant viruses using CRISPR/Cas9 editing and Cre/LoxP system includes multiple steps as previously described (Tang et al., 2019). Targeting gRNAs can be designed using online tools such as CHOPCHOP (https://chopchop.cbu.uib.no/), CRISPOR (http://crispor.tefor.net/), Cas-OFFinder (http://www.rgenome.net/cas-offinder/) and Benchling (https://www.benchling.com/crispr/). Two gRNAs are required, one for targeting the donor plasmid to release the insert and the other to target the viral genome at the desired insertion site to generate DSB. Cas9/gRNA expression vectors such as pX459-v2 (Addgene plasmid # 62988) which express the codon-optimised *Streptococcus pyogenes* Cas9 (Cas9) can be used for gRNA cloning. Detailed steps of donor plasmid construction are outlined in Fig. 1a. The IBDV VP2 expression cassette was placed next to a Red Fluorescent Protein (RFP) expression cassette flanked with LoxP sites (Darteil et al., 1995) to facilitate marker gene excision by Cre recombinase. The two expression cassettes are flanked by the sg-A target sites to generate insert for designed integration into the HVT UL45/46 loci. Concurrent cleavage of donor plasmid DNA and the targeted viral genome by Cas9 will introduce the RFP and VP2 cassettes into UL45/UL46 locus to generate HVT-VP2-RFP.

5.1.3 Generation of recombinant herpesvirus of turkey candidates

Generation of recombinant HVT virus involves multiple steps including (a) co-transfection of donor plasmid together with Cas9/gRNA vectors targeting both donor plasmid and gene junction of UL45/UL46 into chicken embryo fibroblasts (CEF), (b) infection of wild-type HVT, (c) Isolation of recombinant HVT-RFP-VP2 and (d) excision of RFP expression cassette with Cre recombinase. Detailed procedures for the generation and characterisation of the recombinant HVT are shown in Fig. 1B. Primary CEF transfected with the CRISPR components and infected with the virus was used for the generation of recombinant viruses. A fluorescent marker was used for picking plaques to obtain purified virus stocks. Marker gene can then be excised using Cre recombinase, and the successful excision can be confirmed by polymerase chain reaction (PCR). Plaque purified stocks of the recombinant virus candidates were then subjected to validation and characterisation.

5.2 Characterisation of recombinant herpesvirus of turkey-vectored vaccine

Steps for the characterisation of the recombinant HVT viruses include (a) PCR amplification of the inserted and flanking sequences to demonstrate the insertion and orientation of foreign sequence, (b) Sanger sequencing to confirm the insertion and its precise location in the vector genome, (c) examine the expression of the inserted foreign gene using immunological approaches, (d) evaluate the growth of the recombinant virus and (e) confirm the stability of the inserted sequence in the recombinant virus during repeated passages. Insertion of the foreign gene can be tested by PCR with a pair of primers located within the inserted sequence, while the orientation of the insert can be tested by junction PCR with one primer located in the insert sequence and the other one in the flanking region in the viral genome. Following the assessment of recombinant virus having the right sequence and expression of the inserted foreign gene, replication of the recombinant virus will have to be compared with that of the wild-type virus by time course growth curve analysis to make sure that the insertion has not compromised the vector virus replication. The final step in the characterisation of the recombinant virus is to check its stability by continuous passaging of the virus (at least for 15 passages) and assessing the genome integrity and antigen expression. Only the recombinant virus with stable integrated foreign genes needs to be further assessed for immunogenicity and protective efficacy.

6 Future trends in research

Poultry offers a major food source worldwide and continues to expand to meet the rising demands of the global population (Speedy, 2003). Infectious diseases caused by existing and emerging pathogens continue to threaten sustainable growth of the poultry industry (Josefsberg and Buckland, 2012). While vaccines still remain the primary tool for disease prevention in poultry production, continuing diversity and evolution of many viral pathogens, and the immunosuppressive effects of some of them, are threatening the efficacy of many conventional vaccines. Technological advances in the last few years have made it possible to make substantial improvements in genetic manipulation to generate innovative vaccines capable of inducing stronger and broader immune responses against conserved epitopes of multitude of avian pathogens and optimising poultry health, in addition to the ability for differentiating infected and vaccinated animals. For example, CRISPR-Cas9-based gene editing opens a whole new avenue for innovations in the field of vaccine development, and the ability for developing multivalent vaccines, including bespoke vaccines with antigens suited for different geographical regions. In addition to the vaccines, gene editing approaches also offer the prospects of inducing *de novo* genetic resistance to diseases to improve welfare (Khwatenge and Nahashon, 2021; Panda and McGrew, 2021), which is also helpful in the fight against diseases for improved flock health.

7 Where to look for further information

More details of poultry diseases published by the World Organisation of Animal Health (WOAH) can be found at: https://www.woah.org/en/what-we-do/animal-health-and-welfare/animal-diseases/.

Chapters in the recent textbooks related to poultry diseases, immunology and vaccines will give more information.

- Poultry Health – A Guide for Professionals: https://www.cabi.org/bookshop/book/9781789245059/.
- Diseases of Poultry: https://www.wiley.com/en-us/Diseases+of+Poultry%2C+2+Volume+Set%2C+14th+Edition-p-9781119371168.
- Avian Immunology: https://www.elsevier.com/books/avian-immunology/kaspers/978-0-12-818708-1.
- Avian Virology – Current Research and Future Trends: https://www.amazon.co.uk/Avian-Virology-Current-Research-Future/dp/1912530104.

8 References

Afonso, C. L., Tulman, E. R., Lu, Z., Zsak, L., Kutish, G. F. and Rock, D. L. 2000. The genome of fowlpox virus. *J. Virol.* 74(8), 3815–3831.

Altaras, N. E., Aunins, J. G., Evans, R. K., Kamen, A., Konz, J. O. and Wolf, J. J. 2005. Production and formulation of adenovirus vectors. *Adv. Biochem. Eng. Biotechnol.* 99, 193–260.

Andoh, K., Yamazaki, K., Honda, Y. and Honda, T. 2017. Turkey herpesvirus with an insertion in the UL3-4 region displays an appropriate balance between growth activity and antibody-eliciting capacity and is suitable for the establishment of a recombinant vaccine. *Arch. Virol.* 162(4), 931–941.

Atasoy, M. O., Rohaim, M. A. and Munir, M. 2019. Simultaneous deletion of virulence factors and insertion of antigens into the infectious laryngotracheitis virus using NHEJ-CRISPR/Cas9 and Cre-Lox system for construction of a stable vaccine vector. *Vaccines (Basel)* 7(4): 207.

Babiuk, L. A. and Tikoo, S. K. 2000. Adenoviruses as vectors for delivering vaccines to mucosal surfaces. *J. Biotechnol.* 83(1–2), 105–113.

Baigent, S. J., Petherbridge, L. J., Smith, L. P., Zhao, Y., Chesters, P. M. and Nair, V. K. 2006. Herpesvirus of turkey reconstituted from bacterial artificial chromosome clones induces protection against Marek's disease. *J. Gen. Virol.* 87(4), 769–776.

Baigent, S. J. J. M., Turner, A. V., Gilbert, S. C. and Nair, V. K. 2016. Replication defective adenovirus Serotype 5 expressing Marek's disease virus envelope glycoprotein as a potential Marek's disease vaccine in chicken. *J. Vet. Med. Res.* 3(2): 1049.

Baron, M. D., Iqbal, M. and Nair, V. 2018. Recent advances in viral vectors in veterinary vaccinology. *Curr. Opin. Virol.* 29, 1–7.

Bello, M. B., Yusoff, K., Ideris, A., Hair-Bejo, M., Jibril, A. H., Peeters, B. P. H. and Omar, A. R. 2020. Exploring the prospects of engineered Newcastle disease virus in modern vaccinology. *Viruses* 12(4): 451.

Bi, Y., Sun, L., Gao, D., Ding, C., Li, Z., Li, Y., Cun, W. and Li, Q. 2014. High-efficiency targeted editing of large viral genomes by RNA-guided nucleases. *PLoS Pathog.* 10(5), e1004090.

Boursnell, M. E., Green, P. F., Samson, A. C., Campbell, J. I., Deuter, A., Peters, R. W., Millar, N. S., Emmerson, P. T. and Binns, M. M. 1990. A recombinant fowlpox virus expressing the hemagglutinin-neuraminidase gene of Newcastle disease virus (NDV) protects chickens against challenge by NDV. *Virology* 178(1), 297–300.

Boyd, A. C., Ruiz-Hernandez, R., Peroval, M. Y., Carson, C., Balkissoon, D., Staines, K., Turner, A. V., Hill, A. V., Gilbert, S. C. and Butter, C. 2013. Towards a universal vaccine for avian influenza: protective efficacy of modified vaccinia virus Ankara and adenovirus vaccines expressing conserved influenza antigens in chickens challenged with low pathogenic avian influenza virus. *Vaccine* 31(4), 670–675.

Brun, A. 2016. Vaccines and vaccination for veterinary viral diseases: a general overview. *Methods Mol. Biol.* 1349, 1–24.

Bublot, M., Laplace, E. and Audonnet, J. C. 1999. Non-essential loci in the BamHI-I and -F fragments of the HVT FC126 genome. *Acta virol.* 43(2–3), 181–185.

Bublot, M. S. J. 2004. Vaccination against Marek's disease. In: Nair, T. F. D. V. (Ed.) *Marek's Disease – an Evolving Problem*. Academic Press, London, 168–185.

Calnek, B. W. 2001. Pathogenesis of Marek's disease virus infection. *Curr. Top. Microbiol. Immunol.* 255, 25–55.

Calnek, B. W., Schat, K. A., Peckham, M. C. and Fabricant, J. 1983. Field trials with a bivalent vaccine (HVT and SB-1) against Marek's disease. *Avian Dis.* 27(3), 844–849.

Chang, P., Ameen, F., Sealy, J. E., Sadeyen, J. R., Bhat, S., Li, Y. and Iqbal, M. 2019. Application of HDR-CRISPR/Cas9 and erythrocyte binding for rapid generation of recombinant turkey herpesvirus-vectored avian influenza virus vaccines. *Vaccines (Basel)* 7(4): 192.

Chang, P., Yao, Y., Tang, N., Sadeyen, J. R., Sealy, J., Clements, A., Bhat, S., Munir, M., Bryant, J. E. and Iqbal, M. 2018. The application of NHEJ-CRISPR/Cas9 and Cre-Lox system in the generation of bivalent duck enteritis virus vaccine against avian influenza virus. *Viruses* 10(2).

Chen, S., Xu, N., Ta, L., Li, S., Su, X., Xue, J., Du, Y., Qin, T. and Peng, D. 2020. Recombinant fowlpox virus expressing gB gene from predominantly epidemic infectious Larygnotracheitis virus strain demonstrates better immune protection in SPF chickens. *Vaccines (Basel)* 8(4).

Collet, S. R. and Smith, J. A. 2020. Principles of disease prevention, diagnosis and control. In: Swayne, D. E., Boulianne, M., Logue, C. M., McDougald, L. R., Nair, V. and Suarez, D. L. (Eds) *Diseases of Poultry*. Hoboken, NJ, John Wiley & Sons, 4–40.

Cronenberg, A. M., van Geffen, C. E., Dorrestein, J., Vermeulen, A. N. and Sondermeijer, P. J. 1999. Vaccination of broilers with HVT expressing an Eimeria Acervulina antigen improves performance after challenge with Eimeria. *Acta virol.* 43(2–3), 192–197.

Darteil, R., Bublot, M., Laplace, E., Bouquet, J. F., Audonnet, J. C. and Riviere, M. 1995. Herpesvirus of turkey recombinant viruses expressing infectious bursal disease virus (IBDV) VP2 immunogen induce protection against an IBDV virulent challenge in chickens. *Virology* 211(2), 481–490.

Davison, S., Gingerich, E. N., Casavant, S. and Eckroade, R. J. 2006. Evaluation of the efficacy of a live fowlpox-vectored infectious laryngotracheitis/avian encephalomyelitis vaccine against ILT viral challenge. *Avian Dis.* 50(1), 50–54.

Devaux, C. A. 2012. Emerging and re-emerging viruses: A global challenge illustrated by Chikungunya virus outbreaks. *World J. Virol.* 1(1), 11–22.

Ertl, H. C. 2016. Viral vectors as vaccine carriers. *Curr. Opin. Virol.* 21, 1–8.

Esaki, M., Godoy, A., Rosenberger, J. K., Rosenberger, S. C., Gardin, Y., Yasuda, A. and Dorsey, K. M. 2013a. Protection and antibody response caused by turkey herpesvirus vector Newcastle disease vaccine. *Avian Dis.* 57(4), 750–755.

Esaki, M., Noland, L., Eddins, T., Godoy, A., Saeki, S., Saitoh, S., Yasuda, A. and Dorsey, K. M. 2013b. Safety and efficacy of a turkey herpesvirus vector laryngotracheitis vaccine for chickens. *Avian Dis.* 57(2), 192–198.

Ferreira, T. B., Alves, P. M., Aunins, J. G. and Carrondo, M. J. 2005. Use of adenoviral vectors as veterinary vaccines. *Gene Ther.* 12 (Suppl. 1), S73–83.

Folegatti, P. M., Ewer, K. J., Aley, P. K., Angus, B., Becker, S., Belij-Rammerstorfer, S., Bellamy, D., Bibi, S., Bittaye, M., Clutterbuck, E. A., Dold, C., Faust, S. N., Finn, A., Flaxman, A. L., Hallis, B., Heath, P., Jenkin, D., Lazarus, R., Makinson, R., Minassian, A. M., Pollock, K. M., Ramasamy, M., Robinson, H., Snape, M., Tarrant, R., Voysey, M., Green, C., Douglas, A. D., Hill, A. V. S., Lambe, T., Gilbert, S. C., Pollard, A. J. and Oxford COVID Vaccine Trial Group 2020. Safety and immunogenicity of the

ChAdOx1 nCoV-19 vaccine against SARS-CoV-2: a preliminary report of a phase 1/2, single-blind, randomised controlled trial. *Lancet* 396(10249), 467–478.

Fougeroux, C. and Holst, P. J. 2017. Future prospects for the development of cost-effective adenovirus vaccines. *Int. J. Mol. Sci.* 18(4): 686.

Francois, A., Chevalier, C., Delmas, B., Eterradossi, N., Toquin, D., Rivallan, G. and Langlois, P. 2004. Avian adenovirus CELO recombinants expressing VP2 of infectious bursal disease virus induce protection against bursal disease in chickens. *Vaccine* 22(17–18), 2351–2360.

Fuchs, W., Veits, J., Helferich, D., Granzow, H., Teifke, J. P. and Mettenleiter, T. C. 2007. Molecular biology of avian infectious laryngotracheitis virus. *Vet. Res.* 38(2), 261–279.

Gao, H., Cui, H., Cui, X., Shi, X., Zhao, Y., Zhao, X., Quan, Y., Yan, S., Zeng, W. and Wang, Y. 2011. Expression of HA of HPAI H5N1 virus at US2 gene insertion site of turkey herpesvirus induced better protection than that at US10 gene insertion site. *PLoS ONE* 6(7), e22549.

Gao, W., Soloff, A. C., Lu, X., Montecalvo, A., Nguyen, D. C., Matsuoka, Y., Robbins, P. D., Swayne, D. E., Donis, R. O., Katz, J. M., Barratt-Boyes, S. M. and Gambotto, A. 2006. Protection of mice and poultry from lethal H5N1 avian influenza virus through adenovirus-based immunization. *J. Virol.* 80(4), 1959–1964.

Garcia, M. and Spatz, S. 2020. Infectious laryngotracheitis. In: Swayne, D. E., Boulianne, M., Logue, C. M., McDougald, L. R., Nair, V., Suarez, D. L., de Wit, S., Grimes, T., Johnson, D., Kromm, M., Prajitno, T. Y., Rubinoff, I. and Zavala, G. (Eds) *Diseases of Poultry*. John Wiley and Sons, Ames, IA, 189–209.

Ge, J., Deng, G., Wen, Z., Tian, G., Wang, Y., Shi, J., Wang, X., Li, Y., Hu, S., Jiang, Y., Yang, C., Yu, K., Bu, Z. and Chen, H. 2007. Newcastle disease virus-based live attenuated vaccine completely protects chickens and mice from lethal challenge of homologous and heterologous H5N1 avian influenza viruses. *J. Virol.* 81(1), 150–158.

Gimeno, I. M., Cortes, A. L., Faiz, N., Villalobos, T., Badillo, H. and Barbosa, T. 2016. Efficacy of various HVT vaccines (conventional and recombinant) Against Marek's disease in broiler chickens: effect of dose and age of vaccination. *Avian Dis.* 60(3), 662–668.

Heckert, R. A., Riva, J., Cook, S., McMillen, J. and Schwartz, R. D. 1996. Onset of protective immunity in chicks after vaccination with a recombinant herpesvirus of turkeys vaccine expressing Newcastle disease virus fusion and hemagglutinin-neuraminidase antigens. *Avian Dis.* 40(4), 770–777.

Hines, N. L. and Miller, C. L. 2012. Avian paramyxovirus serotype-1: a review of disease distribution, clinical symptoms, and laboratory diagnostics. *Vet. Med. Int.* 2012, 708216.

Hu, H., Roth, J. P., Estevez, C. N., Zsak, L., Liu, B. and Yu, Q. 2011. Generation and evaluation of a recombinant Newcastle disease virus expressing the glycoprotein (G) of avian metapneumovirus subgroup C as a bivalent vaccine in turkeys. *Vaccine* 29(47), 8624–8633.

Huang, Z., Elankumaran, S., Yunus, A. S. and Samal, S. K. 2004. A recombinant Newcastle disease virus (NDV) expressing VP2 protein of infectious bursal disease virus (IBDV) protects against NDV and IBDV. *J. Virol.* 78(18), 10054–10063.

Johnson, D. I., Vagnozzi, A., Dorea, F., Riblet, S. M., Mundt, A., Zavala, G. and Garcia, M. 2010. Protection against infectious laryngotracheitis by in ovo vaccination with commercially available viral vector recombinant vaccines. *Avian Dis.* 54(4), 1251–1259.

Johnson, M. A., Pooley, C., Ignjatovic, J. and Tyack, S. G. 2003. A recombinant fowl adenovirus expressing the S1 gene of infectious bronchitis virus protects against challenge with infectious bronchitis virus. *Vaccine* 21(21–22), 2730–2736.

Josefsberg, J. O. and Buckland, B. 2012. Vaccine process technology. *Biotechnol. Bioeng.* 109(6), 1443–1460.

Kallel, H. and Kamen, A. A. 2015. Large-scale adenovirus and poxvirus-vectored vaccine manufacturing to enable clinical trials. *Biotechnol. J.* 10(5), 741–747.

Kanabagatte Basavarajappa, M., Kumar, S., Khattar, S. K., Gebreluul, G. T., Paldurai, A. and Samal, S. K. 2014. A recombinant Newcastle disease virus (NDV) expressing infectious laryngotracheitis virus (ILTV) surface glycoprotein D protects against highly virulent ILTV and NDV challenges in chickens. *Vaccine* 32(28), 3555–3563.

Kawamura, H., King, D. J., Jr. and Anderson, D. P. 1969. A herpesvirus isolated from kidney cell culture of normal turkeys. *Avian Dis.* 13(4), 853–863.

Khwatenge, C. N. and Nahashon, S. N. 2021. Recent advances in the application of CRISPR/Cas9 gene editing system in poultry species. *Front. Genet.* 12, 627714.

Kim, S. H. and Samal, S. K. 2019. Innovation in Newcastle disease virus vectored avian influenza vaccines. *Viruses* 11, 300; 1–12.

Kortekaas, J., Dekker, A., de Boer, S. M., Weerdmeester, K., Vloet, R. P., de Wit, A. A., Peeters, B. P. and Moormann, R. J. 2010. Intramuscular inoculation of calves with an experimental Newcastle disease virus-based vector vaccine elicits neutralizing antibodies against Rift Valley fever virus. *Vaccine* 28(11), 2271–2276.

Kúdelová, M., Belvončíková, P., Halásová, Z., Košovský, J., Lapuníková, B., Pančík, P., Režuchová, I., Supolíková, M. and Zelník, V. 2013. Recombinant herpesviruses as tools for the study of herpesvirus biology. *Acta virol.* 57(2), 149–159.

Li, K., Liu, Y., Liu, C., Gao, L., Gao, Y., Zhang, Y., Cui, H., Qi, X., Zhong, L. and Wang, X. 2017. Evaluation of two strains of Marek's disease virus serotype 1 for the development of recombinant vaccines against very virulent infectious bursal disease virus. *Antiviral Res.* 139, 153–160.

Li, K., Liu, Y., Liu, C., Gao, L., Zhang, Y., Cui, H., Gao, Y., Qi, X., Zhong, L. and Wang, X. 2016a. Recombinant Marek's disease virus type 1 provides full protection against very virulent Marek's and infectious bursal disease viruses in chickens. *Sci. Rep.* 6, 39263.

Li, K., Liu, Y., Liu, C., Gao, L., Zhang, Y., Gao, Y., Cui, H., Qi, X., Zhong, L. and Wang, X. 2016b. Effects of different promoters on the protective efficacy of recombinant Marek's disease virus type 1 expressing the VP2 gene of infectious bursal disease virus. *Vaccine* 34(47), 5744–5750.

Li, K., Liu, Y., Xu, Z., Zhang, Y., Yao, Y., Nair, V., Liu, C., Zhang, Y., Gao, Y., Qi, X., Cui, H., Gao, L. and Wang, X. 2020. Prevention of avian retrovirus infection in chickens using CRISPR-Cas9 Delivered by Marek's Disease Virus. *Mol. Ther. Nucleic Acids* 21, 343–353.

Li, Y., Reddy, K., Reid, S. M., Cox, W. J., Brown, I. H., Britton, P., Nair, V. and Iqbal, M. 2011. Recombinant herpesvirus of turkeys as a vector-based vaccine against highly pathogenic H7N1 avian influenza and Marek's disease. *Vaccine* 29(46), 8257–8266.

Liang, X., Sun, L., Yu, T., Pan, Y., Wang, D., Hu, X., Fu, Z., He, Q. and Cao, G. 2016. A CRISPR/Cas9 and Cre/Lox system-based express vaccine development strategy against re-emerging pseudorabies virus. *Sci. Rep.* 6, 19176.

Liao, Y., Bajwa, K., Reddy, S. M. and Lupiani, B. 2021. Methods for the manipulation of herpesvirus genome and the application to Marek's disease virus research. *Microorganisms* 9(6): 1260.

Liu, Y., Li, K., Gao, Y., Gao, L., Zhong, L., Zhang, Y., Liu, C., Zhang, Y. and Wang, X. 2016. Recombinant Marek's disease virus as a vector-based vaccine against avian leukosis virus Subgroup J in chicken. *Viruses* 8(11): 301.

Liu, Y., Xu, Z., Zhang, Y., Yu, M., Wang, S., Gao, Y., Liu, C., Zhang, Y., Gao, L., Qi, X., Cui, H., Pan, Q., Li, K. and Wang, X. 2020. Marek's disease virus as a CRISPR/Cas9 delivery system to defend against avian leukosis virus infection in chickens. *Vet. Microbiol.* 242, 108589.

Logunov, D. Y., Dolzhikova, I. V., Zubkova, O. V., Tukhvatulin, A. I., Shcheblyakov, D. V., Dzharullaeva, A. S., Grousova, D. M., Erokhova, A. S., Kovyrshina, A. V., Botikov, A. G., Izhaeva, F. M., Popova, O., Ozharovskaya, T. A., Esmagambetov, I. B., Favorskaya, I. A., Zrelkin, D. I., Voronina, D. V., Shcherbinin, D. N., Semikhin, A. S., Simakova, Y. V., Tokarskaya, E. A., Lubenets, N. L., Egorova, D. A., Shmarov, M. M., Nikitenko, N. A., Morozova, L. F., Smolyarchuk, E. A., Kryukov, E. V., Babira, V. F., Borisevich, S. V., Naroditsky, B. S. and Gintsburg, A. L. 2020. Safety and immunogenicity of an rAd26 and rAd5 vector-based heterologous prime-boost COVID-19 vaccine in two formulations: two open, non-randomised phase 1/2 studies from Russia. *Lancet* 396(10255), 887–897.

Morgan, R. W., Gelb, J., Jr., Schreurs, C. S., Lutticken, D., Rosenberger, J. K. and Sondermeijer, P. J. 1992. Protection of chickens from Newcastle and Marek's diseases with a recombinant herpesvirus of turkeys vaccine expressing the Newcastle disease virus fusion protein. *Avian Dis.* 36(4), 858–870.

Omar, A. R., Schat, K. A., Lee, L. F. and Hunt, H. D. 1998. Cytotoxic T lymphocyte response in chickens immunized with a recombinant fowlpox virus expressing Marek's disease herpesvirus glycoprotein B. Vet. *Vet. Immunol. Immunopathol.* 62(1), 73–82.

Panda, S. K. and McGrew, M. J. 2021. *Genome Editing of Avian Species: Implications for Animal Use and Welfare*. Lab Anim, 23677221998400.

Park, M. S., Steel, J., Garcia-Sastre, A., Swayne, D. and Palese, P. 2006. Engineered viral vaccine constructs with dual specificity: avian influenza and Newcastle disease. *Proc. Natl Acad. Sci. U. S. A.* 103(21), 8203–8208.

Pavlova, S. P., Veits, J., Keil, G. M., Mettenleiter, T. C. and Fuchs, W. 2009a. Protection of chickens against H5N1 highly pathogenic avian influenza virus infection by live vaccination with infectious laryngotracheitis virus recombinants expressing H5 hemagglutinin and N1 neuraminidase. *Vaccine* 27(5), 773–785.

Pavlova, S. P., Veits, J., Mettenleiter, T. C. and Fuchs, W. 2009b. Live vaccination with an H5-hemagglutinin-expressing infectious laryngotracheitis virus recombinant protects chickens against different highly pathogenic avian influenza viruses of the H5 subtype. *Vaccine* 27(37), 5085–5090.

Reddy, S. K., Sharma, J. M., Ahmad, J., Reddy, D. N., McMillen, J. K., Cook, S. M., Wild, M. A. and Schwartz, R. D. 1996. Protective efficacy of a recombinant herpesvirus of turkeys as an in ovo vaccine against Newcastle and Marek's diseases in specific-pathogen-free chickens. *Vaccine* 14(6), 469–477.

Ross, P. J. and Parks, R. J. 2009. Construction and characterization of adenovirus vectors. *Cold Spring Harb. Protoc.* 2009(5), pdb.prot5011.

Sadigh, Y., Powers, C., Spiro, S., Pedrera, M., Broadbent, A. and Nair, V. 2018. Gallid herpesvirus 3 SB-1 strain as a recombinant viral vector for poultry vaccination. *npj Vaccines* 3, 21.

Sakaguchi, M., Nakamura, H., Sonoda, K., Okamura, H., Yokogawa, K., Matsuo, K. and Hira, K. 1998. Protection of chickens with or without maternal antibodies against both

Marek's and Newcastle diseases by one-time vaccination with recombinant vaccine of Marek's disease virus type 1. *Vaccine* 16(5), 472–479.

Schijns, V. E. J. C., van de Zande, S., Lupiani, B. and Reddy, S. M. 2014. Practical aspects of poultry vaccination. In: Schat, K. A., Kaspers, B. and Kaiser, P. (Eds) *Avian Immunology* (2nd edn.). Academic Press, Boston, 345–362.

Shao, Y., Sun, J., Han, Z. and Liu, S. 2018. Recombinant infectious laryngotracheitis virus expressing Newcastle disease virus F protein protects chickens against infectious laryngotracheitis virus and Newcastle disease virus challenge. *Vaccine* 36(52), 7975–7986.

Silva, A. C., Fernandes, P., Sousa, M. F. and Alves, P. M. 2014. Scalable production of adenovirus vectors. *Methods Mol. Biol.* 1089, 175–196.

Skinner, M. A., Laidlaw, S. M., Eldaghayes, I., Kaiser, P. and Cottingham, M. G. 2005. Fowlpox virus as a recombinant vaccine vector for use in mammals and poultry. *Expert Rev. Vaccines* 4(1), 63–76.

Sonoda, K., Sakaguchi, M., Okamura, H., Yokogawa, K., Tokunaga, E., Tokiyoshi, S., Kawaguchi, Y. and Hirai, K. 2000. Development of an effective polyvalent vaccine against both Marek's and Newcastle diseases based on recombinant Marek's disease virus type 1 in commercial chickens with maternal antibodies. *J. Virol.* 74(7), 3217–3226.

Speedy, A. W. 2003. Global production and consumption of animal source foods. *J. Nutr.* 133(11), 4048S–4053S.

Steitz, J., Wagner, R. A., Bristol, T., Gao, W., Donis, R. O. and Gambotto, A. 2010. Assessment of route of administration and dose escalation for an adenovirus-based influenza A virus (H5N1) vaccine in chickens. *Clin. Vaccine Immunol.* 17(9), 1467–1472.

Stewart-Brown, B. 2015. *MSD Veterianry Manual*. Merck & Co., Inc., Rahway, NJ, USA

Stobart, C. C. and Moore, M. L. 2014. RNA virus reverse genetics and vaccine design. *Viruses* 6(7), 2531–2550.

Suarez, D. L. and Pantin-Jackwood, M. J. 2017. Recombinant viral-vectored vaccines for the control of avian influenza in poultry. *Vet. Microbiol.* 206, 144–151.

Swayne, D. E., Beck, J. R. and Kinney, N. 2000. Failure of a recombinant fowl poxvirus vaccine containing an avian influenza hemagglutinin gene to provide consistent protection against influenza in chickens preimmunized with a fowl pox vaccine. *Avian Dis.* 44(1), 132–137.

Tang, N., Zhang, Y., Pedrera, M., Chang, P., Baigent, S., Moffat, K., Shen, Z., Nair, V. and Yao, Y. 2018. A simple and rapid approach to develop recombinant avian herpesvirus vectored vaccines using CRISPR/Cas9 system. *Vaccine* 36(5), 716–722.

Tang, N., Zhang, Y., Pedrera, M., Chang, P., Baigent, S., Moffat, K., Shen, Z., Nair, V. and Yao, Y. 2019. Generating recombinant avian herpesvirus vectors with CRISPR/Cas9 gene editing. *J. Vis. Exp.* Jan 7; (143).

Tang, N., Zhang, Y., Sadigh, Y., Moffat, K., Shen, Z., Nair, V. and Yao, Y. 2020. Generation of A triple insert live avian herpesvirus vectored vaccine using CRISPR/Cas9-based gene editing. *Vaccines (Basel)* 8(1): 97.

Taylor, J., Christensen, L., Gettig, R., Goebel, J., Bouquet, J. F., Mickle, T. R. and Paoletti, E. 1996. Efficacy of a recombinant fowl pox-based Newcastle disease virus vaccine candidate against velogenic and respiratory challenge. *Avian Dis.* 40(1), 173–180.

Tischer, B. K., von Einem, J., Kaufer, B. and Osterrieder, N. 2006. Two-step red-mediated recombination for versatile high-efficiency markerless DNA manipulation in Escherichia coli. *BioTechniques* 40(2), 191–197.

Tong, G. Z., Zhang, S. J., Wang, L., Qiu, H. J., Wang, Y. F. and Wang, M. 2001. Protection of chickens from infectious laryngotracheitis with a recombinant fowlpox virus expressing glycoprotein B of infectious laryngotracheitis virus. *Avian Pathol.* 30(2), 143–148.

Toro, H., Tang, D. C., Suarez, D. L., Sylte, M. J., Pfeiffer, J. and Van Kampen, K. R. 2007. Protective avian influenza in ovo vaccination with non-replicating human adenovirus vector. *Vaccine* 25(15), 2886–2891.

Toro, H., Zhao, W., Breedlove, C., Zhang, Z., Yu, Q. and Van Santen, V. 2014. Infectious bronchitis virus S2 expressed from recombinant virus confers broad protection against challenge. *Avian Dis.* 58(1), 83–89.

Tsukamoto, K., Kojima, C., Komori, Y., Tanimura, N., Mase, M. and Yamaguchi, S. 1999. Protection of chickens against very virulent infectious bursal disease virus (IBDV) and Marek's disease virus (MDV) with a recombinant MDV expressing IBDV VP2. *Virology* 257(2), 352–362.

Tsukamoto, K., Saito, S., Saeki, S., Sato, T., Tanimura, N., Isobe, T., Mase, M., Imada, T., Yuasa, N. and Yamaguchi, S. 2002. Complete, long-lasting protection against lethal infectious bursal disease virus challenge by a single vaccination with an avian herpesvirus vector expressing VP2 antigens. *J. Virol.* 76(11), 5637–5645.

Vagnozzi, A., Zavala, G., Riblet, S. M., Mundt, A. and Garcia, M. 2012. Protection induced by commercially available live-attenuated and recombinant viral vector vaccines against infectious laryngotracheitis virus in broiler chickens. *Avian Pathol.* 41(1), 21–31.

van Zijl, M., Quint, W., Briaire, J., de Rover, T., Gielkens, A. and Berns, A. 1988. Regeneration of herpesviruses from molecularly cloned subgenomic fragments. *J. Virol.* 62(6), 2191–2195.

Weli, S. C. and Tryland, M. 2011. Avipoxviruses: infection biology and their use as vaccine vectors. *Virol. J.* 8, 49.

Witter, R. L., Nazerian, K., Purchase, H. G. and Burgoyne, G. H. 1970. Isolation from turkeys of a cell-associated herpesvirus antigenically related to marek's disease virus. *Am. J. Vet. Res.* 31(3), 525–538.

Witter, R. L. 2001. Protective efficacy of Marek's disease vaccines. *Curr. Top. Microbiol. Immunol.* 255, 57–90.

Witter, R. L. and Lee, L. F. 1984. Polyvalent Marek's disease vaccines: safety, efficacy and protective synergism in chickens with maternal antibodies. *Avian Pathol.* 13(1), 75–92.

Yang, T. C., Dayball, K., Wan, Y. H. and Bramson, J. 2003. Detailed analysis of the CD8+ T-cell response following adenovirus vaccination. *J. Virol.* 77(24), 13407–13411.

Yokoyama, N., Maeda, K. and Mikami, T. 1997. Recombinant viral vector vaccines for the veterinary use. *J. Vet. Med. Sci.* 59(5), 311–322.

Zeshan, B., Zhang, L., Bai, J., Wang, X., Xu, J. and Jiang, P. 2010. Immunogenicity and protective efficacy of a replication-defective infectious bronchitis virus vaccine using an adenovirus vector and administered in ovo. *J. Virol. Methods* 166(1–2), 54–59.

Zhang, X., Wu, Y., Huang, Y. and Liu, X. 2012. Protection conferred by a recombinant Marek's disease virus that expresses the spike protein from infectious bronchitis virus in specific pathogen-free chicken. *Virol. J.* 9, 85.

Zhao, W., Spatz, S., Zhang, Z., Wen, G., Garcia, M., Zsak, L. and Yu, Q. 2014. Newcastle disease virus (NDV) recombinants expressing infectious laryngotracheitis virus (ILTV) glycoproteins gB and gD protect chickens against ILTV and NDV challenges. *J. Virol.* 88(15), 8397–8406.

Zhu, F. C., Li, Y. H., Guan, X. H., Hou, L. H., Wang, W. J., Li, J. X., Wu, S. P., Wang, B. S., Wang, Z., Wang, L., Jia, S. Y., Jiang, H. D., Wang, L., Jiang, T., Hu, Y., Gou, J. B., Xu, S. B., Xu, J. J., Wang, X. W., Wang, W. and Chen, W. 2020. Safety, tolerability, and immunogenicity of a recombinant adenovirus type-5 vectored COVID-19 vaccine: a dose-escalation, open-label, non-randomised, first-in-human trial. *Lancet* 395(10240), 1845–1854.

Zou, Z., Huang, K., Wei, Y., Chen, H., Liu, Z. and Jin, M. 2017. Construction of a highly efficient CRISPR/Cas9-mediated duck enteritis virus-based vaccine against H5N1 avian influenza virus and duck Tembusu virus infection. *Sci. Rep.* 7(1), 1478.

Part 3

Optimising health at differing stages in poultry production

Chapter 7

Incubation and chick health

Ron Meijerhof, Poultry Performance Plus, The Netherlands

1 Introduction

Artificial incubation is an important part of the poultry production chain, both for broilers and layers. The success of incubation influences not only the hatchability and with it the costs of a day-old chick (DOC) but also the quality of that chick. This quality is a crucial factor for obtaining optimum technical performance and immune competence, both through the health status of the DOC and through the development of the embryo. Before we focus on the factors that can influence the health status, we look at the factors that are involved in the development of the embryo during incubation, as they are crucial for chick quality. In this chapter, we will focus on chicken embryos, unless stated otherwise.

Artificial incubation is a very delicate process, which requires an excellent control to maximize results such as hatchability and chick quality. Traditionally, multistage machines were commonly used in the industry: once or twice a week the eggs that are at that moment 17 or 18 days in the setters are removed and transferred to the hatchers, and an equal number of fresh eggs are brought into the machine. In this way, the eggs further in the incubation process can warm

http://dx.doi.org/10.19103/AS.2022.0104.10

up the fresh eggs, which saves energy. As in this situation the machine has to cater for eggs at all stages of incubation, settings of ventilation, temperature and relative humidity (RH) are maintained at a constant level. Nowadays, single-stage machines are getting increasingly popular, where all eggs in the machine are placed and removed at the same time. With this method, the machines can be cleaned regularly and machine settings like temperature, RH and ventilation can be adjusted to the requirements of the eggs at each moment of incubation.

Over time, the technology and control equipment used in hatcheries have improved significantly. At this moment, hatcheries setting up to 2 million eggs weekly are no exception anymore, as well as incubators setting over 100 000 eggs per machine. The technology installed to make this process run constantly and reliably is complex. Modern hatcheries are able to read and follow all the different machines remotely, control and adjust the settings from central computers, receive alarm warnings remotely by telephone and store and process all relevant data automatically. Modern hatchery management aims at creating optimal environmental conditions for eggs and chicks from egg storage at the breeder farm until the moment of chick delivery at the broiler or layer farm. However, in spite of all these technical improvements, it must be questioned if we are actually controlling the key factors for the embryo to the level that we think we are. Controlling the set points on the computer screen does not automatically translate into optimal conditions for the embryo itself. To understand these key factors, we have to look into what happens during incubation.

2 What happens during incubation?

From a biological point of view, incubation, that is, changing a fertile egg into a live chicken, is of a complexity far beyond our imagination and nothing less than a miracle. However, from a technical point of view, we can simplify the process to a very straightforward model. In essence, the embryo builds up the body from the content of the egg. For this process, energy is needed. The energy is derived from carbohydrates during approximately the first 6 days of incubation and after that from the fat sources in the yolk (Romanoff, 1960; Noble and Cocchi, 1990; Moran, 2007). To utilize the fat, the embryo needs oxygen (O_2) which has to come from the environment around the egg. During this process, which is controlled by temperature, three waste products are produced, namely carbon dioxide (CO_2), metabolic water and metabolic heat, which have to be removed from the egg by ventilation, evaporation and cooling. To summarize, it means that for optimal incubation we need temperature, ventilation for O_2 and CO_2 exchange and ventilation for moisture loss. When we arrange this adequately, the embryo will utilize the yolk and change it into body tissue in the most optimal way. Besides these conditions, we also need to turn the eggs regularly, to prevent the embryo from sticking to the membranes inside the egg, to build up the chorioallantoic membrane and to enable the embryos to access the nutrients from the egg.

During the incubation process, many factors can lead to sub-optimal development of the embryo or embryonic death. To determine the causes for incubation failures, it is a good practice to open eggs that do not contain living or viable embryos, either during the incubation process (determined by candling of the eggs) or after the hatch (by opening the unhatched eggs). Numerous scientific publications and instruction materials from breeding companies, incubation manufacturers and other sources are available to support these procedures. Especially the work of Hamburger and Hamilton (1951) is often used as a reference for evaluating embryonic development.

Temperature is by far the most important factor in determining chick quality and hatch success (see also Section 6.1). As an embryo reacts poikilothermic, the speed of its metabolism depends on the temperature inside the egg. The speed of its metabolism also determines the need for energy, which after approximately 6 days of incubation needs to be converted from the fat in the yolk. And as the conversion of fat into energy is limited by the amount of O_2 available for the embryo, especially a too high temperature will harm the development, as the embryo will not be able to match the energy availability with the speed of its development. This happens especially in the later stages of the incubation process but can already be noticed as early as incubation day 14. Until day 14, the heat production which is a result of the development is increasing rapidly. After day 14, the heat production is still increasing but the increase is not happening that fast anymore, indicating that the O_2 supply to the embryo is becoming a limiting factor. At approximately 17–18 days of incubation, the heat production is not increasing any further, the so-called plateau phase.

With an increasing temperature, there is an increase in demand for energy and with it for O_2 to match with the increased embryonic development. If the temperature in the egg becomes too high in the last week of the incubation period (above approximately 38.3–38.5°C when measured as temperature of the surface of the shell) and as a result the O_2 becomes severely limited, the embryo will not be able to utilize all the needed energy from the fat in the yolk and will start using protein as an alternative energy source. This has a detrimental effect on hatch results and chick quality, as proteins that were intended for building up the embryonic tissues are utilized for energy. Research has shown that an internal egg temperature of approximately 1°C above the expected optimum results in a significant reduction in heart size of approximately 20–25% at hatch (Leksrisompong et al., 2007; Molenaar et al.,

2011; Nangsuay et al., 2016) but also in a change in bone formation (Oviedo-Rondon et al., 2008, 2009), development of the intestinal tract (Barri et al., 2011), yolk utilization (Dayan et al., 2020) and response of the immune system (Oznurlu et al., 2010). This has been shown to result in an increase in mortality due to ascites (Molenaar et al., 2011) but also in alterations in the bursa at hatch (Wijnen et al., 2020) and a reduced growth and increased duodenum lesion score after an Eimeria challenge 5 weeks post hatch (Walstra et al., 2012).

Hatchery-related health aspects are influenced by pathogens that are infecting the eggs or the DOC. However, we should realize that the quality of the incubation process itself can influence the health status and the ability of the chick to withstand disease challenges as well.

3 Day-old chick quality

Although judgment of chick quality is subjective, some criteria can be given. A quality DOC should be dry, clean and free of dirt and contamination. It should have clear and bright eyes and be free from deformities. The navel should be clean and sealed, and no yolk sac or membrane should protrude from the navel area. The residual yolk sac should be not more than 10% of the chick's weight. The body should be firm without signs of respiratory distress or dehydration. The chick should be alert, stand up easily and quick after being placed on its back and being interested in its environment and responding to sounds. The legs should have a normal conformation without swelling or lesions and with straight toes. The beak should be well formed and the feathers should be long and fluffy (Funk and Irwin, 1955).

4 Factors that influence the chick quality and health status of the day-old chick

Chick quality and health status are influenced by a lot of different factors, including the health status of the breeders, the quality of the hatching egg, the handling of the hatching egg before setting, the conditions during the incubation period and the handling of the chicks during and after hatching.

4.1 Breeder influences

The breeder flock has a significant influence on the success of the incubation process and the quality of the resulting DOC. The hatching result is of course highly dependent on the fertilization rate of the produced eggs but also on factors such as egg hygiene, shell quality, nutritional aspects and health status (Picture 1).

Picture 1 Poor hatching egg hygiene. Picture provided by Ron Meijerhof.

4.1.1 Diseases

Several diseases in breeder hens can influence the quality of the hatching egg and the DOC. A wide range of viral and bacterial challenges (reovirus, adenovirus, Newcastle disease virus, chicken infectious anemia virus, astrovirus, avian encephalomyelitis virus, *Salmonella pullorum/gallinarum*, *S. enteritidis*, *Escherichia coli*, etc.) are known to affect chick quality, but probably the viral disease that causes severe problems worldwide is infectious bronchitis (IB). Although the disease does not cause excessive mortality in the breeder hen, it influences production, shell quality, internal egg quality, hatchability and DOC quality (Jackwood and de Wit, 2020).

Especially in broiler chicks, lesions in the gizzard can sometimes be found. These lesions are often referred to as gizzard erosion and can be observed in DOCs but sometimes already in the unhatched chick. Gizzard erosion (Picture 2) has already been reported in the 1930s (Holst and Halbrook, 1933). The effect of the lesions on the birds is not clear, as flocks with high incidences

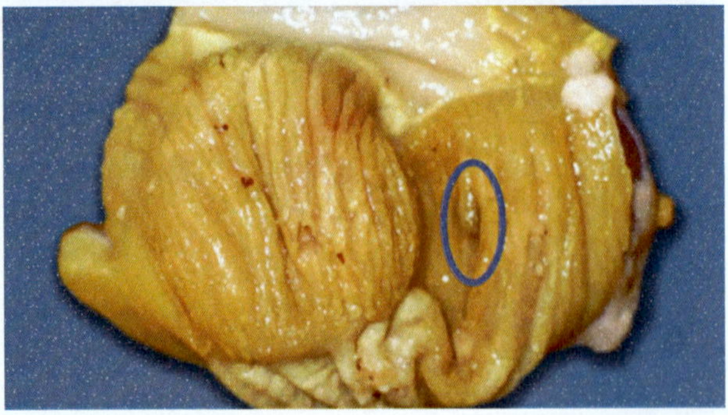

Picture 2 Gizzard erosion. Picture courtesy of Ron Meijerhof.

can be well performing. The cause for these lesions in DOC is largely unknown and can be multifactorial. A clear relation with incubation is not known.

4.1.2 Nutritional factors

As the developing embryo gets all the nutrients from the egg, the breeder diet must fulfill the needs of the embryo. In this respect, especially minerals and vitamins are important. Insufficient levels of vitamins can cause a range of problems with respect to chick quality. Deficiencies in fat-soluble vitamins can cause problems with hatchability (vitamin A and K), eggshell quality and bone strength (vitamin D) and crazy chick disease (vitamin E) although the breeder birds normally have enough body reserves to withstand temporary fluctuations in the feed. Water-soluble vitamins need to be present on a regular base as the breeder bird has limited ability to store them. Especially the B vitamin complex is important as it is involved in many developmental systems in the embryo and can cause chick abnormalities like 'stargazers' (vitamin B1) (Picture 3), 'clubbed down feathers' (vitamin B2) and a severe loss in hatchability (riboflavin, biotin, thiamine, niacin, cobalamin). Shortage of vitamins is not necessarily caused by insufficient dosing in the feed. Storage time and storage condition of the feed or premix as well as heat treatment during manufacturing of the feed can cause a vitamin shortage as well (Leeson and Summers, 2000).

Shortages of minerals like iodine and manganese in the breeder diet affect embryo survival rates. The deficiency of zinc is probably the most striking for embryo development. Zinc deficiency leads to gross skeletal malformations in embryos, and in DOCs that cannot stand, eat or drink. Shortage of calcium or an imbalance between calcium and phosphorus in the diet can create shell problems. Also high environmental temperatures can create shell problems.

Picture 3 'Stargazers'. Picture courtesy of Ron Meijerhof.

When the birds are panting to dispatch the heat, the increased loss of CO_2 can lead to a shift in the acid/base balance of the blood, influencing the calcium metabolism. A comprehensive overview of the influence of breeder nutrition on DOC quality is given by Leeson and Summers (2000, 2009).

4.1.3 Chemical intoxication

Developing embryos are very sensitive to chemical intoxication, especially at a young age. Cross-contamination of breeder feed with traces of coccidiostats and then especially nicarbazin can cause severe embryonic mortality in the first few days of incubation (Sherwood et al., 1956; Leeson et al., 1989). Also the use of antibacterial drugs like florfenicol can cause early embryonic mortality (Hu et al., 2020). As the eggs need to be opened and examined carefully to determine the embryonic mortality associated with it, it is often classified as infertility. Indications for nicarbazin intoxication are the appearance of white eggshells and mottled yolks (Polin and Porter, 1956; Polin et al., 1957). Also other chemicals can cause mottled yolks and increased embryo mortality, for instance, certain deworming compounds (Beane et al., 1965) and the use of cottonseed meal in breeder diets due to the presence of gossypol (Lyon et al., 1972).

4.1.4 Biological intoxication

Mycotoxins are toxins produced by various molds belonging primarily to the Aspergillus, Penicillium and Fusarium species and can cause serious health problems like leg problems (Krough et al., 1989), immunosuppression (Corrier, 1991), kidney and liver damage (Filazi et al., 2017) and hepatic injury (Doupnik and Peckham, 1970). The acceptable maximum levels that are suggested in the feed are very low, for instance, 5 ppb for aflatoxins and ochratoxins, although from other mycotoxins much higher levels seem to be tolerated (fumonisins: 1000 ppb, deoxynivalenol: 500 ppb; Leeson and Summers, 2009; Filazi et al., 2017). In breeder feeds, they create a risk not only for the animal itself but also for the offspring. Some mycotoxins can be the cause for poor eggshell quality (for instance, zearalenone), embryo malformations (aflatoxins, sterigmatocystin) and embryo mortality (aflatoxins). Also the uptake of nutrients and vitamins (for instance, vitamin D3 and B12) by the breeder hen can be impaired, influencing the nutritional status of the hatching egg. Next to that, mycotoxins like fumonisins and deoxynivalenol can result in diarrhea and poor litter quality, creating more bacterial challenges for the produced eggs (Filazi et al., 2017).

4.2 Egg quality

A good-quality egg is needed to produce a good-quality DOC. Not only the internal quality of the egg is important but also the quality of the shell and the hygiene of the egg are important. Many factors can influence the quality of an egg.

4.2.1 Laying house period

When the eggs are produced, they have the body temperature of the hen (41°C) and the shell is wet and fragile. After being produced, the egg starts to cool down and dries up, which makes the shell more rigid. This means that immediately after the egg is produced the chance of hair cracks and other cracks is high, so the laying nest should be designed in such a way that the eggs are handled very carefully. The type of nest and the temperature in the house will influence the cooling process of the eggs. In roll-away nests, the eggs change relatively quickly to the environmental temperature, whereas in litter nests they will keep a higher temperature for a longer time due to the isolating effect of the litter and the warming effect of the visiting birds. As a result of the drop in temperature the egg content will shrink, and a small air cell is formed in the egg. This results in a negative pressure inside the egg and air will enter through the pores to equalize the pressure. When an egg is produced in a dirty environment (for instance, in the litter area of the breeder house or in a dirty laying nest), bacteria can enter the pores with that airflow, and will have a chance during the incubation process to pass the membranes and enter the egg, resulting in bacterial contamination, embryonic death and reduced chick quality. This means that the environment during the cooling process is crucial for egg quality and chick quality. A good chick quality starts with a good egg quality in the laying nests. As bacterial contamination has a negative influence on hatchability and first week mortality, this is also an economic concern for the hatchery/integration.

As mentioned before, elevated house temperatures can influence the deposition of calcium and with it the quality of the shell but can also influence the survival rate of the embryo. Frequent collection, even in roll-away nests, is advisable in heat stress situations.

4.2.2 Washing of eggs

Depending on maintenance and type of nest, litter conditions, intestinal health of the chicks and management practices, a certain amount of dirty eggs will be produced, either from the floor or from the nests. In some operations, dirty eggs are washed or brushed and then used in the normal incubation process. Although after washing or brushing the eggs are visually clean, the contamination from the pores will not be fully removed and these eggs will still produce a reduced hatchability and chick quality (van den Brand et al., 2016). This is why these eggs should still be considered as dirty eggs and set separate from the clean nest eggs, to avoid further contamination. As washing can spread out the bacterial contamination even more and brushing removes the protective cuticle from the eggshell, these procedures cannot replace good hygiene in the egg production period.

4.2.3 Disinfection of eggs

Disinfection of hatching eggs can be done with a range of products. Most commonly used are formaldehyde (H_2CO) and hydrogen peroxide (H_2O_2). Usually, a concentration of 3–7 g of formaldehyde per m^3 of the disinfection room is applied. Hydrogen peroxide is usually applied as a spray or foam in a 3% solution. As formaldehyde is a gas, it will be more effective in the pores of the eggshell than liquid products like hydrogen peroxide. The use of formaldehyde before the start of the incubation process can cause embryonic mortality, especially when the exposure time is longer than 20 minutes (Cardirci, 1997; Elibol et al., 2003) or when the fumigation is done too frequently (Proudfoot and Stewart, 1970). In field situations, a minimum period of 12 h but preferably 24 h between two consecutive disinfections is advisable. When the incubation has started, the embryo is very sensitive to formaldehyde until approximately 96 h of incubation (Marcellus et al., 1930). Also other disinfectant gases like ozone (O_3) can cause embryonic death, as Whistler and Sheldon (1989) demonstrated by applying a concentration of 3% ozone for 2 h. Liquid disinfectants normally do not have this problem as a liquid will not enter the pores of the eggshell.

Although both washing and disinfection of the eggs can reduce the bacterial load on the eggs, it cannot restore the quality of the eggs. Once the egg quality is compromised, the options for repair are limited, as the bacteria in the pores of the egg are very difficult to remove. Good egg quality should therefore be a priority during the production of the eggs.

4.2.4 Egg position

During incubation, embryos orientate themselves by gravity. At the end of the process, they turn in the egg with the head up to open the air cell (internal pipping) and get O_2 to complete the process of opening (external pipping) and breaking free from the eggshell. This means that the egg must be placed blunt end up to allow the embryo to find the air cell. In eggs placed small end up, the embryonic mortality will increase by approximately 15–25% (Bauer et al., 1990). As the eggs are usually placed in their final position already during collection, correct setting procedures should be monitored at the breeder farm.

4.2.5 Egg storage

In a commercial operation, eggs need to be stored for a number of days to supply the required number of chicks for each hatch day. Mismatch of egg production and setting due to, for instance, market situations can lead to non-optimal storage times. In an optimal situation, eggs are stored for a minimum of 2 days and a maximum of 5–7 days (Meijerhof, 1992). Very young (Gucbilmez et al, 2013) and older flocks (Meijerhof, 1992; Nasri et al., 2020) are more sensitive to the negative impact of storage than prime flocks. Depending on the age of the flock, breed, temperature during storage, temperature treatment and egg position during storage and warming profile at the start of incubation, the hatchability of the eggs will go down after 5–7 days with approximately 0.5–1% per day. When the storage period is further extended, the decrease in hatchability will go faster. With every day of extra storage, the incubation period will also get longer (Shiranjang et al., 2018), with approximately 0.5–1 h per day of storage (depending on the warming profile at the start of incubation). Temporary warming during storage for a few hours as well as turning the eggs during storage or storing them upside down will slow down, but not stop, the decrease in hatchability. Not only hatchability is influenced by egg storage but chick quality as well. Extended egg storage results in a decrease in chick quality in terms of malformations (Mather and Laughlin, 1977, 1979), culls (Byng and Nash, 1962) and reduced growth rate (Becker, 1960; Merrit, 1964).

When the eggs are set within 48 h after being produced, the hatchability is decreased as well. After egg production, the albumen increases in pH due to the escape of CO_2 from the egg, and a certain difference in pH between albumen and yolk needs to be present to direct the ion pumps in the embryo (Meijerhof, 1992)

5 Biological risk factors during incubation

5.1 *Bacterial challenges*

As incubation takes place in high temperatures and relatively high humidity in the presence of organic material, the multiplication of bacteria during the incubation process is a continuous concern, not only because bacteria can form a threat to the health of the DOC but can also be a concern for later performance and food safety. It is good to realize that the majority of the bacteria will enter the incubation process through the incoming hatching eggs, which means that quality and hygiene during the production of the eggs are crucial for control of the hygiene status in the hatchery. Although this section is not about hatching egg production, the quality of the hatching egg is so essential for the quality of the incubation process and the DOC that some aspects should be highlighted.

5.1.1 *Omphalitis*

In broiler production, the mortality that occurs in the first week of life is relatively high compared to the mortality in the rest of the production cycle. A significant part of this mortality is due to omphalitis (navel/yolk sac infection). The hatching process is a severe stress for the chicks, not only because of high noise and continuous darkness (Archer and Mench, 2014) but also because of dust, pathogen pressure and sometimes the use of formaldehyde or other disinfectants during the hatching phase (de Gouw et al., 2017). In the period before emerging from the shell, the embryo takes up the remaining yolk sac through the navel, which is closing during hatching. As the navel is not always completely closed at the moment of hatching (Pictures 4 and 5), bacteria can

Picture 4 Unhealed navel. Picture provided by Ron Meijerhof.

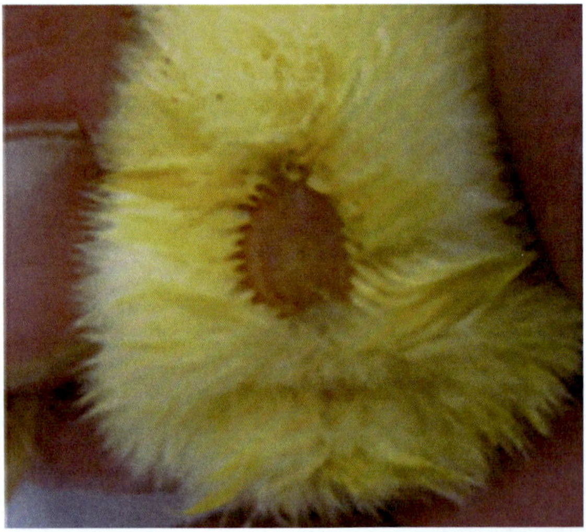

Picture 5 Clean well-closed navel. Picture courtesy of Petersime.

enter the body cavity at this moment and can cause mortality approximately 3–4 days after hatch. DOCs and especially embryos are extremely sensitive to bacterial contamination. Meijerhof and Hulet (1997) demonstrated that introducing a commercial competitive exclusion bacteria culture (used for DOCs to protect them against *Salmonella* infection) at 18 days of incubation into the embryo resulted in almost 100% mortality within 24–36 h after injection. Injecting the culture at day 18 in the air cell of the hatching egg resulted in a hatchability reduction of approximately 10% but also in extra mortality in the first week after hatch (Meijerhof and Hulet, 1997).

5.1.2 Exploders or bangers

The sensitivity of embryos and chicks to bacterial contamination requires a good control on the hygiene of the eggs used in the incubation process. Eggs that are produced in a dirty environment (dirty nests, floor eggs) have an increased risk to get contaminated. This increases the risk for that egg and chick to get contaminated, but it can also result in 'exploders' or 'bangers' (eggs that explode during the incubation process due to the presence of gas-forming bacteria). These 'exploders' will multiply the bacterial pressure during the hatching phase and increase the risk for bacterial contamination of the DOCs and mortality in the first week of life. To reduce the bacterial load on the eggs they are disinfected and sometimes even washed or brushed ('sanded'), but as the bacteria will be present in the pores of the eggshell and are difficult to remove, the effect is limited. Washing of the dirty eggs can increase the problem by spreading out the bacterial load over more eggs. Dry scraping of the eggs to clean them does not spread the bacterial load further but will not remove the bacteria from the pores and therefore is not a real solution.

5.1.3 On-farm hatching

In Western Europe, there is an increasing trend to transport the incubated eggs at 18 days to the broiler farm to let them hatch in the house (Picture 6). The main reason is related to welfare (availability of feed and water directly after hatching), but field results suggest a trend toward a decrease in first week mortality as well. Although the scientific evidence at this moment is limited, the extra space and ventilation during hatching in the house compared to hatching

Picture 6 In-house hatching. Picture courtesy of Nestborn.

in a hatching cabinet perhaps result in more 'relaxed' hatching conditions, avoiding high bacterial pressure, limited ventilation and overheating, resulting in less mortality.

5.1.4 Leg problems

Bacterial contamination of eggs and chicks and then especially with *Staphylococcus* spp. can lead to leg problems early and later in the life of the bird (Andreasen, 2020; McNamee and Smyth, 2000). In case of leg problems, often bacteria can be found in bones and joints. When floor eggs are used or when the hygiene of the eggs is below standard, an increase in leg problems like lameness is often observed.

Splayed legs observed at hatch (Picture 7) are not related to bacterial contamination but are the result of the inability of a chicken to keep its legs under its body. This can be caused by a slippery surface and/or a lack of energy due to sub-optimal incubation conditions.

5.1.5 Bacteria involved

5.1.5.1 E. coli

E. coli can be found in large quantities during the hatching process and in birds that died during and after the hatching process, as well as in the first week of life. Although sometimes pathogenic varieties of *E. coli* can be found, the presence of *E. coli* does not automatically means that there is a problem.

Picture 7 Splayed legs. Picture courtesy of Ron Meijerhof.

5.1.5.2 *Salmonella* control

Within the large group of *Salmonella* serovars, especially *S. pullorum* and *S. gallinarum* cause a health threat to the chick, although also species like *Salmonella enteritidis*, *S. typhimurium* and *S. arizona* can cause health problems. However, in most cases, *Salmonella* serovars found in hatcheries do not create a problem for the chicks but cause a concern from a food safety point of view. *Salmonella* in hatcheries can spread very easily, especially during hatching. Bailey et al. (1998) showed that a few *Salmonella*-positive eggs in a single hatcher can cause a massive outbreak of *Salmonella*-positive chicks from that hatcher. As the presence and persistency of *Salmonella* contamination in hatcheries cause a high colonization risk for the chick at the day of hatch (Cox et al., 1990), strict hygiene practices following Hazard Analysis and Critical Control Points procedures are essential to control the situation.

5.1.5.3 *Streptococcus* and *Enterococcus*

Streptococcus spp and/or *Enterococcus* spp can often be isolated from 'poor-quality chicks' and the presence can lead to problems later in life as well (Borst, 2020). The bacteria are often 'pseudo-vertically' transmitted through dirty eggshells and can sometimes be isolated from unhatched eggs (Dolka et al., 2017). The presence of especially *Enterococcus* can be an indication that dirty eggs (floor eggs) have been introduced in the incubation process.

5.2 Fungal challenges

Except for the production of mycotoxins, fungus in a hatchery operation also causes a direct risk to the quality of the DOC. Especially *Aspergillus* is a risk factor, as it will obstruct the ventilation of the lungs of the DOC (Chute and Richard, 1991). Due to the climatic conditions and the presence of biological material, an *Aspergillus* infection can sometimes be difficult to avoid or remove (Wright and Epps, 1958). Especially the use of litter nest boxes, paper egg trays, carton chick boxes, etc. can increase the risk of *Aspergillus*, but also condensation and the presence of organic material in air ducts can create problems. Hatchery sanitation is crucial for the control of aspergillosis in hatcheries (Wright et al., 1960)

6 Physical risk factors during incubation

During embryonic development, several disorders can occur, usually caused by either temperature issues or turning. Gaseous conditions during incubation (O_2 concentration, CO_2 concentration and RH) can cause problems as well but are usually having less influence with a wider range of tolerance.

6.1 Temperature

An optimal internal egg temperature, reflected in eggshell temperature (Meijerhof and van Beek, 1993), is crucial for hatching success, chick quality and broiler performance (Lourens et al., 2005; Hulet et al., 2007; Molenaar et al., 2011).

A too high temperature during the first days of incubation results in an increase of malformations. Peterka et al. (1996) showed that a temperature of 40°C in the first days of incubation resulted in almost all embryos having malformations. In field conditions, we see the number of malformations such as brain hernia, posterior duplications, double spines, double heads and double beaks tends to increase at initial temperatures above approximately 38–38.1°C. These malformations are often recognized only during break out of the hatch debris, as the chicks will often not be able to emerge from the eggshell.

When the temperature at the start of incubation is too low or the incubation is stopped for a period of time in the first days, embryonic mortality will increase (Romanoff, 1949). After approximately 4–5 days of incubation, a lower temperature is not that critical anymore, and between 14 and 18 days, there is no problem if the incubation process is even temporarily stopped by cooling down the eggs to 22–25°C for a period as long as 24 h (Lancaster and Jones, 1988). During the hatching phase, the temperature of the eggs increases due to the activity of the hatching chick. This increase can be as high as 1–1.5°C. A too low temperature will reduce the hatch because the chicks will be slow in hatching and emerge too late from the eggshell. Low hatching temperatures (eggshell temperatures below 38.8–39°C during internal pipping) also result in a dragging hatch, a large

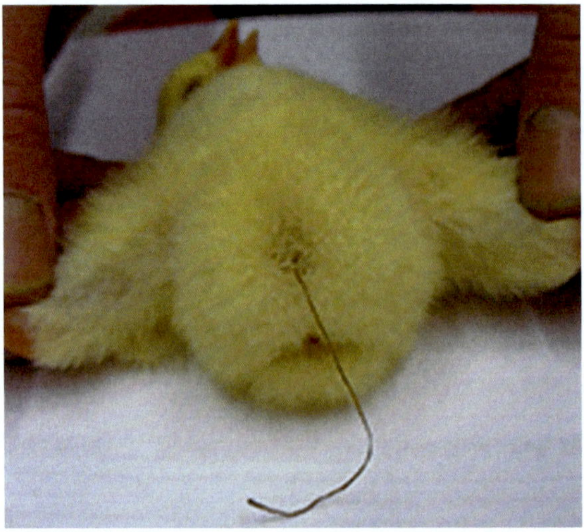

Picture 8 String navel. Picture courtesy of Petersime BV, Belgium.

Pictue 9 Red hocks, black button navel. Picture provided by Ron Meijerhof.

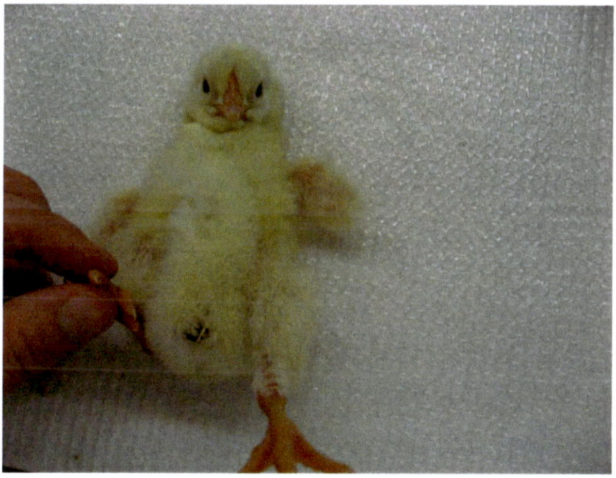

Picture 10 Black button navel. Picture provided by Ron Meijerhof.

hatch window (time between first and last chick hatching) (Willemsen et al., 2010) and more unhealed navels, both button navels and string navels (Pictures 8, 9 and 10). The actual level of the air temperature at which this will happen depends on the type of machine, as the resulting egg temperature is a function of air temperature, evaporation of water and air velocity and is influenced by, for instance, the level of ventilation during the hatching process.

At the end of incubation (until internal pipping), especially high temperatures result in poor chick quality. High temperatures from approximately day 13 to 14 onward will result in chicks with poorly closed navels, enlarged residual yolk sacs, less development of chick and organs, and more weak chicks that will have difficulty finding feed and water (Hulet, 2007; Molenaar et al., 2010). An

Picture 11 Red dots on the beak. Picture courtesy of Ron Meijerhof.

indicator for high temperatures at the end of incubation is the occurrence of red dots on the beak (Picture 11).

6.2 Turning

Turning of the eggs is important during the first half of the incubation period. Especially the period between day 3 and day 7 is critical (Deeming, 1989). Turning supports the uptake and use of nutrients from albumen and yolk (Eycleshymer, 1906) and prevents the embryo from sticking to the inner shell membrane (Romanoff, 1960). Lack of turning results in a decrease in arterial O_2 pressure of late embryos (Tazawa, 1980) through a lack of gas exchange through the incomplete development of the chorioallantoic membrane, as unabsorbed albumen is interposed between the chorioallantois and the inner shell membranes, thus reducing the gas exchange. Failing to turn results in problems with hatch but also in a poor chick quality (Robertson, 1961). After day 12, there is no need to turn anymore, as the chorioallantoic membrane is

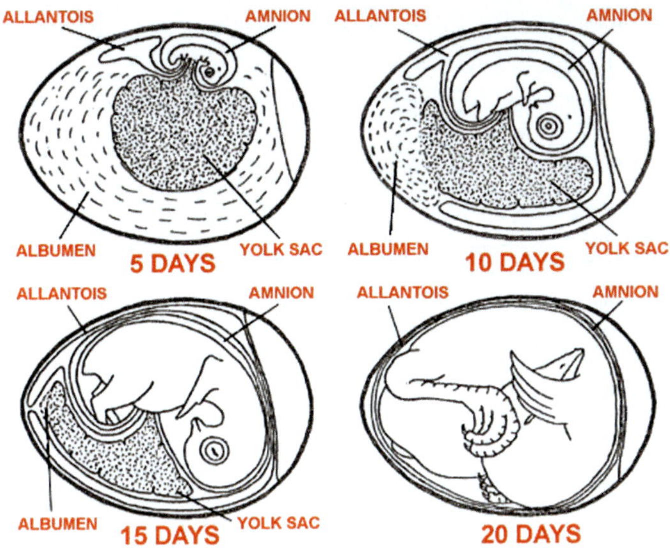

Figure 1 Development of embryo and extraembryonic membranes. Source: Romanoff (1939). Cornell Rural School Leaflet (Fig 9).

fully closed (Figure 1). Placing the egg trays in a horizontal position after day 12 can sometimes even have beneficial effects in some machines, as this position allows more airflow and therefore better cooling of the eggs at the end of incubation (Meijerhof, 2009).

In most commercial machines, the eggs are turned over a 38–45° angle to both sides once per hour. Some research (Wilson, 1991; Elibol and Brake, 2003) suggests that optimal turning frequencies might be higher than currently used, but this could also be influenced by a better distribution of airflow and therefore a more uniform temperature distribution.

6.3 Gaseous conditions

Next to temperature and turning, also gaseous conditions (O_2 and CO_2) and RH can influence the hatch results and chick quality. Based on the heat production of the eggs, the production of CO_2 and the consumption of O_2 can be calculated (Meijerhof and van Beek, 1993). If the requirements for O_2 and CO_2 in the air are known, ventilation of the machines can be calculated based on the number of viable embryos in the machine. As the eggs will lose moisture through evaporation, water in the form of humidity will be removed from the machine through ventilation as well. Due to the increasing heat production and with it O_2 requirement and CO_2 production, the ventilation requirements will be much higher at the end of the incubation process than at the start, which will influence the RH as well. In multistage incubation systems, the ventilation is almost always more than sufficient due to the limited average weight of the embryo in the machine, but in single stage the level of ventilation can be impaired.

6.3.1 Carbon dioxide

Although the embryos produce hardly any CO_2 at the start of incubation, a high level of CO_2 in the first week of incubation is reported to have beneficial effects. Willemsen et al. (2008) showed that CO_2 concentrations above 1% in the first half of incubation stimulate embryonic development, induce early hatching and increase hatchability, perhaps due to a stimulation of the development of the cardiovascular system (Verhoelst et al., 2011; de Smit et al., 2006).

Embryos at the end of the incubation process seem to tolerate high levels of CO_2. Experiments with CO_2 levels above 1% showed little effect on embryo development and hatch results. Tong et al. (2015) increased CO_2 concentration to 1.0% from E18 till hatching and found some effects on hatching moment but not on chick quality. Everaert et al. (2007) increased CO_2 levels up to 4% (40 000 ppm) in the second half of incubation without significant influences on embryonic growth or hatchability. However, Buys et al. (1998) found that a relatively small increase from 0.2% to 0.4% CO_2 by limiting the ventilation between day 14 and day 19 of incubation already created differences in chicken characteristics. This might be due to the way the CO_2 increase is realized by adding CO_2 to the machine or by closing the ventilation. In case of closing the ventilation, temperature patterns in the machine might be influenced as well, for instance, through an alteration of the humidity control.

In practical situations, the CO_2 concentrations in single-stage machines are often maximized at a level of approximately 0.4%, although higher concentrations during hatching are often used as they are believed to stimulate hatch. This might be due to the effect of the limited ventilation of the machines, which will affect the temperature of the eggs.

6.3.2 Oxygen

O_2 concentration is usually not a limiting factor for the normal development of the embryos, unless the temperature is increased. Lourens et al. (2007) showed that a level of 17% O_2 had more negative influence on heat production at 38.9°C than at 37.8°C eggshell temperature, where at 25% O_2 an eggshell temperature of 38.9°C had the highest heat production. At higher altitudes, the partial pressure of O_2 (PO_2) is decreased and with it hatchability will decrease (Visschedijk, 1985). As incubation at high altitudes will limit the availability of O_2 for the embryo, a higher eggshell temperature will be more detrimental. An altitude of more than 1200–1300 m above sea level is usually not advisable for hatcheries, although small negative effects can already be observed at altitudes of 600–700 m above sea level. If the hatchery needs to be situated at higher altitude, it is recommended to put the breeder farms at least at that level, as birds adjust their eggshell conductance based on altitude (Rahn et al., 1977). Injecting O_2 into the setters and hatchers will compensate for the negative effect but is usually too expensive for practical use in broiler hatcheries.

6.3.3 Relative humidity

RH influences the moisture loss of the eggs as it creates a difference in water vapor pressure inside and outside of the egg. A moisture loss (weight loss) of on average 12–14% during the total incubation period is required for optimal hatching results (Ar and Rahn, 1980), as it creates an air cell for the embryo to get sufficient O_2 during hatching and with it energy to escape from the eggshell. The distribution of the moisture loss over the incubation period has no influence on the embryo, as long as the total required moisture loss is achieved at the end of incubation. Usually, the moisture loss is determined at transfer (18 days of incubation) and should be 11–13% at that moment, but preferably not less than 10%. It is often assumed that a high weight loss during incubation will lead to dehydration of the embryo. However, as the moisture is lost from the allantois, which surrounds the amnion in which the embryo is positioned, dehydration of the embryo due to water loss from the egg is not likely to happen. The allantois acts as a buffer between the direct environment of the embryo and the outside of the shell.

Traditionally, RH during incubation is set at a constant level of approximately 55–60%, especially in multistage incubation systems. Single-stage incubation systems offer the opportunity to change the RH settings over the period of incubation, matching the ventilation level of the machines. In modern single-stage machines, the ventilation is based on the level of CO_2. This means that at the start of incubation the ventilation can be kept at zero for a number of days, while at the end of incubation a minimum level of approximately 600–650 m^3 per hour per 100 000 eggs needs to be maintained.

If a constant level of RH is set over the incubation period, it means that the machines have to ventilate at the start of incubation to remove the excess moisture from the eggs, where at the end of incubation the machine needs to evaporate water as moisture is removed with the increased ventilation. This is costing extra energy at the start (warming up cold incoming air) and creates hot and cold spots at the end of incubation (local cooling by the sprayer in the machine). On top of that, spraying water in the machine will increase the risk of bacterial growth. An option to avoid this is so-called 'non-linear moisture loss'. With this system, the RH increases in the first days of incubation and is at a lower level at the end, resulting in a correct amount of total moisture loss at 18 days of incubation. An advantage of this system is that the machine is more even in temperature distribution (no local cooling by the sprayer at the end of incubation) but also that the machines remain relatively dry. This helps to decrease the bacterial growth on the eggs and with it can help to reduce the first week mortality in the field. Downside of no or limited water spray in the machines is that the cooling effect of evaporation will not be there.

6.4 Hatching period

During hatching, the embryo will first open the air cell to start lung ventilation and then externally pip and free itself from the eggshell. During this process, the activity is high, the metabolic heat production approximately doubles (Janke et al., 2004) and as a result the temperature of the egg will go up. This means that during this process the temperature of the machine cannot be steered by measuring the eggshell temperatures. As the embryo before internal pipping is in the so-called plateau phase (a period where the heat production cannot increase anymore due to the limitation of O_2), the air temperature can be kept at a constant level.

A parameter to observe during hatching is the so-called hatch window, the time between the first and last chick hatching. Usually, this period is a minimum of 24 h but often may be up to 36 h. When the period is longer than 36 h, there is a 'dragging' hatch, which can result in poor chick quality and dehydrated chicks. This is usually the result of a too low temperature for the last chicks hatching. This can be due to a too early decrease in temperature but, for instance, also when the ventilation of the machine is increased significantly to comfort the hatched chicks.

Once the hatch starts, the opened eggs and hatched chicks will produce moisture and the RH will go up. This is a clear indication that the hatch has started. Also, the CO_2 production will go up due to the activity of the chicks. When the ventilation is increased to remove the CO_2 or the moisture, more cold air will be introduced into the hatcher and some of the chicks can become too cold to hatch in time, resulting in a dragging hatch. These slow-hatching chicks often have a string navel, a piece of membrane in the navel, which is a sign of too low temperatures during the hatching phase. As the last chicks are slow and late, one can also observe chicks that are still alive in the shell in a normal position with the yolk taken up but do not have the strength to escape timely from the egg. Also button navels (black dried-up 'buttons' on the navel) might be a result of too low temperature but also occur when the temperatures are too high.

When the body temperature of the chicks is too high, they will be panting in the hatchers, to release heat through evaporation. This panting can cause dehydration. Optimal chick body temperatures (cloaca temperatures) are approximately 40°C. When the body temperature rises above 41°C, the bird will start panting and can dehydrate quickly. Not only will the birds dehydrate but they will also lose energy, having more difficulties standing and walking, resulting in a slow start after placement in the broiler house.

During hatching, there is an enormous increase in bacterial pressure, as there is organic material in the presence of high temperature and high humidity. This high pressure can cause omphalitis due to bacterial penetration through the navel. For this reason, sometimes formaldehyde is used during the hatching process. Although this will reduce the bacterial pressure, there is also a risk of damaging the cells in the trachea, making the birds more susceptible to infection or vaccination reactions (Sander et al., 1995: Fauziah et al., 1996) and influence performance negatively (Zulkifli et al., 1999).

As mentioned in Section 5.1.3, recently alternative hatching systems have become available. Some of these systems transport the eggs at 18 days of incubation to the broiler farm and let them hatch on the litter. Other systems supply space, light, feed and water to the hatchers during the hatching process. Although data about the results with these systems are still limited, the first impression is that the birds in some of these systems need less medication in the first week (De Jong et al., 2017), perhaps due to a lower level of bacterial pressure during hatching or due to the early availability of feed and water activating the intestinal tract.

7 Processing, storage and transport of the day-old chick

Once the chicks are hatched and dried up, they can be removed from the hatchers and be processed. This processing consists of different steps like selection, counting, packing and vaccination.

Important during hatching but also during processing and storage is the structure of the floor on which the birds stand. Slippery floors can lead to splayed legs as the birds cannot keep their legs in the right position. Modern hatching and transport baskets are designed to prevent this, but the problem can occur when, for instance, newspapers are used as paper lining. Especially when carton boxes are used, leg and toe damage can occur when the cracks and folds of the boxes are not completely tight and bird feet get stuck underneath them.

The most important thing for chick health to control during storage and transport is the body temperature. When the body temperature is optimal (39-40°C), the bird will lose approximately 1-2 g of water per 24 h through breathing, but the digestion of the residual yolk produces water and will prevent dehydration. When the chick temperature is elevated (above 41°C), the birds will pant to evaporate moisture and cool themselves, resulting in extra weight loss. This can be quite severe, in extreme cases up to 10 g of moisture loss (weight loss) per 24 h. In broiler production, the storage and transport times usually are limited, but even on short periods of a few hours severe overheating and loss of condition can be experienced. The correct transport temperature depends on the heat production of the chicks and the heat loss. The heat production is not always equal. Transportation feed or early feeding applied already in the hatchery will increase the heat production and the transport temperature will need to be reduced. Smaller chicks but especially chicks from young breeder flocks produce less heat and need an adjusted transport temperature. As heat loss is not only dependent on temperature but also on air velocity, the construction of the transport vehicle and the shape of the transport boxes need to be taken into consideration to determine the right transport temperature. To check if the temperature is correct, the body temperature of the chicks should be determined.

As chicks react poikilothermic (no or limited regulation of heat production) for the first 4–5 days of life (Weytjens et al., 1999), a low environmental temperature will result in a rapid loss of body temperature. By themselves, the birds can handle this for a limited time, but they will become inactive and not move to feed and water, and the risk of navel/yolk sac infections will increase.

8 Conclusion

Chick health depends on more than just the absence of diseases. A healthy chick is capable of utilizing its genetic potential when it is provided with the correct nutrition, climate, light and management. To utilize that potential, chick quality needs to be optimal, which starts with good-quality hatching eggs with low bacterial pressure and good shell quality and hygiene. As embryos and DOCs are sensitive to bacterial contamination, bacterial pressure should be as low as possible during the incubation process and hatching. The chick needs to be free of diseases, but also needs to have an optimal development during the embryonic stage resulting in an optimal development of the body, organs and immune system. The conditions during incubation, hatching, processing and transportation need to be managed properly to ensure that the chicks will have a good start, can quickly adapt to the situation on the farm and find feed and water.

9 Acknowledgement

I'd like to thank Dr Ampai Nangsuay and Prof Sjaak de Wit for their valuable and constructive suggestions.

10 Where to look for further information

- Romanoff, A. L. 1960. *The Avian Embryo: Structural and Functional Development*. New York: Macmillan.
- Romanoff A. L. and Romanoff, A. J. 1949. *The Avian Egg*. New York: John Wiley and Sons, Inc.
- www.hatchability.com.
- www.poultryperformanceplus.com.

11 References

Andreasen, C. B. (2020). Staphylococcosis. In: *Diseases of Poultry* (14th edn.) Swayne, D. E., Boulianne, M., Logue, C. M., McDougald, L. R., Nair, V. and Suarez, D. L. (Eds). Wiley-Blackwell, Hoboken, NJ, 995–1003.

Ar, A. and Rahn, H. (1980). Water in the avian egg: overall budget of incubation. *Am. Zool.* 20(2), 373–384.

Archer, G. S. and Mench, J. A. (2014). Natural incubation patterns and the effects of exposing eggs to light at various times during incubation on post-hatch fear and stress responses in broiler (meat) chickens. *Appl. Anim. Behav. Sci.* 152, 44–51.

Bailey, J. S., Cason, J. A. and Cox, N. A. (1998). Effect of *Salmonella* in young chicks on competitive exclusion treatment. *Poult. Sci.* 77(3), 394–399.

Barri, A., Honaker, C. F., Sottosanti, J. R., Hulet, R. M. and McElroy, A. P. (2011). Effect of incubation temperature on nutrient transporters and small intestine morphology of broiler chickens. *Poult. Sci.* 90(1), 118–125.

Bauer, F., Tullett, S. G. and Wilson, H. R. (1990). Effects of setting eggs small end up on hatchability and posthatching performance of broilers. *Br. Poult. Sci.* 31(4), 715–724.

Beane, W. L., Siegel, P. B. and Siegel, H. S. (1965). Piperazine compounds and yolk discoloration. *Poult. Sci.* 44, 666–668.

Becker, W. A. (1960). The storage of hatching eggs and the post-hatching body weights of chickens. *Poult. Sci.* 39(3), 588–590.

Borst, L. B. (2020). Streptococcus and enterococcus. In: *Diseases of Poultry* (14th edn.) Swayne, D. E., Boulianne, M., Logue, C. M., McDougald, L. R., Nair, V. and Suarez, D. L. (Eds). Wiley-Blackwell, Hoboken, NJ, 1003–1010.

Buys, N., Dewil, E., Gonzales, E. and Decuypere, E. (1998). Different carbon dioxide levels during incubation interact with hatching time and ascites susceptibility in two broiler lines selected for different growth rate. *Av. Pathol.* 27(6), 605–612.

Byng, A. J. and Nash, D. (1962). The effects of egg storage on hatchability. *Br. Poult. Sci.* 3(2), 81–87.

Cardirci, S. (1997). The effect of fumigation regimens on shell structure and embryo viability. M.Sc Thesis. University of Glasgow.

Chute, H. L. and Richard, J. L. (1991). Fungal infections. In: *Diseases of Poultry* (9th edn.) Calnek, B. W., Barnes, H. J., Beard, C. W., Reid, W. M. and Yoder, H. W., Jr. (Eds). Iowa State University Press, Ames, Iowa, 326–339.

Corrier, D. E. (1991). Mycotoxicosis: mechanisms of immunosuppression. *Vet. Immunol. Immunopathol.* 30(1), 73–87.

Cox, N. A., Bailey, J. S., Maulding, J. M. and Blankenship, L. C. (1990). Presence and impact of *Salmonella* contamination in commercial broiler hatcheries. *Poult. Sci.* 69(9), 1606–1609.

Dayan, J., Reicher, N., Melkman-Zhavi, T. and Uni, Z. (2020). Incubation temperature affects yolk utilization through changes in expression of yolk sac tissue functional genes. *Poult. Sci.* 99(11), 6128–6138.

Deeming, D. C. (1989). Failure to turn eggs during incubation: development of the area vasculosa and embryonic growth. *J. Morphol.* 201(2), 179–186.

de Gouw, P., van de Ven, L. J. F., Lourens, S., Kemp, B. and van den Brand, H. (2017). Effects of dust, formaldehyde and delayed feeding on early postnatal development of broiler chickens. *Res. Vet. Sci.* 112, 201–207.

De Jong, I. C., van Riel, J., Bracke, M. B. M. and van den Brand, H. (2017). A 'meta-analysis' of effects of post-hatch food and water deprivation on development, performance and welfare of chickens. *PLoS ONE* 12(12), e0189350.

de Smit, L., Bruggeman, V., Tona, J. K., Debonne, M., Onagbesan, O., Arckens, L., De Baerdemaker, J. and Decuypere, E. (2006). Embryonic developmental plasticity of the chick: increased carbon dioxide during early stages of incubation changes the developmental trajectories during prenatal and postnatal growth. *Comp. Biochem. Physiol. A Mol. Integr. Physiol.* 145(2), 166–175.

Dolka, B., Gotebiewska-Kosakowska, M., Krajewski, K., Kwiecienski, P., Nowak, T., Szubstarski, J., Wilczynski, J. and Szeleszczuk, P. (2017). Occurrence of Enterococcus spp. in poultry in Poland based on 2014–2015 data. *Med. Weter.* 73(4), 220–224.

Doupnik, B. Jr. and Peckham, J. C. (1970). Mycotoxicity of *Aspergillus ochraceus* to chicks. *Appl. Microbiol.* 19(4), 594–597.

Elibol, O. and Brake, J. (2003). Effect of frequency of turning from three to eleven days of incubation on hatchability of broiler hatching eggs. *Poult. Sci.* 82(3), 357–359.

Elibol, O., Uysal, A. and Ertas, S. (2003). The effect of preincubation fumigation with different formaldehyde fumigation concentrations and periods on hatchability of chicken eggs. *J. Agr. Sci.* 9, 9–12.

Everaert, N., Kamers, B., Witters, A., de Smit, L., Debonne, M., Decuypere, E. and Bruggeman, V. (2007). Effect of four percent carbon dioxide during the second half of incubation on embryonic development, hatching parameters, and posthatch growth. *Poult. Sci.* 86(7), 1372–1379.

Eycleshymer, A. C. (1906). Some observations and experiments on the natural and artificial incubation of the common fowl. *Biol. Bull.* 12, 360–374.

Fauziah, O., Purton, M. D. and Solomon, S. E. (1996). Scanning electron microscopy of the respiratory epithelium of chicks fumigated with formaldehyde vapour. *Br. Poult. Sci.* 37(3), 563–570.

Filazi, A., Yurakok-Dikmen, B., Kuzukiran, O. and Tansel Sireli, U. (2017). Mycotoxins in poultry. In: *Poultry Science* Manafi, M. (Ed.). IntechOpen Limited, London, UK, 73-92

Funk, M. E. and Irwin, M. R. (1955). *Hatchery Operation and Management*. John Wiley & Sons Inc., New York.

Gucbilmez, M., Özlü, S., Shiranjang, R., Elibol, O. and Brake, J. (2013). Effects of preincubation heating of broiler hatching eggs during storage, flock age, and length of storage period on hatchability. *Poult. Sci.* 92(12), 3310-3313.

Hamburger, V. and Hamilton, H. L. (1951). A series of normal stages in the development of the chick embryo. *Dev. Dyn.* 195(4), 231–272.

Holst, W. F. and Halbrook, E. R. (1933). A 'scurvy-like' disease in chicks. *Science* 77(1997), 354.

Hu, D., Meng, F., Cui, Y., Yin, M., Ning, H., Yin, Z., Chen, L., Ge, Y. and Liu, S. (2020). Growth and cardiovascular development are repressed by florfenicol exposure in early chicken embryos. *Poult. Sci.* 99(5), 2736–2745.

Hulet, R. M., Gladys, G., Hill, D., Meijerhof, R. and El-Shiekh, T. (2007). Influence of egg shell embryonic incubation temperature and broiler breeder flock age on post hatch growth performance and carcass characteristics. *Poult. Sci.* 86(2), 408–412.

Hulet, R. M. (2007). Symposium: managing the embryo for performance, managing incubation: where are we and why? *Poult. Sci.* 86(5), 1017–1019.

Jackwood, M. and de Wit, J. J. (2020). Infectious bronchitis virus. In: *Diseases of Poultry* (14th edn.) Swayne, D. E., Boulianne, M., Logue, C. M., McDougald, L. R., Nair, V. and Suarez, D. L. (Eds). Wiley-Blackwell, Hoboken, NJ, 167–188.

Janke, O., Tzschentke, B. and Boerjan, M. (2004). Comparative investigations of heat production and body temperature in embryos of modern chicken breeds. *Avian Poult. Biol. Rev.* 15(3), 191–196.

Krough, P., Christensen, D. H., Hald, B., Halou, B., Larson, C., Pederson, E. J. and Thrane, U. (1989). Natural occurrence of the mycotoxin fusarochromanone, a metabolite of Fusarium equiseti, in cereal feed associated with tibial dyschondroplasia. *Appl. Environ. Microbiol.* 55(12), 3184–3188.

Lancaster, F. M. and Jones, D. R. (1988). Cooling of broiler hatching eggs during incubation. *Br. Poult. Sci.* 29(3), 597–604.

Leeson, S. and Summers, J. D. (2000). *Broiler Breeder Production*. Nottingham University Press, Nottingham, England.

Leeson, S. and Summers, J. D. (2009). *Commercial Poultry Nutrition*. Nottingham University Press, Nottingham, England.

Leeson, S. L. J., Caston, L. J. and Summers, J. D. (1989). The effect of graded levels of nicarbazin on reproductive performance of laying hens. *Can. J. Anim. Sci.* 69(3), 757–764.

Leksrisompong, N., Romero-Sanchez, H., Plumstead, P. W., Brannan, K. E. and Brake, J. (2007). Broiler incubation. 1. Effect of elevated temperature during late incubation on body weight and organs of chicks. *Poult. Sci.* 86(12), 2685–2691.

Lourens, A., van den Brand, H., Meijerhof, R. and Kemp, B. (2005). Effect of eggshell temperature during incubation on embryo development, hatchability and posthatch development. *Poult. Sci.* 84(6), 914–920.

Lourens, A., van den Brand, H., Heetkamp, M. J. W., Meijerhof, R. and Kemp, B. (2007). Effects of eggshell temperature and oxygen concentration on embryo growth and metabolism during incubation. *Poult. Sci.* 86(10), 2194–2199.

Lyon, C. E., Newell, G. W. and Morrison, R. D. (1972). Vitelline membrane strength and egg yolk mottling. *Poult. Sci.* 51(2), 480–487.

Marcellus, F. N., Gwatking, R., and Glover, J. S. (1930). Incubator disinfection in the control of Salmonella Pullorum. Proceedings of the 4th World's Poultry Congress, 373-378.

Mather, C. M. and Laughlin, K. F. (1977). Storage of hatching eggs: the effect on early embryonic development. *Br. Poult. Sci.* 18(5), 597–603.

Mather, C. M. and Laughlin, K. F. (1979). Storage of hatching eggs: the interaction between parental age and early embryonic development. *Br. Poult. Sci.* 20(6), 595–604.

McNamee, P. T. and Smyth, J. A. (2000). Bacterial chondronecrosis with osteomyelitis (femoral head necrosis) of broiler chickens: a review. *Avian Pathol.* 29(4), 253–270.

Meijerhof, R. (1992). Pre-incubation holding of hatching eggs. *Worlds Poult. Sci. J.* 48(1), 57–68.

Meijerhof, R. (2009). Principles and practice of incubator design. *Avian Biol. Res.* 2(1-2), 81–86.

Meijerhof, R. and van Beek, G. (1993). Mathematical modelling of temperature and moisture loss of hatching eggs. *J. Theor. Biol.* 165(1), 27–41.

Meijerhof, R. and Hulet, R. M. (1997). In ovo injection of competitive exclusion culture in broiler hatching eggs. *J. Appl. Poult. Res.* 6(3), 260–266.

Merrit, E. S. (1964). Pre-incubation storage effects on subsequent performance of chickens 1. *Br. Poult. Sci.* 5(1), 67–73.

Molenaar, R., van Roovert-Reijrink, I. A. M., Meijerhof, R. and van den Brand, H. (2010). Meeting embryonic requirements of broilers throughout incubation: a review. *Br. Poult. Sci.* 12(3), 243–250.

Molenaar, R., Hulet, R., Meijerhof, R., Maatjens, C. M., Kemp, B. and van den Brand, H. (2011). High eggshell temperatures during incubation decrease growth performance and increase the incidence of ascites in broiler chickens. *Poult. Sci.* 90(3), 624–632.

Moran, E. T. Jr. (2007). Nutrition of the developing embryo and hatchling. *Poult. Sci.* 86(5), 1043–1049.

Nangsuay, A., Meijerhof, R., van den Anker, I., Heetkamp, M. J. W., De Souza Morita, V. S., Kemp, B. and van den Brand, H. (2016). Effect of breeder age, broiler strain, and eggshell temperature on development and physiological status of embryos and hatchlings. *Poult. Sci.* 95(7), 1666–1679.

Nasri, H., van den Brand, H., Najjar, T. and Bouzouaia, M. (2020). Egg storage and breeder age impact on egg quality and embryo development. *J. Anim. Physiol. Anim. Nutr. (Berl)* 104(1), 257–268.

Noble, R. C. and Cocchi, M. (1990). Lipid metabolism and the neonatal chicken. *Prog. Lipid Res.* 29(2), 107–140.

Oviedo-Rondon, E. O., Small, J., Wineland, M. J., Christensen, V. L., Mozdziak, P. S., Koci, M. D., Funderburk, S. V. L., Ort, D. T. and Mann, K. M. (2008). Broiler embryo bone development is influenced by incubator temperature, oxygen concentration and eggshell conductance at the plateau stage in oxygen consumption. *Br. Poult. Sci.* 49(6), 666–676.

Oviedo-Rondon, E. O., Wineland, M. J., Small, J., Cutchin, H., McElroy, A., Barri, A. and Martin, S. (2009). Effect of incubation temperatures and chick transportation conditions on bone development and leg health. *J. Appl. Poult. Sci. Res.* 18(4), 671–678.

Oznurlu, Y., Celik, I., Telatar, T. and Sur, E. (2010). Histochemical and histological evaluations of the effects of high incubation temperature on embryonic development of thymus and bursa of Fabricius in broiler chickens. *Br. Poult. Sci.* 51(1), 43–51.

Peterka, M., Peterkova, R. and Likovsky, Z. (1996). Teratogenic and lethal effects of long-term hyperthermia and hypothermia in the chick embryo. *Reprod. Toxicol.* 10(4), 327-332.

Polin, D. and Porter, C. C. (1956). The effect of nicarbazin on porphyrin and yolk formation. *Poult. Sci.* 35, 1165.

Polin, D., Ott, W. H. and Siegmund, O. H. (1957). The incidence and degree of yolk mottling in eggs from hens fed diets with and without nicarbazin. *Poult. Sci.* 36(3), 524–528.

Proudfoot, F. G. and Stewart, D. K. R. (1970). Effect of pre-incubation fumigation with formaldehyde on the hatchability of chicken eggs. *Can. J. Anim. Sci.* 50(3), 453–465.

Rahn, H., Carey, C., Balmas, K., Bhatia, B. and Paganelli, C. (1977). Reduction of pore area of the avian eggshell as an adaptation to altitude. *Proc. Nat. Acad. Sci.* 74(7), 3095–3098.

Robertson, I. S. (1961). The influence of turning on the hatchability of hens' eggs. II. The effect of turning frequency on the pattern of mortality, the incidence of malpositions, malformations and dead embryos with no somatic abnormality. *J. Agric. Sci.* 57(1), 57–69.

Romanoff, A. L. (1949). Critical periods and causes of death in avian embryonic development. *Auk* 66(3), 264–270.

Romanoff, A. L. (1960). *The Avian Embryo. Structural and Functional Development*. MacMillan Company, New York, 1041–1140.

Sander, J. E., Wilson, J. L., Rowland, G. N. and Middendorf, P. J. (1995). Formaldehyde vaporization in the hatcher and the effect on tracheal epithelium of the chick. *Avian Dis.* 39(1), 152–157.

Sherwood, D. H., Milby, T. T. and Higgins, W. A. (1956). The effect of nicarbazin on reproduction white rock breeder hens. *Poult. Sci.* 35(5), 1014–1019.

Shiranjang, R., Ozlu, S. and Elibol, O. (2018). The effect of egg storage period, hatching time, and initial brooding litter temperature on performance of chicks from young breeder. *International Poultry Science Congress of WPSA Turkey Branch*.

Tazawa, H. (1980). Adverse effect of failure to turn the avian egg on embryo oxygen exchange. *Respir. Physiol.* 41(2), 137–142.

Tong, Q., McGonnell, I. M., Roulston, N., Bergoug, H., Romanini, C. E. B., Garain, P., Eterradossi, N., Exadaktylos, V., Bahr, C., Berckmans, D. and Demmers, T. G. M. (2015). Higher levels of carbon dioxide during late incubation alter the hatch time of chicken embryos. *Br. Poult. Sci.* 56(4), 503–509.

van den Brand, H. M. P., Sosef, M. P., Lourens, A. and van Harn, J. (2016). Effects of floor eggs on hatchability and later life performance in broiler chickens. *Poult. Sci.* 95(5), 1025–1032.

Verhoelst, E., De Ketelaere, B., Decuypere, E. and De Baerdemaeker, J. (2011). The effect of early prenatal hypercapnia on the vascular network in the chorioallantoic membrane of the chicken embryo. *Biotechnol. Prog.* 27(2), 562–570.

Visschedijk, A. H. (1985). Gas exchange and hatchability of chicken eggs incubated at simulated high altitude. *J. Appl. Physiol. (1985)* 58(2), 416–418.

Walstra, I., Ten Napel, J., Kemp, B. and van den Brand, H. (2012). Adaptive response to Eimeria Acervulina in rearing hens is affected by suboptimal incubation temperature and heat exposure in later life. *Animal* 6(1), 137–144.

Weytjens, S., Meijerhof, R., Buyse, J. and Decuypere, E. (1999). Thermoregulation in chicks originating from breeder flocks of two different ages. *J. Appl. Poult. Res.* 8(2), 139–145.

Whistler, P. E. and Sheldon, B. W. (1989). Bactericidal activity, eggshell conductance and hatchability effects of ozone versus formaldehyde disinfection. *Poult. Sci.* 68(8), 1074–1077.

Wijnen, H. J., van den Brand, H., Lammers, A., van Roovert-Reijrink, I. A. M., van der Pol, C. W., Kemp, B. and Molenaar, R. (2020). Effects of eggshell temperature pattern during incubation on primary immune organ development and broiler immune response later in life. *Poult. Sci.* 99(12), 6619–6629.

Willemsen, H., Tona, K., Bruggeman, V., Onagbesan, O. and Decuypere, E. (2008). Effects of high CO2 level during early incubation and late incubation in ovo dexamethasone injection on perinatal embryonic parameters and post-hatch growth of broilers. *Br. Poult. Sci.* 49(2), 222–231.

Willemsen, H., Kamers, B., Dahlke, F., Han, H., Song, Z., Ansari Pirsareaei, Z., Tona, K., Decuypere, E. and Everaert, N. (2010). High- and low-temperature manipulation during late incubation: effect on embryonic development, the hatching process, and metabolism in broilers. *Poult. Sci.* 89(12), 2678–2690.

Wilson, H. R. (1991). Physiological requirements of the developing embryo: temperature and turning. In: *Avian Incubation* Tullett, S. G. (Ed.). Butterworth-Heinemann, London, 145–156.

Wright, M. L. and Epps, N. A. (1958). Hatchery sanitation. *Can. J. Comp. Med. Vet. Sci.* 22(11), 396–399.

Wright, M. L., Anderson, G. W. and Epps, N. A. (1960). Hatchery sanitation as a control measure for Aspergillosis in fowl. *Avian Dis.* 4(4), 369–379.

Zulkifli, I., Fauziah, O., Omar, A. R., Shaipullizan, S. and Siti Selina, A. H. (1999). Respiratory epithelium, production performance and behavior of formaldehyde exposed broiler chicks. *Vet. Res. Commun.* 23(2), 91–99.

Chapter 8

Optimizing the health of broilers

Edgar Orlando Oviedo-Rondón, North Carolina State University, USA

1 Introduction

The broiler sector keeps growing in most countries worldwide at an average annual rate of 3.4% from 2009 to 2019 (IndexBox, 2021). Global broiler production reached 137 million tons of meat in 2020 (IndexMundi, 2021) (Fig. 1). Broiler meat is still one of the most well-accepted and affordable animal protein sources in all markets. In 2019, the highest levels of poultry meat per capita consumption were observed in Malaysia (63 kg per person), the United States (58 kg per person), and Brazil (57 kg per person). But, the highest volumes of poultry consumption were in China (20 M tons), the United States (19 M tons), and Brazil (12 M tons), compiling a 40% share of global consumption (Berkhout, 2020). The FAO has forecasted that poultry meat will decelerate with a compound annual growth rate of 2.3% between 2019 and 2030. Thus, it is projected that the poultry meat market worldwide will reach 166 M tons by 2030 (FAO, 2003). The top three broiler producers in the world continue to be the United States (23 M tons), China (20 M tons), and Brazil (16 M tons), with a combined 45% share of global production. They were followed by the 27 countries of the European Union, the Russian Federation, and India. The top five broiler-producing countries have accounted for almost 59% of global broiler production in the past 5 years.

Broiler production systems have been adapting to consumer demands related to the size and weight of birds, the final products required, food safety issues, welfare concerns, environmental impact, use of antibiotics, and other

http://dx.doi.org/10.19103/AS.2022.0104.11

Figure 1 Estimated broiler production in the top 30 country producers in thousands of MT during 2020. (https://www.indexmundi.com/agriculture/?commodity=broiler-meat&graph=production).

related problems. The necessary changes in the production systems to meet these market demands represent some challenges. As the broiler industry develops, some health issues have become significant concerns with emerging and re-emerging diseases (Hafez and Attia, 2020). Despite using similar genetics worldwide, the infrastructural and organizational biosecurity, the technological level of housing, and environmental control vary greatly (Gelaude et al., 2014; Van Limbergen et al., 2018). Consequently, disease incidence, prevalence, and mortality rates fluctuate among countries and regions. The resources lost in mortality, disease prevention, nutritional and management interventions, carcass defects, and condemnations are considerable. Optimizing broiler health is a top priority for this industry due to the high cost of the major diseases, impacts on welfare, condemnations, and food safety issues at the processing plant (Jones et al., 2019; Blake et al., 2020). Control of broiler diseases has been demanding for several countries. For example, a recent study (Swidan et al., 2020) indicated that the losses due to mortalities resulted mainly from avian influenza (AI), Newcastle disease (ND), Gumboro, and coccidiosis diseases between 2008 and 2019 constituted about 52% of the total income from animal production in Egypt.

This chapter will present the critical infectious disease challenges faced by the top five broiler-producing countries. The current and future strategies to control these diseases will be discussed. These strategies include biosecurity, vaccination, surveillance, diagnostics, environmental management, nutritional interventions, and genetic selection. In addition, the potential strategies to enhance and optimize resistance to infectious diseases will be discussed while reducing the reliance on antibiotics or only on vaccination.

2 Key infectious disease challenges faced by broilers around the world

The disease challenges that broilers face change according to the type of housing and level of environmental control, weather conditions, poultry population density, endemic diseases in the area, final body weight, downtime between flocks, reuse of litter, water quality, diet composition and feed additives, health management plans, and biosecurity level, among other factors (Gaucher et al., 2015; Luyckx et al., 2015; Van Limbergen et al., 2018; Smith, 2019; Fancher et al., 2020; Abioja and Abiona, 2020; Van Limbergen et al., 2020; Swelum et al., 2021). The ranking of broiler diseases varies from country to country and within countries from region to region or company to company. There are major concerns around the world to preserve the food safety of poultry meat. Consequently, the prevention of diseases and contamination of meat and other edible poultry products caused by *Salmonella* and *Campylobacter* spp. are always top priorities due to the zoonotic implications (Hafez and Attia, 2020). This topic will be covered in other chapters and consequently will not be covered here. However, there are good sources of information to seek strategies to control *Salmonella* (Gast and Porter, 2020), *Campylobacter* (Zhang and Sahin, 2020), and other pathogens related to food safety.

With the fantastic collaboration of veterinarian colleagues, poultry scientists, producers, managers, and diverse experts located in the top broiler-producing countries or production areas globally, this chapter summarizes their informed opinion about the ranking of current broiler diseases. These rankings illustrate how specific disease occurrences may change depending on the production system used. These examples also help to discuss local challenges where they are more relevant. The broiler disease rankings were organized according to what experts surveyed and considered important in 2020, but data since 2017 are presented for the four bigger broiler producer countries. Thus, readers can conclude which are the critical points of control in each production system to minimize the incidence of diseases and optimize health. For the sake of brevity, only the most common diseases will be commented on for each country, but the complete list obtained in each country may serve as a future reference, and keep in mind that other health challenges must be addressed.

2.1 United States

In the United States, the Association of Veterinarians in Broiler Production (AVBP) produces an annual ranking of major broiler diseases and significant concerns of the industry included within the Committee on Poultry and other Avian Species of the United States Animal Health Association (USAHA, 2021). The AVBP ranking of broiler diseases during the past 4 years is presented in Table 1. This ranking indicates that coccidiosis, necrotic enteritis (NE), chick quality and early mortality, infectious bronchitis virus (IBV), and gangrenous dermatitis have been among the five most common diseases.

Intestinal diseases related to coccidia infection, dysbacteriosis, and NE have been the major challenges in the past 4 years. One factor could be the common utilization of reused litter in broiler houses generally for more than a year in several flocks. However, the greater relevance of gut health issues could be due to the rapid increase of antibiotic-free production. Currently,

Table 1 Broiler disease ranking in the United States from 2017 to 2020

Infectious disease category	Ranking			
Year	2017	2018	2019	2020
Coccidiosis	1	1	1	1
Necrotic enteritis	2	2	2	2
Chick quality and early mortality	4	3	3	3
Infectious bronchitis, IBV, respiratory	3	4	4	4
Gangrenous dermatitis	6	6	5	5
Novel reovirus	9	7	6	6
Bacterial osteomyelitis of the legs	11	8	8	7
General polyserositis - *Escherichia coli*	10	12	7	8
Infectious laryngotracheitis (ILT)	5	5	9	9
Infectious bursal disease	12	11	10	10
Vertebral osteomyelitis/kinky back	14	13	11	11
Infectious bronchitis, IBV, nephropathogenic	16	14	12	12
Histomoniasis	13	10	14	13
Avian influenza (AI)	7	9	13	14
Mycoplasma	15	15	15	15
Cholera			16	16
Newcastle disease (ND)		16	17	17
Marek's disease	17	17		

Source: Opinion of the Association of Veterinarians in Broiler Production (AVBP) included in the annual report of the United States Animal Health Association (USAHA), compiled and organized by Oviedo-Rondón (2019).

over 60% of the broilers in the United States do not receive growth-promoting antibiotics, and a growing percentage will meet requirements for no antibiotics ever programs (Smith, 2019; Fancher et al., 2020). In 2018, less than 8% of broilers were raised with the full spectrum of any feed antibiotics approved by the FDA.

Less than 15% of the chicks received an antibiotic in the hatchery in 2018, down from about 90% in 2013. Consequently, chick early mortality frequently includes bacterial contamination. Under this production system, chick quality and early survival are the vital aspects of success (Smith, 2019). Chick quality is related to the good development of all organs and physiological systems, including the immune system during incubation. Therefore, hatchery infrastructure, maintenance, management, and low contamination are pivotal in broiler production. Other bacterial diseases like gangrenous dermatitis, bacterial osteomyelitis, general polyserositis caused by *Escherichia coli*, and vertebral osteomyelitis could be related to dysbacteriosis that facilitates these pathogenic commensal bacteria (Smith, 2019; Fancher et al., 2020).

The most common viral challenges are IBV (Jackwood and de Wit, 2020; Legnardi et al., 2020), novel reovirus, and infectious laryngotracheitis (ILT) (Maekawa et al., 2019). These diseases have become either endemic or seasonal in some regions of the leading broiler-producing states. In some areas, the high population density, decades of using live vaccines, ILT chicken embryo origin vaccines, and the consequent appearance of IBV variants contribute to this higher incidence.

2.2 China

The author compiled in Table 2 the opinions of poultry veterinarians and live production managers from China and the incidence of diseases diagnosed in private laboratories recorded by the three largest broiler producers. These data include only the production of white-feathered chickens comparable to those produced in other countries. However, the housing systems for these broilers are mainly in cages and slats. Therefore, this table also includes the percentage of positive flocks to each one of the diseases.

The reader can observe the changes in disease categories reported in China compared to the United States and other regions of the world. Housing systems and the use of antibiotics may influence these results. In China, three housing systems are used for white-feathered broilers: the net-flooring system (NFS), the normal cage system, and the high-standard cage system. The main difference between the two cage systems is whether they use automatic feces-scraping equipment. Farmers raising white-feathered broilers prefer the NFS and cage systems over the standard flooring system with litter (Xin et al., 2016; Anderson, 2017; Chen et al., 2020). These housing conditions bring disease

Table 2 Ranking of diseases observed in white-feathered broilers in China from 2017 to 2020

Year	2017		2018		2019		2020	
Infectious disease category	Ranking	Positive[a] %	Ranking	Positive[a] %	Ranking	Positive[a] %	Ranking	Positive[a] %
General *Escherichia coli*	1	30.80	1	35	2	14	1	56.60
Staphylococcus	6	3.80	3	10	1	18	2	14.50
Infectious bronchitis, IBV (nephropathogenic and respiratory)	4	7.70	5	5	4	10%	3	11.50
Salmonella spp			10	2	6	6	4	4.00
Avian influenza (AI; H9N2)	2	19.20	5	5	8	4	5	2.90
Fowl adenovirus infections	4	7.70	12	2	13	1	6	2.60
Mycoplasma synoviae (MS)	2	19.20	5	5	7	5	7	1.70
Chicken infectious anemia (CIA)			2	13	8	4	8	1.20
Avian leukosis			10	2	3	13	9	1.20
Marek's disease	6	3.80	4	7	5	9	10	0.70
Reticuloendotheliosis					13	1	11	0.60
AI (H5N1)					8	4	12	0.40
Infectious laryngotracheitis (ILT)			9	3	11	3	13	0.20
Newcastle disease (ND)	6	3.80	8	4	11	3	14	0.10
Avian reovirus			12	2				
Infections bursal disease	6	3.80	12	2				

Source: Poultry veterinarians and live production managers of the Republic of China.
[a]Percentage of positive flocks for a specific disease over the total evaluated in diagnostic laboratories of the largest three broiler integrator companies.

challenges different from those observed in flooring systems with litter or reused material common in other countries. In addition, the low contact of birds with excreta reduces the incidence of coccidiosis and NE.

In contrast, Chinese veterinarians and experts reported a high incidence of *E. coli* and *Staphylococcus* spp. infections that also have been described in publications (Xu et al., 2020). These are some of the most prevalent bacteria in the intestines and the skin of poultry and humans. If these groups of commensal bacteria and *Salmonella* spp., especially *Salmonella pullorum*, cause persistent diseases in broilers, subclinical dysbiosis could be one of the main unforeseen issues (Ducatelle et al., 2018). It is important to notice that in the opinion of Chinese veterinarians working with broiler production in flooring systems and also probably some yellow-feathered chickens, coccidiosis, NE, and mycoplasma are the top three diseases together with viral infections like ND, AI, IBV, infectious bursal disease, and ILT.

Regarding dysbacteriosis, it is necessary to consider the type of feed ingredients, feed processing, water quality, and feed additives used. The use of growth-promoting antibiotics in China was elevated in the past years (Krishnasamy et al., 2015; Schoenmakers, 2020). In 2013, China consumed nearly half the world's antibiotics, 162 000 tons, 52% administered to animals. These volumes raised concerns in the public and governmental institutions.

Krishnasamy et al. (2015) commented that coccidiostats and arsenicals were the most common antimicrobials used in Chinese poultry production by weight. However, macrolides, penicillins, and tetracyclines were also widely used, resulting in potential antimicrobial resistance (AMR) and public health. In 2020, Xu et al. reported that the misuse of antibiotics in medium- and small-sized farms is still high in some regions of China. These authors also indicated that amoxicillin, oxytetracycline, and ceftriaxone were the most commonly used antibiotics, while ofloxacin and norfloxacin were the second most common antibiotics employed. This report confirms that penicillin, tetracyclines, cephalosporins, fluoroquinolones, and quinolones utilized in human medicine were misused in poultry. As a result, China has had one of the highest frequencies of AMR in the world (Schoenmakers, 2020). For instance, colistin resistant *E. coli* (CREC) infections in humans and animals have been widely reported in China (Wang et al., 2018).

The situation may be changing at this moment (Schoenmakers, 2020). In 2016, the Chinese government announced the National Action Plan to Combat AMR from Animal Resources (2017–2020). For example, colistin use in animal feed was banned on April 30, 2017. Consequently, colistin sulfate premix production decreased from 27 170 tons in 2015 to 2497 tons in 2018. A recent evaluation by Wang et al. (2020) of CREC and *mcr*-1 bacterial gene abundance, the bacterial marker for this AMR, from animals and humans between 2015 and 2019, indicated that after the ban, colistin residue in chicken farms decreased

96%, and the median relative abundance of *mcr*-1 per 16S RNA decreased 77%. The CREC identification went from 18.1% to 5%. However, clinical CREC infections in humans from 26 provinces and municipalities comprised 1.7% of *E. coli* infections in 2015–16 and only reduced to 1.3% in 2018–19. This data and other evidence indicate that AMR infections may persist in human and poultry populations for a few more years (Schoenmakers, 2020). This situation in China is a good example of how the problems related to antibiotic use may evolve in other countries that still are initiating this process.

The Ministry of Agriculture and Rural Affairs has pointed out Chinese medicine and probiotics as likely candidates for new product research among antibiotic substitutes. However, like in any other part of the world, plans for antibiotic substitution must also include improving chick quality, farm sanitation, biosecurity, enhancing drinking water and feed quality, integral programs of disease prevention, training about appropriate antibiotic use to farmers, and awareness of the risks of misusing them. Other key factors include educating farmers, veterinarians, and pharmacists in disease diagnostics, providing technological tools for faster and better diagnostics, restricting free availability, and enforcing the national regulations.

The IBV disease is also a key challenge for China's broiler industry like it is in almost the whole world. Both respiratory and nephropathogenic IBV strains are observed. The AI subtype H9N2 of low pathogenicity (LPAI) and the highly pathogenic subtype H5 (HPAI) occurrence is more common in China than in other countries (Parvin et al., 2020; Bo et al., 2021). Xu et al. (2020) also indicated that ILT and ND had been diagnosed in approximately 42% of 88 broiler farms surveyed in their study, which could mean a high incidence in some regions of China.

2.3 Brazil

The opinions of renowned Brazilian poultry veterinarians working in diverse integrations, vaccine companies, and diagnostic centers were compiled in Table 3. Currently, coccidiosis ranks again as the number one disease, and chick quality and early mortality are considered even more critical than nonspecific enteritis observed (Gazoni et al., 2020, 2021).

In surveys between 2012 and 2019, Gazoni et al. (2021) evaluated 13 648 broilers between 9 and 49 days of age in 13 Brazilian States. This group detected subclinical coccidiosis evaluated by microscopy due to *Eimeria maxima* in 34.8% of the flocks. In another study conducted between 2017 and 2018, the same group (Gazoni et al., 2020) observed *E. maxima* in 45% of mucosa scrapings in 32 broiler integrator companies. However, *Eimera acervulina* had the highest prevalence in macroscopic lesions, positively correlated with cellular desquamation, bedding ingestions, and feed passage. In addition,

Table 3 Ranking of broiler diseases in Brazil from 2017 to 2020

Disease category (infectious disease)	Ranking			
Year	2017	2018	2019	2020
Coccidiosis	1	1	1	1
Chick quality and early mortality	5	5	2	2
Enteritis (unspecific)	3	2	4	3
Infectious bronchitis, IBV (respiratory)	2	3	3	4
Escherichia coli (general polyserositis)	4	4	5	5
Mycoplasma (MG and MS)	6	8	7	6
Novel reovirus	9	10	8	7
Salmonellosis (*Salmonella pullorum* and *Salmonella gallinarum*)	8	7	6	8
Infectious bursal disease (IBD)	7	6	9	9
Infectious coryza	14	13	13	10
Infectious bronchitis, IBV (nephropathogenic)	11	9	10	11
Cholera	10	11	11	12
Bacterial osteomyelitis in legs	12	12	12	13
Infectious laryngotracheitis	17	17	16	14
Necrotic enteritis	16	16	15	15
Chicken infectious anemia (CIA)	13	14	14	16
Avian encephalomyelitis	15	15	17	17
Gangrenous dermatitis	18	18	18	18
Vertebral osteomyelitis	19	19	19	19
Marek's disease[a]	20	20	20	20
Histomoniasis[a]	21	21	21	21
Newcastle disease[b]	22	22	22	22

Source: Brazilian poultry veterinarian specialists.
[a]Extremely low incidence. Avian Influenza is still considered an exotic disease in Brazil.
[b]Latest outbreak limited to wild or backyard birds.

these studies identified factors that correlated with a reduction in intestinal health and negative effects on performance.

The fourth disease is the respiratory form of IBV (Colvero et al., 2015; Bande et al., 2017), and *E. coli* infection (Gonzalez and Cerqueira, 2019) is the fifth disease in this ranking. Colvero et al. (2015) calculated that IBV causes total losses of US$3563 per 1000 broiler breeders and $266.4 per 1000 broilers. There are several reports of multiple occurrences of outbreaks caused by IBV genotypes that differ (>25% divergent) from the Massachusetts type, which is the only type of live IB vaccine for vaccination in Brazil (Bande et al., 2017). Consequently, the economic losses could be attributed to poor vaccine protection (Colvero et al., 2015). This could be due to rapid mutations during replication of H120 and Ma5 viral strains used for vaccination in Brazil (Saraiva et al., 2018).

Mycoplasma infection is still the sixth most common disease (Reck et al., 2019). AI is still exotic in Brazil, and ND has occurred only in backyard flocks in the past decades. In Brazil, the housing, management, feed ingredients, feed additives, and the whole broiler production system are similar to those observed in the United States. Therefore, the critical disease challenges are very similar.

Brazil is the largest exporter of broiler meat globally, with a market share of almost 40%, exporting to over 150 countries. Consequently, broiler producers must meet Brazilian regulations and those from their main markets in China, Saudi Arabia, South Korea, the European Union, South Africa, and Japan. Some of these regulations include exporting products free of residues and even metabolites of antibiotics and coccidiostats. As a result, part of the Brazilian broiler production is antibiotic-free or has more extended withdrawal periods.

2.4 The European Union

The author used two sources of information to formulate a view of Europe's key infectious disease challenges. First, the author was given access to a disease prevalence data set of the Poultry Veterinary Study Group of the European Union (PVSGEU) and the most recent quarterly reports from the European Food Safety Authority related to AI (EFSA, 2021b).

The PVSGEU members are an informal group of poultry veterinarians from countries across the continent who meet twice each calendar year to share experiences on poultry disease, poultry welfare, and food safety issues in the poultry sector that have arisen during the previous 6 months. The PVSGEU has provided a summary document to the author of the trends in poultry disease incidence reported by the country members over the last two decades. However, this data set should be interpreted with caution. It is not an official record of the EU poultry disease incidence but reports individual veterinarians' views that can be influenced by various factors related to their work area.

The author has noted the following findings from the PVSGEU data set (Table 4). The use of antibiotic growth promotant in poultry diets was banned from January 1, 2006. However, despite the antibiotic ban, coccidiosis and NE do not feature the top three diseases of concern. Clostridium-associated diseases rank third, suggesting that overt NE is less of an issue for European poultry producers, but other clostridium-associated conditions are not uncommon. Together with leg problems, chick quality and early mortality are the two most important disease conditions that the PVSGEU encounter. This report probably reflects the complex etiology of these conditions and, therefore, why they

Table 4 Ranking of broiler diseases based on PVSGEU data set from 2017 to 2020

Infectious disease category	Ranking			
Year	2017	2018	2019	2020
Chick quality and early mortality	7	2	1	1
Leg problems	2	6	5	2
Clostridium	3	3	3	3
Enterococcus	4	4	1	3
Bacterial osteomyelitis of the legs (femoral head necrosis)	3	5	8	4
Enterococcus caecorum	5	8	4	4
Infectious bronchitis, respiratory	8	6	14	5
Escherichia coli infections	1	1	2	5
Dysbacteriosis	10	6	8	5
Necrotic enteritis	13	7	7	6
Marek's disease	10	14	15	6
Infectious bronchitis, IBV (nephropathogenic)	13	9	10	7
Coccidiosis	6	4	4	8
Infectious bursal disease	13	12	9	9
Campylobacter	8	6	6	9
Inclusion body hepatitis (IBH)	13	9	14	10
Gizzard erosion	10	10	13	11
Novel reovirus	14	13	11	12
Mycoplasma synoviae	11	11	16	12
Ruptured tendon	13	11	15	12
Avian influenza (H5N8) broilers	5			12
Eimeria necatrix	15	14	15	13
Vertebral osteomyelitis/kinky back	9	8	15	14
Enterococcus hirae	11	10	16	14
Gangrenous dermatitis	18		17	15
Mycoplasma gallisepticum	14	15	17	15
Avian encephalomyelitis	17	15	17	16
Infectious laryngotracheitis	11	14	15	17
Chicken anemia virus (CAV)	16	13	17	17
Black bone	17	16	17	18
Newcastle disease	17	16	18	

Source: Poultry Veterinarians Study Group from Europe (PVSGEU). Data were compiled and adapted from Oviedo-Rondón (2019).

continue to be reported. *Enterococcus* infections rank high in more recent years, which undoubtedly reflects the significant reduction in the use of therapeutic antibiotics in poultry production across the EU (Jung et al., 2018).

E. coli infections regularly rank high in the list, but this is unsurprising as *E.coli* is a common bacterial finding at postmortem examination in poultry (Nolan et al., 2020). Dysbacteriosis ranked fifth in 2020, but this is a syndromic condition related to intestinal integrity and stability that can be influenced by multiple factors, including feed ingredient quality, which often reflects feed pricing.

AI was not ranked high in the 2020 PVSGEU summary (Table 4). Still, the data set was collated 6 months before the 2020/2021 winter HPAI incursion and, therefore, the spike in case numbers related to the latest incursions. The author has collated the information about the HPAI status from the *EFSA Journal* quarterly reports, which are available online. According to *EFSA Journal* (2017), since 2005, at least 10 major incursions of HPAI H5 viruses through wild migratory waterbirds have occurred in Europe, almost one every year. The HPAI outbreak 2016–2017 affected 109 thousand birds categorized as chickens, which was only 3.9% of the total birds' population found positive for HPAI (*EFSA Journal*, 2017). The great majority of chicken cases were backyard flocks, breeders, and layers. Most of the outbreaks were reported in wild birds and domestic waterfowl (Foie Gras production).

From November 26, 2020, to May 31, 2021, 1247 cases had been reported in poultry, 65 detections in captive birds, and 2243 HPAI in wild birds (EFSA Report February–May, 2021). Until March 2020, most affected flocks had less than 1000 broilers, mainly backyard flocks with mixed species. But, between February 12 and April 30, 2021, 297 poultry farms and more than 10 million birds were affected only in Poland. Among 560 outbreaks in this second peak and short period, 98.4% were positive for HPAI subtype H5N8, six outbreaks were H5N1, and three outbreaks were H5N5. In these outbreaks, 35.2% of the poultry facilities had more than 10 000 birds, 18% had between 1000 and 10 000 birds, and 46.8% had less than 1000 birds.

Over 22.4 million birds have been affected, which is already eight-fold more than in the previous biggest outbreak that occurred between 2016 and 2017 (EFSA, 2017, 2021a). Like in the last epidemic, France has had 95.91% of the cases in domestic poultry, Germany has reported 83.05% of the cases in wild birds, but this time, Poland had 78.75% of the outbreaks in poultry. However, these reports also depend on the importance given in each country to wild bird surveillance. The risks for zoonosis are from very low to negligible, but monitoring is undergoing. No human infection with AI viruses has been identified until May 31, 2021, in the EU (EFSA, 2021b).

Biosecurity is crucial for AI control and all outdoor farming, movement of poultry should be minimized, and compulsory indoor confinement of free-range poultry must be implemented (EFSA, 2021a,b). The main issue with HPAI and even LPAI in Europe is the growing tendency for free-range poultry and outdoor farming (EFSA, 2021a,b). Under these conditions, domestic waterfowl get closer to migratory birds that bring viruses common in Asia in their migration cycles (EFSA, 2021a,b). Other aspects of current and future AI diagnosis, surveillance, epidemiology, and control are discussed in detail by Swayne et al. (2020).

2.5 Other countries

A survey was conducted with at least 12 poultry veterinarians from Russia, India, Colombia, and Chile to obtain their informed opinion of the most prevalent diseases in these countries. The results are presented in Table 5. The complex respiratory disease, *E. coli* infections, and IBV ranked high in almost all countries. However, AI is exotic in Colombia, and no cases have been observed in broilers in Chile.

Table 5 Ranking of broiler diseases in other countries in 2020

Infectious disease category	Russia	India	Colombia	Chile
Complex respiratory disease		1	1	
Escherichia coli	3	2	3	4
Infectious bronchitis (respiratory), IBV	1	5	2	5
Avian influenza	5	3	[a]	[a]
Coccidiosis	7	6	4	1
Dysbacteriosis	8	7	5	2
Infectious bursal disease	4	10	7	8
Newcastle disease (ND)[b]	9	4	8	11
Bacterial osteomyelitis of the legs	2		10	14
Mycoplasma	14		6	6
Necrotic enteritis	10	8	14	3
Infectious bronchitis (nephropathogenic)	6		12	12
Novel reovirus			13	7
Chick quality and early mortality	11	11	9	10
Infectious laryngotracheitis	12		11	9
Inclusion body hepatitis		9	15	
Vertebral osteomyelitis/kinky back	13		16	
Gangrenous dermatitis			17	13

[a] Avian influenza is considered an exotic disease in Colombia, and in Chile, no cases have been observed in broilers.
[b] Latest outbreaks are limited to wild or backyard birds.

AI is of great concern in India and Russia. India has had outbreaks of HPAI H5N8 since the end of 2020 in poultry and wild birds (EFSA Report February-May, 2021). Russia has had reports of H5N8. This time, seven poultry workers were infected in Russia with the subtype H5N8 presenting mild or no symptoms. These human infections were very unusual and required further precautions (*EFSA Journal*, 2021). Additionally, the wide expansion of H9N2 in Russia may be causing greater economic damage to poultry production than H5 and H7 subtypes that are generally more localized (Parvin et al., 2020). H9N2 viruses are also of greater concern for zoonosis due to their human receptor binding and risk of virus reassortant with human influenza (Peacock et al., 2019; Parvin et al., 2020).

Coccidiosis, non-specific dysbacteriosis, and NE are also important. Intestinal and respiratory health are again the main aspects to optimize broiler health independently of latitude, housing technologies, and environmental control capabilities. However, issues become worst when extreme weather conditions occur everywhere in the world.

3 Present and potential strategies to optimize control of infectious diseases

In the final section of this chapter, the current and future strategies for preventing and controlling the five most important broiler infectious diseases in the great majority of countries will be discussed. Disease prevention and control involve the following three interconnected processes: bioexclusion, surveillance, and biocontainment (Collett et al., 2020). The first step is limiting the challenge by biosecurity, sanitation, water and feed quality, and house environmental control. The second is enhancing resistance by immunization, immunomodulation, or genetic selection. The third is preventing the spread of diseases by boosting biosecurity and surveillance and practicing quarantine. These strategies include housing infrastructure, equipment maintenance, management, nutritional, immunological, and genetic selection factors that could be used to optimize broiler health.

Additionally, new technologies with potential benefits in prevention will be briefly discussed. The options to reduce dependence on antibiotics and reliance on only vaccination will be emphasized. Finally, this discussion will illustrate how other diseases could be addressed.

3.1 General strategies to optimize broiler health

3.1.1 A multifactoral approach

There is a global agreement that a multifactorial approach is required to prevent and control all infectious diseases (Williams, 2005; Oviedo-Rondón,

2019; Smith, 2019; Hafez and Attia, 2020; Jackwood and de Wit, 2020; Collett et al., 2020; Swelum et al., 2021). Diagnostics is key to identifying the exact etiology of the health issue to control. A variety of laboratory and field diagnostic techniques and tools are available for each broiler disease (Collett et al., 2020). However, the occurrence of a single disease caused only by one pathogen is not the most common case, and often, there are multiple factors and even pathogens involved (Smith, 2019; Collett et al., 2020).

Then, diagnostics, prevention, and control of infectious disease involve understanding the whole production system. The design of strategies to optimize health requires identifying and understanding the presenting illness and considering the big picture of other contributing diseases and influential aspects (Owen, 2017). In this way, we can articulate complete pathogenesis that indicates points of action for control and prevention. These diagnostic exercises require accurate and complete information that may integrate the disease *per se* and production over time (Owen, 2017). Many infectious diseases are subclinical and could be detected on the production records, meat inspection services at the processing plants, welfare audits, or monitoring systems (Huneau-Salaün et al., 2015).

Data analytics is becoming more vital to optimize broiler health, to summarize multiple environmental and performance parameters, and to detect trends, correlations, and causation (Collett et al., 2020). In diagnostics, the conventional microbiological and histological methods are being replaced by serology, mass spectroscopy (matrix-assisted laser desorption ionization time-of-flight mass spectrometry), polymerase chain reaction-based techniques, and other DNA techniques to detect pathogens. These techniques can also identify changes in the microbiome that may need to be modulated to achieve good results in the prevention. Diagnostic procedures are becoming more portable, rapid, sensitive, accurate, and cost affordable (Umesha and Manukumar, 2018). The development and adoption of these tools have good value to optimize health.

In the prevention and control of infectious diseases, biosecurity is the first line of defense and the most economical in the long term (Gelaude et al., 2014; Owen, 2017; Van Limbergen et al., 2018; Smith, 2019; Hafez and Attia, 2020). Van Limbergen et al. (2020) conducted a study with 2309 flocks from 358 broiler farms in 2016 to investigate the risk and protective factors that may affect health, performance, dead on arrival, and condemnation rates in conventional broiler farms. This study was conducted in seven EU member states that included Belgium, Cyprus, Finland, Greece, Poland, Spain, and the United Kingdom as a part of the EU-funded project PROHEALTH. The factors found to be significant in each case are summarized in Table 6 by their order of importance. It is essential to highlight the relevance of ventilation in the majority of the health aspects evaluated. It is also remarkable that intestinal

Table 6 Order of importance of factors associated with mortality live performance, dead on arrival, and condemnation rates in 2309 conventional broiler flocks from 358 broiler farms in 7 EU member states[a]

Independent variables	Mortality	Live performance	Dead on arrival	Condemnation rate
Litter quality[b]	1			
Neonatal septicemia[c]	2			
Ventilation type[d, e]	3	6		4
Farmer management activities	4	7		
Chick sex[f]		1		
Coccidiosis		2		
Necrotic enteritis		3		
Dysbacteriosis		4		
Light intensity adaptations[g]		5	2	
Daily growth rate			1	2
Type of drinker system[h]			3	1
Feed withdrawal time				3
House size				5
Septicemia after 7 days				6
Type of feed[i]		8		7

Source: Adapted from Van Limbergen et al. (2020).
[a] Only the statistically significant risk factors in the final models are presented.
[b] The reference used for floor quality was a floor in perfect conditions without cracks.
[c] This was only the case in tunnel-ventilated broiler houses.
[d] The reference used for the type of ventilation was roof ventilation.
[e] Interpreted as a trend, as it was not significant ($P > 0.05$).
[f] If only male chicks were housed, a higher EPI was present.
[g] The reference for light adaptation was dimming the light intensity.
[h] The reference used for drinking system was the nipple drinking system.
[i] Only the case when dysbacteriosis was absent in the flock.

health diseases were significant for live performance and involve water and feed factors.

3.1.2 Broiler housing

Proper house ventilation is critical to guarantee optimal temperatures and maintain air and litter quality (Dunlop et al., 2016). House ventilation is the primary management factor affected when weather conditions become extreme. Either heat or cold stress, often observed under extreme weather conditions, plays a vital role in several health and welfare issues (Das et al., 2011; Part et al., 2016; Saeed et al., 2019; Kpomasse et al., 2021). As global climate change circumstances become more frequent, broiler housing must be more prepared to deal with extreme conditions. Increased temperatures may increase cooling energy demand and reduce heating energy demand in broiler houses (Izar-Tenorio et al., 2020; Abioja and Abiona, 2020). However, these tendencies may change depending on season and location. Increasing the size of the houses and adding insulation may help mitigate the energy demand variability (Izar-Tenorio et al., 2020). Other strategies to ameliorate heat stress could be related to diet manipulations, use of phytochemicals, breeding for thermo-tolerant strains, and thermal conditioning during embryo development or early life (Das et al., 2011; He et al., 2018; Abioja and Abiona, 2020; Ncho et al., 2021).

3.1.3 Feed and water

At least 3 of the top 10 critical diseases for broilers worldwide are related to gut health. Coccidiosis and clostridiosis are frequently reported as the most common and costly (Jones et al., 2019). Control strategies for these intestinal diseases share common roots. The most successful nutritional and management plans have been discussed in previous publications (Williams, 2005; Yegani and Korver, 2008; Cowieson and Kluenter, 2019; Oviedo-Rondón, 2019; Smith, 2019). However, it is essential to emphasize the importance of monitoring water quality and implementing technologies to maintain it or improve it wherever is necessary (Barton, 1996; Watkins et al., 2005; Manning et al., 2007a,b; Abbas et al., 2008; Elsaidy et al., 2015; Jacobs et al., 2020). Distance monitoring with sensors of water pH, oxidation-reduction potential, temperature, and flow can help broiler growers to control water quality (Astill et al., 2020; Neethirajan, 2020).

Water is the primary nutrient ingested by broilers and sometimes the most variable among farms and even daily in a farm (Manning et al., 2007a,b; Williams et al., 2013; McCreery, 2015). Water line height, flow rates or water pressure, and drinker types may impact broiler water and feed consumption and consequently growth (Lott et al., 2001; Ipek et al., 2002; Brown, 2013; Quilumba et al., 2015). Additionally, water drinker management impacts relative humidity, ammonia concentration, and litter quality (Dunlop et al., 2016). High moisture in the house environment contributes to increasing not only *Eimeria* spp. and *Clostridium* spp. populations but also all kinds of microorganisms, including viruses and insects, which can negatively affect broiler health or propagate vectors for transmission to other locations.

Consequently, in most cases, water management has a more significant impact than any feed factor in gut health. Water properties may alter physicochemical conditions in the gastrointestinal tract, affect gut motility, digestion, and microbiota. Water alkalinity, magnesium, sulfates, nitrates, and nitrites have the most harmful effects (Barton, 1996; Abbas et al., 2008; Elsaidy et al., 2015). Water acidification with acetic acid at 3% in drinking water has been proven to be as or more effective under coccidia infection than amprolium to improve body weight gain, feed conversion, reduce lesion scores, and oocyst counts in broilers (Abbas et al., 2011).

Feed quality assurance is pivotal in gut health (Kiarie and Mills, 2019). However, quality can never be limited to optimizing nutrient utilization, reducing prevalent mycotoxins, reducing contamination and oxidation. Plant feed ingredients may contain numerous antinutritional factors, proteins, and fibers of difficult digestion by birds. The adverse effects of those feed components that alter digestibility, absorption, and even metabolism of nutrients can be effectively minimized by adequate pre-processing of feed ingredients during drying and storage, proper uniform particle size post-grinding, and proper temperature settings in the pelletization process (Lilly et al., 2011; Kiarie and Mills, 2019). In addition, feed hygienic status and efficacy of feed additives have significant effects on animal health and food safety. The addition of exogenous enzymes is another effective methodology to improve gut health. Enzymes improve digestion by mitigating the impact of antinutritional factors such as phytate, non-starch polysaccharides, proteins of difficult digestion, providing substrate for symbiotic bacteria, stimulating beneficial microbial fermentation, and modulating gut microbial communities (Keerqin et al., 2017; Cowieson and Kluenter, 2019).

3.1.4 Data and information for decision-makers

Data-driven technologies, modeling, smart sensors, machine vision, big data, and the internet of things may help to model and understand the epidemiological factors, predict the onset of infectious diseases, and improve the welfare and health of broilers (Brooks-Pollock et al., 2015; Astill et al., 2020; Li et al., 2020; Neethirajan, 2020). Sensoring data together with advanced data analytics techniques can provide early warnings for intervention to control diseases when they start and reduce the use of antimicrobials and anticoccidial drugs (Corkery et al., 2013).

Borgonovo et al. (2020) presented the development of a data-driven machine learning algorithm that uses specific critical values of volatile organic compounds in the air that are associated with abnormal levels of oocysts count at the early stage of coccidiosis. Applications of early detection technologies for sick broilers using sensors and machine vision have been growing in the past 10 years, and implementations are starting worldwide (Aydin, 2017; Zhuang et al., 2018; Okinda et al., 2019).

Training in data analytics for the poultry industry professionals will become more important to optimize broiler health. It is important to notice that the rankings of key broiler disease challenges were done with educated opinions because diagnostic data were never properly recorded, stored, consolidated, or analyzed in any country to become information available for decision-makers.

One important source of information about broiler health is the information given to the processing plant (Manning et al., 2007b). The final bird counts culled broilers, and dead-on-arrival are obtained (Jacobs et al., 2017). This information can be associated with health risk factors and welfare indicators. Data from the processing plant can include reports of septicemia, skin lesions, cellulitis, ascites, edema, emaciation, bruising, pericarditis, pericarditis, air sacculitis (USAHA, 2021; Coldebella et al., 2018). The information about the average rate of condemnations observed in the past 5 years in the United States and Brazil is presented in Table 7.

Additionally, processing plants collect data on carcass and meat yields and paw quality. All these are markers of health status. This information could be linked to identifying specific pathogens and antibody titers from blood samples frequently obtained before processing. These data need to be used regularly, and alert systems were created to investigate specific cases that indicate health

Table 7 Plant condemnations of broilers in the United States and Brazil from 2016 to 2020

	2016	2017	2018	2019	2020
Plant condemnations			%		
The United States[a]					
Wooden breast, farm + parts condemns	0.53	0.49	0.42	0.35	0.31
Septox condemns	0.14	0.13	0.13	0.11	0.09
Airsac condemns	0.10	0.09	0.07	0.05	0.04
IP condemns	0.02	0.02	0.01	0.01	0.01
Leukosis condemns	0.00	0.00	0.00	0.00	0.00
Brazil[b]					
Myopathies	0.07	0.06	0.09	0.10	0.01
Septox condemns	0.00	0.00	0.00	0.00	0.03
Cellulitis	0.13	0.13	0.13	0.14	0.11
Colibacillosis	0.12	0.14	0.10	0.12	0.01
Air sacculitis	0.05	0.05	0.05	0.05	0.08

Source: [a] USAHA (2021).
[b] Dr. José Mauricio França based on data from Ministério da Agricultura e Instituto Brasileiro de Geografia e Estatística IBGE and Coldebella et al. (2018). The Brazilian system does not consider leukosis condemns.

risks for a future flock in a house or farm or the flocks in the area (Huneau-Salaün et al., 2015).

3.2 Coccidiosis

Coccidiosis ranked high as one of the key disease challenges among the top broiler-producing countries presented herein and in other reports (Jones et al., 2019; Blake et al., 2020; McDougald et al., 2020). This disease has a substantial economic impact. Blake et al. (2020) updated Williams' (1999) compartmentalized model to estimate coccidiosis cost in chickens. These authors estimated a global annual cost of £10.36 billion (US$14.05 billion) at prices of 2016, equivalent to £0.16 or $0.22 per chicken produced. Williams (1999) had calculated over US$3 billion annually, but the broiler population has doubled, and production systems have evolved in a direction that facilitates the infection with this parasite or increases its harmful effects, and the currency values have also changed. In addition, the withdrawal of growth-promoting antibiotics has increased the costs related to coccidiosis (Smith, 2019).

Eimerias may cause disruptions in the gut mucosa and epithelium that exacerbate the dysbacteriosis, malabsorption, and enteritis observed in antibiotic-free broiler production (Fatoba and Adeleke, 2018; Gautier, 2019; Smith, 2019; Fancher et al., 2020; Gazoni et al., 2020; McDougald et al., 2020). In many countries, reused litter became more common in the past 20 years, and oocysts may persist more flock after flock (Barbour et al., 2015; Gazoni et al., 2021). The current models used to calculate costs do not include the additional charges of dysbiosis that increase NE occurrence, the impacts on wet litter, and the consequential issues on pododermatitis, culling, condemnations, and downgrades (Blake et al., 2020; Williams, 1999; Gazoni et al., 2020).

Existing control methods rely on chemical and ionophore coccidiostats, live non-attenuated and attenuated anticoccidial vaccines, and in some countries, phytobiotic products are becoming popular (Williams, 2005; Barbour et al., 2015; Abebe and Gugsa, 2018; Fatoba and Adeleke, 2018; Soutter et al., 2020). Phenylarsonic acid compounds, primarily roxarsone (3-nitro-4-hydroxyphenylarsonic acid) and *p*-arsanilic acid (4-aminobenzenearsonic acid), have been prohibited or withdrawn from the market worldwide by most manufacturers. However, their use in some countries is still allowed. China became the latest country to ban them in May 2019 (Hu et al., 2019).

Ionophores, which are natural metabolites of fungi species such as monensin, lasalocid, salinomycin, narasin, or maduramicin, are common worldwide. Some *Eimeria* strains have gained resistance, but appropriate shuttle or rotation programs still allow the successful application of these products for most broiler producers (Fatoba and Adeleke, 2018; Abebe and Gugsa, 2018; McDougald et al., 2020). However, there are regulatory constraints to use ionophores. In the EU, ionophores were reclassified as feed additives, and they are still in use despite the ban of antibiotic growth promotants. In contrast, the US FDA and Canadian regulators classify coccidiostats as antibiotics and cannot be used in the labeled antibiotic-free production (Smith, 2019).

Chemical products, in contrast, do not have known antibacterial activity and are not used in human medicine. Consequently, they are ideal for broiler antibiotic-free production. Unlike ionophores, each product has a unique way of action. This group includes nicarbazin, żoalene, decoquinate, clopidol, robenidine, and diclazuril (Abebe and Gugsa, 2018; Kadykalo et al., 2018). They are generally safe and have a zero-day withdrawal period in America, except for Robenz with 6 days. But, coccidia develops resistance against these chemical coccidiostats. The exception to some of the last affirmations is nicarbazin. A higher dose than recommended can cause broiler mortality under elevated environmental temperatures (Da Costa et al., 2017). However, there are no reports of resistance against nicarbazin in broilers. Despite the issues previously discussed, anticoccidial use remains a valuable tool for coccidiosis control worldwide (Kadykalo et al., 2018; McDougald et al., 2020).

The new economic model from Blake et al. (2020) included the cost of anticoccidial vaccines that have become the dominant strategy of anticoccidial prophylaxis in layers and breeding stock worldwide. Anticoccidial vaccines are not commonly applied to broilers in all countries (Barbour et al., 2015; McDougald et al., 2020). In the United States, more than 50% of broiler-producing companies have adopted the application of anticoccidial vaccines. Live vaccines have been applied in at least one flock per year, and more than 40% of the chickens produced in the United States have received anticoccidial live non-attenuated vaccines during the past decade (Blake et al., 2020). The bioshuttle programs with a live vaccine applied in the hatchery, followed by an anticoccidial drug, are popular in the United States and many Latin-American countries to reduce drug-resistant *Eimeria* populations (Dalloul and Lillehoj, 2006; Chapman and Jeffers, 2014; Kimminau and Duong, 2019).

The production of traditional live anticoccidial vaccines is limited by their low reproductive index and high production costs, among other factors. Recombinant, subunit, or DNA-based vaccines overcome these limitations by eliciting undesired contaminants and preventing the reversal of toxoids back to their original toxigenic form (Venkatas and Adeleke, 2019; McDougald et al., 2020). Recombinant vaccines are produced using defined *Eimeria* antigens that include specific genes that encode immunogenic protective proteins and harmless adjuvants. Vectors are ligated to these genes to ensure their penetration inside the chicken cells, resulting in efficient translation of the protective antigenic proteins. The search for identifying potent novel *Eimeria* antigens that stimulate both cell-mediated and humoral immune responses in chickens is under development (Dalloul and Lillehoj, 2006; Barbour et al., 2015).

Additionally, improving the efficacy of mass delivery methods of live vaccines and the future DNA-based vaccines is another factor to optimize coccidia control (Williams, 2005; Venkatas and Adeleke, 2019). Nutrient levels of protein and specific amino acids (Keerqin et al., 2017; Gautier, 2019) like glutamine, threonine, and arginine, nutrient digestibility, and feed additives like essential oil blends (Oviedo-Rondón et al., 2006; Muthamilselvan et al., 2016) and probiotics (Miranda et al., 2016) should be revised to achieve optimal immunological responses to vaccines and modulate microbiota to avoid dysbiosis issues (Bryan et al., 2020).

Genetic selection against *Eimeria* has been unsuccessful so far (Smith et al., 2002). However, there is a substantial variance for weight gain or interleukin-10 induction in chickens infected with *Eimeria tenella* (Boulton et al., 2018a) or *E. maxima* (Boulton et al., 2018b). Still, breeding for increased weight gain under challenge alone would not distinguish between resilience/resistance and tolerance (Boulton et al., 2018a,b). The detection of appropriate biomarkers for enhanced immunological responses is under development.

3.3 Necrotic enteritis

According to veterinarians' reports and other studies, NE is the second key disease challenge for broilers (Jones et al., 2019; Boulianne et al., 2020). Due to this disease, the worldwide economic losses are now estimated to be over US$6 billion per year, calculating with the prices of 2015 (Wade and Keyburn, 2015). Furthermore, its impact may be growing as antibiotic-free production systems become more popular (Dahiy et al., 2006; Smith, 2019; Fancher et al., 2020). However, the main difficulty in controlling this disease has been the subclinical presentation before signs are observed. Consequently, treatment is not effective (Timbermont et al., 2011; Boulianne et al., 2020).

NE can be minimized by improving feed quality assurance programs, water quality, digestibility, reducing digesta viscosity with enzymes, house environmental control, and cocci control (M'Sadeq et al., 2015; Boulianne et al., 2020). New strategies include the immunization of breeders (Keyburn et al., 2013). Additionally, it has been proposed to conduct genetic selection to enhance immunity and resistance to NE. There are some candidate genes for this selection. The substantial genetic variations among different chicken strains indicate that it is possible to develop broiler lines with genetic resistance against NE (Zahoor et al., 2018). A better understanding of the netB gene encoding for the pore-forming toxin in *Clostridium perfringens* has been important to understand this disease's pathogenesis. The gene is present in many isolates of *C. perfringens* in healthy or sick broilers (Rood et al., 2016).

The expression of the netB gene in *C. perfringens* is necessary for the development of the disease, and it occurs when broilers are immunologically stressed by *Eimeria* infection (Timbermont et al., 2011; Yang et al., 2019; Dierick et al., 2021), digesta with high viscosity due to the presence of non-starch polysaccharides, undigested material, and ingredients that enhance mucus production (M'Sadeq et al., 2015). Studies have been centered on ingredients such as wheat, barley, meat and bone meal, and fish meal (Zanu et al., 2020). However, broilers fed corn-soybean diets without animal byproducts in the United States or Brazil also present this disease.

Diets with excessively high protein content (40% CP) containing three times more glycine than commercial diets have been proven to increase *C. perfringens* overgrowth (Wilkie et al., 2005). There is some influence of high-protein levels in intestinal health, but diets with protein content higher than 25% are seldom observed under commercial conditions in the starter phases. However, protein digestibility during the first 10 days of life in poultry species is low and can be further reduced by alkaline water (pH > 8). Non-digested protein could play a similar role in increasing *Clostridium* populations and many other pathobiont microbiota of broiler's ceca (Keerqin et al., 2017; Chen and Vitetta, 2020; Zanu et al., 2020). Excessive calcium levels, small particle size (dgw < 75 μm), and high solubility of some sources could increase digesta pH, chelation of Ca by phytic acid, decreasing phytase efficacy, and nutrient digestibility. Excessive nutrients in the hindgut may trigger an overgrowth of enteric pathogens, including *Clostridium*, changes in the microflora and mucosa that facilitate *Eimeria* infection. They could cause the onset of NE incidence (Zanu et al., 2020). Feed additives like tannins and essential oil blends have been tested extensively to control *C. perfringens*, coccidian, and modulate microflora leading to the lesser impact of NE in broilers (Muthamilselvan et al., 2016; Diaz-Carrasco et al., 2016).

The environmental factors previously discussed are also essential to control NE. For example, heat stress, high relative humidity, and high stocking density have been proven to increase NE incidence (M'Sadeq et al., 2015; Tsiouris, 2016; Dunlop et al., 2016; Boulianne et al., 2020). However, reducing stocking density from 0.23 m^2 to 0.27 m^2 per bird had minimal effects in reducing production issues in broilers raised without antibiotics (McKeith et al., 2020). Accordingly, litter management is also important to minimize NE (Hermans and Morgan, 2007).

3.4 Chick quality and early mortality

Chicken early mortality is a major concern for the broiler industry. The main cause is contamination with bacteria and occasionally fungi (Swelum et al., 2021). Embryo or chick contamination not only increases total mortality, but also serves as a marker of other health, performance, and food safety issues that the flock may present later on (Yassin et al., 2009; Yerpes et al., 2020). Chick quality and livability can be influenced by breeder management, nutrition, and health (de Jong and van Riel, 2020). The hen gut microbiome could be transmitted vertically to the progeny (Shterzer et al., 2020) or throughout the eggshell (Maki et al., 2020). Improving broiler breeder nutrition and health has a tremendous impact on broiler health and productivity. The feeding strategies and nutrient levels in breeder nutrition can influence livability, immunity, and progeny performance (Leandro et al., 2011a,b; Oviedo-Rondón et al., 2013; Chang et al., 2016). de Jong and van Riel (2020) connected records from 2174 broiler flocks at the house level in 74 farms with 88 broiler breeder farms and 209 breeder flocks in the Netherlands. The authors observed that breeder flock effects affected carcass condemnations.

Eggshell integrity is essential to avoid contamination. Eggs with hairline cracks usually incubated can reduce hatchability, increase contamination likelihood, and cause more significant mortality (Barnett et al., 2004). In addition, proper egg disinfection with various products (Tebrün et al., 2020) may fail in eggs with hairline cracks. Identifying and minimizing factors that cause eggshell hairline cracks may help to reduce this issue.

Grochowska et al. (2019) conducted a field study seeking to determine factors that affect chick mortality. These authors concluded that breeder flock age, egg storage time, hatcher, and setter type affected chick mortality in the first 7 days of life. Incubation conditions have been related to chick quality and first-week mortality in several studies (Tona et al., 2004; Yassin et al., 2009; Oviedo-Rondón, 2019). Egg storage increases embryo mortality and has several deleterious effects in embryo development, reduces chick quality, the abundance of $CD3^+$, $CD4^-$, $CD8^-$ T cells in chicks, and the humoral response to booster Newcastle disease virus immunization of the birds (Goliomytis et al., 2015). Incubation conditions such as elevated temperatures or suboptimal conditions in the last phase of incubation, especially in the hatchers, also affect embryonic development, immunity, and health post-hatch (Oznurlu et al., 2010; Simon, 2016; Wijnen et al., 2020).

Finally, chick reception and brooding conditions are critical for chick survival. Yerpes et al. (2020) identified the risk factors associated with first-week mortality in chickens using data from one hatchery in Spain. Factors were classified either as internal or individual-dependent and external that included management or environmental factors. Breeder age, chick gender, and breed were the internal factors significantly related to chick mortality. Among the 21 external factors considered, the type of broiler house, the presence or absence of drip cup, egg storage, study year, and season were related to chick mortality. These authors concluded that the identified housing factors and management routines should reduce the first-week mortality rate. de Jong and van Riel

(2020) determined that broiler health variation was mainly explained by broiler farm and day-old chick batch.

3.5 Infectious bronchitis

IBV is one of the main challenges for broiler health worldwide (Jackwood, 2012; Colvero et al., 2015; Jackwood and de Wit, 2020; Legnardi et al., 2020). The avian coronavirus causes the IBV, and depending on the tissue tropism of each strain, its infection can cause acute upper respiratory tract disease or nephritis in broilers (Bande et al., 2017). The most classic IBV includes the Massachusetts serotype, which infects the respiratory tract and is observed worldwide. Nephropathogenic strains are observed more in Asia and Middle Eastern countries. But, specific IBV strains affect other tissues like the gastrointestinal tract (Moroccan IBV-G), proventriculus (QX in China). In addition, the QX IBV present in Asia, Europe, the Middle East, and Africa can affect kidneys and reproductive tract, causing false layers and high mortality (Bande et al., 2017).

Transmission occurs by inhalation or contact with contaminated objects. Morbidity is usually 100%, but mortality varies depending on host factors, strains, and secondary infections (Bande et al., 2017). The current control of this disease includes strict biosecurity and the application of attenuated live vaccines for broilers and live and killed vaccines for broiler breeders (Legnardi et al., 2020). However, the worldwide spread of many antigenic types and emerging new varieties of this IBV make it very difficult to obtain vaccines to protect and prevent transmission (Jackwood and de Wit, 2020).

There are two schools of thought for IBV vaccination that have often opposed each other. Some are in favor of homologous vaccination, others of heterologous one (Legnardi et al., 2020). This dichotomy is based on the principle that a homologous vaccine to the field strain is more likely to generate neutralizing immunity against it. In contrast, heterologous vaccination usually seeks a broader but less specific immunity toward different and potentially unknown circulating strains, in line with the 'protectotype' concept. Nevertheless, 'hybrid' solutions are often applied, including one heterologous strain Mass-based vaccine and one homologous vaccine strain. Several vaccine combinations have been tested so far in line with the different epidemiological scenarios, with particular attention to Mass-like (GI-1) and 793B-like (GI-13) strains, American strains such as Ark (GI-9), Conn (GI-1), DE072, and GA98 (GIV-1), QX (GI-19), Q1 (GI-16), and Var2 (GI-23), suggesting a variable degree of protection depending on the challenge virus. However, based on Mass-like and 793B-like strains, the most common combinations usually ensure good protection against new variants (Legnardi et al., 2020).

Several studies have shown that live IBV, ND, ILT, and avian metapneumovirus vaccines might interfere with each other's replication, humoral response, and induced level of protection when administered together or in relatively short intervals of each other (Jackwood and de Wit, 2020). However, mixing two viruses in a vaccine is the standard practice in the broiler industry, and to improve the efficacy of IBV vaccination, this practice may need to change. Improving the mass vaccine application methods for broilers is another priority in the years to come (Colvero et al., 2015). Failures in immunization can lead to an increased susceptibility to secondary infections and reversion of the vaccine virus to virulence by mutations leading to amino acid substitutions in proteins key for immunogenicity like the S1 glycoprotein (Saraiva et al., 2018). At this moment, all commercially available vaccines cannot be applied *in ovo*. The development of new vaccines may take time. Consequently, the effective prevention method is based on biosecurity with sanitation of all equipment, early and proper vaccination, control of other diseases like mycoplasma and *E. coli* infections, and adequate management of the house environment temperatures, relative humidity, and good air quality (Van Limbergen et al., 2018).

3.6 Escherichia coli *infection*

Bacterial infections have become more common in many countries, especially those with *E. coli* in countries that adopted antibiotic-free production (Smith, 2019; Fancher et al., 2020; Swelum et al., 2021). Omphalitis, respiratory tract infection, and septicemia are the most common diseases caused by *E. coli* that frequently cause mortality. Pericarditis, airsacculitis, and perihepatitis are observed in subacute forms that often lead to culling or condemnations at the processing plant. Most isolates are from O1, O2, and O75 serogroups (Nolan et al., 2020). There are intestinal and extraintestinal pathogenic *E. coli*. Among the intestinal isolates, there are enteropathogenic, enterotoxigenic, enteroinvasive, enterohemorrhagic, and enteroaggregative *E. coli*. Three extraintestinal isolates include the avian pathogenic *E. coli* (APEC), causing neonatal meningitis, and the uropathogenic *E. coli* (Nolan et al., 2020).

Drug resistance, in many cases, is a big issue. The average resistance rates in *E. coli* to diverse antibiotic classes are higher than 40% in all countries, except for ampicillin in the United States. The resistance rates to fluoroquinolones and quinolones in the United States, where fluoroquinolones are not registered for use, are below 5%. The average of resistant *E. coli* is above 40% in Brazil, China, and the EU, where fluoroquinolones are legalized. However, the banning of fluoroquinolones and quinolones has not eliminated the occurrence of resistant populations (Roth et al., 2019).

The alternative strategies to antibiotics of *E. coli* control include probiotics, prebiotics, synbiotics, and postbiotics. Plant extracts containing essential oil blends, phenols, alkaloids, and lectins have shown some immunomodulatory and antimicrobial activity. However, most of them have variable efficacy (Nolan et al., 2020). Postbiotics were recently introduced in the broiler industry. These products are the soluble factors, products, or metabolic byproducts, secreted by live bacteria or released after bacterial lysis, such as enzymes, peptides, teichoic acids, peptidoglycan-derived muropeptides, polysaccharides, cell surface proteins, and organic acids. The potential beneficial effects as anti-inflammatory, immunomodulatory, antihypertensive, antiproliferative, hepatoprotective, and antioxidant agents are still under evaluation (Swelum et al., 2021).

There are three commercially available vaccines against colibacillosis in the market worldwide: Nobilis *E. coli* (MSD Animal Health, Summit, NJ), Poulvac *E. coli* (Zoetis, Florham Park, NJ), and Nisseiken Avian Colibacillosis Vaccine CBL (Nisseiken Co., Ltd., Tokyo, Japan). Nobilis is an inactivated subunit vaccine consisting of fimbrial antigen F11 and a flagellar toxin. Poulvac is a live attenuated *E. coli* EC34195 serotype O78 with deleted aroA gene as the active substance. The immune response is mainly cell-mediated by CD 4+TCRVβ1+ on the mucosal surfaces producing immunoglobulin A (IgA) and CD8 cells. The Nisseiken is a live mutant vaccine that contains 107–109 colony-forming units/dose of AESN1331 O78 APEC strain, which has a deleted crp gene (Swelum et al., 2021).

Recent reports indicate that a recombinant or subunit multiantigen vaccine comprising a combination of common surface proteins can produce significant levels of IgA and IgY against specific antigens and impose immune responses favorable for killing the bacteria. The recombinant vaccine has reduced bacterial growth of multiple APEC serotypes, reduced bacterial load in organs, and reduced air sac lesions (Ebrahimi-Nik et al., 2018). However, mass evaluations need to be conducted. Vaccination against *E. coli* still faces several challenges, like inducing cross-protection against various APEC serogroups and massive delivery methods. Generally, studies revealed that the inactivated vaccines protected homologous challenges only (Swelum et al., 2021).

Other alternative control strategies include biosecurity, proper water sanitation, and good air quality (Nolan et al., 2020). In addition, the application of bacteriophages has been demonstrated to be successful against several APEC (Żbikowska et al., 2020). Finally, the use of toll-like receptor ligands and agonists has been evaluated to activate signaling cascades leading to the expression of various innate immune responses. These include cell wall components such as lipopolysaccharide, peptidoglycans, bacterial deoxyribonucleic acid, and double-stranded viral ribonucleic acid. However, the synthetic agonist products like CpG-oligodeoxynucleotides and polynosinic:polycytidylic acid can be easier to produce and apply (Swelum et al., 2021).

4 Conclusion

The economic impact of all broiler infectious diseases indicates a strong need to develop more work in this area. The optimization of broiler health comes with a better understanding and monitoring of the whole production system in each location. The data presented in this chapter indicate that intestinal and respiratory diseases are the main aspects to improve worldwide. There are big challenges to improving immunity against coccidiosis, *Clostridium* spp., IBV, and *E. coli*. These four pathogens cause major infectious diseases for broilers. Proper incubation and broiler-house environmental management play an important role in preventing and controlling diseases caused by these and other pathogens. Improving chick quality and reducing early mortality is a priority to enhance broiler health. Feed and water quality are necessary to control all main pathogens except for IBV. Feed composition and additives may help modulate the microflora that influence the pathogenicity or virulence of the three intestinal pathogens. Broiler breeder management, nutrition, and immunization play a big role in controlling these and other diseases. Genetic selection to develop more host resistance against *Eimeria* and *Clostridium* species is possible, but it may take a long time to achieve. Genetic resistance against *E. coli* is very difficult since this is one of the most common commensal species. Enhanced immunity and genetic resistance against IBV are problematic due to the high genetic variability of this virus. Diagnostic methods should include more information from each live production phase, including hatchery, transportation, farms, and processing plant. Data from serology and PCR-DNA-based techniques are needed for better pathogen identification and surveillance of present diseases and emerging ones. The whole toolbox of data analytics should be applied more frequently to optimize broiler health.

5 Where to look for further information

Readers interested in learning more about the current status of broiler or other poultry infectious diseases should visit the United States Animal Health (USAHA) website and obtain their annual reports and other publications. There is one yearly report on transmissible diseases of poultry and avian species from the Committee on poultry and other avian species. https://www.usaha.org/transmissible-diseases-of-poultry-avian-species.

Another thorough material to learn about all infectious diseases of broilers is the book *Diseases of Poultry* (eds D. E. Swayne, M. Boulianne, C. M. Logue, L. R. McDougald, V. Nair, D. L. Suarez, S. Wit, T. Grimes, D. Johnson, M. Kromm, T. Y. Prajitno, I. Rubinoff and G. Zavala). The extensive lists of references will provide further information to readers.

6 Acknowledgements

The author would like to express his gratitude to all people who collaborated to contribute or gather information for this Chapter in diverse countries. Some preferred to remain anonymous, but I will list a few of the participants of this global project who accepted to be named. My acknowledgments to The members of the Poultry Veterinary Study Group of the EU (PVSGEU), especially to Dr. Paul Cornelissen and Dr. Juan Carlos Abad from Cobb-Vantress Spain; Poultry Veterinarian Specialists and Production personnel of broiler companies in Brazil and China; Dr. José Mauricio França from UTP Brazil; Dr. Luiz Sesti from CEVA Brazil; Dr. Dmitry Spiridonov from DSM Russia; Dr. Tigran Papazyan from Alltech Russia; Dr. Dinabandhu Joardar from Cargill; Dr. Pablo Campino from Chile; and from Colombia: Dr. Luis Carlos Bernal, Dr. Gustavo Martinez, Dr. Juan Carlos Leyton, Dr. Carlos Ardila, Dr. Luis Carlos Monroy, Dr. Juan Carlos Cardenas, Dr. Diego Aldana, Dr. Leonardo Cotamo, Dr. Juan E. Ibarra, Dr. Marco Augusto Gutierrez, and Dr. Andres Rodriguez. Thanks for your comments and suggestions.

7 References

Abbas, R. Z., Munawar, S. H., Manzoor, Z., Iqbal, Z., Khan, M. N., Saleemi, M. K., Zia, M. A. and Yousaf, A. (2011). Anticoccidial effects of acetic acid on performance and pathogenic parameters in broiler chickens challenged with *Eimeria tenella*. *Pesq. Vet. Bras.* 31(2), 99–103. doi.org/10.1590/S0100-736X2011000200001.

Abbas, T. E. E., Elzubeir, E. A. and Arabbi, O. H. (2008). Drinking water quality and its effects on broiler chicks performance during winter season. *Int. J. Poult. Sci.* 7(5), 433–436.

Abebe, E. and Gugsa, G. (2018). A review on poultry coccidiosis. *Abyss. J. Sci. Technol.* 3(1), 1–12.

Abioja, M. O. and Abiona, J. A. (2020). Impacts of climate change to poultry production in Africa: Adaptation options for broiler chickens. In: *African Handbook of Climate Change Adaptation* Leal Filho, W., Ogugu, N., Adelake, L., Ayal, D. and da Silva, I. (Eds). Cham: Springer. doi.org/10.1007/978-3-030-42091-8_111-1.

Anderson, L. (2017). Poultry and products semi-annual in China. *United States Department of Agriculture Foreign Agricultural Service* 1, 1–11.

Astill, J., Dara, R. A., Fraser, E. D. G., Roberts, B. and Sharif, S. (2020). Smart poultry management: Smart sensors, big data, and the internet of things. *Comput. Electron. Agric.* 170, 105291. doi.org/10.1016/j.compag.2020.105291.

Aydin, A. (2017). Development of an early detection system for lameness of broilers using computer vision. *Comput. Electron. Agric.* 136, 140–146. doi.org/10.1016/j.compag.2017.02.019.

Bande, F., Arshad, S. S., Omar, A. R., Hair-Bejo, M., Mahmuda, A. and Nair, V. (2017). Global distributions and strain diversity of avian infectious bronchitis virus: A review. *Anim. Health Res. Rev.* 18(1), 70–83. doi.org/10.1017/S1466252317000044.

Barbour, E. K., Ayyash, D. B., Iyer, A., Harakeh, S. and Kumosani, T. (2015). A review of approaches targeting the replacement of coccidiostat application in poultry production. *Rev. Bras. Cienc Avic* 17(4), 405–418. doi.org/10.1590/1516-635x1704405-418.

Barnett, D. M., Kumpula, B. L., Petryk, R. L., Robinson, N. A., Renema, R. A. and Robinson, F. E. (2004). Hatchability and early chick growth potential of broiler breeder eggs with hairline cracks. *J. Appl. Poult. Res.* 13(1), 65–70. doi.org/10.1093/japr/13.1.65.

Barton, T. L. (1996). Relevance of water quality to broiler and turkey performance. *Poult. Sci.* 75(7), 854-856.

Berkhout, N. (2020). Popularity of poultry continues globally. *Poult. World*. Accessed July 20, 2020. Misset Uitgeverij B.V. Available at: https://www.poultryworld.net/Meat/Articles/2020/7/Popularity-of-poultry-continues-globally-615517E/#:~:text=In%202019%2C%20the%20global%20poultry,40%25%20share%20of%20global%20consumption. Accessed June 18, 2021.

Blake, D. P., Knox, J., Dehaeck, B., Huntington, B., Rathinam, T., Ravipati, V., Ayoade, S., Gilbert, W., Adebambo, A. O., Jatau, I. D., Raman, M., Parker, D., Rushton, J. and Tomley, F. M. (2020). Re-calculating the cost of coccidiosis in chickens. *Vet. Res.* 51(1), 115. doi.org/10.1186/s13567-020-00837-2.

Bo, H., Zhang, Y., Dong, L. B., Dong, J., Li, X. Y., Li, Z., Shu, Y. L. and Wang, D. Y. (2021). Distribution of avian influenza viruses according to environmental surveillance during 2014-2018. *China. Infect. Dis. Pover.* 10, 60. doi.org/10.1186/s40249-021-00850-3.

Borgonovo, F., Ferrante, V., Grilli, G., Pascuzzo, R., Vantini, S. and Guarino, M. A. (2020). A Data-driven prediction method for an early warning of coccidiosis in intensive livestock systems: A preliminary study. *Animals (Basel)* 10(4), 747. doi.org/10.3390/ani10040747.

Boulianne, M., Uzal, F. A. and Opengart, K. (2020). Clostridial diseases. In: *Diseases of Poultry* Swayne, D. E., Boulianne, M., Logue, C. M., McDougald, L. R., Nair, V., Suarez, D. L., Wit, S., Grimes, T., Johnson, D., Kromm, M., et al. (Eds). doi.org/10.1002/9781119371199.ch22.

Boulton, K., Nolan, M. J., Wu, Z., Psifidi, A., Riggio, V., Harman, K., Bishop, S. C., Kaiser, P., Abrahamsen, M. S., Hawken, R., Watson, K. A., Tomley, F. M., Blake, D. P. and Hume, D. A. (2018a) Phenotypic and genetic variation in the response of chickens to *Eimeria tenella* induced coccidiosis. *Genet. Sel. Evol.* 50(1), 63. doi.org/10.1186/s12711-018-0433-7.

Boulton, K., Nolan, M. J., Wu, Z., Riggio, V., Matika, O., Harman, K., Hocking, P. M., Bumstead, N., Hesketh, P., Archer, A., Bishop, S. C., Kaiser, P., Tomley, F. M., Hume, D. A., Smith, A. L., Blake, D. P. and Psifidi, A. (2018b). Dissecting the genomic architecture of resistance to *Eimeria maxima* parasitism in the chicken. *Front. Genet.* 9, 528. doi.org/10.3389/fgene.2018.00528.

Brooks-Pollock, E., de Jong, M. C. M., Keeling, M. J., Klinkenberg, D. and Wood, J. L. N. (2015). Eight challenges in modelling infectious livestock diseases. *Epidemics* 10, 1-5. doi.org/10.1016/j.epidem.2014.08.005.

Brown, C. C. (2013). The influence of drinker water pressure on broiler water usage and performance. Master's Thesis. 51 pp.

Bryan, D. D. S. L., Abbott, D. A., Van Kessel, A. G. and Classen, H. L. (2020). The influence of indigestible protein on broiler digestive tract morphology and caecal protein fermentation metabolites. *J. Anim. Physiol. Anim. Nutr. (Berl)* 104(3), 847- 866. doi.org/10.1111/jpn.13256.

Chang, A., Halley, J. and Silva, M. (2016). Can feeding the broiler breeder improve chick quality and offspring performance? *Anim. Prod. Sci.* 56(8), 1254-1262. doi.org/10.1071/AN15381.

Chapman, H. D. and Jeffers, T. K. (2014). Vaccination of chickens against coccidiosis ameliorates drug resistance in commercial poultry production. *Int. J. Parasitol. Drugs Drug Resist.* 4(3), 214-217. doi.org/10.1016/j.ijpddr.2014.10.002.

Chen, J. and Vitetta, L. (2020). The role of butyrate in attenuating pathobiont-induced hyperinflammation. *Immune Network* 20(2), e15. doi.org/10.4110/in.2020.20.e15.

Chen, Q., Saatkamp, H. W., Cortenbach, J. and Jin, W. (2020). Comparison of Chinese broiler production systems in economic performance and animal welfare. *Animals (Basel)* 10(3), 491. doi.org/10.3390/ani10030491.

Coldebella, A., Caron, L., Albuquerque, E. R. and Viana, A. L. (2018). *Avaliação dos dados de abate e condenações de aves registrados no Sistema de Informações Gerenciais do Serviço de Inspeção Federal nos anos de 2012 a 2015*. Embrapa: Concórdia Suínos e Aves. 44 p.; 21 cm. ISSN 01016245; 195.

Collett, S. R., Smith, J. A., Boulianne, M., Owen, R. L., Gingerich, E., Singer, R. S., Johnson, T. J., Hofacre, C. L., Berghaus, R. D. and Stewart-Brown, B. (2020). Principles of disease prevention, diagnosis, and control. In: *Diseases of Poultry* Swayne, D. E., Boulianne, M., Logue, C. M., McDougald, L. R., Nair, V., Suarez, D. L., Wit, S., Grimes, T., Johnson, D., Kromm, M., et al. (Eds). doi.org/10.1002/9781119371199.ch1.

Colvero, L. P., Villarreal, L. Y., Torres, C. A. and Brañdo, P. E. (2015). Assessing the economic burden of avian infectious bronchitis on poultry farms in Brazil. *Rev. Sci. Tech.* 34(3), 993 -999. doi.org/10.20506/rst.34.3.2411.

Corkery, G., Ward, S., Kenny, C. and Hemmingway, P. (2013). Incorporating smart sensing technologies into the poultry industry. *J. World's Poult. Res.* 3(4), 106–128.

Cowieson, A. J. and Kluenter, A. M. (2019). Contribution of exogenous enzymes to potentiate the removal of antibiotic growth promoters in poultry production. *Anim. Feed Sci. Technol.* 250, 81–92. doi.org/10.1016/j.anifeedsci.2018.04.026.

Da Costa, M. J., Bafundo, K. W., Pesti, G. M., Kimminau, E. A. and Cervantes, H. M. (2017). Performance and anticoccidial effects of nicarbazin-fed broilers reared at standard or reduced environmental temperatures. *Poult. Sci.* 96(6), 1615–1622. doi.org/10.3382/ps/pew475.

Dahiy, J. P., Wilkie, D. C., Van Kesse, A. G. and Drew, M. D. (2006). Potential strategies for controlling necrotic enteritis in broiler chickens in post-antibiotic era. *Anim. Feed Sci. Technol.* 129(1–2), 60–88.

Dalloul, R. A. and Lillehoj, H. S. (2006). Poultry coccidiosis: Recent advancements in control measures and vaccine development. *Expert Rev. Vaccines* 5(1), 143–163. doi.org/10.1586/14760584.5.1.143.

Das, S., Palai, T. K., Mishra, S. R., Das, D. and Jena, B. (2011). Nutrition in relation to diseases and heat stress in poultry. *Vet. World* 4(9), 429–432. doi.org/10.5455/vetworld.2011.429-432.

de Jong, I. C. and van Riel, J. W. (2020). Relative contribution of production chain phases to health and performance of broiler chickens: A field study. *Poult. Sci.* 99(1), 179–188. doi.org/10.3382/ps/pez562.

Diaz-Carrasco, J. M., Redondo, L. M., Redondo, E. A., Dominguez, J. E., Chacana, A. P. and Fernandez Miyakawa, M. E. (2016). Use of plant extracts as an effective manner to control *Clostridium perfringens* induced necrotic enteritis in poultry. *BioMed Res. Int.* 2016, 3278359. doi.org/10.1155/2016/3278359.

Dierick, E., Ducatelle, R., Immerseel, F. V. and Goossens, E. (2021). Research note: The administration schedule of coccidia is a major determinant in broiler necrotic enteritis models. *Poult. Sci.* 100(3), 100806. doi.org/10.1016/j.psj.2020.10.060.

Ducatelle, R., Goossens, E., De Meyer, F., Eeckhaut, V., Antonissen, G., Haesebrouck, F. and Van Immerseel, F. (2018). Biomarkers for monitoring intestinal health in poultry:

Present status and future perspectives. *Vet. Res.* 49(1), 43. doi.org/10.1186/s13567-018-0538-6.

Dunlop, M. W., Moss, A. F., Groves, P. J., Wilkinson, S. J., Stuetz, R. M. and Selle, P. H. (2016). The multidimensional causal factors of 'wet litter' in chicken-meat production. *Sci. Total Environ.* 562, 766–776. doi.org/10.1016/j.scitotenv.2016.03.147.

Ebrahimi-Nik, H., Bassami, M. R., Mohri, M., Rad, M. and Khan, M. I. (2018). Bacterial ghost of avian pathogenic *E. coli* (APEC) serotype O78:K80 as a homologous vaccine against avian colibacillosis. *PLoS ONE* 13(3), e0194888.

EFSA (2017). Avian influenza overview October 2016–August 2017. *EFSA J.* 15(10), 5018. doi.org/10.2903/j.efsa.2017.5018.

EFSA (2021a). Avian influenza overview December 2020 – February 2021. *EFSA J.* 19(3), 6497. doi.org/10.2903/j.efsa.2021.6497.

EFSA (2021b). Avian influenza overview February – May 2021. *EFSA J.*

Elsaidy, N., Mohamed, R. A. and Abouelenien, F. (2015). Assessment of variable drinking water sources used in Egypt on broiler health and welfare. *Vet. World* 8(7), 855–864. doi.org/10.14202/vetworld.2015.855-864.

Fancher, C. A., Zhang, L., Kiess, A. S., Adhikari, P. A., Dinh, T. T. N. and Sukumaran, A. T. (2020). Avian pathogenic *Escherichia coli* and *Clostridium perfringens*: Challenges in no antibiotics ever broiler production and potential solutions. *Microorganisms* 8(10), 1533. doi.org/10.3390/microorganisms8101533.

FAO (2003). *World Agriculture: Towards 2015/2030 An FAO Perspective*. Edited J. Bruinsma. London: EarthScan publications Ltd. 444 p. Available at: http://www.fao.org/3/y4252e/y4252e07.htm. Accessed June 18, 2021.

Fatoba, A. J. and Adeleke, M. A. (2018). Diagnosis and control of chicken coccidiosis: A recent update. *J. Parasit. Dis.* 42(4), 483–493.

Gast, R. K. and Porter, R. E., Jr. (2020). Salmonella infections. In: *Diseases of Poultry* Swayne, D. E., Boulianne, M., Logue, C. M., McDougald, L. R., Nair, V., Suarez, D. L., Wit, S., Grimes, T., Johnson, D., Kromm, M., et al. (Eds). doi.org/10.1002/9781119371199.ch16.

Gaucher, M. L., Quessy, S., Letellier, A., Arsenault, J. and Boulianne, M. (2015). Impact of a drug-free program on broiler chicken growth performances, gut health, *Clostridium perfringens*, and *Campylobacter jejuni* occurrences at the farm level. *Poult. Sci.* 94(8), 1791–1801.

Gautier, A. E. (2019). Coccidiosis vaccination and nutrient utilization in broiler chickens. Ph.D. Dissertation. Available at: https://scholarworks.uark.edu/etd/3469. Accessed June 18, 2021.

Gazoni, F. L., Adorno, F. C., Matte, F., Alves, A. J., Campagnoni, I. D. P., Urbano, T., Zampar, A., Boiago, M. M. and da Silva, A. S. (2020).Correlation between intestinal health and coccidiosis prevalence in broilers in Brazilian agroindustries. *Parasitol. Int.* 76, 102027. doi.org/10.1016/j.parint.2019.102027.

Gazoni, F., Matte, F., Chiarelli-Adorno, F., Mariely-Jaguezeski, A., Tellez-Isaias, G. and Schafer-da-Silva, A. (2021). Coccidiosis in commercial broilers in Brazil between 2012 and 2019: Main species and degrees of injury. *Abanico Vet.* 11, e101. Available at: http://www.scielo.org.mx/scielo.php?script=sci_arttext&pid=S2448-61322021000100101&lng=es. doi.org/10.21929/abavet2021.2.

Gelaude, P., Schlepers, M.,Verlinden, M., Laanen, M. and Dewulf, J. (2014). Biocheck. UGent: A quantitative tool to measure biosecurity at broiler farms and the relationship with technical performances and antimicrobial use. *Poult. Sci.* 93(11), 2740–2751. doi.org/10.3382/ps.2014-04002.

Goliomytis, M., Tsipouzian, T. and Hager-Theodorides, A. L. (2015). Effects of egg storage on hatchability, chick quality, performance and immunocompetence parameters of broiler chickens. *Poult. Sci.* 94(9), 2257–2265. doi.org/10.3382/ps/pev200.

Gonzalez, G. M. A. and Cerqueira, M. F. A. (2019). Shiga toxin-producing Escherichia coli in the animal reservoir and food in Brazil. *J. Appl. Microbiol.* 128, 1568–1582. doi.org/10.1111/jam.14500.

Grochowska, E., Kinal, A., Sobek, Z., Siatkowski, I. and Bednarczyk, M. (2019). Field study on the factors affecting egg weight loss, early embryonic mortality, hatchability, and chick mortality with the use of classification tree technique. *Poult. Sci.* 98(9), 3626-3636. doi.org/10.3382/ps/pez180.

Hafez, H. M. and Attia, Y. A. (2020). Challenges to the poultry industry: Current perspectives and strategic future after the COVID-19 outbreak. *Front. Vet. Sci.* 7, 516. doi.org/10.3389/fvets.2020.00516.

He, S. P., Arowolo, M. A., Medrano, R. F., Li, S., Yu, Q. F., Chen, J. Y. and He, J. H. (2018). Impact of heat stress and nutritional interventions on poultry production. *Worlds Poult. Sci. J.* 74(4), 647–664. doi.org/10.1017/S0043933918000727.

Hermans, P. G. and Morgan, K. L. (2007). Prevalence and associated risk factors of necrotic enteritis on broiler farms in the United Kingdom; a cross-sectional survey. *Avian Pathol.* 36(1), 43–51. doi.org/10.1080/03079450601109991.

Hu, Y., Cheng, H., Tao, S. and Schnoor, J. L. (2019). China's ban on phenylarsonic feed additives, a major step toward reducing the human and ecosystem health risk from arsenic. *Environ. Sci. Technol.* 53(21), 12177–12187. doi.org/10.1021/acs.est.9b04296.

Huneau-Salaün, A., Stärk, K. D., Mateus, A., Lupo, C., Lindberg, A. and Le Bouquin-Leneveu, S. (2015). Contribution of meat inspection to the surveillance of poultry health and welfare in the European Union. *Epidemiol. Infect.* 143(11), 2459–2472. doi.org/10.1017/S0950268814003379.

IndexBox (2021). World - poultry - market analysis, forecast, size, trends and insights update: COVID-19. *Impact*. Available at: https://www.indexbox.io/store/world-poultry-market-report-analysis-and-forecast-to-2020/. Accessed June 18, 2021.

IndexMundi (2021). Broiler meat production annual growth rate by country. Available at: https://www.indexmundi.com/agriculture/?commodity=broiler-meat&graph=production-growth-rate. Accessed June 18, 2021.

Ipek, A., Sahan, U. and Yilmaz, B. (2002). The effect of drinker type and drinker height on the performance of broiler cockerels. *Czech J. Anim. Sci.* 47, 460–466.

Izar-Tenorio, J., Jaramillo, P., Griffin, W. M. and Small, M. (2020). Impacts of projected climate change scenarios on heating and cooling demand for industrial broiler chicken farming in the Eastern U.S. *J. Clean. Prod.* 255. doi.org/10.1016/j.jclepro.2020.120306, 120306.

Jackwood, M. W. (2012). Review of infectious bronchitis virus around the world. *Avian Dis.* 56(4), 634–641. doi.org/10.1637/10227-043012-Review.1.

Jackwood, M. W. and de Wit, S. (2020). Infectious bronchitis. In: *Diseases of Poultry* Swayne, D. E., Boulianne, M., Logue, C. M., McDougald, L. R., Nair, V., Suarez, D. L., Wit, S., Grimes, T., Johnson, D., Kromm, M., et al. (Eds). doi.org/10.1002/9781119371199.ch4.

Jacobs, L., Delezie, E., Duchateau, L., Goethals, K. and Tuyttens, F. A. M. (2017). Broiler chickens dead on arrival: Associated risk factors and welfare indicators. *Poult. Sci.* 96(2), 259–265. doi.org/10.3382/ps/pew353.

Jacobs, L., Persia, M. E., Siman-Tov, N., McCoy, J., Ahmad, M., Lyman, J. and Good, L. (2020). Impact of water sanitation on broiler chicken production and welfare parameters. *J. Appl. Poult. Res.* 29(1), 258–268.

Jones, P. J., Niemi, J., Christensen, J.-P., Tranter, R. B. and Bennett, R. M. (2019). A review of the financial impact of production diseases in poultry production systems. *Anim. Prod. Sci.* 59(9), 1585–1597. doi.org/10.1071/AN18281.

Jung, A., Chen, L. R., Mitsu Suyemoto, M. M., Barnes, H. J. and Borst, L. B. (2018). A review of *Enterococcus cecorum* infection in poultry. *Avian Dis.* 62(3), 261–271. doi.org/10.1637/11825-030618-Review.1.

Kadykalo, S., Roberts, T., Thompson, M., Wilson, J., Lang, M. and Espeisse, O. (2018). The value of anticoccidials for sustainable global poultry production. *Int. J. Antimicrob. Agents* 51(3), 304–310. doi.org/10.1016/j.ijantimicag.2017.09.004.

Keerqin, C., Wu, S. B., Svihus, B., Swick, R., Morgan, N. and Choct, M. (2017). An early feeding regime and a high-density amino acid diet on growth performance of broilers under subclinical necrotic enteritis challenge. *Anim. Nutr.* 3(1), 25–32. doi.org/10.1016/j.aninu.2017.01.002.

Keyburn, A. L., Portela, R. W., Sproat, K., Ford, M. E., Bannam, T. L., Yan, X., Rood, J. I. and Moore, R. J. (2013). Vaccination with recombinant NetB toxin partially protects broiler chickens from necrotic enteritis. *Vet. Res.* 44(1), 54. doi.org/10.1186/1297-9716-44-54.

Kiarie, E. G. and Mills, A. (2019). Role of feed processing on gut health and function in pigs and poultry: Conundrum of optimal particle size and hydrothermal regimens. *Front. Vet. Sci.* 6, 19. doi.org/10.3389/fvets.2019.00019.

Kimminau, E. A. and Duong, T. (2019). Longitudinal response of commercial broiler operations to bio-shuttle administration. *J. Appl. Poult. Res.* 28(4), 1389–1397. doi.org/10.3382/japr/pfz092.

Kpomasse, C. C., Oke, O. E., Houndonougbo, F. M. and Tona, K. (2021). Broiler production challenges in the tropics: A review. *Vet. Med. Sci.* 00, 1– 12. doi.org/10.1002/vms3.435.

Krishnasamy, V., Otte, J. and Silbergeld, E. (2015). Antimicrobial use in Chinese swine and broiler poultry production. *Antimicrob. Resist. Infect. Control* 4, 17. doi.org/10.1186/s13756-015-0050-y.

Leandro, N. M., Ali, R., Koci, M., Moraes, M. R. D., Wineland, M. J. and Oviedo-Rondón, E. O. (2011a). Effects of broiler breeder genetic, diet type, and feeding program on maternal antibody transfer and development of lymphoid tissues of chicken progeny. *J. Appl. Poult. Res.* 20, 474–484.

Leandro, N. M., Ali, R., Koci, M., Moraes, V., Eusebio-Balcazar, P. E., Jornigan, J., Malheiros, R. D., Wineland, M. J., Brake, J. and Oviedo-Rondón, E. O. (2011b). Maternal antibody transfer to broiler progeny varies among strains and is affected by grain source and cage-density. *Poult. Sci.* 90(12), 2730–2739.

Legnardi, M., Tucciarone, C. M., Franzo, G. and Cecchinato, M. (2020). Infectious bronchitis virus evolution, diagnosis and control. *Vet. Sci.* 7(2), 79. doi.org/10.3390/vetsci7020079.

Li, N., Ren, Z., Li, D. and Zeng, L. (2020). Review: Automated techniques for monitoring the behaviour and welfare of broilers and laying hens: Towards the goal of precision livestock farming. *Animal* 14(3), 617–625. doi.org/10.1017/S1751731119002155.

Lilly, K. G. S., Gehring, C. K., Beaman, K. R., Turk, P. J., Sperow, M. and Moritz, J. S. (2011). Examining the relationships between pellet quality, broiler performance, and bird sex. *J. Appl. Poult. Res.* 20(2), 231–239.

Lott, B. D., May, J. D., Simmons, J. D. and Branton, S. L. (2001). The effect of nipple height on broiler performance. *Poult. Sci.* 80(4), 408–410.

Luyckx, K. Y., Van Weyenberg, S., Dewulf, J., Herman, L., Zoons, J., Vervaet, E., Heyndrickx, M. and De Reu, K. (2015). On-farm comparisons of different cleaning protocols in broiler houses. *Poult. Sci.* 94(8), 1986–1993.

Maekawa, D., Riblet, S. M., Newman, L., Koopman, R., Barbosa, T. and García, M. (2019). Evaluation of vaccination against infectious laryngotracheitis (ILT) with recombinant herpesvirus of turkey (rHVT-LT) and chicken embryo origin (CEO) vaccines applied alone or in combination. *Avian Pathol.* 48(6), 573–581. doi.org/10.1080/03079457.2019.1644449.

Maki, J. J., Bobeck, E. A., Sylte, M. J. and Looft, T. (2020). Eggshell, and environmental bacteria contribute to the intestinal microbiota of growing chickens. *J. Anim. Sci. Biotechnol.* 11, 60. doi.org/10.1186/s40104-020-00459-w.

Manning, L., Chadd, S. A. and Baines, R. N. (2007a). Water consumption in broiler chicken: A welfare indicator. *Worlds Poult. Sci. J.* 63(1), 63–71. doi.org/10.1017/s0043933907001274.

Manning, L., Chadd, S. A. and Baines, R. N. (2007b). Key health and welfare indicators for broiler production. *Worlds Poult. Sci. J.* 63(1), 46–62. doi.org/10.1017/s0043933907001262.

McCreery, D. (2015). Water consumption behavior in broilers. Ph.D. Dissertation, University of Arkansas. 72 pp.

McDougald, L. R., Cervantes, H. M., Jenkins, M. C., Hess, M. and Beckstead, R. (2020). Protozoal infections. In: *Diseases of Poultry* Swayne, D. E., Boulianne, M., Logue, C. M., McDougald, L. R., Nair, V., Suarez, D. L., Wit, S., Grimes, T., Johnson, D., Kromm, M., et al. (Eds). doi.org/10.1002/9781119371199.ch28.

McKeith, A., Loper, M. and Tarrant, K. J. (2020). Research note: Stocking density effects on production qualities of broilers raised without the use of antibiotics. *Poult. Sci.* 99(2), 698–701. doi.org/10.1016/j.psj.2019.09.004.

Miranda, R. M., Abdelrahman, W., van-Heerden, K. and Mohnl, M. (2016). Combination of probiotics and coccidiosis vaccine enhances protection against an *Eimeria* challenge. *Vet. Res. (Paris)* 47, 1, 111.

M'Sadeq, S. A., Wu, S., Swick, R. A. and Choct, M. (2015). Towards the control of necrotic enteritis in broiler chickens with in-feed antibiotics phasing-out worldwide. *Anim. Nutr.* 1(1), 1–11. doi.org/10.1016/j.aninu.2015.02.004.

Muthamilselvan, T., Kuo, T. F., Wu, Y. C. and Yang, W. C. (2016). Herbal remedies for coccidiosis control: A review of plants, compounds, and anticoccidial actions. *Evid. Based Complement. Alternat. Med.* 2016, 2657981. doi.org/10.1155/2016/2657981.

Ncho, C. M., Gupta, V. and Goel, A. (2021). Effect of thermal conditioning on growth performance and thermotolerance in broilers: A systematic review and meta-analysis. *J. Therm. Biol.* 98, 102916. doi.org/10.1016/j.jtherbio.2021.102916.

Neethirajan, S. (2020). The role of sensors, big data and machine learning in modern animal farming. *Sensing and Bio-Sensing Research* 29. doi.org/10.1016/j.sbsr.2020.100367, 100367.

Nolan, L. K., Vaillancourt, J.-P., Barbieri, N. L. and Logue, C. M. (2020). Colibacillosis. In: *Diseases of Poultry* Swayne, D. E., Boulianne, M., Logue, C. M., McDougald, L. R., Nair, V., Suarez, D. L., Wit, S., Grimes, T., Johnson, D., Kromm, M., et al. (Eds). doi.org/10.1002/9781119371199.ch18.

Okinda, C., Lu, M., Liu, L., Nyalala, I., Muneri, C., Wang, J., Zhang, H. and Shen, M. (2019). A machine vision system for early detection and prediction of sick birds: A broiler chicken model. *Biosyst. Eng.* 188, 229–242. doi.org/10.1016/j.biosystemseng.2019.09.015.

Oviedo-Rondón, E. O., Hume, M. E., Hernández, C. and Clemente-Hernández, S. (2006). Intestinal microbial ecology of broilers vaccinated and challenged with mixed Eimeria species, and supplemented with essential oil blends. *Poult. Sci.* 85(5), 854–860. doi.org/10.1093/ps/85.5.854.

Oviedo-Rondón, E. O., Leandro, N. M., Ali, R., Koci, M., Moraes, V. and Brake, J. (2013). Broiler breeder feeding programs and trace minerals on maternal antibody transfer and broiler humoral immune response. *J. Appl. Poult. Res.* 22(3), 499–510.

Oviedo-Rondón, E. O. (2019). Holistic view of intestinal health in poultry. *Anim. Feed Sci. Technol.* 250, 1–8. doi.org/10.1016/j.anifeedsci.2019.01.009.

Owen, R. L. (2017). *A Practical Guide for Managing Risk in Poultry Production*. 2nd edn. Jacksonville, FL: AAAP, Inc.

Oznurlu, Y., Celik, I., Telatar, T. and Sur, E. (2010). Histochemical and histological evaluations of the effects of high incubation temperature on embryonic development of thymus and bursa of Fabricius in broiler chickens. *Br. Poult. Sci.* 51(1), 43–51. doi.org/10.1080/00071660903575558.

Part, C. E., Edwards, P., Hajat, S. and Collins, L. M. (2016). Prevalence rates of health and welfare conditions in broiler chickens change with weather in a temperate climate. *R. Soc. Open Sci.* 3(9), 3160197160197. doi.org/10.1098/rsos.160197.

Parvin, R., Schinkoethe, J., Grund, C., Ulrich, R., Bönte, F., Behr, K. P., Voss, M., Samad, M. A., Hassan, K. E., Luttermann, C., Beer, M. and Harder, T. (2020). Comparison of pathogenicity of subtype H9 avian influenza wild-type viruses from a wide geographic origin expressing mono-, di-, or tri-basic hemagglutinin cleavage sites. *Vet. Res.* 51(1), 48. doi.org/10.1186/s13567-020-00771-3.

Peacock, T. H. P., James, J., Sealy, J. E. and Iqbal, M. A. (2019). A Global perspective on H9N2 avian influenza virus. *Viruses* 11(7), 620. doi.org/10.3390/v11070620.

Quilumba, C., Quijia, E., Gernat, A., Murillo, G. and Grimes, J. (2015). Evaluation of different water flow rates of nipple drinkers on broiler productivity. *J. Appl. Poult. Res.* 24(1), 58–65.

Reck, C., Menin, Á., Canever, M. F., Pilatic, C. and Miletti, L. C. (2019). Molecular detection of *Mycoplasma synoviae* and avian reovirus infection in arthritis and tenosynovitis lesions of broiler and breeder chickens in Santa Catarina State, Brazil. *J. S. Afr. Vet. Assoc.* 90, e1–e5. doi.org/10.4102/jsava.v90i0.1970.

Rood, J. I., Keyburn, A. L. and Moore, R. J. (2016). NetB and necrotic enteritis: The hole movable story. *Avian Pathol.* 45(3), 295–301. doi.org/10.1080/03079457.2016.1158781.

Roth, N., Käsbohrer, A., Mayrhofer, S., Zitz, U., Hofacre, C. and Domig, K. J. (2019). The application of antibiotics in broiler production and the resulting antibiotic resistance in *Escherichia coli*: A global overview. *Poult. Sci.* 98(4), 1791–1804. doi.org/10.3382/ps/pey539.

Saraiva, G. L., Santos, M. R., Pereira, C. G., Vidigal, P. M. P., Fietto, J. L. R., de Oliveira Mendes, T. A., Bressan, G. C., Soares-Martins, J. A. P., de Almeida, M. R. and Silva-Júnior, A. (2018). Evaluation of the genetic variability found in Brazilian commercial vaccines for infectious bronchitis virus. *Virus Genes* 54(1), 77–85. doi.org/10.1007/s11262-017-1515-2.

Saeed, M., Abbas, G., Alagawany, M., Kamboh, A. A., Abd El-Hack, M. E., Khafaga, A. F. and Chao, S. (2019). Heat stress management in poultry farms: A comprehensive overview. *J. Therm. Biol.* 84, 414–425. doi.org/10.1016/j.jtherbio.2019.07.025.

Schoenmakers, K. (2020). How China is getting its farmers to kick their antibiotics habit. *Nature* 586(7830), S60–S62. doi.org/10.1038/d41586-020-02889-y.

Shterzer, N., Rothschild, N., Sbehat, Y., Stern, E., Nazarov, A. and Mills, E. (2020). Large overlap between the intestinal and reproductive tract microbiomes of chickens. *Front. Microbiol.* 11, 1508. doi.org/10.3389/fmicb.2020.01508.

Simon, K. (2016). Effects of early life conditions on immunity in broilers and layers. Wageningen University Ph.D. Dissertation. ISBN 9789462576711 - 188.

Smith, A. L., Hesketh, P., Archer, A. and Shirley, M. W. (2002). Antigenic diversity in *Eimeria maxima* and the influence of host genetics and immunization schedule on cross-protective immunity. *Infect. Immun.* 70(5), 2472–2479. doi.org/10.1128/iai.70.5.2472-2479.2002.

Smith, J. A. (2019). Broiler production without antibiotics: United States field perspectives. *Anim. Feed Sci. Technol.* 250, 93–98. doi.org/10.1016/j.anifeedsci.2018.04.027.

Soutter, F., Werling, D., Tomley, F. M. and Blake, D. P. (2020). Poultry coccidiosis: Design and interpretation of vaccine studies. *Front. Vet. Sci.* 7, 101. doi.org/10.3389/fvets.2020.00101.

Swayne, D. E., Suarez, D. L. and Sims, L. D. (2020). Influenza. In: *Diseases of Poultry* Swayne, D. E., Boulianne, M., Logue, C. M., McDougald, L. R., Nair, V., Suarez, D. L., Wit, S., Grimes, T., Johnson, D., Kromm, M., et al. (Eds). doi.org/10.1002/9781119371199.ch6.

Swelum, A. A., Elbestawy, A. R., El-Saadony, M. T., Hussein, E. O. S., Alhotan, R., Suliman, G. M., Taha, A. E., Ba-Awadh, H., El-Tarabily, K. A. and El-Hack, M. E. A. (2021). Ways to minimize bacterial infections, with special reference to *Escherichia coli*, to cope with the first-week mortality in chicks: An updated overview. *Poult. Sci.* 100(5), 101039. doi.org/10.1016/j.psj.2021.101039.

Swidan, A. N., Atallah, S. T. and El-Ktany, E. M. (2020). Economic impact of broiler diseases on national income of Egypt. *Alex. J. Vet. Sci.* 67(1), 39–46.

Tebrün, W., Motola, G., Hafez, M. H., Bachmeier, J., Schmidt,V., Renfert, K., Reichelt, C., Brüggemann-Schwarze, S. and Pees, M. (2020). Preliminary study: Health and performance assessment in broiler chicks following application of six different hatching egg disinfection protocols. *PLoS ONE* 15(5), e0232825. doi.org/10.1371/journal.pone.0232825.

Timbermont, L., Haesebrouck, F., Ducatelle, R. and Van Immerseel, F. (2011). Necrotic enteritis in broilers: An updated review on the pathogenesis. *Avian Pathol.* 40(4), 341–347. doi.org/10.1080/03079457.2011.590967. Available at: org.prox.lib.ncsu.edu/10.1080/03079457.2011.590967 .

Tona, K., Onagbesan, O. M., Jego, Y., Kamers, B., Decuypere, E. and Bruggeman, V. (2004). Comparison of embryo physiological parameters during incubation, chick quality, and growth performance of three lines of broiler breeders differing in genetic composition and growth gate. *Poult. Sci.* 83(3), 507–513. doi.org/10.1093/ps/83.3.507.

Tsiouris, V. (2016). Poultry management: A useful tool for the control of necrotic enteritis in poultry. *Avian Pathol.* 45(3), 323–325. doi.org/10.1080/03079457.2016.1154502.

Umesha, S. and Manukumar, H. M. (2018). Advanced molecular diagnostic techniques for detection of food-borne pathogens: Current applications and future challenges. *Crit. Rev. Food Sci. Nutr.* 58(1), 84–104. doi.org/10.1080/10408398.2015.1126701.

USAHA (2021). United States animal health. Transmissible diseases of poultry and avian species. Committee on Poultry and other avian species. Available at: https://www.usaha.org/transmissible-diseases-of-poultry-avian-species.

Van Limbergen, T., Dewulf, J., Klinkenberg, M., Ducatelle, R., Gelaude, P., Mendez, J., Heinola, K., Papasolomontos, S., Szeleszczuk, P., Mae, D. and PROHEALTH consortium (2018). Scoring biosecurity in European conventional broiler production. *Poult. Sci.* 97(1), 74–83. doi.org/10.3382/ps/pex296.

Van Limbergen, T., Sarrazin, S., Chantziaras, I., Dewulf, J., Ducatelle, R., Kyriazakis, I., McMullin, P., Méndez, J., Niemi, J. K., Papasolomontos, S., Szeleszczuk, P., Erum, J. V., Maes, D. and PROHEALTH consortium (2020). Risk factors for poor health and performance in European broiler production systems. *BMC Vet. Res.* 16(1), 287. doi.org/10.1186/s12917-020-02484-3.

Venkatas, J. and Adeleke, M. A. (2019). A review of Eimeria antigen identification for the development of novel anticoccidial vaccines. *Parasitol. Res.* 118(6), 1701–1710. doi.org/10.1007/s00436-019-06338-2.

Wade, B. and Keyburn, A. (2015). The true cost of necrotic enteritis. *World Poult.* 31, 16–17.

Wang, R., van Dorp, L., Shaw, L. P., Bradley, P., Wang, Q., Wang, X., Jin, L., Zhang, Q., Liu, Y., Rieux, A., Dorai-Schneiders, T., Weinert, L. A., Iqbal, Z., Didelot, X., Wang, H. and Balloux, F. (2018). The global distribution and spread of the mobilized colistin resistance gene mcr-1. *Nat. Commun.* 9(1), 1179. doi.org/10.1038/s41467-018-03205-z.

Wang, Y., Xu, C., Zhang, R., Chen, Y., Shen, Y., Hu, F., Liu, D., Lu, J., Guo, Y., Xia, X., Jiang, J., Wang, X., Fu, Y., Yang, L., Wang, J., Li, J., Cai, C., Yin, D., Che, J., Fan, R., Wang, Y., Qing, Y., Li, Y., Liao, K., Chen, H., Zou, M., Liang, L., Tang, J., Shen, Z., Wang, S., Yang, X., Wu, C., Xu, S., Walsh, T. R. and Shen, J. (2020). Changes in colistin resistance and mcr-1 abundance in *Escherichia coli* of animal and human origins following the ban of colistin-positive additives in China: An epidemiological comparative study. *Lancet Infect. Dis.* 20(10), 1161–1171. doi.org/10.1016/S1473-3099(20)30149-3.

Watkins, S. E., Fritts, C. A., Yan, F., Wilson, M. L. and Waldroup, P. W. (2005). The interaction of sodium chloride levels in poultry drinking water and the diet of broiler chickens. *J. Appl. Poult. Res.* 14(1), 55–59.

Wijnen, H. J., van den Brand, H., Lammers, A., van Roovert-Reijrink, I. A. M., van der Pol, C. W., Kemp, B. and Molenaar, R. (2020). Effects of eggshell temperature pattern during incubation on primary immune organ development and broiler immune response in later life. *Poult. Sci.* 99(12), 6619–6629. doi.org/10.1016/j.psj.2020.09.088.

Wilkie, D. C., Van Kessel, A. G., White, L. J., Laarveld, B. and Drew, M. D. (2005). Dietary amino acids affect intestinal *Clostridium perfringens* populations in broiler chickens. *Can. J. Anim. Sci.* 85(2), 185–193.

Williams, C. L., Tabler, G. T. and Watkins, S. E. (2013). Comparison of broiler flock daily water consumption and water-to-feed ratios for flocks grown in 1991, 2000–2001, and 2010–2011. *J. Appl. Poult. Res.* 22(4), 934–941.

Williams, R. B. (1999). A compartmentalised model for the estimation of the cost of coccidiosis to the world's chicken production industry. *Int. J. Parasitol.* 29(8), 1209–1229.

Williams, R. B. (2005). Intercurrent coccidiosis and necrotic enteritis of chickens: Rational, integrated disease management by maintenance of gut integrity. *Avian Pathol.* 34(3), 159–180. doi.org/10.1080/03079450500112195.

Xin, X. F., Zhang, Y., Wang, J. M. and Nuetah, J. A. (2016). Effects of farm size on technical efficiency in China's broiler sector: A stochastic meta-frontier approach. *Can. J. Agr. Econ.* 64(3), 493–516.

Xu, J., Sangthong, R., McNeil, E., Tang, R. and Chongsuvivatwong, V. (2020). Antibiotic use in chicken farms in northwestern China. *Antimicrob. Resist. Infect. Control* 9(1), 10. doi.org/10.1186/s13756-019-0672-6.

Yang, W. Y., Lee, Y. J., Lu, H. Y., Branton, S. L., Chou, C. H. and Wang, C. (2019). The netB-positive *Clostridium perfringens* in the experimental induction of necrotic enteritis with or without predisposing factors. *Poult. Sci.* 98(11), 5297–5306. doi.org/10.3382/ps/pez311.

Yassin, H., Velthuis, A. G. J., Boerjan, M. and van Riel, J. (2009). Field study on broilers' first-week mortality. *Poult. Sci.* 88(4), 798–804. doi.org/10.3382/ps.2008-00292.

Yegani, M. and Korver, D. R. (2008). Factors affecting intestinal health in poultry. *Poult. Sci.* 87(10), 2052–2063.

Yerpes, M., Llonch, P. and Manteca, X. (2020). Factors associated with cumulative first-week mortality in broiler chicks. *Animals (Basel)* 10(2), 310. doi.org/10.3390/ani10020310.

Zahoor, I., Ghayas, A. and Basheer, A. (2018). Genetics and genomics of susceptibility and immune response to necrotic enteritis in chicken: A review. *Mol. Biol. Rep.* 45(1), 31–37.

Zanu, H. K., Kheravii, S. K., Bedford, M. R. and Swick, R. A. (2020). Dietary calcium and meat and bone meal as potential precursors for the onset of necrotic enteritis. *Worlds Poult. Sci. J.* 76(4), 743–756. doi.org/10.1080/00439339.2020.1831419.

Żbikowska, K., Michalczuk, M. and Dolka, B. (2020). The use of bacteriophages in the poultry industry. *Animals (Basel)* 10(5). doi.org/10.3390/ani10050872.

Zhang, Q. and Sahin, O. (2020). Campylobacteriosis. In: *Diseases of Poultry* Swayne, D. E., Boulianne, M., Logue, C. M., McDougald, L. R., Nair, V., Suarez, D. L., Wit, S., Grimes, T., Johnson, D., Kromm, M., et al. (Eds). doi.org/10.1002/9781119371199.ch17.

Zhuang, X., Bi, M., Guo, J., Wu, S. and Zhang, T. (2018). Development of an early warning algorithm to detect sick broilers. *Comput. Electron. Agric.* 144, 102–113. doi.org/10.1016/j.compag.2017.11.032.

Chapter 9

Optimizing the health of poultry layers

Yuko Sato, Iowa State University, USA

1 Introduction

Commercial egg production is an important part of poultry production worldwide. Table egg production is facing times of change and transition with a focus on animal well-being. Several major global trends have impacted egg production in many ways: (1) housing systems, (2) restrictions on beak treatments, (3) increasing focus on sustainability, and (4) consumer demand for speciality eggs [1]. In some countries, conventional cage systems are being phased out completely, while in other countries, cages and floor spaces are subjected to stringent requirements. Although the majority of the US egg production still remains in cages, increasing numbers of table egg layer producers transition into cage-free/aviary-type production or organic egg production, which involves raising birds cage-free with outdoor access or free-range. The choice of the management and housing system is up to the poultry producer and depends on regulations, personal preference, financial returns, and environmental options. Producers are constantly looking at different challenges with egg production based on customer preference, in which case these alternative systems demand better and different management skills in order to keep birds in good health and productivity [2].

Approaches to laying hen health largely depend on management styles based on housing systems. In general, a simple mnemonic of 'F-L-A-W-S', which stands for feed, light, air, water, and space/sanitation/(bio)security, is used to look at management best practices [3, 4]. Great strides have been made with bird genetics with respect to overall health and resistance to diseases. Since

http://dx.doi.org/10.19103/AS.2022.0104.12

laying hens have a long production cycle, strong emphasis is made on good pullet production to prepare hens in good health through egg production. On average, 1-2% mortality is standard in the first few weeks of age, with the total end of flock mortality around 2-5%. Average mortality is typically 0.1% per week, or <0.5% per month during late pullet stages or during production [4]. With limited antimicrobial options in laying hens along with challenges of non-antibiotic treatments for organic layers, the focus on optimizing laying hen health comes down to genetic selection, high-quality nutrition, sound biosecurity, and good management practices.

2 Key disease issues faced by pullets

2.1 Vertically transmitted diseases of egg-type chickens

There are several infectious diseases that are known to transmit vertically from infected breeders to their progeny. In the United States, the National Poultry Improvement Plan (NPIP) programme, established in 1935, aims to eradicate bacillary white diarrhoea, caused by *Salmonella enterica* serovar Pullorum, from commercial breeding stock. The NPIP is a cooperative between industry, state, and federal programmes in which diagnostic protocols and technologies work synergistically to the improvement of poultry and poultry products and to preventing, controlling, and eradicating breeding stock diseases in commercial poultry [5]. Other disease-causing agents of significance to egg-type chickens covered by the programme include *Salmonella enterica* serovar Gallinarum (fowl typhoid), *Salmonella enterica* var. *enteritidis*, *Mycoplasma gallisepticum*, *Mycoplasma synoviae*, and avian influenza (fowl plague) viruses [6]. Individual diseases and the causative agents will be discussed in detail in subsequent portions of this section. Under the NPIP programme, it is the responsibility of the breeding companies to maintain disease-free breeders to reduce the risk of horizontal transmission of these diseases within commercial flocks.

Vaccination at the pullet stage is critical for both ensuring the health of the birds and addressing food safety concerns, especially vaccination against *Salmonella enterica* var. *enteritidis* (SE) is important, for its ability of egg transmission, which is responsible for several foodborne illnesses in humans from poultry products. For example, a common vaccine schedule in the United States against SE includes two rounds of live *Salmonella typhimurium* vaccines followed by one or two boosters by inactivated SE vaccines. In Europe, live SE vaccines are given three times. Some other major differences between the commercial broiler vaccines include (1) the lack of *in ovo* vaccination of agents such as Marek's due to the selection of pullets versus cockerel as compared to straight-run operations in broilers, (2) routine use of salmonella vaccines in commercial birds for food safety purposes, and (3) the use of spray and injection technology versus water-based vaccines due to the majority of housing, especially in the United States, being in cages. Coccidiosis vaccines have been routine in commercial broilers and not in commercial layers; however, with the increase in cage-free layers, this practice has become much more standard in pullet vaccination at the hatchery. In addition, several studies show the difference in immunity between broilers and layers, and thus a higher immunogenic vaccine using chicken-embryo origin (CEO) vaccines for ILT is more routinely used in layer production than in broilers. To account for the long lifespan of a commercial layer, numerous vaccines that are important to sustain egg production and protect from disease are given during the pullet stage.

Table 1 An example of a commercial layer pullet vaccine schedule in the United States

Age	Disease	Vaccine	Route
Hatchery	Marek's/IBD	rHVT-IBD[a], Classic Intermediate	Subcutaneous
	Marek's	Rispens (CVI988)	Subcutaneous
	Salmonella	Live *Salmonella typhimurium (ST)*	Very coarse spray/ gel
	IBV	MA5 (optional)	Spray
	Coccidia	Live oocysts (optional)	Very coarse spray/ gel
3-4 weeks	ND-IBV	B1-B1/Massachusetts + Arkansas	Spray
	Live *Escherichia* Coli (*E. coli*)	homologous O78 strain and heterologous O2	Spray
5 weeks	ND-IBV	B1-B1/Massachusetts + Connaught	Spray
	Salmonella	Live ST	Spray
	Live *E. coli*	homologous O78 strain and heterologous O2	Spray
8 weeks	MG	F strain, given with ILT	Eye drop
	AE (add to Pox)	AE CEO Calneck CEO Hitchner (pigeon pox) and CEO Fowlpox	Wing web
	ILT	CEO, given with MG	Eye drop
10 weeks	ND-IBV	LaSota/Holland + Massachusetts/ Connaught	Intramuscular
14 weeks	ND-IBV-SE	Three-way inactivated	Subcutaneous or intramuscular
	Live *E. coli*	homologous O78 strain and heterologous O2	Spray

r = recombinant DNA vector vaccines, two main vectors are fowlpox (rPox) virus and turkey herpesvirus that are used to construct recombinant vaccines for poultry [7].
[a] rHVT= recombinant herpesvirus of turkeys (Marek's), insert could be IBD, ND, ILT, IBV, MG, etc.

Table 1 shows an example of a vaccination schedule in the United States for a typical layer pullet production.

Vaccinations are largely based on disease pressures regionally, as well as the type of production system (e.g. shell egg versus liquid egg, organic versus conventional, and caged versus cage-free) these birds will be in. Not only do vaccinations cost in terms of unit cost and labour for administration; the stress associated with immunogenicity can interfere with growth and production. Thus, vaccination in the lay house is not performed routinely due to it being detrimental to egg production. Vaccination schedules must be reassessed by veterinarians on a regular basis to examine the dynamic disease pressures, recognize and control emerging diseases, and troubleshoot management strategies based on the goals needed to achieve success in production.

2.2 Viral diseases of pullets

2.2.1 Avian adenovirus infections

Aviadenoviruses are DNA viruses of the family *Adenoviridae* that includes diseases of concern to poultry. They are divided into three genera: *Aviadenovirus*, *Siadenovirus*, and *Atadenovirus*. Viruses affecting poultry, which will be referred to hereinafter as Fowl aviadenoviruses (FAdV) are further subdivided into 5 species (A–E), numerous genotypes, and 12 serotypes [8]. Although many of the avian adenoviruses are subclinical, exceptions include Quail Bronchitis, Egg Drop Syndrome (EDS), Turkey Haemorrhagic Enteritis, Pheasant Marble Spleen Disease, and Inclusion Body Hepatitis (IBH), to name a few. Of significance to laying hens and pullets include EDS in the genus *Atadenovirus*, which causes significant drops in egg production (which will be discussed in 3.1.1) and IBH in the genus *Aviadenovirus*, which might cause disease and mortality [9].

In the last two decades, IBH has been mainly due to species D or E [10]. In most cases, outbreaks of IBH are introduced vertically and thus control at the breeder level is very important for transmission. Once the virus is introduced in the flocks, it spreads horizontally via the faecal–oral route. Clinical signs of IBH are characterized by the sudden onset of mortality, especially in chicks less than 3 weeks of age, and by relatively low morbidity. In many cases, clinical signs are not as pronounced as in broilers, with the most common presentation being reduced laying performance and a slight increase in mortality (<0.5%) [11, 12]. Gross lesions are characterized by swollen livers with small pinpoint foci and lesions in immune organs (bursal atrophy, thymic atrophy, bone marrow aplasia). Diagnosis using histopathologic lesions is pathognomonic, with clear basophilic intranuclear inclusion bodies being found in areas of necrosis within the liver, +/− pancreas, kidneys, and gizzards, and in enterocytes. Confirmatory diagnosis is performed by virus isolation, the use of polymerase-chain reaction

(PCR) with further subtyping to characterize the virus. Prevention and control largely depends on controlling at the primary breeder level, as FAdV are fairly hardy and resistant to inactivation [9]. Vaccination against IBH and other FAdV is not routinely done in commercial layers.

2.2.2 Avian encephalomyelitis virus

Also known as epidemic tremors, avian encephalomyelitis virus (AEV) is caused by a Tremovirus of the family *Picornaviridae*. Clinical AEV is characterized by iconic fine tremors of the head and neck, and ataxia progressing to paralysis and prostration [13]. Clinical signs from vertically transmitted AEV can start as early as 1 day of age. Horizontal transmission occurs in flocks seeded by vertical transmission and affected birds start to show clinical signs as early as 10 days post-infection. Chickens typically become resistant to neurological signs if the infection occurred after 3 weeks of age and would manifest as egg production drops in adult laying hens [14]. Morbidity can be as high as 60% and mortality up to 25% in naïve flocks. There are typically no gross lesions except for rare whitish discolourations in the muscular layer of the gizzard. Histopathologic lesions are often pathognomonic, characterized by degenerating neurons with central chromatolysis in the brain and spinal cord, and nodular aggregates of lymphocytes within muscles of the heart, gizzard, proventriculus, and also in the pancreas. Diagnosis can otherwise be done via PCR and comparison of acute and convalescent titers using serological assays. Differentials in young pullets include Newcastle disease, nutritional encephalitis (vitamin E and riboflavin deficiency), Marek's disease, and other causes of bacterial and viral encephalitis. Control of AEV is by vaccination of breeder flocks and commercial flocks to prevent drops in egg production. Routinely, AEV vaccines in layer pullets are given after 8 weeks of age and at least 4 weeks before egg production so as not to interfere with maternal antibodies and egg production, respectively [13, 14].

2.2.3 Avian influenza

Avian influenza (AI) is caused by type A influenza virus, further classified into 16 hemagglutinin and 9 neuraminidase subtypes. The disease is further classified by pathotype as low pathogenicity AI (LPAI) and high pathogenicity AI (HPAI) and is an important transboundary disease of poultry. LPAI H5 and H7 subtypes have the potential to mutate into HPAI strains and are monitored globally for disease outbreaks, as H5/H7 LPAI and HPAI are of interest to the World Organization for Animal Health (WOAH) [15]. AI viruses, like other respiratory viruses, spread via aerosol transmission from respiratory secretions and contamination of fomites, and also via infected faeces, making oropharyngeal swabs the primary sample for detection, but also via cloacal swabs, since the virus can replicate in respiratory, intestinal, renal, and/or reproductive organs. Potential sources of infection into commercial layers include (1) other commercial poultry, (2) migratory waterfowl and other wild birds (direct or indirect transmission), (3) domestic swine, and (4) companion or pet birds. In commercial layer operations where biosecurity is elevated, transmission is likely due to airborne viruses from infected commercial sites, movement of infected pullets or infection during transit, mechanical transmission from contaminated equipment, clothing, human and vehicular traffic, and broadly through breaches in biosecurity [16]. Although, experimentally, incubation periods of AI viruses are as short as a few hours to 2–5 days naturally, realistic incubation periods are longer. The WOAH has recently updated the incubation period of HPAI at the flock level to 14 days [17]. Clinical signs can vary greatly from severe mortality due to HPAI infections to mild respiratory signs of LPAI infections. In many cases, a decrease in feed or water consumption is the first noticeable sign at the flock level [16]. As such, gross lesions are also extremely variable in location and severity. LPAI lesions are typical of mild respiratory lesions, which may or may not be complicated by secondary bacterial infections. HPAI lesions are typically a range of haemorrhages, oedema, and necrotizing lesions in visceral organs and skin. Histopathologic lesions consist of multi-organ necrosis, inflammation, and haemorrhage. Definitive diagnosis is via detection of the virus via antigen assays such as PCR and virus isolation. AI is an NPIP-monitored disease, and serological samples are collected routinely from seemingly negative birds for active surveillance. Differentials depend on the pathotype, with virulent Newcastle disease, fowl cholera, heat stress, water deprivation, and acute toxicosis for HPAI infection, and lentogenic Newcastle disease, ILT, Infectious bronchitis virus (IBV), mycoplasmosis, and other respiratory bacterial infections for LPAI. Biosecurity is the primary prevention against LPAI/HPAI, and virus elimination is the preferred control strategy in outbreak situations. In select countries, vaccination is also used as a preventative and emergency control measure [15].

2.2.4 Infectious bronchitis virus

IBV is a highly contagious, economically important disease of both pullets and layers caused by avian gammacoronavirus. Transmission is through inhalation or direct contact with infected birds or contaminated fomites, but vertical transmission is not reported. IBV has tropism of the respiratory, reproductive, and renal (nephropathogenic strains) tissues, and its effects on egg production will be discussed later in Section 3.1.4. IBV is consistently one of the top five health concerns in pullet production within the United States and globally, due to disease and death loss, but also its long-lasting impact on reproductive health and thus egg production in laying hens [18].

Clinical signs thus depend on the pathotype of virus that is infecting the birds – i.e. respiratory tropic virus will cause respiratory clinical signs, and nephropathogenic strains might also affect the urogenital organs, etc. Incubation period of IBV could be as short as 18 h and it spreads rapidly through a flock. IBV replicates readily in the respiratory and enteric cells and disseminates via respiratory secretions and faecal droppings. Airborne spread by contaminated dusts is the most common route of transmission, especially in areas with dense chicken populations. As with other coronaviruses in veterinary and human medicine, it is also well established that IBV has high genetic diversity and mutation rates, resulting in numerous variants and subtypes of the virus, making diagnosis and control of the disease extremely challenging. In addition, usually there is moderate to no cross-protection amongst different serotypes [19].

Clinical signs in young pullets include coughing, sneezing, rales, nasal and ocular discharge, and sometimes voice changes due to inflammation in the upper respiratory tract. Mortality is usually minimal unless it is complicated by other infectious agents such as *E. coli* or *Mycoplasma gallisepticum*, or unless the virus is a nephropathogenic strain (e.g. Holte, Gray, Italian I1731, Chinese QX, Australian T) to which there might be high mortality associated with birds dying from urolithiasis and renal gout [20]. In the United States and Canada, periodic outbreaks can happen from isolates related to Delmarva (DMV)/1639 which can result in explosive mortality in pullets [21]. Early infection of pullets around 5–14 days of age with a variant strain of IBV may result in damaged or hypoplastic oviducts, which causes false layer syndrome in layers at the onset of production [19]. There has been anecdotal reported success of reducing incidence of false layer syndrome with the use of MA5 vaccines given at the hatchery to help with early protection against IBV. Nonspecific signs could manifest in growth retardation, respiratory signs, and enteritis. Histopathologic lesions are usually acute inflammatory lesions followed by necrosis of affected tissues, such as tracheitis, air sacculitis, and pneumonia. Differential diagnosis in the respiratory form includes other acute respiratory diseases such as Newcastle disease, infectious laryngotracheitis (ILT), LPAI, avian metapneumovirus, infectious coryza, and *Mycoplasma gallisepticum* and *E. coli* infections [18] Changes in the kidney with the nephropathogenic IBV can also resemble those of astroviruses, infectious bursal disease (IBD), mycotoxicosis, and other toxicities [22].

Diagnosis is through clinical history, lesions, seroconversion using acute and convalescent titers, and antigen detection through PCR or virus isolation. Since vaccination for IBV is routine, complete diagnosis of IBV must include identification and characterization of the serotype or genotype of the virus to examine both its relevance to the disease and control strategies, including the appropriate vaccines and comparing sequencing and mutations of this highly variable virus [23]. Sample types used include tracheal swabs or fresh trachea in early infection, and also lung, kidney, and oviduct depending on the pathotype of the virus. In late infection, the virus could be found persisting in caecal tonsils for several weeks to months after acute disease [24, 25]. If a predominant genotype or variant that is not currently covered by the routine vaccines is rampant in an operation, virus isolation necessary to manufacture autogenous vaccines may be considered to control future outbreaks to reduce mortality and production losses.

2.2.5 Infectious laryngotracheitis

Infectious laryngotracheitis (ILT) virus is a highly contagious upper respiratory disease of chickens caused by the Gallid herpesvirus type of the subfamily *Alphaherpesvirinae*. High flock density, raising of multiage chickens within dense poultry-producing areas, and improper vaccination and breach in the biosecurity contribute to the increased ILT outbreaks across the world [26]. Pullets above the age of 4 weeks are reported to have ILT; however, flocks typically do not have outbreaks until about 7 weeks of age. Many of the recent outbreaks around the world are largely suspected to have emerged from CEO vaccines and are often termed CEO vaccine-like or vaccinal ILT infections. Clinical signs are characterized by acute respiratory disease with dyspnoea and expectoration of bloody mucus. Characteristic gross lesions include severe haemorrhages of the laryngeal and tracheal mucosa with obstruction of airways by caseous casts or plugs in the laryngotracheal junction (Fig. 1), predisposing chickens to asphyxiation. Milder symptoms are characterized by general unthriftiness, conjunctivitis, and swelling of the infraorbital sinuses. Due to the virus' slow spread, it may take 10–14 days for birds to recover after the onset of clinical signs [27]. The virus becomes latent carriers and can cause problems later in lay. Histopathologic lesions in early stages of the infection, which is often pathognomonic with haemorrhage and necrosis of the tracheal mucosa with epithelial cell debris containing intranuclear eosinophilic inclusion bodies forming syncytia, are observed in the first 1–5 days

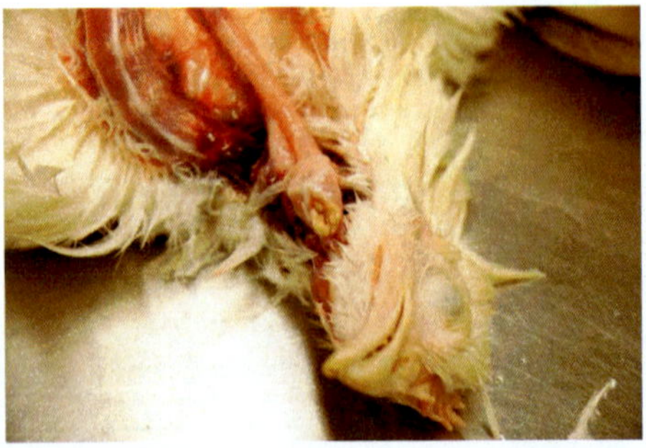

Figure 1 Caseous tracheal plug formed at the laryngotracheal junction characteristic of ILT infection in a layer pullet.

post-infection. Inclusions may also be seen in the conjunctiva and in the lungs. Subacute to chronic infections are characterized by a hyperplastic tracheitis with proliferation of regenerating epithelial cells that are hard to distinguish. After early stages of infection, diagnosis is commonly facilitated by the use of PCR and other antigen-detection methods. ILT must be distinguished from other pathogens for appropriate control measures. The most common rule-out is the diphtheritic form of the fowlpox virus, which could be differentiated easily via histopathology. Other differentials include *Mycoplasma gallisepticum*, Newcastle disease, AI, IBV, severe *E. coli* infections, and infectious coryza [26, 27]. Control of ILT generally involves the use of routine vaccinations and good biosecurity practices. Three major types of vaccines have been used historically: modified live CEO vaccines, modified live tissue culture origin (TCO) vaccines, and recombinant viral vector ILT vaccines, which are used alone or in combination with live attenuated vaccines. CEO is a much more immunogenic vaccine and is used extensively due to its ability of mass vaccination; however, improper use may lead to unwanted spread of the vaccine virus and potential for disease (i.e. vaccinal ILT). The first dose of CEO is recommended to be administered between 3 weeks and 8 weeks of age, followed by a second one at 9–14 weeks of age, prior to moving to lay house. TCO is primarily used in operations where ILT challenges are relatively low, given 10 weeks apart. Cell-mediated and local immunity versus humoral immunity provides resistance to the disease, and due to this nature the disease spreads much slower as compared to agents such as IBV, making vaccination during an outbreak a practical option to control the disease. Although both CEO and TCO have shown protection against mortality, clinical signs, and reducing replication and shedding of the virus, vaccination via live attenuated CEO vaccines has been demonstrated to reduce mortality rates and reduce rolling infections within a flock during an outbreak [28]. Recombinant ILT (either rPox-ILT or rHVT-ILT) followed by a booster of TCO or CEO ILT has been used in high ILT challenge areas since the protection with recombinant vaccine alone may not be adequate. Modified live vaccines should not be administered prior to 3 weeks of age. Since re-vaccinating a flock can be time consuming and costly, definitive diagnosis must be made before making decisions. Heat treatment during clean up stages of 100 h at 100°F (38°C) may be used in outbreak situations to reduce virus burdens in hopes of reducing virus spread in subsequent flocks.

2.2.6 Fowlpox virus

Fowlpox is a common viral disease in commercial layers and pullets and is characterized as one of two forms: cutaneous lesions in unfeathered skin (dry pox) or proliferative, diptheritic lesions in mucosal membranes (wet pox).

Fowlpox is a slow-spreading DNA virus that is often transmitted via mechanical methods through injured skin or direct contact with mucosal membranes. Insects such as mosquitoes and mites can serve as biological vectors, and others like flies and gnats can serve as mechanical vectors, which explains the seasonality/climate preference (most commonly in the spring and summer, more frequently diagnosed in warm climates such as the southern United States) of the disease correlated with insect activity. Clinical signs can depend on the form of the virus and location of the lesions, especially with the cutaneous form [29]. Cutaneous lesions around the eye, nares, and mouth can interfere with feed and water consumption, breathing, and eating, respectively, and result in high mortality (Fig. 2). Wet form of fowlpox could resemble gross lesions from ILT and could result in partial or full obstruction of the respiratory tract and in death from asphyxiation. Skin lesions can start in 4–5 days, followed by a formation of a scab, which may last for another 1–2 weeks. Mild cutaneous form of the disease runs for about 3–4 weeks, but can prolong with complications. Microscopic lesions are characterized by marked hyperplasia of the epithelium with ballooning degeneration of infected cells, as well as associated inflammatory changes. Characteristic eosinophilic intracytoplasmic inclusion bodies (Bollinger bodies) with viral particles (Borrell bodies) are pathognomonic within infected cells [30]. Respiratory lesions and clinical signs can be very similar to ILT and histopathology will be a great way to rule out, although co-infections are not uncommon and should be considered in each case for control options. Other differentials are viscerotropic velogenic Newcastle disease (vvND), AI, *Mycoplasma gallisepticum* (MG), infectious coryza, and fowl cholera. Cutaneous lesions can resemble a pantothenic acid or biotin deficiency in young chicks, as well as *Trichomonas gallinae* and T-2 toxin lesions [30, 31]. In areas where pox challenge happens early, chicks can be vaccinated as early as 1 day of age with a TCO vaccine at hatch with a recombinant (rPox) HVT (Marek's disease) vaccine given as rPox-HVT. This early vaccination does not provide long-term

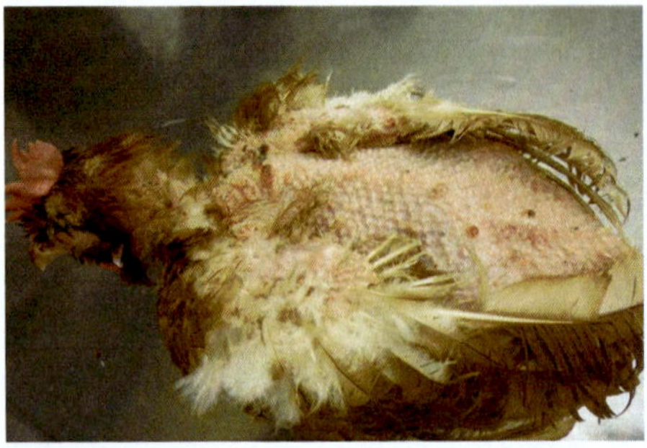

Figure 2 Cutaneous lesions of fowlpox with scabs on unfeathered skin and around the eyes of a laying hen.

fowlpox protection, and thus a second vaccination at 8–10 weeks of age using a live CEO vaccine is recommended. For long-term protection, pullets should be vaccinated with a CEO vaccine after 6 weeks of age, which is often in combination with AE and/or pigeon pox vaccine, typically given in the wing web. The wing web vaccine produces a vaccine 'take', which causes a small scab or skin swelling 5–10 days post-vaccination. A check to look and see if 99–100% of the bird have vaccine takes will ensure appropriate vaccine application, as many fowlpox outbreaks are a result of improper vaccine application. Pigeon pox given together with fowlpox vaccines provides better overall protection by stimulating broader immune responses, especially in high challenge areas where fowlpox vaccine alone may not be adequate to prevent outbreaks [31]. Similar to ILT, long incubation periods and slow spread within a flock make re-vaccination during an outbreak a practical option to control the disease. The use of iodine disinfectants through the waterlines may help inactivate the virus and help heal dry scab lesions, especially in managing wet pox outbreaks [32]. In addition, controlling insect vectors and proper biosecurity will help prevent the spreading of the virus and outbreaks in the future [30, 31].

2.2.7 Newcastle disease virus and avian metapneumovirus

Of significance to layers include Newcastle disease (ND), now reclassified as avian avulavirus 1 in 2016 and avian metapneumovirus (AMPV), which will be described in 3.1.7. Similar to AI, virulent ND, formerly known as exotic Newcastle disease (END), infection is of interest to WHOA [33]. ND affects the respiratory, nervous, gastrointestinal, and reproductive systems and can cause up to 100% mortality in unvaccinated chickens. There are four pathotypes of the disease based on clinical signs, in order of decreasing pathogenicity: (1) velogenic, which creates high mortality and lesions in all systems; (2) mesogenic, which creates neurologic signs in young chicks; (3) lentogenic, which creates mild respiratory signs and are commonly used in commercial vaccines; and (4) asymptomatic enteric, which has a tropism of the gastrointestinal system [33, 34]. Since the virus can infect over 236 species of birds and even vND is endemic in some species of birds, a breach in biosecurity and transmission from wild birds could be a way for introduction into commercial flocks, followed by horizontal spread [35]. Similar to AI, neither the clinical signs nor the gross lesions are pathognomonic for ND. Virulent neurotropic ND can cause clinical signs such as torticollis and head tremors, while the typical lentogenic form may be respiratory signs such as dyspnoea, facial swelling, and oculonasal discharges. Likewise, gross lesions vary greatly and may be absent in vaccinated pullets. Necrosis and haemorrhages in lymphoid tissues such as caecal tonsils and Peyer's patches have been frequently described with virulent ND. Gross and microscopic lesions in the respiratory form of the virus are rare [33]. As such, ND should be on the differential list with any high mortality events, any neurologic diseases, and nonspecific respiratory diseases. Trachea, oropharyngeal, and/or cloacal swabs are good samples to take for PCR diagnosis followed by possible virus isolation. Serology is often not helpful as vaccination is common and vaccination schedule varies. Differential diagnosis can include anything from aspergillosis, mycoplasmosis, ILT, fowl cholera, IB, avian metapneumovirus, and HPAI/LPAI depending on clinical signs. Strict biosecurity, vaccination, and culling infected birds are often necessary to contain ND outbreaks [33, 35]. Vaccines are used to prevent losses from morbidity and mortality and will not prevent infection. The most widely used vaccines are the live lentogenic vaccines, using B1 and La Sota strains. These viruses belong to genotype II and are antigenically and genetically similar. The La Sota strain is the most antigenic and results in higher antibody levels, and is most widely used in countries where vND is endemic. The B1 strain is the most attenuated and is often used in cases with low exposure or in very young chicks. The VG/GA (Villegas-Glisson/University of Georgia) strain is typically sold as an enterotropic vaccine and has been used internationally for 20+ years, whereas it is not as widely used in the United States [36]. The live vaccine can be delivered as a recombinant (rHVT-ND) vaccine, through drinking water, nasal sprays, and coarse sprays, of which coarse sprays are most widely used in layer pullets. In the United States, the recombinant vaccine is most commonly used in broilers and extra-label in turkeys but not as mainstream in egg-type pullets. The coarse spray, often given early, can cause the most significant vaccine reactions but also is the most immunogenic [36, 37]. This is boostered by inactivated vaccines mixed with other agents such as SE and IB and given at least once or twice before birds go into lay.

2.2.8 Infectious bursal disease virus

IBD, also known as Gumboro disease, is a birnavirus that is highly resistant to inactivation. The disease is characterized by high mortality and severe immunosuppression due to its targeting of lymphoid tissue, especially the bursa of Fabricius, from which the disease name is derived. There are two serotypes of IBD, but only serotype 1 is pathogenic, although there are several variants of serotype-1 IBD which are antigenically distinct from the classic strains of IBD [37]. The virus is in the United States endemic in most, if not all, chicken farms, due to the highly infectious nature and hardy properties of this virus and its persistent survival in houses even when thorough cleaning and disinfection procedures are followed [38]. Outbreaks of very virulent strains of IBD (vvIBD), which causes higher severity of clinical signs and mortality, have been reported in Europe, Africa, Asia, South America, and the United States

[37, 39]. The period of greatest susceptibility to clinical disease is between 2 weeks and 6 weeks of age. Susceptible chickens younger than 2 weeks do not exhibit clinical signs but have subclinical infections resulting in severe immunosuppression, which is catastrophic for all ages. Clinical signs and gross lesions are characterized by dehydrated birds with haemorrhages in the thigh and pectoral muscles, and markedly swollen, oedematous cloacal bursa with necrosis and haemorrhages on the mucosal surface. Histopathologic lesions are not pathognomonic but highly suggestive of IBD infection with degeneration and necrosis of lymphoid cells, some with haemorrhages, followed by atrophy and fibroplasia of damaged bursal follicles (Fig. 3). Microscopic lesions in the bursa can appear as early as within 24 h post-infection. Differential diagnoses based on clinical signs are typical of coccidiosis or other conditions that can cause nephrosis, such as nephropathogenic IBV infection and dehydration due to water deprivation. Diagnosis is typically made by molecular characterization of IBD using bursa as a sample type, and subclinical IBD is diagnosed by either histopathologic surveillance or serology to look at titers [37]. Since the virus is ubiquitous, vaccination is the primary method for prevention and control of IBD. Most US vaccine protocols will include the classic strain of IBD and one or more of the variant strains of IBD. Inactivated vaccines are usually not practical unless for use in breeders for transferring maternally derived antibodies to progeny, especially because there is little cross-protection across variant strains of IBD. Vaccination via live vaccines or recombinant vectored vaccines is mainstream in layer production, with the preference for vectored vaccines, as it does not seem to interfere with even high levels of neutralizing maternal antibodies [40]. A universal vaccination programme cannot be offered because of the variability in maternal immunity, management, and regional disease pressures, and thus

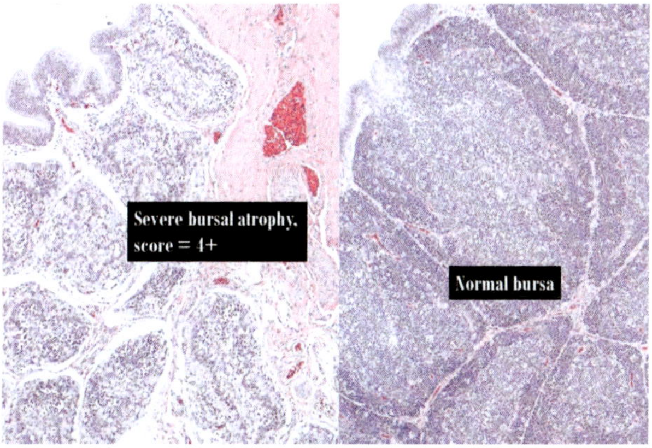

Figure 3 Histopathologic images of severe bursal atrophy (left) compared to normal bursa (right) H&E 10x magnification.

must be evaluated on a case-by-case basis [37]. Careful observation of vaccine titers, not limited to IBD but to other disease agents which may be affected by immunosuppression, monitoring for potential variant strains, and periodic gross and histopathologic evaluation of bursas are routinely done to evaluate the status of IBD in given flocks. With vvIBD, thorough cleaning and disinfection with an all-in-all-out system can be essential to successfully eliminate the virus from infected premises.

2.2.9 Neoplastic viruses (Marek's disease, avian leukosis and reticuloendotheliosis virus)

The three most economically important virus-induced transmissible neoplastic diseases of concern to layers include (1) Marek's disease (MD), (2) Avian leukosis/lymphoid leukosis (LL), and (3) reticuloendotheliosis (RE). These agents are easily confused in diagnosis, although there are characteristics based on gross lesions, microscopic features, and age distributions that help in differentiating between the three agents (Tables 2 and 3). Different methods of differential diagnosis may or may not be available depending on the global distribution and availability of these assays. Prevention and control strategies vary greatly between MD, where the primary strategy is through vaccination, and retroviruses (LL and RE), where control is via eradication from infected

Table 2 Comparison of the three tumour viruses on gross lesions, histopathologic lesions, and age predilections.

		MD	LL	RE
Gross lesions	Liver	+++	+++	+++
	Spleen	+++	+++	+++
	Peripheral nerves	+++	–	++
	Skin	+++	+	+
	Gonads	+++	+++	+++
	Heart	+++	+	+++
	Bursa	+	+++	+++/––
	Intestines	+	++	+++
	Lung	+++	++	++
	Kidney	+++	+++	+++
Histopathology	Pleomorphic cells	+	–	–/+
	Uniform blast cells	–	+	+/–
Age	Peak time (in months)	2-7	4-10	2-6
	Limits (in months)	>1	>3	>1

Source: Modified from Ojkic et al. [19].The number of + and – correlates to likelihood. The +/– means more likely than not, and vice versa.

Table 3 List of tests and procedures for differential diagnosis.

Criteria	Test of procedure	Criteria/assay	Value	Present use
Standard	Epidemiological data	Age	Mod	Yes
	Clinical signs	Neurological	Mod	Yes
	Gross pathology	Nerves, bursa, skin	High	Yes
	Microscopic pathology (H&E stain)	Location and type of lesions	High	Yes
Advanced	Quantitative PCR	Oncogenic MD Ct value	High	Yes
	Immunohistochemistry	T-cell (CD3/CD4) and B-cell markers (CB1/PAX5/IgM)	High	Yes
		Meq antigen (encoded by MD *meq* oncogene)	High	Rare
	Southern blotting	Clonal insert, *c-myc* alt.	High	Rare
	PCR-retrovirus/*c-myc* junctions	Signature fragment	High	Rare
Viral assays	Immunohistochemistry	MD antigens – pp38, gB	Low	Rare
		RE or LL antigens	Mod	Rare
	Virus isolation	RE-7d culture/FA or PCR	High	Yes
		LL-7d culture/FA or ELISA	Mod	Yes
		MD-7d culture/FA or ELISA	Low	Yes
		LL-VN or ELISA test	Low	Yes
		MD antigen detection in skin by AGP	Low	Rare
	PCR assays	Qualitative assay, RE and LL	Mod	Yes
		Qualitative assay for MD	Low-mod	Yes

Source: Modified from Dunn et al. [41] and Witter et al. [42].FA, fluorescent assay; VN, virus neutralization; AGP, agar-gel precipitin.

flocks, supplemented by selection for genetic resistance. Below, MD will be discussed in pullets. LL and RE will be discussed in 3.1.9 due to the age predilection of the disease.

MD is a highly contagious, ubiquitous alphaherpesvirus characterized by the rapid onset of lymphoid tumours, immunosuppression, and paralysis. The virus replicates readily in feather follicles and stays latent in birds for weeks to months prior to development of clinical signs. Infective MD virus particles are shed in feather dander and chicks as early as day of age are exposed to the virus. As such, detection of MD virus alone is not sufficient to make a confirmatory diagnosis unless there are supportive gross and microscopic lesions, potentially with the help of additional ancillary tests [42]. There are three different species and four different pathotypes, ranging from mild to very virulent plus (vvMDV+) which could influence vaccination strategies based on disease pressures within a region [43, 44]. MD is characterized by four different pathologic syndromes: (1) lymphoproliferative form, characterized by visceral organ tumours and neuropathy of peripheral nerves; (2) neural form or fowl paralysis, characterized by paralysis and lesions in the peripheral nerves; (3) integumentary form, characterized by swollen feather follicles in young birds; and (4) ocular form, characterized by blindness and grey iridal lesions. The lymphoproliferative form is most common in layers and can happen at any age after ~4 weeks of age. The neural form is occasionally seen in layers, mostly between the ages of 8 weeks and 20 weeks of age, characterized by paralysis and subsequent death in up to 20% of the flock. Both the integumentary and ocular forms of MD are rare in layers, and usually there is no mortality associated with this form of disease, let alone could be gone unnoticed in large flocks [42, 45]. The incidence of MD is quite variable in commercial layer flocks and in general low since the worldwide introduction of vaccines, although occasional outbreaks can happen due to several factors, with improper vaccine administration among the most frequently identified issues. Definitive diagnosis is often complicated (see Tables 2 and 3), as antemortem diagnosis is not reliably done. Due to the variety of gross and microscopic lesions, there is lack of truly pathognomonic lesionsthat can definitively distinguish between other neoplastic viruses. The other differentials for the neural form that doesn't involve the brain include peripheral neuropathy, a disease of unknown aetiology, characterized by paralysis and nerve enlargement in a low proportion of commercial chickens between 6 weeks and 12 weeks of age [42]. Histopathologic examination of the brain to distinguish the two, although not always the case, may be beneficial for distinction.

Vaccination against MD is among the most costly for live pullet vaccines based on the use of vectored vaccine technology, which requires specific equipment, as well as the maintenance of the Rispens and other MD viruses

Table 4 Current vaccine strains licensed in the United States according to serotype and virus properties [43].

Vaccine	Serotype	Oncogenic potential	Horizontal transmission	Current use
HVT (FC126)	III	No	No	Worldwide
SB-1	II	No w/ HVT	Yes	Combined
301 B/1	II	No w/ HVT	Yes	Combined
CVI988 clone C	I	No	Yes	No
CVI988/C/R6	I	No	Yes	No
Rispens (CVI988)	I	No	Yes	Worldwide
R2/23 (Md11/75)	I	No	Yes	No

under liquid nitrogen. Several types of vaccines (Table 4) are commonly used, some given alone, and some used in combination due to their synergistic activities, depending on the virus challenges [42, 45]. For example, Rispens (CVI988) is a must in high challenge areas where there are dense populations of chickens, where double-dosing and multiple dosages may be given in areas, and where vvMDV+ is endemic and higher thresholds of the virus are necessary for adequate protection [45]. In comparison, immunity induced by HVT may be excellent against low virulence challenges and help reduce cost of vaccination. In recent years, recombinant vectored MD (rHVT) vaccines have been used extensively in layer pullets, with the added benefit of vectoring other disease agents along with HVT. The most recent development has been the use of trivalent rHVT-IB-ND vaccines [46], which have been developed to control vND worldwide. Although impractical, strict biosecurity practices to limit the extent of early MD virus exposure are a cost-effective strategy in conjunction with sound vaccination programmes. This could be somewhat problematic in multiage layer sites and thus a more biosecure pullet site located remotely from lay barns may be more successful in controlling early exposure. Heat or formalin inactivation of barns to knock down virus burden in facilities with MD outbreaks, or multiple serotype I vaccine applications, is used, similar to controlling ILT.

2.3 Bacterial diseases of pullets

In Section 2.3.1, primary agents of bacterial diseases (Clostridial diseases, mycoplasmosis, Pullorum-typhoid salmonellosis) will be discussed. In Section 2.3.2, opportunistic pathogens (*E. coli, Staphylococcus, Streptococcus, Pseudomonas*, paratyphoid *Salmonella* spp.) will be discussed. Other bacteria of concern to layers will be discussed in 3.2

2.3.1 Primary bacterial diseases of pullets

2.3.1.1 Clostridial diseases (botulism, necrotic enteritis and ulcerative enteritis)

Clostridia are gram-positive, spore-forming anaerobic rod-shaped bacteria. Although many of them are considered commensal microflora and are present ubiquitously within the environment, select species produce several toxins that are extremely pathogenic to poultry. The three diseases of concern are botulism, caused by *Clostridium botulinum*; necrotic enteritis, caused by *C. perfringens*; and ulcerative enteritis, caused by *C. colinum* [47]. Ulcerative enteritis is rarely reported in layer pullets and is often confused with necrotic enteritis, with the only difference being the etiologic agent. However, the bacteria are extremely hard to isolate by bacterial culture, and unless PCR confirmation is done, presumptive diagnosis is via gross and microscopic lesions.

Botulism is a disease of humans and other animals caused by clostridia-produced botulinum neurotoxins, which causes high mortality and flaccid paralysis that progresses cranially and leads to acute death by cardiac and respiratory failure. Botulism occurs sporadically in commercial layers, although with the increase in cage-free and free-range birds in the last decade, it is considered an emerging disease in Europe [48]. Cases of botulism in poultry have predominantly been due to type C toxins, although there have been reports of infection by mosaic C/D or D/C toxins as well. Mature hens may be less susceptible to botulism due to a potentially high threshold of tolerance to toxins as compared to younger pullets, although poultry of all ages are susceptible. Rotting vegetation and hardy Clostridial spores in the environment, as well as decaying poultry carcasses, can contain enough exotoxin to be highly lethal. Over 2000 minimum lethal doses of type C toxin per gram of carcass tissue have been found. Maggots are known to concentrate the toxin, and may contain up to 10^4–10^5 minimal lethal doses of toxin [49]. Botulinum toxins are produced when the spores germinate, which require higher temperatures. This is evident where most botulism outbreaks take place during the summer and fall when ambient temperatures are high [47]. The incubation period is usually very short, ranging from a few hours to 13 days, depending on the amount of toxin ingested. Clinical signs are characterized by flaccid paralysis of the legs, wings, neck, and eyelids, which progresses cranially to paralyse the head and neck (limberneck), with feathers easily epilating due to paralysis of the piloerector muscles. Prior to death, birds typically gasp due to cardiac and respiratory failure. Gross and microscopic lesions are absent, and confirmation is typically done by using mouse serum toxin bioassay and/or by molecular methods detecting the neurotoxin gene. Diagnosis is by collecting serum samples from acutely affected birds and collecting crop or gizzard contents, respectively. Differentials are through process of ruling out other diseases that cause acute mortality and/or neurologic signs, such as vND, HPAI, AE, fowl cholera, MD, and drug or chemical toxicities [47]. Treatment is usually impractical in commercial birds, although antibiotic treatment may help with secondary infections. In organic birds, the use of prophylactic phytobotanicals such as oregano oil have minimal reported success in controlling mortality. Emphasis is put on management strategies to reduce exposure to source(s) of infection, such as prompt removal of mortality, distancing compost piles from ranging birds, and elimination of shallow waters in outdoor access that can harbour *C. botulinum* spores.

Necrotic enteritis, also known as enterotoxemia, is one of the major gastrointestinal diseases of poultry caused by pathogenic strains of *Clostridium perfringens* which leads to acute mortality with characteristic necrotic lesions of the mucosa and distension of the small intestines. Although the pathogenesis is not clearly understood, necrotic enteritis is often associated with predisposing factors such as enteric coccidiosis, certain feed ingredients and/or feed quality issues, or underlying immunosuppressive diseases such as IBD that allow for proliferation of the *C. perfringens* within the intestines [47]. Necrotic enteritis is commonly caused by *C. perfringens* types A (α or CPA toxin), C (β or CPB, and α toxins), and G (Netβ toxin), with increasing reports of the recently discovered netB toxin-related outbreaks produced by a subset of *C. perfringens* type A strains [50–53]. Clinical signs are characterized by the acute onset of depressed, ruffled birds progressing to death in a couple of hours, with a sharp increase in flock mortality. Gross lesions are characterized by markedly distended intestines with brown, fowl-smelling fluid and friable mid-small intestines with diptheritic pseudomembranes (Fig. 4) [47]. Scattered pinpoint foci may be present in the livers due to bacterial showering. Histopathologic lesions include extensive mucosal necrosis of the upper third of the villous epithelium, with sloughed, necrotic coagulum and abundant large rods at the submucosal and necrotic

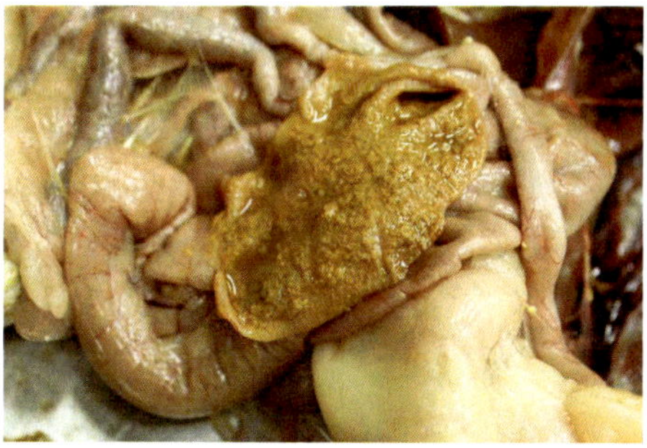

Figure 4 Jejunum from a layer pullet with markedly distended intestines with diptheritic pseudomembranes (necrotic enteritis).

margins. Diagnosis is often based on typical history and gross diagnosis and does not require ancillary testing. Differentials include coccidiosis, ulcerative enteritis, and histomoniasis [53]. Treatment in acute cases includes the use of feed or water-based antimicrobials that target gram-positive bacteria such as bacitracin and penicillin. Prophylactic use of bacitracin and/or coccidiostats that help reduce clostridial load such as ionophores, may be used throughout the pullet grow-out period to control necrotic enteritis. However, the key to success is managing predisposing factors. Effective coccidiosis control with the use of vaccines and/or coccidiostats; appropriate litter amendments and ventilation to reduce environments conducive to clostridial overload; pre- and probiotics to improve overall gut health and enrich with healthy microflora; improvement of feed quality; and vaccination protocols against agents such as IBD. This is increasingly important as higher percentages of layer production are progressing to antibiotic-free strategies.

2.3.1.2 Mycoplasmosis (Mycoplasma gallisepticum and Mycoplasma synoviae)

Mycoplasmosis is a bacterial disease comprising respiratory disease, synovitis, and general poor performance in pullets. The bacteria are vertically and horizontally transmitted and the clinical signs vary greatly with the species, the pathogenicity of the strain, co-infections with other agents, and the age and immune status of the infected bird. Vaccination and antimicrobials have been used to control infections and prevent disease. However, larger success in commercial poultry involves elimination of specific avian mycoplasma species from breeding stock using test and cull programmes such as via the NPIP in the United States and the MG eradication programme in the Netherlands [54, 55]. The two species of concern in commercial pullets include *Mycoplasma gallisepticum* and, to a lesser degree, *M. synoviae*.

MG, also known as chronic respiratory disease, is the most pathogenic and economically significant mycoplasma of poultry. MG is characterized by respiratory rales, coughing, nasal discharge, conjunctivitis, and infraorbital sinusitis which results in swollen heads. In adult laying hens, MG infection can result in egg production drops and decreased hatchability in breeders. Despite negative chick sources, MG is relatively common in commercial multiage egg-layer farms in many countries, and endemically infected flocks present as potential reservoirs of infection [56]. In addition, the prevalence of MG in backyard poultry flocks is relatively high, and can also be a source of infection in proximity to commercial flocks. Although poultry of any age can technically be infected, pullets less than 4 weeks of age almost never get infected horizontally. MG is highly contagious, and can spread by aerosol or droplets to susceptible birds, or via contaminated fomites as it can survive for several days on materials such as dust, feathers, and faeces [54, 56]. Clinical signs with MG alone can be very subtle, ranging from general growth retardation to keratoconjunctivitis with facial swelling, increased lacrimation, and conjunctival congestion [57]. Gross lesions may also be very subtle with mucus in nasal and tracheal passages, and minimal without complications with other agents such as *E. coli*, which then can cause severe air sac infections or pneumonia, especially in cold months. Histopathologic lesions are characterized by marked thickening of the mucosal membranes of respiratory tissue with primarily lymphocytic infiltrates, often resulting in nodular lymphoid hyperplasia [54]. Diagnosis is typically through PCR, as the organism is fastidious in nature and can take upwards of 2–3 weeks for bacterial culture. Molecular characterization can be done directly from PCR samples to look at wild-type strains versus vaccine strains, as well as investigate epidemiologic links in outbreak scenarios. Serological surveillance using serum plate agglutination (SPA) and/or enzyme-linked immunosorbent assay (ELISA) followed by confirmation via hemagglutination inhibition (HI) assay is useful for monitoring flock status. For early infection, SPA can detect antibodies to MG as early as 7–10 days post-infection due to the target being IgM versus IgG in ELISAs [56]. Differential diagnosis includes a list of respiratory diseases such as lentogenic ND, IB, LPAI, colibacillosis, ILT in older pullets, and keratoconjunctivitis with high ammonia levels in poorly ventilated barns. Despite its short-term use to ameliorate clinical disease, treatment is usually impractical in commercial pullets since antibiotics cannot eliminate MG from infected flocks (hence the term chronic respiratory disease). Tylosin and tetracycline are commonly used for treatment, but require prolonged treatment duration (3–4 weeks) with a risk of reversion to disease and reinfection in MG positive flocks. Historically and perhaps in other countries, enrofloxacin and other fluoroquinolones have been used for treatment but are now subsequently banned in the United States since 2005 [58] and similarly in other countries. Vaccination is used in some commercial pullets, especially those destined for multiage lay sites that are positive for MG, to ameliorate clinical signs, to reduce issues in egg production in layers, and to displace virulent wild-type strains with vaccine strains. Inactivated, live attenuated, and recombinant MG vaccines are available commercially [54]. Inactivated vaccines are more standard in replacement breeder pullets since there is no risk of transmission or reversion to virulence; however, there are issues including high cost of administration and local vaccine reactions from injections as well as limited protection due to only stimulating humoral immunity. Three commercial live MG vaccines used worldwide are F strain, ts-11, and 6/85, all of which can revert to virulence. The original F strain had moderate virulence and posed problems with vaccine-related infection when administering to pullet sites close to turkey operations. The current F strain is a relatively mild MG strain and has been used extensively in layer pullets to replace field strains. It is also more virulent and thus more reactive than ts-11 and 6/85 and produces good antibody responses. Pullet sites may choose to use ts-11 for a lower reaction and to replace the F strain in facilities. Although it has good local immunity and colonization, ts-11 has a weak humoral antibody response when compared to F strain and is perhaps useful in sites where wild-type strains of MG are controlled. The 6/85 is the least virulent and immunogenic of the live MG vaccines with minimal virulence and immunogenicity. Therefore, it has weak local colonization and little to no serologic conversion and is likely not a good choice in endemic areas. The recombinant fowlpox-MG vaccine alone has limited protection against disease and perhaps could be useful in sites where wild-type MG strains are controlled [59]. Since maintaining flocks free of MG infection is only possible by obtaining day-old chicks from MG-free breeders and hatcheries, followed by good biosecurity to prevent introduction of the organism, routine monitoring to evaluate MG stats is important in maintaining MG-free pullet sites.

Mycoplasma synoviae (MS) causes subclinical upper respiratory infection and usually causes MG-like air sac infection via co-infection with ND or IB, or both. Although there are variations among isolates in its pathogenicity, strains in the United States typically cause little to no clinical disease in pullets alone. Production of eggs with abnormal shells (egg-shell apex abnormalities) has been reported in Italy, the Netherlands, Germany, Australia, and Brazil in strains of MS with tropism to the oviduct. This is characterized by roughening of the shell surface at the top cone of the egg, with thin, translucent shell surfaces with cracks and breaks in the egg shell [60]. Similar to MG, MS infection in multiage layer sites is not uncommon, although serological and molecular detection of MS does not happen often in pullet sites. Swollen hock joints and foot pads with viscous exudates around tendon sheaths and synovial membranes may happen with MS infection coupled with high morbidity, subclinical respiratory infection. Similar to other mycoplasmas, *in ovo* and horizontal infection is possible with MS. Diagnosis is also by molecular methods via PCR and serological monitoring using ELISA/SPA followed by HI. Differentials include agents that have the potential to cause both synovitis and respiratory disease, such as *Staphylococcus aureus, E. coli*, fowl cholera, and salmonella. Since clinical symptoms are mostly subclinical, treatment is usually not provided [54]. Inactivated vaccines and live MS vaccines are available in many other countries but are pending registration in the United States [61] The live vaccines reduce the shedding of wild-type MS as long as the flock is vaccinated prior to exposure to field strains, and are given via eye drop as early as 3 weeks of age. Care must be taken for administrating antibiotics after vaccination since it can kill vaccine strains that are replicating in the respiratory tract [62].

2.3.1.3 *Pullorum disease and fowl typhoid*

There are two species of *Salmonella*: *S. enterica* and *S. bongori*. There are over 2500 serovars of *Salmonella*, most of which are in *S. enterica* ssp. *enterica,* which is associated with disease in warm-blooded animals [63]. Of concern to poultry are two serovars, *Salmonella* Pullorum (SP), the cause of Pullorum disease, and *Salmonella* Gallinarum (SG), the cause of fowl typhoid, are host-specific, non-motile Salmonellas that cause septicemic diseases in affected poultry [64]. Both of them have been responsible for devastating economic losses and are the reason for the establishment of NPIP in 1935 in the United States to eliminate them from commercial breeding stock [6]. Although outbreaks may appear periodically in backyard flocks, Pullorum and typhoid are eliminated from commercial poultry in the United States as one example, and continue to be monitored as a reportable disease in several economically developed countries [64, 65]. Mortality associated with SP infection is generally limited to 2–3 weeks of age, although birds that survive early infection may become inapparent carriers and shed horizontally and vertically through the egg. Vertical transmission from infected breeders is one of the primary routes of spread of Pullorum disease, along with contamination of egg-shell surfaces that are a more significant issue in SG. The bacteria are then transmitted horizontally via the faecal–oral route and target the intestines, especially the ceca, for colonization; however, SP has been shown to preferentially target sites such as the bursa of Fabricius [66]. Birds infected *in ovo* may have dead or dying chicks upon hatching, while the majority will be live chicks with signs of acute septicaemia including depression, weakness, droopy wings, huddling, dehydration, and ruffled feathers. Pullorum disease is also called 'bacillary white diarrhoea' based on the clinical sign of white and viscous diarrhoea and faecal pasting around the vent [65]. The chicks may have lameness caused by swelling of the hock joint [63]. SG can affect all ages of birds and can cause similar clinical signs and even mortality in adult birds [65]. Chicks infected with SP sometimes die without gross lesions, but classic lesions include white to grey nodules (granulomas) in the heart, liver, and gizzard; caseous, white to yellow exudates in the caecal lumen; and white to yellow, caseous exudate within the joints of the legs and wings [63]. Typical gross lesions of fowl typhoid include a dark, enlarged, friable liver that often has a bronze to copper colour, as well as catarrhal enteritis with bile-tinged digesta in the small intestines [65]. Differentials in young chicks may include other enteric diseases such as coccidiosis, and diseases that cause acute death and septicaemia such as AI and ND. Management issues such as temperature extremes causing white diarrhoea, as well as nutritional imbalances and acute toxicities, are also considered. Diagnosis is mainly through bacterial culture and isolation of affected organs and/or faeces, although shedding of the bacteria is intermittent and thus asymptomatic carriers may be harder to detect through culture. Serological surveillance programmes are used for health certificates for interstate and international movement, as well as screening breeding stock to reduce risks of vertical infection. SPA, although not specific for Pullorum-typhoid, have been used as a screening test, followed by tube agglutination or microagglutination tests, which have been used in the US NPIP programme. Prevention and control is based on obtaining eggs and chicks from SP and SG free breeders and clean hatcheries, as medicating with antibiotics to prevent or treat Salmonella infections in poultry has been debated for many years as inconsistent and not cost effective [63]. Pullorum-typhoid is a test and cull disease in countries where it has been eradicated from commercial poultry.

2.3.2 Opportunistic bacterial diseases of pullets

2.3.2.1 E. coli, Staphylococcus, Streptococcus, Enterococcus, Pseudomonas and paratyphoid Salmonella

Escherichia coli infection, or colibacillosis, is the most common infectious bacterial disease of poultry and is caused by a gram-negative, non-acid-fast, non-spore-forming bacillus which is a normal bacterium found in the gastrointestinal tract and respiratory tract of birds [67, 68]. *E. coli* can manifest in many forms of localized and systemic disease including colisepticemia, coligranuloma (Hjarre's disease), airsacculitis, swollen-head syndrome, coliform cellulitis (inflammatory or infectious process), peritonitis, salpingitis in layers, osteomyelitis/synovitis, panophthalmitis, meningitis/encephalitis, omphalitis/yolk sac infection, and less frequently, enteritis [67]. There is a lot of debate whether or not *E. coli* is a primary pathogen (avian pathogenic *E. coli*, or APEC) or a secondary, opportunistic pathogen since much of the clinical cases of *E. coli* infection is subsequent to increased host susceptibility to stress, immunosuppression, co-infections/predisposing infections, or poor environmental conditions. In comparison with broilers and turkeys, APEC strains in layers appear to be much more homogenous [68–70]. Thus, vaccination strategies in the pullet growth period have rendered some success over the years to reduce salpingitis in adult layers, reducing mortality associated with *E. coli* peritonitis. Incubation period and clinical signs vary greatly between the susceptible host and the disease manifestation. Since *E. coli* is a normal commensal organism of chickens, disease is most common 5–7 days after an insult such as a predisposing infection or adverse conditions such as high ammonia levels in the barns. Birds with colisepticemia are often moribund or very lethargic, resulting in acute death after these clinical signs. Bacterial omphalitis is one of the most common problems in layer pullets. In one study, *E. coli* and *Enterococcus faecalis* infections accounted for approximately half of the mortality in layer chicks during the first week after hatch [71]. Differentials vary depending on the type of clinical signs and manifestation of the gross lesions. Respiratory infections can resemble MG, IB, lentogenic ND, to name a few, whereas septicemia lesions can resemble salmonellosis, fowl cholera, erysipelas, LPAI, and others. Diagnosis is often by process of elimination, based on potential underlying factors. If *E. coli* was cultured from lesions such as peritonitis, investigation and controlling the underlying primary agent/issue is often necessary to address the problem at hand. Antibiotic treatment may be complicated since *E. coli* is commonly resistant to many antimicrobials. In general, antimicrobials against gram-negative bacteria such as tetracyclines and sulpha drugs are used in pullets, although treatment options are limited. Antimicrobial sensitivity testing is recommended to look at susceptible antimicrobial options. Vaccination by using live attenuated vaccines has been commonly used in the United States, both for prevention and in the face of an outbreak, with anecdotal success with multiple doses prior to the onset of colibacillosis. Vaccines for *E. coli* have been based on the most common types of serotypes that are identified from poultry with *E. coli* septicemia, which are based on the O (somatic) antigens O1, O2, O18, O35, O36, O78, and O111 [67, 72]. There have been anecdotal reports of *E. coli* vaccination reducing mortality associated with colisepticemia by up to 50% in layers. In cases where commercial vaccination has not been effective, some operations have used autogenous vaccination based on isolates of *E. coli* from their own operations. Control measures include minimizing contamination of egg shells, hatcheries, pullet barns, etc., by vigorous sanitation programmes, including better ventilation and reducing faecal-contaminated dust, as well as continuous water sanitation programmes to reduce coliforms in drinking water. Diseases, parasitism, and other stressors that cause immunosuppression are key to reducing risks of elevated mortality due to secondary *E. coli* infections.

Less common pathogens such as *Staphylococcus, Streptococcus, Enterococcus, Pseudomonas*, and paratyphoid *Salmonella* may contribute to morbidity and mortality in pullets, especially in the first few weeks of life.

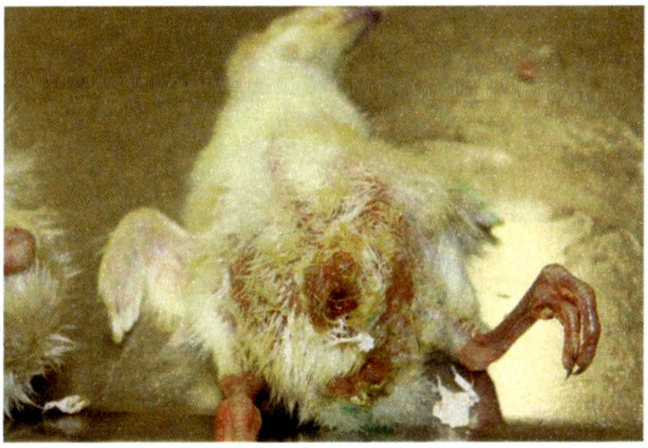

Figure 5 Navel ill in a 4-day-old chick with dirty, unhealed navel. Notice the shanks are dry and darkened, which is a sign of dehydration.

These organisms are opportunists or are co-pathogens with another agent in order to cause any or significant disease in healthy pullets [69]. Many of these are commonly associated with yolk sac infections in the first week of life, as well as localized or systemic infections in older pullets, resulting in general poor-doers, anorexia, respiratory signs, neurologic issues due to meningitis/meningoencephalitis, lameness due to osteomyelitis/synovitis, stunting, and septicaemia subsequent to overwhelming infection. These organisms are ubiquitous in nature and are found either as an environmental contaminant (e.g. *Pseudomonas*, paratyphoid *Salmonella*) or as normal commensal organisms in the gastrointestinal tract and tissues of chickens (e.g. *Enterococcus, Staphylococcus,* and *Streptococcus*). Disease often manifests due to a breakdown in the natural defence mechanisms of the host, such as skin wounds, inflammation of the mucous membranes, parasitism, immunosuppressive diseases, and open, unhealed navel infections in baby chicks (Fig. 5) [69]. Morbidity and mortality is highly variable, but usually low when uncomplicated. Infections and mortality are usually sporadic, with horizontal spread of high bacterial load due to unsanitary conditions. Diagnosis and antimicrobial treatment is done via bacterial culture and antimicrobial sensitivity testing [69]. Select antibiotics based on the spectrum of usage, targeting either gram positive (*Staphylococcus, Streptococcus, Enterococcus*) or gram negative (*Pseudomonas, Salmonella*), and associated lesions can dictate the choice of antimicrobials. There is little value in further diagnostic testing due to relatively low genetic diversity of these organisms. For example, only about 10% of the known paratyphoid *Salmonella* serovars have ever been found in poultry, and an even smaller subset of these is consistently common [64]. However, in the case of *Salmonella* and other organisms such as *Campylobacter*, less focus is on pullet health and more on human public health concerns as contaminated poultry and poultry products are a significant source of foodborne human illness. Prevention and control measures are consistent with colibacillosis, with emphasis on reducing predisposing factors and improving environmental conditions to minimize breaches in the host natural defence mechanisms.

3 Key disease issues faced by laying hens

3.1 Viral diseases of layers

3.1.1 Avian adenovirus infections

EDS-76, also known as duck atadenovirus 1, was first described in 1976 by a group of Dutch workers as characterized by a sudden, severe but not permanent fall in egg production accompanied by the production of soft-shelled eggs and shell-less eggs with watery albumin [73, 74]. The virus has been detected in many countries, including several countries in Europe, South America, Asia, and Africa. Recently, EDS has been reported in brown egg layer chickens in the United States with reported loss of 30-60% of marketable eggs. Although this is the first detection of the virus associated with clinical disease in the United States, sporadic cases in North America date back to the 1980s [9]. Since EDS is resistant to a lot of disinfectants and is hardy in the environment similar to other adenoviruses, transmission is by contamination of materials such as egg-packers, trucks, and feed by faeces and oviduct exudates [75, 76]. Once in a lay house, lateral spread is usually slow or intermittent, taking up to 11 weeks to spread in a caged lay house. However, spread within a cage-free housing is usually faster [77]. Clinical signs for EDS are characterized by loss of egg-shell pigment, egg-shell abnormalities including thin or soft-shelled eggs to shell-less eggs [73, 74]. Flocks experience dramatic drops in egg production, up to 40%, which can extend on average from 4 weeks to 10 weeks. Mortality is usually non-existent or negligible. Diagnosis is typically done by looking at both antigen detection via PCR and virus isolation from oviduct swabs or oviduct tissue, or by antibody detection via serologic assays such as HI, ELISA, immunofluorescent assay (IFA), to name a few. Negative serological tests prior to development of clinical signs do not guarantee freedom from infection, as vertically transmitted birds do not test positive for EDS during the pullet period [9]. Differentials for EDS include low dietary calcium and IBV infection. Since classic EDS spreads via vertical transmission, biosecurity, especially to limit contact with waterfowl, to keep breeders free of EDS and vaccination to prevent introduction, is used globally or by regional permission within the United States. Vaccination, either given as an inactivated vaccine or autogenous vaccine, is typically done a few weeks prior to move to lay and can last for a year [75].

3.1.2 Avian encephalomyelitis virus

With AEV, adult layers seldom show clinical signs when the infection is going through the flock, and would manifest as temporary egg production drops [14]. An adult flock with AEV might experience a 5–30% drop in egg production that lasts no more than 2 weeks; therefore, any instances with a mild, temporary drop in egg production should include AEV as a differential. There are no gross lesions associated with adults infected by AE, although birds infected as chicks can survive the infection with a bluish opacity in the lens. Experimentally, adult layers infected with AEV did not have viral antigens present in the central nervous system, which explains the lack of neurological signs [78].

3.1.3 Avian influenza virus

Please refer to the AI section in Section 2.2.3.

3.1.4 Infectious bronchitis virus

In layers and breeder chickens, infection by IBV may result in up to a 70% reduction in egg production and negatively impact egg quality. The virus can replicate in the oviduct and can cause permanent damage in young hens or can cause cystic left oviducts, resulting in limited egg production over a prolonged period of time as well as birds that fail to come into production (false layers), respectively [18]. Respiratory clinical signs in adult layers can be mild or minimal, even in the face of severe egg production drops. Production drops are more pronounced if IBV infection happens at the onset of production versus at peak production, with increased numbers of abnormal shelled eggs and longer-lasting effects on egg weight and internal egg quality. A range of egg-quality defects vary from loss of egg-shell pigment; misshapen, thin, soft-shelled, slab-sided, and rough-shelled eggs (wrinkled eggs); and thin to watery albumen (loss of internal egg quality). Microscopically, oviduct epithelium has loss of cilia, dilatation of tubular glands, and mononuclear cell inflammation, as well as oedema and fibroplasia of the mucosa [79]. A study in Australia showed differences between strains of IBV in pathogenicity for the oviduct, including that using real-time RT-PCR, that some strains can persist in reproductive tissue for longer than was first thought, with a peak between 10 days and 14 days post-infection [80]. EDS, LPAI, MG, avian metapneumovirus, ND, infectious coryza, and other ascending bacterial infections can mimic clinical signs of IBV in adult layers [18].

3.1.5 Infectious laryngotracheitis

Similar to pullets, infectious laryngotracheitis (ILT) should be one of the differentials in flocks that experience elevated mortality and respiratory signs. Birds that recover from an ILT infection become long-term carriers of the virus, where the virus lays dormant in the trigeminal nerve ganglia [27]. Stressful events such as moving pullets to the lay house and the onset of laying can stimulate the re-excretion of virus in recovered birds [81]. Due to infection, birds may experience a drop in feed and water consumption, leading to a drop in egg production. However, egg quality is not affected unlike with infection by EDS, which does not typically have mortality issues, and compared to IBV, which causes internal and external egg-quality issues. Diagnosis of ILT requires laboratory confirmation because a diagnosis based on clinical presentation can be misleading, and often a combination of histopathology and virus-detection methods such as PCR or virus isolation is recommended [27].

3.1.6 Fowlpox virus

Please refer to the Fowlpox section in Section 2.2.6.

3.1.7 Newcastle disease and avian metapneumovirus

Gross lesions, especially with virulent ND infection, will closely resemble those in pullets. In addition, egg yolk peritonitis with atrophied follicles and degenerated oviduct may appear during birds in lay [82]. Well-vaccinated layers infected with vND may present with only a decrease in egg production 1 week after infection, with the fewest eggs produced 2–3 weeks post-infection, after which egg production will start to increase [83]. More details regarding general aetiology, pathogenesis, and interventions for ND are discussed in Section *2.2.7*.

Avian metapneumovirus (aMPV) is a paramyxovirus under the genus *Metapneumovirus* that can cause respiratory disease in turkeys but is frequently only associated with a decrease in egg production and potential egg-quality issues in adult layers [84]. Four different subtypes (A–D) have been described globally, and wild migratory birds are thought to be natural reservoirs of the virus, which is consistent with the seasonality of the virus [85]. Experimentally, aMPV infection caused up to 25% egg production drops with high incidence of soft and thin-shelled eggs [86]. Egg-shell abnormalities may be more pronounced in brown egg layers than in white egg layers. Definitive diagnosis of aMPV may be challenging due to the short period of secretion and the labile nature of the virus. Samples can range from oculonasal secretions, to choanal swabs, to tissues scraped from sinuses or nasal turbinates. Screening by serology using ELISA is done in breeders and flocks with a lack of clinical signs. Differentials include other paramyxoviruses, AI, ND, IB, and mycoplasmas, which can also cause mild respiratory disease and egg production issues [84]. The virus is reportable in the United States and requires state and federal approval to use live vaccines in breeders. As such, good biosecurity and vaccination in endemic areas are used to control the disease.

3.1.8 Infectious bursal disease

Please refer to the Infectious bursal disease in Section 2.2.8.

3.1.9 Neoplastic viruses (Marek's disease, avian leukosis and reticuloendotheliosis virus)

LL virus and reticuloendotheliosis virus (REV) are both retroviruses that are associated with neoplastic disease in older birds, above the age of 14–16 weeks. LL has seven exogenous subgroups in chickens - A, B, C, D, E, J, and K - on the basis of differences in their antigenicity, viral interference patterns with subgroups, and host range [87]. Subgroups A (most often) and B (less frequently) are common in the field in countries of North America and Asia.

Endogenous (subgroup E) viruses do not produce tumours and are transmitted genetically (this is different from congenital) as part of the host genome. Endogenous viruses may have complete or defective genomes. If the genome is complete, fully infectious particles may be shed at low levels. Endogenous viruses may be non-defective or replication defective. An average chicken carries on average about five endogenous virus loci. Non-defective viruses can produce exogenous virus particles. The virus is shed vertically from the hen via the yolk at a low rate (1–10%), and congenitally infected birds remain viremic for life. The virus is much more contagious horizontally, and infected chickens have transient viremia. The earlier the infection, the more likely it is to lead to tolerance, persistent viremia, and tumours [45]. The incubation period and pathologic disease for LL are dependent on strain and dose of virus, route of infection, age at exposure, and genetics. In field cases, LL is typically not seen in chickens less than 14 weeks of age (refer to Table 2). Lesions can range from visceral B-cell lymphoid tumours (especially in the bursa of Fabricius), erythroid leukosis which occurs mostly in the liver and kidney, myeloblastosis of the bone marrow, hemangiomas in the skin or visceral organs, nephroblastomas in the kidney, connective tissue tumours resulting in fibroblastomas and osteopetrosis, and other nonspecific tumours [45]. LL in the absence of overt tumours can negatively impact egg production. Compared with non-shedders, hens that shed the virus produced 20–35 fewer eggs per hen housed to 497 days of age; had delayed sexual maturity (i.e. age at first egg); and produced smaller eggs and at a lower rate, with thinner shells. Mortality from non-neoplastic reasons was 5–15% higher, fertility was 2.4% lower, and hatchability was 12.4% lower [88, 89]. Definitive diagnosis is challenging due to the virus being ubiquitous, and requires viral and molecular characterization to rule out other neoplastic viruses and spontaneous neoplasms (See Table 3). No commercial vaccine is available for the protection of chickens from infection with LL, and prevention and control is by eradication of LL from the breeding stock in the form of testing and culling positive birds [45, 89].

Reticuloendotheliosis is caused by a non-defective RE retrovirus, which can result in a runting syndrome or chronic lymphoma syndrome. Although the virus is not ubiquitous in contrast to MD and LL, the infection is fairly widespread in countries where RE is reported, such as the United States, China, Japan, Korea, Argentina, Brazil, and Egypt [45, 90]. However, clinical disease is uncommon. Infection with REV may be transmitted by a horizontal route by direct contact between birds, of which there are a very wide range of host species, and indirectly by some insect vectors like mosquitoes (*Culex pipiens*) or flies (*Musca domestica*), and also via vertical route by eggs and infected semen [91–93]. Disease in young chicks is likely due to contaminated vaccines, most commonly in fowlpox or MD vaccines, although it has been detected in

ND and IB vaccines as well [93]. RE neoplasms are highly variable since they can manifest as either B or T-cell lymphomas and affect birds of any age, In addition, lesions of different types (B cell lymphoma and T cell lymphoma) can be observed in the same flock or even in individual birds [45]. However, definitive diagnosis of RE is simpler than that of MD and LL in that the virus is not ubiquitous and the detection of the virus, viral antigens, or proviral DNA is diagnostic (See Table 3). RE in commercial poultry is typically controlled by testing and elimination of infected breeders and quality control of vaccines to ensure that contamination of RE is not present. Other control strategies are not used due to the sporadic and self-limiting nature of the disease [45].

3.2 Bacterial disease of layers

In addition to those mentioned in the pullet section, adult layers have several other bacterial issues of concern including avian chlamydiosis, campylobacteriosis, erysipelas, fowl cholera, infectious coryza, and other clostridial diseases.

3.2.1 Avian chlamydiosis

Avian chlamydiosis is caused by a gram-negative, obligate intracellular bacteria, *Chlamydia psittaci*. Although chickens are relatively resistant to infection compared to other species such as turkeys and pigeons, there have been increased reports of outbreaks on chicken farms and zoonotic transmissions, especially in Europe [94]. Infected chickens showed respiratory clinical signs and gross lesions characterized by conjunctivitis, rhinitis, pneumonia, fibrinous airsacculitis, fibrinous pericarditis, and hepatosplenomegaly [95, 96]. Diagnosis is through antigen-detection assays such as qPCR versus bacterial culture, which takes a long time, through oropharyngeal swabs or conjunctival swabs from birds with ocular lesions [94]. Smears and histopathology lesions with intracellular bacteria can be visualized by cytological staining such as Giemsa and Macchiavello stains, although not definitive and losing popularity compared to immunohistochemistry and molecular methods [97]. Differentials include respiratory viral infections such as AI, NDV, and AMPV; bacterial infections such as colibacillosis, MG, and fowl cholera; and environmental conditions such as high ammonia. Prevention and control is mainly through vector control, including indirect contact with infected pet birds (especially psittacines), rodents, arthropods, and wild birds. There are no commercial vaccines available, and their practicality is limited. Treatment by antibiotics, most commonly using tetracyclines, is highly effective and used in rare situations where outbreaks are detected [94].

3.2.2 Campylobacteriosis/spotty liver disease

Historically, campylobacteriosis in poultry has been only linked to human public health issues. The bacteria is generally regarded as a commensal organism, and its infection rarely results in clinical disease or significant pathological lesions with the exception of vibrionic hepatitis, thought to have been caused by *Campylobacter coli and C. jejuni*. Recently, there is a re-emerging disease known as 'spotty liver disease' (SLD) caused by a novel species *C. hepaticus*, characterized by multifocal, small white-grey lesions in the liver (Fig. 6) [98, 99]. SLD was first identified in 2015 in the United Kingdom and was characterized in Australia in 2016 as a novel Campylobacter species causing disease in free-ranging laying hens (100–102). Clinical signs with SLD can range from no apparent clinical signs to nonspecific loose stools, and in severe cases, increase in mortality and drops in egg production, which can occur simultaneously but without any effects on external and internal egg quality [98–102]. Flocks with this condition experience a 5-20% drop in egg production over a 3 to 4-week period and have an increase in mortality, ranging from 0.5% to 5% [103]. Gross lesions are characterized by random, miliary, 1-2 mm tan to white foci in the liver that can coalesce. Histopathologic lesions of SLD are characterized by multifocal to coalescing hepatocellular necrosis characterized by discrete margins and necrotic centres mixed with degenerating heterophils in acute to

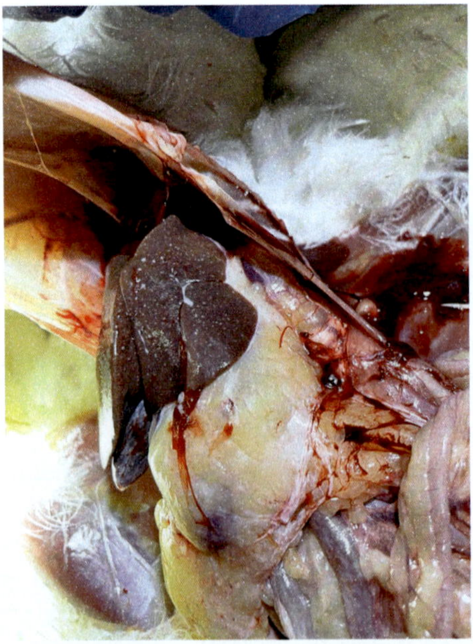

Figure 6 Spotty liver disease (*Campylobacter hepaticus*) in 32 week old laying hens with miliary foci in the liver.

subacute stages. Diagnosis is through bacterial culture, although larger success is through the use of bile as a sample source versus liver, which requires an additional enrichment step and lower recovery success, or through molecular methods using published PCR protocols [98]. Gross lesions can resemble other bacterial hepatitis due to organisms such as salmonella, *Pasteurella multocida* (fowl cholera), and clostridia, as well as Hepatitis E virus infection. Treatment and prevention strategies for SLD are somewhat problematic. Although there are no approved antibiotics against SLD, supportive antibiotic therapy to prevent secondary infections has helped reduce mortality, as well as the use of water acidification using citric acid and equivalent has had limited success. There is a promise of success using autogenous vaccines based on isolates recovered from SLD lesions. The author predicts higher incidences of SLD as commercial egg production moves towards cage-free and free-range production with more access to the outdoors.

3.2.3 Erysipelas

Erysipelas is a septicemic disease caused by a gram-positive bacteria, *Erysipelothrix rhusiopathiae*. Several cases of high mortality with a sudden onset have been increasingly seen in recent years in organic and pastured layers [104] with some flocks losing as much as 10% in a week and sharp drops in egg production [103]. The disease is rarely seen in younger pullets. The bacteria are ubiquitous in the soil and found in other species of livestock, such as pigs, and spread by the faecal-oral route , as well as breaks in the natural defences of the host, such as through skin scratches and feather picking. Gross lesions in adult layers are typical of septicaemia, characterized frequently by hepatomegaly and splenomegaly, sometimes accompanied by air sacculitis and peritonitis, with birds frequently being in lay, in addition to petechial and ecchymotic lesions seen on serosal surfaces, adipose tissue, and mucosal surfaces of select organs. Histopathologic lesions are characterized by vascular changes, fibrinoid necrosis, and a minimal inflammatory response [105]. Diagnosis is through bacterial culture of the lesions, and the bacteria are isolated with relative ease in clinical cases, especially with enrichment media. Differentials should include diseases that cause acute and high mortality such as HPAI, NDV, and septicemic bacterial diseases such as fowl cholera and colibacillosis. In terms of prevention and control, layer flocks are usually vaccinated with inactivated vaccines once or twice at placement, due to the labour-intensive vaccination protocol. However, recent field experience and reports indicate that flocks may suffer from an outbreak despite vaccination [105]. In addition, the increase in pullets raised in cage-free settings makes vaccination less practical. In field settings, live Erysipelas vaccines labelled for swine and turkeys have been used experimentally in flocks/sites with repeated outbreaks with mixed success, although this may be the only control measure possible in organic production where antibiotic treatment is not an option. Mortality is usually controlled effectively with antibiotic therapy in uncomplicated cases.

3.2.4 Fowl cholera

Fowl cholera, caused by gram-negative bacteria *Pasteurella multocida*, is a highly contagious disease resulting in septicaemia and high mortality in all types of poultry, including laying hens. In recent years, outbreaks of fowl cholera resulting in high mortality have been reported more frequently due to more laying hens raised on pastures with outdoor access. The bacteria are classified into 16 different somatic serovars according to the Heddleston serotyping scheme [106]. The most common serovars are 1, 3, 4, 3 × 4, which are included in the commercial vaccines, and immunologically, these serotypes do not cross react. Mortalities range up to 20%, although even higher rates are reported, especially with secondary infections. In one case in Denmark, up to 91% mortality was reported in a co-infection with Erysipelas [107]. After acute infections, localized infections in organs such as ovaries (oophoritis) and joints can maintain chronic and persistent shedders within a flock of chickens, causing smouldering infections resulting in nagging mortality and egg production drops in subsequent flocks. This is especially a problem in multiage flocks, where all-in-all-out practices cannot be done. Clinical signs in acute fowl cholera are usually depression and catarrhal exudates preceding acute death. Gross and microscopic lesions include severe vasculitis, necrotizing lesions with severe heterophilic inflammation (especially in the liver and lung) and organomegaly. In acute bacteraemic cases, impression smears from affected organs can reveal bipolar rods. Chronic fowl cholera can manifest in different localized lesions, including the cranial bones and inner ears, resulting in torticollis [106]. Differentials are similar to Erysipelas in including acute septicemic diseases with high mortality and respiratory disease. Infectious coryza and *Gallibacterium anatis* infection can cause respiratory disease with similar pinpoint foci in the liver. Diagnosis is commonly through bacterial culture or via PCR detection in cases where antibiotic therapy may make isolation difficult. Additional serotyping for future interventions with the use of autogenous vaccines and investigating epidemiologic links may be warranted. Treatment and prevention is usually done by the use of antibiotics and good sanitation practices. In flocks where fowl cholera is endemic, US pullets are typically vaccinated at 8 and 14 weeks with either two live vaccines or one live followed by one inactivated vaccine. In Europe, live vaccines are not available and thus flocks are vaccinated twice using inactivated vaccines. Vaccination in the face of a fowl cholera outbreak with the live vaccine via wing web has met with success in some organic flocks. It is not advised to vaccinate all flocks already in lay in an affected area due to the stress of vaccination, and risk of introducing fowl cholera outweighs the benefits. Antibiotic therapy in conventional flocks has also been successful in the short term; however, chronic mortality and suppression of production usually recur. Increasing the frequency of vaccination during grow and lay is being used preventatively [103].

3.2.5 Infectious coryza

Infectious coryza (IC) is an acute respiratory disease of chickens caused by gram-negative bacteria *Avibacterium paragallinarum*. According to the Page serotyping scheme, *A. paragallinarum* is classified into three serotypes (A, B, C), which have low cross-protection. Chronically ill or healthy, subclinical carriers are known reservoirs of infection. Although all ages of chickens are susceptible, susceptibility increases with age and is especially a problem in multiage layer flocks [108]. IC is prevalent in layer flocks of almost all countries of Latin America, many countries in Asia and Africa, and selected regions of the European Union. Recently, there have been increased cases of IC in the United States, in both layer flocks and broilers/broiler breeders. IC infection is characterized by low mortality and high morbidity with clinical signs of a sharp drop in feed consumption and egg production (10–40%) and respiratory signs including catarrhal and foul-smelling oculonasal discharge, conjunctivitis, swollen infraorbital sinuses, and swollen wattles (Fig. 7). Although mortality and clinical signs can be exacerbated with co-infections or secondary disease, IC alone has been reported to cause up to 28% mortality in layer flocks [109]. Diagnosis is mainly through identification of the organism through bacterial culture, which could be challenging to grow, and/or through detection via molecular methods via qPCR. However, there have been several reports of positive detection of *A. paragallinarium* with the absence of clinical signs. Transmission is primarily through carrier birds, although there are also risks of infection from clinical or

Figure 7 Conjunctivitis and sinusitis with catarrhal exudate from a laying hen. Picture courtesy of Dr. Mohamed El-Gazzar, Iowa State University.

carrier backyard flocks in close proximity to commercial flocks. Differentials for IC include other respiratory diseases such as MG, chronic fowl cholera, fowlpox, AMPV, IB, and ILT. Treatment with antibiotics has mixed success. Commercial inactivated vaccines are available but are not used universally, as it usually requires two injections in the pullet stage for adequate protection [108]. Pullets that have experienced IC during their growing period are generally protected against a later drop in egg production if they are infected with the same strain [110], which lends itself to using vaccination to control IC infection in multiage layer flocks.

3.2.6 Clostridial diseases (focal duodenal necrosis)

In addition to necrotic enteritis, ulcerative enteritis, and botulism that occurs in both layers and pullets, there are other clostridial diseases of significance to adult laying hens.

Focal duodenal necrosis (FDN) is a poorly understood intestinal disease of layers that has been first described in 1997 and is considered a top 10 disease concern in layers in the United States for the last decade [103]. FDN is characterized by a decrease in egg case weight and a drop in egg production in affected flocks aged 15 weeks to the end of lay [111, 112]. The aetiology of this disease is still unknown, but the condition has been found in both the United States and Europe, associated with *Clostridium colinum* and *C. perfringens*. Egg production drops can range from 1% to 10% and lowered egg weights can be as much as 2.5 g per egg or 2 lb per case [113]. Gross lesions of FDN are characterized by grey-green ulcerations in the mucosa of the duodenum, and infrequently in the jejunum. These ulcerations may be single to multiple, irregularly shaped, 5–15 mm diameter areas that disappear rapidly after death. Since these infections can go unnoticed, routine necropsy surveillance of laying hens with observed clinical signs is the best way to identify and diagnose FDN [114]. Microscopic lesions correlate with gross lesions with ulceration of villous tips with numerous rod-shaped gram-positive bacteria. Treatment by antibiotics until clinical signs and gross lesions resolve is usually successful, although relapsing of FDN can happen 6–8 weeks after treatment [113].

3.3 Opportunistic and emerging bacterial diseases of layers

In addition to colibacillosis caused by *E. coli* infection, *Gallibacterium anatis* has been more frequently isolated from diseased layers in countries such as the United States and Canada in the last two decades. However, *Gallibacterium anatis* could also be isolated from the upper respiratory tract and oviducts of clinically healthy birds. In contrast, other *Gallibacterium spp.* have been isolated in pure culture from diseased birds with salpingitis and decreased egg production at the peak of lay. The infection associated with salpingitis includes regressed oviducts, oophoritis or atretic ovarian follicles, deformed or ruptured ovarian follicles, peritonitis, septicaemia, pericarditis, hepatitis, liver necrosis characterized by pinpoint foci, tracheitis, and pneumonia [115]. Similar reports have been reported in other countries such as Morocco and Egypt [116]. Autogenous vaccines have been used in commercial operations to control salpingitis with minimal success.

In recent years, a condition called focal ulcerative dermatitis syndrome (FUDS) has been described in laying hens in the Midwestern states of the United States, originally in cage-free brown layers, but is now described in white laying hens as well [103, 117]. Skin lesions generally present on the dorsum of the birds, cranial to the uropygial gland and are described on a scale from mild to severely exudative and necrotic. Outbreaks result in up to 50% cumulative mortality over 20 weeks due to septicaemia. A commercial company conducted microbiome studies of the skin of affected and non-affected birds and suspected the causative agent to be *Staphylococcus agnetis*. Pre- and probiotic treatment has shown some success in controlling infections, although there is scepticism over validity of this research at this point [117].

4 Other issues of layers and pullets

4.1 Fungal infections

Fungal infections can manifest as either a systemic infection (e.g. Aspergillosis) or toxicities due to mycotoxins. Similar to other bacteria, fungal infections are typically opportunistic and are relatively uncommon. However, once infection spreads, it is often devastating to not only the host, but to the flock and its surrounding environment [118]. Aspergillosis, or brooder pneumonia, is the most important fungal infection in pullets and can result in high mortality in chicks less than 10 days of age. Aspergillosis can manifest in several different ways in layer pullets, ranging from respiratory pneumonia and air sacculitis, to systemic infections via hematogenous routes, to opthalmitis, to encephalitis. Although not transmissible between birds, infection is most common in very young pullets in contaminated barns, feedstuffs, litter, egg, and other hatchery-related reservoirs, such as contaminated vaccination equipment and incubators. Chronic lesions can manifest in peritonitis or localized infections that can cause sporadic deaths in mature layers. The classic gross lesions are yellow-white plaques in air sacs (Fig. 8), meninges of the brain, and caseous plugs within the larynx and trachea, as well as caseous granulomas in the lungs. Microscopically, a caseous, granulomatous reaction with numerous branching, septate fungal hyphae within lesions is diagnostic

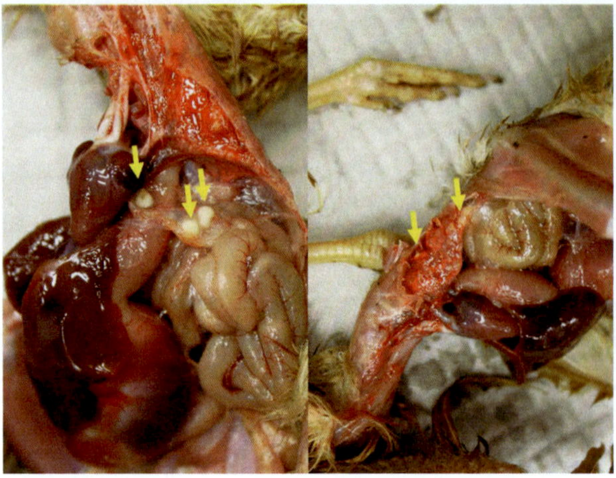

Figure 8 Aspergillosis in young pullets with yellow fungal plaques in the air sacs (left, arrow) and granulomas in the lungs (right, arrow).

of mycotic infections. Since fungal spores are ubiquitous, positive culture from the skin or respiratory tissues without any gross lesions may be frequent and does not equate to diagnosis [118]. Differentials depend on the organ and organ systems affected, where respiratory aspergillosis can resemble colibacillosis, mycoplasmosis, IB, ILT, ND, or high ammonia in barns. Blindness and opthalmitis on top of respiratory distress can also resemble vitamin A deficiency.

Mycotoxins of interest to layers include toxicosis due to aflatoxins, trichothecenes (T-2), and other mycotoxins produced by *Fusarium*, ochratoxins, oosporein, citrinin, and fumonisins [119]. Mycotoxins are commonly detected in poultry feed, and routine feed analysis to look at levels is recommended. In 2020, a survey of over 21 000 feedstuffs from 79 different countries found the three highest mycotoxin contamination by *Fusarium* mycotoxins DON (65% of all samples) and fumonisin (64%), followed by zearalenone (48%) with 87% of the samples containing ten or more mycotoxins and metabolites [120]. Chickens, compared to other poultry counterparts and especially compared to mammals, are fairly resistant to mycotoxicosis, although even subclinical toxicosis can cause loss in production. Table 5 describes the effects of mycotoxins on bird health and levels of concern to layers and layer pullets.

A definitive diagnosis of mycotoxicosis requires quantitative analysis of feedstuffs to look for specific toxins and their levels of contamination. If mycotoxicosis is suspected, toxic feed should be removed and replaced with non-contaminated feed. Birds usually recover from mycotoxicoses soon after an uncontaminated diet is available, but this may not compensate for production losses. As such, combined with the frequency of mycotoxin in the

Table 5 Mycotoxins and levels of concern that can affect health and production performance of layers and pullets.

Mycotoxin	Levels of concern	Effects on birds
Aflatoxins	0.02 ppm (in EU and US)	Liver toxicity can cause immunosuppression.
DON (vomitoxin)	5 ppm	Effect on immunity. Feed refusals.
Fumonisins	30 ppm (laying hens)	Black adhesive diarrhoea, reduction in feed intake, egg production, and body weight, lameness, mortality.
Ochratoxin-A	0.1 ppm	Nephrotoxicity, issues with feed efficiency, decrease in egg production, and shell thickness issues.
T-2 Toxin	0.25 ppm	Suppresses the immune system, can cause gizzard erosion and oral lesions, leading to lower feed intake and egg production.
Zearalenone	0.5 ppm	Egg production issues including numbers and quality, cystic oviducts.
Aflatoxins	0.02 ppm (pullets) 0.10 ppm (layers)	Liver toxicity can cause immunosuppression.

Source: Modified from UGA extension bulletin by J. Fowler [121].ppm = parts per million.

feed, an integrated prevention programme should be used to optimize bird health and production. In layer production, several methods have been used which broadly include: (1) feed management, (2) antifungals, and (3) mycotoxin binding agents [119]. In terms of feed management, layers and layer pullets are only fed a small inclusion percentage of distillers' dry grain solubles (DDGS), which is a byproduct of ethanol that is known to concentrate mycotoxins at three times the level due to its fermentation process. Since poultry have a formulation limit in the finished feed, the level of contamination is kept to a certain limit [122]. Although they have no effect on already formed mycotoxins, antifungal agents such as organic acids (e.g. citric acid) and compounds such as copper sulphate and formaldehyde are commonly added to prevent fungal growth and to also reduce growth of other microbials such as *Salmonella*. Finally, most layer and pullet feed contain mycotoxin binding agents using various clays (e.g. bentonite clay) and silica-containing zeolites [119]. However, prevention through routine surveillance of layer and pullet diets for mycotoxin and other metabolites should be done at the feed mills to monitor mycotoxin levels.

4.2 Nutrition

Feed and water are some of the most fundamental needs for growth and production in pullet and layer health, respectively. As mentioned in the mycotoxin section, routine surveillance of feed and revisiting formulated diets in order to approach target weights and case weights on eggs is important for good bird health. Similarly, water sources should be tested at least once a year to look at water quality, including mineral levels and total solids [123]. In general, birds should have access to feed and water at all times, and monitoring drops in feed and water consumption is very useful in early detection of health issues. Many production systems use phase feeding to optimize intake. The feed consumption rate is governed by several factors including body weight, age of the birds, rate of egg production, egg weight, ambient temperature, feed texture, and dietary energy content. Decisions regarding nutrition should always be in consultation with a poultry nutritionist and input from a veterinarian if health concerns arise. Deficiencies and excess should be monitored at a flock level for each macro- and micronutrient, starting with looking at the diet and analysis of the feed ingredients.

4.3 Parasites

Parasites of concern to egg-laying chickens are broadly divided into external and internal parasites. External parasites of interest (C-1) include mites (northern fowl mites and red mites), lice, flies, and bed bugs. Internal parasites of interest (C-2) include coccidia (*Eimeria* species), roundworms (*Ascaridia galli, Capillaria* spp., *and Heterakis* spp.*)*, protozoal infections (Histomoniasis), and tapeworms. The difficulty in controlling both external and internal parasites is also based on the fact that there are limited treatment options for laying hens and that the withdrawal period on eggs is detrimental to production. Integrated pest management programmes involving identification of the parasite or pest, prevention, use of approved and alternative pesticides or insecticides, and routine monitoring are critical to successful parasite and pest control. Although there are several treatment options available for each of the parasites, regulations surrounding approved products in laying hens may differ per country and are constantly changing. Please consult with product labels and carefully consider treatment options. Small animal topical products such as Frontline (fipronil), Revolution (selamectin), Bistraz (amitraz), and Advantage (imidacloprid/pyriproxyfen) are forbidden for use in food animals, and are considered adulterants if detected in poultry and poultry products, including eggs.

4.4 External parasites of layers

4.4.1 Northern fowl mites

Ornithonyssus sylviarum, the northern fowl mite (NFM), is the most prevalent poultry ectoparasite found in caged layers and breeder flocks. These mites feed on blood and skin cells of the chicken and can cause significant losses of productivity and even death with heavy infestations. Mites can be found on eggs, egg belts, egg pallets, or other equipment, and on poultry workers when infestations are severe, transferring from one house to another. In one study, NFM infestations in caged layer operations at periods of peak mite infestation can reduce profits by 50 cents per hen annually [124]. Egg production can be reduced by ~5% and rates of floor eggs may increase as birds refuse to use nests infested with mites [123].

4.4.2 Red mites

Dermanyssus gallinae, or red mites, are another blood-sucking mite that is very common in many European caged layer operations, and is arguably the primary pest of laying hens in Europe. A study in a lay facility in Korea revealed that prevalence was close to 90% at the farm level [125]. Although the reported prevalence is low in the United States, the author presumes the rate is much higher today due to higher percentages of cage-free and free-range housing for egg-layer facilities. Not only do they affect egg production, but red mites can pose a threat to bird welfare in the form of causing restlessness, lack of sleep, stress, severe feather pecking, aggression, anaemia, and sometimes even death [126]. These mites are nocturnal and do not stay on the host all the time, compared to NFM and control needs extensive environmental treatment. In addition, they can carry pathogens of significance such as Salmonella, *E. coli*, Pasteurella, and Erysipelas.

4.4.3 Lice

Only chewing lice (order *Mallophaga*) infest birds and therefore do not cause anaemia when compared to mites (Fig. 9). These lice feed on feathers, feather

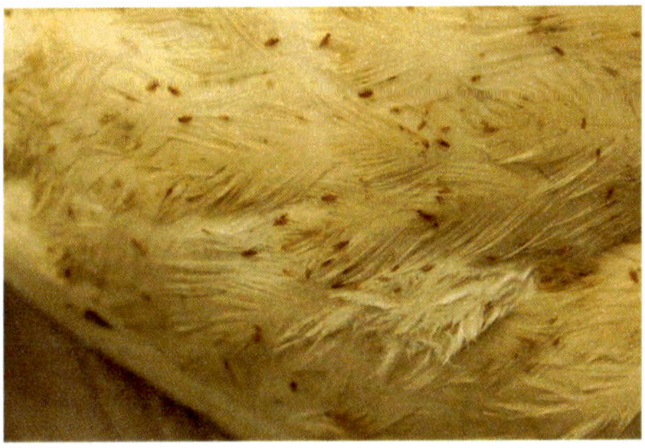

Figure 9 Pediculosis (louse infestation) in a commercial white hen. This hen had mild-moderate weight loss and loss in egg production.

debris, and skin scales, and are highly host dependent [127]. Clinical signs are similar to mites in causing restlessness, weight loss, and local skin irritation.

4.4.4 Flies

Although there are many types of flies that feed on poultry, only a few are of veterinary importance. Black flies (family Simuliidae) are important as vectors for transmission of *Leucocytozoon* spp. and can result in anaphylaxis, in addition to causing egg production loss due to its nuisance [128]. House flies (*Musca domestica*) are the most common pest in egg-layer facilities and serve as mechanical vectors for poultry pathogens such as fowlpox virus and serving as intermediate hosts of other parasites such as tapeworms. In addition, they can carry pathogens of significance to human health (e.g. Salmonella, Campylobacter) and can be a public health issue in terms of egg production [129].

4.4.5 Bed bugs

In addition to impacting bird welfare and effects of decreased egg production, bed bugs are re-emerging as a public health issue. Concerns from bed bugs migrating from poultry facilities, or being carried on clothing or belongings of personnel, to infest homes have been reported [129]. Once established and identified, bed bugs are extremely difficult and expensive to eliminate from a house.

4.5 Internal parasites of layers

4.5.1 Coccidiosis

Coccidiosis is one of the most common and most expensive diseases of laying hens and pullets due to the costs related to performance losses, mortality, and treatment and prevention strategies. Worldwide, cost associated with coccidiosis is estimated to exceed £10.4 billion according to estimated data from 2016 [130]. Nine species of *Eimeria* have been described from chickens, only seven are universally recognized. Co-infections with two or more species are not uncommon, and infections with coccidia are well established as a predisposing factor to several enteric diseases. Immunity is conferred by repeated exposure to low numbers of oocysts through vaccination with live sporulated oocysts or by controlled exposure with step-down doses of an anticoccidial to control field infections until immunity develops [131]. There is no cross-protection between different species, and thus individual species of

Eimeria have to be included in vaccination. Compared to commercial broilers, where the number of species included in a vaccine is limited to three or four species, layer vaccines often contain additional species (approx.six species) due to the long lifespan of production. The three most important *Eimeria* in chickens include *Eimeria acervulina*, *Eimeria maxima*, and *Eimeria tenella*. *Eimeria acervulina* is the most frequently encountered coccidia in commercial poultry, and gross lesions are often limited to the duodenal loop with characteristic ladder-like streaks in the mucosa. Watery and mucoid droppings may be seen as early as 4 days post-infection, due to its shorter life cycle [132]. *Eimeria maxima* is characterized by yellow-orange mucus and bloody fluid, causing ballooning of the midgut close to Meckel's diverticulum. The lifecycle is moderate, with sexual stages developing in deeper tissues by day 5–8 post-infection. Along with *E. acervulina, E. maxima* is correlated to predispose birds to necrotic enteritis [133]. *Eimeria tenella* is a highly pathogenic species of coccidia that causes haemorrhagic and necrotizing lesions (aka 'cores') in the ceca. *E. tenella* has a long lifecycle (6–7 days) comparatively. *E. brunetti* and *E. necatrix* can occasionally cause problems especially in adult layers, resulting in ileocaecal and midgut lesions, respectively, with haemorrhagic droppings. Outbreaks most often happen around 3–6 weeks, although clinical coccidiosis can happen in layers as well, especially in situations where pullets raised in cages get moved to a cage-free lay house. Once birds are in lay, there are limited options for treatment without egg withdrawal. In the United States, amprolium is the only approved anticoccidial available once birds are laying. There are phytobotanicals and other alternative therapies available for organic layers, although efficacy is somewhat limited. Drug resistance is often a problem, and with the added benefit of attenuated vaccine strains of *Eimeria* spp. requiring a lower dose and shorter length of treatment, the use of coccidiosis vaccine is increasingly more popular in layers [132]. Adequate moisture levels at the pullet farms, including the use of chick papers in the first 10–21 days of life, to allow for adequate cycling of oocysts, are critical for the success of vaccination programmes. Routine faecal sampling to look at oocysts counts may be warranted to make sure the coccidia is cycling properly.

4.5.2 *Nematodes (*Ascaridia galli, Capillaria, Heterakis gallinarum*)*

The most prevalent nematodes reported in free-range layers are *Ascaridia galli, Heterakis gallinarum, and Capillaria* spp. All of these nematodes have direct life cycles and, although many of them cause little to no clinical signs, can result in direct or indirect issues with bird welfare and health [134]. Roundworm infection, caused by *Ascaridia galli*, in cage-free laying hens has been a problem with egg quality and can contribute to production issues. *A. galli* is often the most prevalent (22–84%) gastrointestinal nematode in laying production systems with access to outdoor areas [134]. The move towards cage-free housing has increased the reporting of *A. galli*. According to the Association of Veterinarians in Egg Production (AVEP) 2020 survey, roundworms ranked #5 in the top 10 diseases in the cage-free layer industry (US Animal Health Association national meeting, 2020). Although subclinical causes are frequent, there is variability in clinical signs experienced by layers with *A. galli* infection dependent on the infectious dose and immune status of the host, including loss of appetite and body weight, ruffled feathers, drooped wings, delayed development, altered hormone levels, anorexia, depression, and increased mortality [135]. The occasional finding of adult worms in table eggs has been a significant concern, especially in brown eggs where worms are harder to identify even with candling [136, 137]. Comparatively, the threadworm, or *Capillaria* spp. has a higher pathogenicity, where morbidity with catarrhal enteritis and mortality are possible in heavy infections. *Heterakis gallinarum*, or the chicken caecal worm, is abundant in replacement pullets and, although it causes little to no clinical signs in chickens, can be important in its role as an intermediate host for blackhead disease, which can affect laying hens (see Section 4.5.3). Currently, there is only one labelled deworming product (Fenbendazole) against nematodes available on the US market for laying hens; however, its practical usage has been limited with increasing drug resistance, and even more in organic production where treatment options are not possible.

4.5.3 Histomoniasis

Histomonas meleagridis is a protozoal parasite and causative agent of histomoniasis, commonly known as blackhead disease. In chickens, mortality can be up to 10–20%, although the morbidity is very high, as they could serve as asymptomatic reservoirs of disease [138] However, many layers have antibodies against histomonas with absence of clinical signs. Chicken caecal worms, *Heterakis gallinarum*, is a well-established intermediate host, which is commonly found in multiage layers without causing any production loss or clinical signs. Coupled with the earthworm acting as a paratenic host, floor-raised pullets and layers have exposure to the parasite. Thus, due to increase in cage-free and free-range housing, histomoniasis in layer pullets and adult layers have been reported as a re-emerging disease [139]. Gross lesions may or may not involve characteristic hepatic necrosis, but oftentimes may have typhlitis with microscopic lesions including intralesional histomonads (Fig. 10) [134]. Layers and layer pullets may experience depression and loss in egg production, and co-infection with colibacillosis is common [140, 141]. The biggest differential in layers is caecal

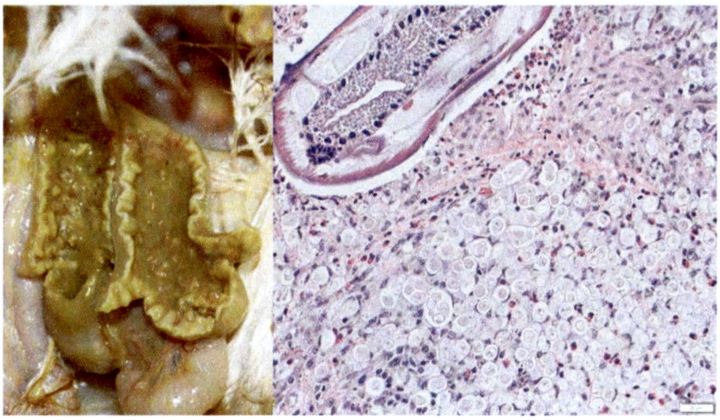

Figure 10 Gross lesions of fibrinonecrotizing typhlitis (left) and histopathologic image (right, H&E 40x magnification) of numerous histomonads and a cross section of *Heterakis gallinarum* within the cecum from the lesion on the left.

coccidiosis caused by *Eimeria tenella*, which could be ruled out by microscopic examination and/or PCR for histomoniasis. Currently, there are no efficacious drugs for treatment nor prevention that are approved for use in poultry in the United States, Europe, and many other countries. There are no vaccines for this disease, and management practices are key in reducing outbreaks.

4.5.4 Tapeworms

Most tapeworms of chickens are often asymptomatic unless parasitic loads are high. Weight loss with a decrease in feed intake, and as a result, a decrease in egg production will be more pronounced in younger birds but are rare [138]. Diagnosis is simply by detection of proglottids or entire segments of worms adhered to the intestinal mucosa. Interventions are by controlling intermediate hosts, which include insects: flies are most common in caged facilities, whereas darkling beetles are more common in cage-free facilities.

5 Surveillance strategies on laying hens to ensure good bird health

It takes a team to manage and monitor layer farms and complexes, and communication is key between each of the parties. Often the team includes field technicians/production managers, hatchery managers, nutritionists, veterinarians, quality assurance officers, and good connections to allied industry partners including, but not limited to: genetics, pharmaceuticals, nutrition, and others. Outside resources from regulatory agencies and diagnostic support are often warranted in investigating disease and other topics of research. Compliance with animal welfare programmes and audits, biosecurity protocols, food safety programmes, as well as speciality programmes such as with organic egg production, takes significant time, effort, and research by a team of experts in order to produce safe and wholesome egg and egg products to be marketed. For the purpose of bird health, this section will focus on the role of the veterinarian relative to laying hen production.

Routine farm visits with walk through of the facility, including assessment of bird well-being, is the first step. Well-being and behaviour are assessed both at the individual bird level, whether they need to be culled due to welfare/health reasons, and at the flock level, to assess overall conditions. Collection of metrics such as body condition, keel bone palpation, routine faecal examinations, as well as looking at production records with a focus on feed and water consumption, egg production, and mortality is essential in benchmarking for normal and abnormal parameters. Veterinarians may choose to perform post-mortem evaluations of daily mortality to investigate issues with morbidity and mortality, as well as collection of diagnostic samples, both ante- and post-mortem, to investigate subclinical conditions as well as production issues that may not necessarily be related to bird health. Serological and faecal samples may be a minimally invasive sample collected during routine visits, whereas sacrificing clinically normal birds may yield additional useful information. Some examples include assessing immunity through looking at bursal scores; measuring thickness of subcutaneous fat pads to look at accurate body conditions for nutritional status; and collecting organ samples for histopathologic examination and viral/molecular diagnostics to investigate subclinical conditions, such as nonspecific enteritis and false laying, respectively. Routine serologic monitoring at specific time points, especially prior to and after administration of inactivated vaccines or at the point of lay, may be a useful tool in assessing vaccine application and benchmarking of expected titers levels required to confer protection against diseases. In addition, significant changes in titer levels can indicate a disease challenge or missed vaccinations that can be useful in making accurate clinical interpretations and adjustments in the future. Additional strategies include auditing bird handling and animal welfare audits, which could lead to opportunities for training and re-training staff; biosecurity audits to ensure producers stay vigilant in their biosecurity measures in the face of disease outbreaks such as Highly Pathogenic AI; and euthanasia training to improve culling of appropriate birds to minimize bird suffering and optimize production.

In summary, optimizing layer health is no cookie-cutter protocol, with different approaches necessary in navigating the most efficient and appropriate husbandry methods based on the type of production, type of bird housing, and type of birds. Routine health assessments are key to observing differences and looking for troubleshooting strategies. Recognizing that there will be emerging and re-emerging health issues and staying vigilant in monitoring those issues will help stay ahead of the problem and in being proactive in responding to the ever-changing demands of the industry.

6 Where to look for further information

Optimizing layer health is a complex and dynamic topic. There are ongoing research in all sectors of animal agriculture regarding precision livestock farming, which will undoubtedly be part of monitoring layer health in the future. In addition, advances in animal well-being and objective measurements of bird welfare is an ongoing trend in the industry. To name a few, the American Association of Avian Pathologists (AAAP), International Poultry Welfare Alliance (IPWA), and Poultry Science Association (PSA) hold annual meetings publish educational materials including textbooks, slide decks, and white papers, and house publications (e.g. *Avian Diseases, Avian Pathology, Journal of Poultry Science*) for those interested in keeping up with the latest trends and upcoming new research regarding egg layer health.

7 References

1. Van de Braak, T. 2020. Global trends in egg production. *Hendrix Genetics*. Available at: https://layinghens.hendrix-genetics.com/en/news/Egg_production-Laying_hens-Poultry-sustainable_production-agriculture-chicken_eggs/. Published November 6, 2020.
2. Bestman, M., Marko, R., Heijmans, J. and van Middelkoop, K. 2012. *Poultry Signals: Laying Hens*. Roodbont Publishers, Zutphen, The Netherlands. ISBN: 9789087401245.
3. Gingerich, E. and Section, I. 2015. Chapter 4: commercial table egg production: manual of poultry diseases. In: Brugere-Picoux, J. and Valliancourt, J.-P. (Eds), Association Française pour l'Avancement des Sciences (AFAS), Paris. ISBN: 2-908014-04-1 978-2-908014-04-4. pp. 25–31.
4. Boulianne, M. 2013. How do we investigate a sick flock? In: Boulianne, M. (Ed.), *Avian Disease Manual* (7th edn.). Omni Press, Jacksonville, FL. pp 1–5.
5. Schat, K. A., Nagaraja, K. V. and Saif, Y. M. 2021. Pullorum disease: evolution of the eradication strategy. *Avian Dis.* 65(2):227–236. 10.1637/aviandiseases-D-21-00024.
6. National poultry improvement plan website. Available at: http://www.poultryimprovement.org/default.cfm. Accessed November 11, 2021.
7. Hein, R., Koopman, R., Garcia, M., Armour, N., Dunn, J. R., Barbosa, T. and Martinez, A. 2021. Review of poultry recombinant vector vaccines. *Av. Dis.* 65(3):438–452.
8. Haıyılatı, A., Li, X. and Zheng, S. J. 2021. Fowl adenovirus: pathogenesis and control. *Int. J. Plant Anim. Environ. Sci.* 11(3):566–589.
9. Fitzgerald, S. D., Hess, M. 2020. Chapter 9: adenovirus infections. In: Swayne, D. E., Boulianne, M., Logue, C. M., McDougald, L. R., Nair, V., Suarez, D. L., de Wit, S., Grimes, T., Johnson, D., Kromm, M., Prajitno, T. Y., Rubinoff, I. and Zavala, G. (Eds), *Diseases of Poultry* (14th edn.). John Wiley & Sons, Inc, Hoboken, NJ. pp. 321–363.
10. Schachner, A., Matos, M., Grafl, B. and Hess, M. 2018. Fowl Aviadenovirus (FAdV) induced diseases and strategies for their control – a review on the current global situation. *Av. Path.* 47(2):111–126.

11. Lim, T. H., Kim, B. Y., Kim, M. S., Jang, J. H., Lee, D. H., Kwon, Y. K., Lee, J. B., Park, S. Y., Choi, I. S. and Song, C. S. 2012. Outbreak of gizzard erosion associated with fowl adenovirus infection in Korea. *Poult. Sci.* 91(5):1113–1117.
12. Matczuk, A. K., Niczyporuk, J. S., Kuczkowski, M., Wozniakowski, G., Nowak, M. and Wieliczko, A. 2017. Whole genome sequencing of Fowl Aviadenovirus A - a causative agent of gizzard erosion and ulceration, in adult laying hens. *Infect. Genet. Evol.* 48:47–53.
13. Suarez, D. L. 2020. Avian encephalomyelitis. In: Swayne, D. E., Boulianne, M., Logue, C. M., McDougald, L. R., Nair, V., Suarez, D. L., de Wit, S., Grimes, T., Johnson, D., Kromm, M., Prajitno, T. Y., Rubinoff, I. and Zavala, G. (Eds), *Diseases of Poultry* (14th edn.). John Wiley & Sons, Inc, Hoboken, NJ. pp 520–527.
14. Tannock, G. A. and Shafren, D. R. 1994. Avian encephalomyelitis: a review. *Avian Pathol.* 23(4):603–620.
15. Swayne, D. E., Suarez, D. L. and Sims, L. D. 2020. Infectious laryngotracheitis. pp. 189-209. In: Swayne, D. E., Boulianne, M., Logue, C. M., McDougald, L. R., Nair, V., Suarez, D. L., de Wit, S., Grimes, T., Johnson, D., Kromm, M., Prajitno, T. Y., Rubinoff, I. and Zavala, G. (Eds), *Diseases of Poultry* (14th edn.). John Wiley & Sons, Inc, Hoboken, NJ. pp. 210–243.
16. Spickler, A. R., Trampel, D. W. and Roth, J. A. 2008. The onset of virus shedding and clinical signs in chickens infected with high-pathogenicity and low-pathogenicity avian influenza viruses. *Avian Pathol.* 37(6):555–577.
17. OIE Terrestrial Code. Chapter 10.4: infection with high pathogenicity avian influenza viruses. Available at: https://www.oie.int/en/what-we-do/standards/codes-and-manuals/terrestrial-code-online-access/?id=169&L=1&htmfile=chapitre_avian_influenza_viruses.htm. Accessed December 13, 2021.
18. Jackwood, M. W. and de Wit, S. 2020. Infectious bronchitis. pp. 167-188. In: Swayne, D. E., Boulianne, M., Logue, C. M., McDougald, L. R., Nair, V., Suarez, D. L., de Wit, S., Grimes, T., Johnson, D., Kromm, M., Prajitno, T. Y., Rubinoff, I. and Zavala, G. (Eds), *Diseases of Poultry* (14th edn.). John Wiley & Sons, Inc, Hoboken, NJ. pp. 167–187.
19. Ojkic, D., Brash, M. L., Jackwood, M. W. and Shivaprasad, H. L. 2013. Viral diseases. In: Boulianne, M. (Ed.), *Avian Disease Manual* (7th edn.). Omni Press, Jacksonville, FL. pp. 6–73.
20. Albassam, M. A., Winterfield, R. W. and Thacker, H. L. 1986. Comparison of the nephropathogenicity of four strains of infectious bronchitis virus. *Avian Dis.* 30(3):468–476.
21. Martin, E. A., Brash, M. L., Hoyland, S. K., Coventry, J. M., Sandrock, C., Gueri`n, M. T. and Ojkic, D. 2014. Genotyping of infectious bronchitis viruses identified in Canada between 2000 and 2013. *Avian Pathol.* 43(3):264–268.
22. Gelb, J. Jr. 1989. Infectious bronchitis. In: Purchase, H. G., Arp, L. H., Domermuth, C. H. and Pearson, J. E. (Eds) *A Laboratory Manual for the Isolation and Identification of Avian Pathogens* (3rd edn.). Kendall/Hunt Publishing Co, Dubuque, IA. pp. 124–127.
23. Gelb, J. J. and Jackwood, M. W. 2016. Infectious bronchitis. In: Williams, S. M., Dufour-Zavala, L., Jackwood, M. W., Lee, M. D., Lupiani, B., Reed, W. M. and Woolcock, P. (Eds), *A Laboratory Manual for the Isolation, Identification, and Characterization of Avian Pathogens* (6th edn.). American Association of Avian Pathologists, John Wiley and Sons, Inc, Jacksonville, FL. pp. 146–149.

24. Cook, J. K. A. 1968. Duration of experimental infectious bronchitis in chickens. *Res. Vet. Sci.* 9(6):506–514.
25. Alexander, D. J. and Gough, R. E. 1977. Isolation of avian infectious bronchitis virus from experimentally infected chickens. *Res. Vet. Sci.* 23(3):344–347.
26. Gowthaman, V., Kumar, S., Koul, M., Dave, U., Murthy, T. R. G. K., Munuswamy, P., Tiwari, R., Karthik, K., Dhama, K., Michalak, I. and Joshi, S. K. 2020. Infectious laryngotracheitis: etiology, epidemiology, pathobiology, and advances in diagnosis and control - a comprehensive review. *Vet. Q.* 40(1):140–161. 10.1080/01652176.2020.1759845.
27. Garcia, M. and Spatz, S. 2020. Infectious laryngotracheitis. In: Swayne, D. E., Boulianne, M., Logue, C. M., McDougald, L. R., Nair, V., Suarez, D. L., de Wit, S., Grimes, T., Johnson, D., Kromm, M., Prajitno, T. Y., Rubinoff, I. and Zavala, G. (Eds), *Diseases of Poultry* (14th edn.). John Wiley & Sons, Inc, Hoboken, NJ. pp. 189-203.
28. García, M. and Zavala, G. 2019. Commercial vaccines and vaccination strategies Against infectious laryngotracheitis: what we have learned and knowledge gaps that remain. *Avian Dis.* 63(2):325–334. 10.1637/11967-090218-Review.1. PMID: 31251534.
29. Tripathy, D. N. and Hanson, L. E. 1978. Pathogenesis of fowlpox in laying hens. *Avian Dis.* American Association of Avian Pathologists, Allen Press 22(2):259–265, 10.2307/1589537.
30. Tripathy, D. N. and Reed, W. M. 2020. Pox. In: Swayne, D. E., Boulianne, M., Logue, C. M., McDougald, L. R., Nair, V., Suarez, D. L., de Wit, S., Grimes, T., Johnson, D., Kromm, M., Prajitno, T. Y., Rubinoff, I. and Zavala, G. (Eds), *Diseases of Poultry* (14th edn.). John Wiley & Sons, Inc, Hoboken, NJ. pp. 364–378.
31. Technical update: fowl pox in layers: an overview. Available at: https://www.hyline.com/ViewFile?id=2d379dc0-599a-417c-8237-bcd92869a71d. Accessed December 17, 2021.
32. Chambers, A. E., Dixon, M. M. and Harvey, S. P. 2009. Studies of the suitability of fowlpox as a decontamination and thermal stability simulant for Variola major. *Int. J. Microbiol.* 2009:158749. 10.1155/2009/158749.
33. Suarez, D. L., Miller, P. J. and Koch, G. 2020. Newcastle disease, other avian paramyxoviruses, and avian Metapneumovirus infections. In: Swayne, D. E., Boulianne, M., Logue, C. M., McDougald, L. R., Nair, V., Suarez, D. L., de Wit, S., Grimes, T., Johnson, D., Kromm, M., Prajitno, T. Y., Rubinoff, I. and Zavala, G. (Eds), *Diseases of Poultry* (14th edn.). John Wiley & Sons, Inc, Hoboken, NJ. pp. 111–166.
34. Alexander, D. J., Aldous, E. W. and Fuller, C. M. 2012. The long view: a selective review of 40 years of Newcastle disease research. *Avian Pathol.* 41(4):329-335. 10.1080/03079457.2012.697991. PMID: 22834545.
35. Kaleta, E. F. and Baldauf, C. 1988. Newcastle disease in free-living and pet birds. In: Alexander, D. J. (Ed.), *Newcastle Disease*. Kluwer Academic Publishers, Dordrecht, Netherlands; Boston, MA. pp. 197–246.
36. Choi, K. S. 2017. Newcastle disease virus vectored vaccines as bivalent or antigen delivery vaccines. *Clin. Exp. Vaccin. Res.* 6(2):72–82.
37. Landman, W. J. M., Vervaet, C., Remon, J. P., Huyge, K. and van Eck, J. H. H. 2017. Primary Newcastle disease vaccination of broilers: comparison of the antibody seroresponse and adverse vaccinal reaction after eye-nose drop or coarse spray application, and implication of the results for a previously developed coarse dry powder vaccine. *Avian Pathol.* 46(4):451–461.

38. Eterradossi, N. and Saif, Y. M. 2020. Infectious bursal disease. In: Swayne, D. E., Boulianne, M., Logue, C. M., McDougald, L. R., Nair, V., Suarez, D. L., de Wit, S., Grimes, T., Johnson, D., Kromm, M., Prajitno, T. Y., Rubinoff, I. and Zavala, G. (Eds), *Diseases of Poultry* (14th edn.). John Wiley & Sons, Inc, Hoboken, NJ. pp. 257–283.
39. Ignjatovic, J. 2004. Very virulent infectious bursal disease. *Australia and New Zealand Standard Diagnostic Procedures*. Available at: https://www.awe.gov.au/sites/default/files/sitecollectiondocuments/animal/ahl/ANZSDP-Very-virulent-infectious-bursal-disease-virus-VVIBV.pdf. Accessed December 29, 2021.
40. Goutebroze, S., Curet, M., Jay, M. L., Roux, C. and Le Gros, F. X. 2003. Efficacy of a recombinant vaccine HVTVP2 against Gumboro disease in the presence of maternal antibodies. *Br. Poult. Sci.* 44(5):824–825.
41. Dunn, J. R., Gimeno, I. M., Wakenell, P. and Sato, Y. in press. AAAP Slide set: differential diagnosis of lymphoid and myeloid tumors in the chicken. Available at: https://www.aaap.info/slide-study-sets. Accessed December 31, 2021.
42. Witter, R. L., Gimeno, I. M., Pandiri, A. R. and Fadly, A. M. 2013. *Tumor Diagnosis Manual* (1st edn.). Omni Press, Jacksonville FL.
43. Witter, R. L. 1997. Increased virulence of Marek's disease virus field isolates. *Avian Dis.* 41(1):149–163.
44. Witter, R. L., Calnek, B. W., Buscaglia, C., Gimeno, I. M. and Schat, K. A. 2005. Classification of Marek's disease viruses according to pathotype: philosophy and methodology. *Avian Pathol.* 34(2):75–90.
45. Nair, V., Gimeno, I. and Dunn, J. 2020. Neoplastic diseases. In: Swayne, D. E., Boulianne, M., Logue, C. M., McDougald, L. R., Nair, V., Suarez, D. L., de Wit, S., Grimes, T., Johnson, D., Kromm, M., Prajitno, T. Y., Rubinoff, I. and Zavala, G. (Eds), *Diseases of Poultry* (14th edn.). John Wiley & Sons, Inc, Hoboken, NJ. pp. 548–657.
46. Lemiere, S., Delveccio, A., Herrmann, A., Alloin, H., Prandini, F. and Crussard, S. *ThePoultrySite*. Available at: https://www.thepoultrysite.com/articles/protection-against-various-velogenic-newcastle-disease-challenges-of-a-trivalent-hvt-vectored-vaccine-against-mareks-disease-infectios-bursal-disease-and-newcastle-disease. Accessed December 31, 2021.
47. Boulianne, M., Opengart, K. and Uzal, F. A. 2020. Clostridial diseases: ulcerative enteritis, necrotic enteritis, botulism. In: Swayne, D. E., Boulianne, M., Logue, C. M., McDougald, L. R., Nair, V., Suarez, D. L., de Wit, S., Grimes, T., Johnson, D., Kromm, M., Prajitno, T. Y., Rubinoff, I. and Zavala, G. (Eds), *Diseases of Poultry* (14th edn.). John Wiley & Sons, Inc, Hoboken, NJ. pp. 967–985.
48. Souillard, R., Le Maréchal, C., Ballan, V., Rouxel, S., Léon, D., Balaine, L., Poëzevara, T., Houard, E., Robineau, B., Robinault, C., Chemaly, M. and Le Bouquin, S. 2017. Investigation of a type C/D botulism outbreak in free-range laying hens in France. *Avian Pathol.* 46(2):195–201, 10.1080/03079457.2016.1240355.
49. Swayne, D. E., Glisson, J. R., McDougald, L. R., Nolan, L. K., Suarez, D. L. and Nair, V. 2013. *Diseases of Poultry* (13th edn.). Wiley-Blackwell, Hoboken, NJ.
50. Rood, J. I., Adams, V., Lacey, J., Lyras, D., McClane, B. A., Melville, S. B., Moore, R. J., Popoff, M. R., Sarker, M. R., Songer, J. G., Uzal, F. A. and Van Immerseel, F. 2018. Expansion of the Clostridium perfringens toxin-based typing scheme. *Anaerobe* 53:5-10.
51. Thompson, D. R., Parreira, V. R., Kulkarni, R. R. and Prescott, J. F. 2006. Live attenuated vaccine-based control of necrotic enteritis of broiler chickens. *Vet. Microbiol.* 113(1-2):25–34.

52. Rood, J. I., Keyburn, A. L. and Moore, R. J. 2016. NetB and necrotic enteritis: the hole movable story. *Avian Pathol.* 45(3):295–301. 10.1080/03079457.2016.1158781. PMID: 27009522.
53. Prescott, J. F., Parreira, V. R., Mehdizadeh Gohari, I., Lepp, D. and Gong, J. 2016. The pathogenesis of necrotic enteritis in chickens: what we know and what we need to know: a review. *Avian Pathol.* 45(3):288–294. 10.1080/03079457.2016.1139688. PMID: 26813023.
54. Ferguson-Noel, N., Armour, N. K. and Noormohammadi, A. H. 2020. Mycoplasmosis. In: Swayne, D. E., Boulianne, M., Logue, C. M., McDougald, L. R., Nair, V., Suarez, D. L., de Wit, S., Grimes, T., Johnson, D., Kromm, M., Prajitno, T. Y., Rubinoff, I. and Zavala, G. (Eds), *Diseases of Poultry* (14th edn.). John Wiley & Sons, Inc, Hoboken, NJ. pp. 907–929.
55. Landman, W. J. 2014. Is Mycoplasma synoviae outrunning Mycoplasma gallisepticum? A viewpoint from the Netherlands. *Avian Pathol.* 43(1):2–8.
56. Kleven, S. H. 2008. Control of avian mycoplasma infections in commercial poultry. *Avian Dis.* 52(3):367–374.
57. Nunoya, T., Yagihashi, T., Tajima, M. and Nagasawa, Y. 1995. Occurrence of keratoconjunctivitis apparently caused by Mycoplasma gallisepticum in layer chickens. *Vet. Pathol.* 32(1):11–18.
58. FDA Website. Available at: https://www.fda.gov/animal-veterinary/recalls-withdrawals/withdrawal-enrofloxacin-poultry. Accessed December 31, 2021.
59. Ferguson-Noel, N., Cookson, K., Laibinis, V. A. and Kleven, S. H. 2012 Jun. The efficacy of three commercial Mycoplasma gallisepticum vaccines in laying hens. *Avian Dis.* 56(2):272–275. 10.1637/9952-092711-Reg.1. PMID: 22856181.
60. Kursa, O., Pakula, A., Tomczyk, G., Pasko, S., Sawicka, A. 2019. Egg-shell apex abnormalities caused by two different Mycoplasma synoviae genotypes and evaluation of egg-shell anomalies by full-field optical coherence tomography. *BMC Vet. Res.* 15:1.
61. Noormohammadi, A. H., Jones, J. F., Harrigan, K. E. and Whithear, K. G. 2003. Evaluation of the nontemperature-sensitive field clonal isolates of the Mycoplasma synoviae vaccine strain MS-H. *Avian Dis.* 47(2):355–360.
62. Hy-line Technical bulletin. Mycoplasma synoviae (MS). https://www.hyline.com/Upload/Resources/TU%20MS%20ENG.pdf
63. Brenner, F. W., Villar, R. G., Angulo, F. J., Tauxe, R. and Swaminathan, B. 2000. Salmonella nomenclature. *J. Clin. Microbiol.* 38(7):2465–2467. 10.1128/JCM.38.7.2465-2467.2000.
64. Gast, R. K. and Porter, R. E. 2020. Salmonella infections. In: Swayne, D. E., Boulianne, M., Logue, C. M., McDougald, L. R., Nair, V., Suarez, D. L., de Wit, S., Grimes, T., Johnson, D., Kromm, M., Prajitno, T. Y., Rubinoff, I. and Zavala, G. (Eds), *Diseases of Poultry* (14th edn.). John Wiley & Sons, Inc, Hoboken, NJ. pp. 719–753.
65. Fowl Typhoid and Pullorum Disease. Updated 2019. Center for Food Security and Public Health, Iowa State University. Available at: https://www.cfsph.iastate.edu/Factsheets/pdfs/fowl_typhoid.pdf. Accessed January 3, 2021.
66. Henderson, S. C., Bounous, D. I. and Lee, M. D. 1999. Early events in the pathogenesis of avian salmonellosis. *Infect. Immun.* 67(7):3580–3586.
67. Nolan, L. K., Barbieri, N. and Logue, C. M. 2020. Colibacillosis. In: Swayne, D. E., Boulianne, M., Logue, C. M., McDougald, L. R., Nair, V., Suarez, D. L., de Wit, S.,

Grimes, T., Johnson, D., Kromm, M., Prajitno, T. Y., Rubinoff, I. and Zavala, G. (Eds), *Diseases of Poultry* (14th edn.). John Wiley & Sons, Inc, Hoboken, NJ. pp. 770–880.
68. Collingwood, C., Kemmett, K., Williams, N. and Wigley, P. 2014. Is the concept of avian pathogenic Escherichia coli as a single pathotype fundamentally flawed? *Front. Vet. Sci.* 1:5. 10.3389/fvets.2014.00005.
69. Logue, C. M. and Andreasen, C. B. 2020. Other bacterial diseases. In: Swayne, D. E., Boulianne, M., Logue, C. M., McDougald, L. R., Nair, V., Suarez, D. L., de Wit, S., Grimes, T., Johnson, D., Kromm, M., Prajitno, T. Y., Rubinoff, I. and Zavala, G. (Eds), *Diseases of Poultry* (14th edn.). John Wiley & Sons, Inc, Hoboken, NJ. pp. 995–1085.
70. Guabiraba, R. and Schouler, C. 2015. Avian colibacillosis: still many black holes. *FEMS Microbiol. Lett.* 362(15):fnv118, 10.1093/femsle/fnv118.
71. Olsen, R. H., Frantzen, C., Christensen, H. and Bisgaard, M. 2012. An investigation on first-week mortality in layers. *Avian Dis.* 56(1):51–57.
72. Barbieri, N. L., de Oliveira, A. L., Tejkowski, T. M., Pavanelo, D. B., Matter, L. B., Pinheiro, S. R., Vaz, T. M., Nolan, L. K., Logue, C. M., de Brito, B. G. and Horn, F. 2015. Molecular characterization and clonal relationships among Escherichia coli strains isolated from broiler chickens with colisepticemia. *Foodborne Pathog. Dis.* 12(1):74–83.
73. McFerran, J. B. 1979. Egg drop syndrome, 1976 (EDS'76). *Vet. Q.* 1(4):176–180, 10.1080/01652176.1979.9693744.
74. Van Eck, J. H. H., Davelaar, F. G., Van den Heuvel-Plesman, T. A. M., Van Kol, N., Kouwenhoven, B. and Guldie, F. H. M. 1976. Dropped egg production, soft shelled and shell-less eggs associated with appearance of precipitins to adenovirus in flocks of laying fowls. *Avian Pathol.* 5(4):261–272.
75. Smyth, J. A. 2013. Atadenovirus (egg drop syndrome and related infections). In: Swayne, D. E., Glisson, J. R., McDougald, L. R., Nolan, L. K., Suarez, D. L. and Nair, V. (Eds), *Diseases of Poultry* (13th edn.). Wiley-Blackwell, Hoboken, NJ. pp. 301–309.
76. Gay, L. and Mundt, E. 2010. Testing of a new disinfectant process for poultry viruses. *Avian Dis.* 54(Suppl. 1):763–767.
77. Darbyshire, J. H. and Peters, R. W. 1980. Studies on EDS-76 virus infection in laying chickens. *Avian Pathol.* 9(3):277–290.
78. Miyamae, T. 1981. Localization of viral protein in avian-encephalomyelitis-virus-infected hens. *Avian Dis.* 25(4):1065–1069.
79. Sevoian, M. and Levine, P. P. 1957. Effects of infectious bronchitis on the reproductive tracts, egg production, and egg quality of laying chickens. *Avian Dis.* 1(2):136–164.
80. Cook, J. K. A., Jackwood, M. and Jones, R. C. 2012. The long view: 40 years of infectious bronchitis research. *Av. Pathol.* 41(3):239–250.
81. Hughes, C. S., Gaskell, R. M., Jones, R. C., Bradbury, J. M. and Jordan, F. T. 1989. Effects of certain stress factors on the re-excretion of infectious laryngotracheitis virus from latently infected carrier birds. *Res. Vet. Sci.* 46(2):274–276.
82. Bwala, D. G., Clift, S., Duncan, N. M., Bisschop, S. P. and Oludayo, F. F. 2012. Determination of the distribution of lentogenic vaccine and virulent Newcastle disease virus antigen in the oviduct of SPF and commercial hen using immunohistochemistry. *Res. Vet. Sci.* 93(1):520–528.
83. Bertran, K., Susta, L. and Miller, P. J. 2017. Avian influenza virus and Newcastle disease virus. In: Hester, P. (Ed.), *Egg Innovation and Strategies for Improvement*. Academic Press, Oxford, New York. pp. 547–560.
84. Rautenschlein, S. 2020. Avian metapneumovirus. In: Swayne, D. E., Boulianne, M., Logue, C. M., McDougald, L. R., Nair, V., Suarez, D. L., de Wit, S., Grimes, T., Johnson,

D., Kromm, M., Prajitno, T. Y., Rubinoff, I. and Zavala, G. (Eds), *Diseases of Poultry* (14th edn.). John Wiley & Sons, Inc, Hoboken, NJ. pp. 135–143.
85. Shin, H. J., Cameron, K. T., Jacobs, J. A., Turpin, E. A., Halvorson, D. A., Goyal, S. M., Nagaraja, K. V., Kumar, M. C., Lauer, D. C., Seal, B. S. and Njenga, M. K. 2002. Molecular epidemiology of subgroup C avian pneumoviruses isolated in the United States and comparison with subgroup A and B viruses. *J. Clin. Microbiol.* 40(5):1687–1693.
86. Cook, J. K. A., Chesher, J., Orthel, F., Woods, M. A., Orbell, S. J., Baxendale, W. and Huggins, M. B. 2000. Avian Pneumovirus infection of laying hens: experimental studies. *Avian Pathol.* 29(6):545–556.
87. Zhao, Z., Rao, M., Liao, M., Cao, W. 2018. Phylogenetic analysis and pathogenicity assessment of the emerging recombinant subgroup K of avian leukosis virus in South China. *Viruses* 10(4):194. doi:10.3390/v10040194.
88. Davidson, I., Maray, T., Malkinson, M. and Becker, Y. 1986. Detection of Marek's disease virus antigens and DNA in feathers from infected chickens. *J. Virol. Meth.* 13(3):231–244.
89. Morgan, R. W., Gelb, J., Jr., Schreurs, C. S., Lutticken, D., Rosenberger, J. K. and Sondermeijer, P. J. A. 1992. Protection of chickens from Newcastle and Marek's diseases with a recombinant herpesvirus of turkeys vaccine expressing the Newcastle disease virus fusion protein. *Avian Dis.* 36(4):858–870.
90. Woźniakowski, G., Frant, M. and Mamczur, A. 2018. Avian reticuloendotheliosis in chickens - an update on disease occurrence and clinical course. *J. Vet. Res.* 62(3):257–260. 10.2478/jvetres-2018-0036.
91. Davidson, I. and Braverman, Y. 2005. Insect contribution to horizontal transmission of reticuloendotheliosis virus. *J. Med. Entomol.* 42(2):128–133.
92. Motha, M. X. J. and Egerton, J. R. 1987. Vertical transmission of reticuloendotheliosis virus in chickens. *Avian Pathol.* 16(1):141–147.
93. Wei, K., Sun, Z., Zhu, S., Guo, W., Sheng, P., Wang, Z., Zhao, C., Zhao, Q. and Zhu, R. 2012. Probable congenital transmission of reticuloendotheliosis virus caused by vaccination with contaminated vaccines. *PLoS ONE* 7(8):e43422.
94. Vanrompay, D. 2020. Avian chlamydiosis. In: Swayne, D. E., Boulianne, M., Logue, C. M., McDougald, L. R., Nair, V., Suarez, D. L., de Wit, S., Grimes, T., Johnson, D., Kromm, M., Prajitno, T. Y., Rubinoff, I. and Zavala, G. (Eds), *Diseases of Poultry* (14th edn.). John Wiley & Sons, Inc, Hoboken, NJ. pp. 1086–1101.
95. Yin, L., Kalmar, I. D., Lagae, S., Vandendriessche, S., Vanderhaeghen, W., Butaye, P., Cox, E. and Vanrompay, D. 2013. Emerging Chlamydia psittaci infections in the chicken industry and pathology of Chlamydia psittaci genotype B and D strains in specific pathogen free chickens. *Vet. Microbiol.* 162(2–4):740–749.
96. Yin, L., Lagae, S., Kalmar, I. D., Borel, N., Pospischil, A. and Vanrompay, D. 2013. Pathogenicity of low and highly virulent Chlamydia psittaci isolates for specific-pathogen-free chickens. *Avian Dis.* 57(2):242–247.
97. Campbell, T. W. 2015. Normal avian cytology. In: Campbell, T. W. and Wiley-Blackwell, O. (Eds), *Exotic Animal Hematology and Cytology* (4th edn.). Wiley-Blackwell, Hoboken, NJ. pp. 219–227.
98. Gregory, M., Klein, B., Sahin, O. and Girgis, G. 2018. Isolation and characterization of Campylobacter hepaticus from layer chickens with spotty liver disease in the United States. *Avian Dis.* 62(1):79–85.

99. Sahin, O. and Zhang, Q. 2020. Campylobacteriosis. In: Swayne, D. E., Boulianne, M., Logue, C. M., McDougald, L. R., Nair, V., Suarez, D. L., de Wit, S., Grimes, T., Johnson, D., Kromm, M., Prajitno, T. Y., Rubinoff, I. and Zavala, G. (Eds), *Diseases of Poultry* (14th edn.). John Wiley & Sons, Inc, Hoboken, NJ. pp. 754–769.
100. Crawshaw, T. R., Chanter, J. I., Young, S. C., Cawthraw, S., Whatmore, A. M., Koylass, M. S., Vidal, A. B., Salguero, F. J. and Irvine, R. M. 2015. Isolation of a novel thermophilic Campylobacter from cases of spotty liver disease in laying hens and experimental reproduction of infection and microscopic pathology. *Vet. Microbiol.* 179(3-4):315-321.
101. Van, T. T. H., Elshagmani, E., Gor, M. C., Scott, P. C. and Moore, R. J. 2016. Campylobacter hepaticus sp. nov., isolated from chickens with spotty liver disease. *Int. J. Syst. Evol. Microbiol.* 66(11):4518–4524.
102. Van, T. T., Elshagmani, E., Gor, M. C., Anwar, A., Scott, P. C. and Moore, R. J. 2017. Induction of spotty liver disease in layer hens by infection with Campylobacter hepaticus. *Vet. Microbiol.* 199:85-90.
103. Gingerich, E. 2021. Layer health report for October 1, 2020 to October 1, 2021. Proceedings 125th Annual Meeting of the United States Animal Heath Association, Virtual; Committee on Poultry and Other Avian Species. Pending Publication. Presented October 14, 2021.
104. Eriksson, H., Nyman, A. K., Fellstrom, C. and Wallgren, P. 2013. Erysipelas in laying hens is associated with housing system. *Vet. Rec.* 173(1):18.
105. Eriksson, H. 2020. Other bacterial diseases. In: Swayne, D. E., Boulianne, M., Logue, C. M., McDougald, L. R., Nair, V., Suarez, D. L., de Wit, S., Grimes, T., Johnson, D., Kromm, M., Prajitno, T. Y., Rubinoff, I. and Zavala, G. (Eds), *Diseases of Poultry* (14th edn.). John Wiley & Sons, Inc, Hoboken, NJ. pp. 995–1085.
106. Blackall, P. J. and Hofacre, C. L. 2020. Fowl cholera. In: Swayne, D. E., Boulianne, M., Logue, C. M., McDougald, L. R., Nair, V., Suarez, D. L., de Wit, S., Grimes, T., Johnson, D., Kromm, M., Prajitno, T. Y., Rubinoff, I. and Zavala, G. (Eds), *Diseases of Poultry* (14th edn.). John Wiley & Sons, Inc, Hoboken, NJ. pp. 831–889.
107. Stokholm, N. M., Permin, A., Bisgaard, M. and Christensen, J. P. 2010. Causes of mortality in commercial organic layers in Denmark. *Avian Dis.* 54(4):1241-1250.
108. Blackall, P. J. and Soriano-Vargas, E. 2020. Infectious coryza and related bacterial infections. In: Swayne, D. E., Boulianne, M., Logue, C. M., McDougald, L. R., Nair, V., Suarez, D. L., de Wit, S., Grimes, T., Johnson, D., Kromm, M., Prajitno, T. Y., Rubinoff, I. and Zavala, G. (Eds), *Diseases of Poultry* (14th edn.). John Wiley & Sons, Inc, Hoboken, NJ. pp. 890–900.
109. Bland, M. P., Bickford, A. A., Charlton, B. R., Cooper, G. C., Sommer, F. and Cutler, G. 2002. Case report: a severe infectious coryza infection in a multi-age layer complex in central California. In: 51st Western Poultry Disease Conference/XXVII Convencion Anual ANECA, Puerto Vallarta, Mexico.
110. Sato, S. and Shifrine, M. 1964. Serologic response of chickens to experimental infection with *Hemophilus gallinarum*, and their immunity to challenge. *Poult. Sci.* 43(5):1199–1204.
111. Villegas, A. M., Berghaus, R., Zavala, G., Barrios, M. A. and França, M. 2018. Profiling of egg layer flocks affected with focal duodenal necrosis based on a questionnaire survey. *Avian Dis.* 62(2):241-246.
112. França, M., Barrios, M. A., Stabler, L., Zavala, G., Shivaprasad, H. L., Lee, M. D., Villegas, A. M. and Uzal, F. A. 2016. Association of Beta2-positive Clostridium perfringens

Type A With focal duodenal necrosis in egg-laying chickens in the United States. *Avian Dis.* 60(1):43–49.
113. Hy-Line. 2022. An overview of focal duodenal necrosis (FDN). Available at: https://www.hyline.com/Upload/Resources/TU%20FDN%20ENG.pdf. Accessed Januaruy 16, 2022.
114. Allaart, J. G., de Bruijn, N. D., van Asten, A. J. A. M., Fabri, T. H. F. and Gröne, A. 2012. NetB-producing and beta2-producing Clostridium perfringens associated with subclinical necrotic enteritis in laying hens in the Netherlands. *Avian Pathol.* 41(6):541–546.
115. Elbestawy, A. R., Abd-Ellatieff, H. A., Ellakany, H. F., Abd El-Hamid, H. S., Abou Rawash, A. A., Gado, A. R., Abd El-Aziz, A. H., Eid, A. A. M. and El-Shall, N. A. 2020. Respiratory and reproductive impairment of commercial layer chickens After experimental infection with Gallibacterium anatis Biovar haemolytica. *Avian Dis.* 64(4):536–541.
116. Nassik, S., Tallouzt, S., Karbach, N., Touzani, C., Bidoudan, Y., Aamarine, N. and Hess, C. 2019. First report of isolation of Gallibacterium anatis from layer chickens in morocco with decrease in laying performance. *Avian Dis.* 63(4):727–730.
117. Russo, K., Evans, N., Grum, D., Ayala, D., Kimminau, E. and Karnezos, P. 2021. Karnezos: focal ulcerative dermatitis syndrome in midwestern laying hens. Proceedings from the 2021 American Association of Avian Pathologists annual meeting, Virtual. Available at: https://aaap.memberclicks.net/assets/2021_Annual_Meeting/AAAP_2021_Proceedings.pdf
118. Arné, P. and Lee, M. D. 2020. Fungal infections. In: Swayne, D. E., Boulianne, M., Logue, C. M., McDougald, L. R., Nair, V., Suarez, D. L., de Wit, S., Grimes, T., Johnson, D., Kromm, M., Prajitno, T. Y., Rubinoff, I. and Zavala, G. (Eds), *Diseases of Poultry* (14th edn.). John Wiley & Sons, Inc, Hoboken, NJ. pp. 1111–1136.
119. Hoerr, F. J. 2020. Fungal infections. In: Swayne, D. E., Boulianne, M., Logue, C. M., McDougald, L. R., Nair, V., Suarez, D. L., de Wit, S., Grimes, T., Johnson, D., Kromm, M., Prajitno, T. Y., Rubinoff, I. and Zavala, G. (Eds), *Diseases of Poultry* (14th edn.). John Wiley & Sons, Inc, Hoboken, NJ. pp. 1330–1344.
120. Muccio, M. 2021. BIOMIN world mycotoxin survey. Available at: https://www.biomin.net/science-hub/world-mycotoxin-survey-impact-2021/. Accessed January 4, 2022.
121. Fowler, J. 2018. University of Georgia extension. Available at: https://site.extension.uga.edu/poultrytips/files/2018/08/Fowler-August-2018.pdf. Accessed January 4, 2022.
122. Wu, F. and Munkvold, G. P. 2008. Mycotoxins in ethanol co-products: modeling economic impacts on the livestock industry and management strategies. *J. Agric. Food Chem.* 56(11):3900–3911.
123. Hy-Line. 2022. Management guide. Available at: https://www.hyline.com/filesimages/Hy-Line-Products/Hy-Line-Product-PDFs/Brown/Brown%20Alt/BRN%20ALT%20COM%20ENG.pdf (Other versions available for other commercial varieties on website). Accessed January 19, 2022.
124. Mullens, B. A., Owen, J. P., Kuney, D. R., Szijj, C. E. and Klinger, K. A. 2009. Temporal changes in distribution, prevalence and intensity of northern fowl mite (*Ornithonyssus sylviarum*) parasitism in commercial caged laying hens, with a comprehensive economic analysis of parasite impact. *Vet. Parasitol.* 160(1–2):116–133.
125. Oh, S. I., Do, Y. J., Kim, E., Yi, S. W. and Yoo, J. G. 2020. Prevalence of poultry red mite (Dermanyssus gallinae) in Korean layer farms and the presence of avian pathogens in the mite. *Exp. Appl. Acarol.* 81(2):223–238. 10.1007/s10493-020-00502-5.

126. Pavličević, A., Pavlović, I., Ratajac, R., Popović, D., Davidović, B. and Krnjajić, D. 2019. Poultry welfare in terms of poultry red mite (Dermanyssus gallinae) impact and control. *Biotechnol. Anim. Husbandry* 35(1):1–11. 10.2298/BAH1901001P.
127. Murillo, A. C. and Mullens, B. A. 2016. Diversity and prevalence of ectoparasites on backyard chicken flocks in California. *J. Med. Entomol.* 53(3):707–711.
128. Edgar, S. A. 1953. A field study of the effect of black fly bites on egg production of laying hens. *Poult. Sci.* 32(5):779–780.
129. Hinkle, N. C. and Corrigan, R. M. 2020. External parasites and poultry pests. In: Swayne, D. E., Boulianne, M., Logue, C. M., McDougald, L. R., Nair, V., Suarez, D. L., de Wit, S., Grimes, T., Johnson, D., Kromm, M., Prajitno, T. Y., Rubinoff, I. and Zavala, G. (Eds), *Diseases of Poultry* (14th edn.). John Wiley & Sons, Inc, Hoboken, NJ. pp. 1137–1155.
130. Blake, D. P., Knox, J., Dehaeck, B., Huntington, B., Rathinam, T., Ravipati, V., Ayoade, S., Gilbert, W., Adebambo, A. O., Jatau, I. D., Raman, M., Parker, D., Rushton, J. and Tomley, F. M. 2020. Re-calculating the cost of coccidiosis in chickens. *Vet. Res.* 51(1):115. 10.1186/s13567-020-00837-2.
131. Mathis, G. F., Newman, L. J., Fitz-Coy, S., Lumpkins, B., Charette, R. and Fuller, L. 2018. Comparison of breeder/layer coccidiosis vaccines: part 1 -precocity and pathogenicity. *J. Appl. Poult. Rsch* 27(1):33–37.
132. Cervantes, H. M., McDougald, L. R. and Jenkins, M. C. 2020. Coccidiosis. In: Swayne, D. E., Boulianne, M., Logue, C. M., McDougald, L. R., Nair, V., Suarez, D. L., de Wit, S., Grimes, T., Johnson, D., Kromm, M., Prajitno, T. Y., Rubinoff, I. and Zavala, G. (Eds), *Diseases of Poultry* (14th edn.). John Wiley & Sons, Inc, Hoboken, NJ. pp. 1193–1217.
133. Nicholds, J. F., McQuain, C., Hofacre, C. L., Mathis, G. F., Fuller, A. L., Telg, B. E., Montoya, A. F., Williams, S. M., Berghaus, R. D. and Jones, M. K. 2021. The effect of different species of eimeria with Clostridium perfringens on performance parameters and induction of clinical necrotic enteritis in broiler chickens. *Avian Dis.* 65(1):132–137. 10.1637/aviandiseases-D-20-00106.
134. Liebhart, D., Sulejmanovic, T., Grafl, B., Tichy, A. and Hess, M. 2013. Vaccination against histomonosis prevents a drop in egg production in layers following challenge. *Avian Pathol.* 42(1):79–84. 10.1080/03079457.2012.760841.
135. Dahl, C., Permin, A., Christensen, J. P., Bisgaard, M., Muhairwa, A. P., Petersen, K. M., Poulsen, J. S. and Jensen, A. L. 2002. The effect of concurrent infections with Pasteurella multocida and Ascaridia galli on free range chickens. *Vet. Microbiol.* 86(4):313–324.
136. Reid, W. M., Mabon, J. L. and Harshburger, W. C. 1973. Detection of worm parasites in chicken eggs by candling. *Poult. Sci.* 52(6):2316–2324.
137. Sharma, N., Hunt, P. W., Hine, B. C. and Ruhnke, I. 2019. The impacts of Ascaridia galli on performance, health, and immune responses of laying hens: new insights into an old problem. *Poult. Sci.* 98(12):6517–6526. 10.3382/ps/pez422.
138. Callait-Cardinal, M. P. and Zenner, L. 2015. Histomoniasis. In: Brugere-Picoux, J. and Vaillancourt, J. P. (Eds), *Manual of Poultry Diseases*. AFAS (French Association for the Advancement of Science), Zutpher, The Netherlands. pp. 424–427.
139. Esquenet, C., De Herdt, P., De Bosschere, H., Ronsmans, S., Ducatelle, R. and Van Erum, J. 2003. An outbreak of histomoniasis in free-range layer hens. *Avian Pathol.* 32(3):305–308, 10.1080/0307945031000097903.

140. Hess, M. and McDougald, L. R. 2020. Histomoniasis (histomonosis, blackhead disease). In: Swayne, D. E., Boulianne, M., Logue, C. M., McDougald, L. R., Nair, V., Suarez, D. L., de Wit, S., Grimes, T., Johnson, D., Kromm, M., Prajitno, T. Y., Rubinoff, I. and Zavala, G. (Eds), *Diseases of Poultry* (14th edn.). John Wiley & Sons, Inc, Hoboken, NJ. pp. 1223–1230.
141. Abdelhamid, M. K., Quijada, N. M., Dzieciol, M., Hatfaludi, T., Bilic, I., Selberherr, E., Liebhart, D., Hess, C., Hess, M. and Paudel, S. 2020. Co-infection of chicken layers With Histomonas meleagridis and avian pathogenic Escherichia coli is associated With dysbiosis, cecal colonization and translocation of the bacteria From the gut lumen. *Front. Microbiol.* 11:586437.

Chapter 10

Optimizing the health of broiler breeder birds

Rick van Emous, Wageningen Livestock Research, The Netherlands; and André Steentjes, Veterinary Centre Someren, The Netherlands

1 Introduction

During the last decade, poultry meat has become the most important protein source in human diets, and production is growing worldwide. Global poultry meat production in 2018 was 96 million tons and increased to approximately 101 million tons in 2020 (USDA, 2021). To produce this volume of poultry meat, approximately 70 billion broilers originating from approximately 600 million broiler breeders are required. Thus, a relatively small number of parent stock (PS; broiler breeders) has a major impact on poultry meat production. Decades of intensive genetic selections have given broilers the potential for efficient growth. Since they originate from broiler breeders, PS and offspring share the same genetic material with this high growth potential (Havenstein et al., 2003a,b; Renema et al., 2007; Zuidhof et al., 2014). Broiler breeders need to produce first-class hatching eggs and healthy chicks. Appropriate management is essential to produce a high number of fertile eggs and high-quality broiler chickens (van Emous, 2015). Due to the continuing increase in the genetic potential of the offspring, this is becoming increasingly challenging as well (Zuidhof et al., 2014). Although a high growth rate and efficiency is desirable

http://dx.doi.org/10.19103/AS.2022.0104.13

in broilers, it may lead to health and reproduction problems in their parents during the rearing and laying period. Therefore, in this chapter, we do not limit ourselves only to broiler breeders, but we also discuss the effects on their offspring (broilers).

The objective of the chapter is to provide an overview of the diseases, vaccinations and direct and indirect effects of management and nutritional interventions on health and productivity in broiler breeders (and offspring).

2 Diseases during the rearing period

Actual scientific figures about diseases affecting broiler breeders are scarce, and only some outdated overviews are available (Collett, 2009; Zavala and Barnes, 2012). However, there is a lot of experience from daily practice, and the most important diseases in broiler breeders during the rearing period are described in this section.

There are several moments of risk identified during the rearing period with potentially harmful effects on the birds. During the first days of life, after hatchery manipulation (sexing, vaccinations), transport and delivery, the adaptation to a new environment (body temperature, finding water and feed places) is very important. In this period, the microflora in the intestinal tract is naive and very susceptible to agents like Salmonella. During the first weeks of life when maternally derived immunity will decrease and active immunity has to be built up, birds are susceptible to introductions of all kinds of infectious agents, e.g. infectious bronchitis virus (IBV), reovirus, infectious bursal disease virus (IBDV), Marek's disease virus, coccidiosis. Moreover, introducing (extra) chicks is always a risk for potential introduction of infectious diseases. Well-known examples are coccidiosis, worms, Histomonas, Salmonella and Mycoplasma. Any disease problems or adverse vaccination reactions affecting feed intake and growth can influence body weight (BW) and BW uniformity in a rearing flock. This is especially of importance during the early rearing period when e.g. the immunity for diseases like coccidiosis, Newcastle disease (ND), IB, IBDV, infectious laryngotracheitis (ILT), etc. has to be built up. Injection with inactivated vaccines (which are mostly oil emulsion vaccines containing adjuvantia, e.g. bacterins) can harm flock uniformity especially when they are administered in combination. In some vaccination programmes on multiple ages combinations of several oil emulsion vaccinations, sometimes even in combination with eyedrop and wing web vaccinations are planned. Multiple injection vaccinations and live vaccinations at one time will harm the development of the birds and logically also flock uniformity. Moreover, bird handling by vaccination crews or service companies (chasing, catching, vaccinations, loading or deloading/delivery) is not only a moment of stress but also a risk for the possible introduction of infectious agents. Before any transport from one farm to another farm or from one house to another house should always be representative, reliable samples must be taken and tested to prevent for the introduction of at least: *Salmonella spp., Mycoplasma gallisepticum* (MG) (and preferably *Mycoplasma synoviae* (MS)) and worms. The interval of sampling until transport must be as short as possible. Polymerase chain reaction (PCR) testing of MG, MS and Salmonella spp., especially when the number of samples per house is taken with the purpose to check for early infections, is preferred compared to serology testing, which always is a 'delayed check' because it takes time to build up antibodies after field infection.

2.1 Early chick mortality

The mortality of breeder chicks during the first days after delivery at the rearing farm can be caused by several problems (Collett, 2009; Zavala and Barnes, 2012). Failures in management, such as poor heating and subsequent abnormal temperature of the environment, floor or litter, too high CO_2 concentration, are generally excluded in modern houses and by good management; however, improvements might still be necessary. Nevertheless, chick quality can be impaired, especially due to the hatchery or even Grand Parent Stock (GPS) flock or farm problems. Chick quality (and early chick mortality) is influenced by many factors such as breeder age, egg storage, season, breed, feed company, year, hatchery and breeder nutrition (Lourens and Steentjes, 2008; Yassin et al., 2009). Nowadays, the production of hatching eggs at GPS level is under very high biosecurity standards, and failures from egg collection, egg hygiene (floor eggs), egg storage, hatching egg disinfections, etc. are rare. It is a common practice in modern hatcheries that PS chicks are differentiated in females c.q. males by professional chick sexers, either by cloacal sexing or by feather sexing. Cloacal sexing, especially for chicks from young (grand-) parents, is usually more stressful and harmful. Other bird handling like injection vaccinations, debeaking, detoeing not only are stressful but can also cause wounds that are a potential route (port d'entrée) of infection. Not only bacteria (e.g. *Escherichia coli, Enterococcus spp., Staphylococcus aureus, Pseudomonas aeruginosa* but also moulds (*Aspergillus fumigatus*) can be a cause of increased mortality when biosecurity measures at the GPS farm, during transport or in the hatchery have failed. Not only *in ovo* vaccination (at 18 days of incubation) but also any injection vaccination (subcutaneous or intramuscular) in 1-day-old chicks (DOC) for Marek's disease or other diseases has to be carried out in a clean, disinfected area.

But still under these top biosecurity conditions, hatching eggs are never sterile. Consequently, also PS chicks and broiler chicks from hatching eggs produced at bio-secure GPS and PS farms can carry infectious bacteria. Of those bacteria, *E. coli, Enterococcus* spp., *P. aeruginosa and S. aureus* are the most important bacteria that can cause mortality during the first days of life. Mortality can differ from the same origin flocks (GPS and PS) in successive deliveries. Mortality is usually due to navel infection causing omphalitis/yolk sacculitis and subsequently bacterial sepsis mainly by the mentioned bacteria species and can be observed more prominent in the first days after hatching with inflammations of the peritoneum (peritonitis), the pericard (pericarditis) and the liver capsula (perihepatitis). Hatching egg hygiene is the key factor to prevent these infections (Lourens and Steentjes, 2008).

Numerous possibilities to influence early chick mortality have been researched, especially focussing on egg hygiene measures at production farms and fumigation of hatching eggs at the farm, during transport and before incubation at the hatchery. Also, vaccination of (grand-) parent females with commercial and/or autogenous vaccines containing e.g. *E. coli* or *Enterococcae* (carrying parts of or the whole killed bacteria) has been carried out to protect against vertical transmission and induce maternal protection. However, in daily practice, none of them have proven a convincing vertical transmitted protective effect. Autogenous vaccines with selected strains of *E. coli* bacteria, which seem to have more pathogenic factors (virulence factors), have been produced and are used commonly around the world. However, the real protective effect on their offspring, and consequently reduction of early chick mortality, is still doubtful.

In order to reduce early chick mortality, antibiotics have been (and still are being) used in hatcheries all over the world for PS chicks and broilers either *in ovo* or by subcutaneous injections; substances like gentamycin, lincomycin-spectinomycin and ceftiofur have been used in many countries.

But antimicrobial treatment as a preventive measurement should be avoided in early life. Administrating antibiotics, even first-choice antibiotics, in the first days of a chick's life can create antimicrobial resistance in the normal inhabitants of the intestines, like *E. coli, Proteus, Klebsiella,* etc. (Verrette et al., 2019). In many countries in north-west Europe, there is no longer a legal possibility to use ceftiofur, a third-generation cephalosporin antimicrobial; due to the discovery of increased resistance to extended-spectrum cephalosporins in the broiler poultry production chain, the preventive use of ceftiofur was banned. Due to cross-resistance with other extended-spectrum cephalosporins, such as ceftriaxone and cephamycin, antimicrobials that are used widely in human medicine and classified by the World Health Organization as the highest priority critically important antimicrobials, this is a public health concern (Verrette et al., 2019).

After the cessation, Dutch (Dierikx et al., 2013) and Canadian (Verrette et al., 2019) reports showed a beneficial effect on the sensitivity of *E. coli*, cultured from caeca of broilers at slaughter age.

2.2 Coccidiosis

In practice, coccidiosis is known as a major detrimental disease that affects pullet's performance by reducing digestion efficiency, growth and flock uniformity (Collett, 2009). Some Eimeria species (especially *E. necatrix*, *E. brunetti* and *E. tenella)* can cause serious mortality. To prevent outbreaks of coccidiosis, hygiene measures (all in-all out, thorough cleaning and the use of chemicals) in combination with administration of coccidiostats (by feed) in a shuttle programme with products that still give possibility to get immunization against Eimeria species is one of the possible approaches. As coccidiosis vaccines are not registered in some countries, chemoprophylaxis is widely used.

Vaccination is the other approach: since the introduction of live coccidiosis vaccines (containing selected, precocious or attenuated strains), the clinical problems of coccidiosis changed from severe outbreaks to minor problems that sometimes have to be treated (Collett, 2009).

Vaccination against coccidiosis usually results in well-protected flocks, but failures in administration or treatments (sometimes by incidence) have a negative effect.

Well-known causes of a poor coccidiosis immunity in vaccinated flocks are as follows:

- Administration of sulpha-products (e.g. sulphonamides, sulphadimidine sodium, sulphaquinoxaline, some (but not all) trimethoprim-sulpha combinations);
- Simultaneous and mistaken feed administration of ionophore coccidiostats or chemical anticoccidiosis substances that block building up immunity and obstruct recirculation of oocysts;
- Too short withdrawal of water (water deprivation) before vaccination: a lot of birds would not drink from the vaccine solution;
- Too much remaining water inside the drinking system before administration: a part of the flock will drink water without vaccine;
- Too little water in total: to prevent this problem the day before vaccination the farmer should perform a 'test vaccination' to measure how much water the birds will drink after a planned deprivation of water during several hours;
- Inactivation of the oocysts because of storage in an improper way (too cold, e.g. in the freezer instead of the refrigerator) or because of administration in combination with or after administration with acid products or other detrimental disinfectants.

If outbreaks in rearing broiler breeders occur, these are usually caused by *E. necatrix* (the Eimeria species that is known to produce slow and low protection) and *E. brunetti*. Eimeria species are present in some vaccines, not in all, e.g. vaccines that are specifically developed for broilers. Usually outbreaks of coccidiosis (*E. necatrix*, *E. brunetti*) occur in pullets, especially after stress moments, such as change of feed (charges), handling of birds (catching, transport) or after vaccinations by vaccination crews (injections, eyedrop). Outbreaks can also happen at the end of the rearing period or in adult birds, e.g. in the period of onset of production.

Sulpha products (sulphadimidine sodium, sulphaquinoxaline, etc.) need to be administered with care when the morbidity is high: to avoid an overdose in healthy birds, the dosage needs to be adjusted since the healthy birds will have a higher water intake due to the drop of drinking by sick birds. If birds stop drinking due to severe infection, treatment via drinking water with anticoccidial drugs is not sufficient and individual injections with sulphonamides could save birds. For administration via drinking water, the choice of the right product must be discussed with the farmers' veterinarian since some drugs kill the parasite immediately and therefore also stop the development of immunity. Others (especially sulphonamides) are risky because they are toxic when overdosed.

2.3 Leg problems

In this subsection, leg problems due to bacterial infections (*S. aureus, Enterococcus spp., E. coli*) are described. Since prohibitions of mutilations such as beak trimming, toe cutting and spur cutting are more common, the entrance of bacteria via these wounds has become less important.

Mutilations such as beak trimming, toe cutting and spur cutting are still common practices in many countries and can result in an increase of lame birds due to the entrance of *S. aureus* bacteria via these wounds. *S. aureus* can also infect birds via other wounds (skin scratches, foot pad lesions) causing inflammation of joints (Aviagen, 2015a). Especially the hock joint (synovitis) and foot joint (bumble foot) can be involved. Although outbreaks of severe synovitis due to *S. aureus* infections are rare, they still can occur suddenly causing major problems. It seems that other factors, like compromising immunity problems, contribute to severe synovitis as well.

Biosecurity measures to prevent other concurrent infections (IBDV, reovirus, chicken infectious anaemia (CIA)) and proper vaccination against these viruses are important as are preventative measures to minimize leg problems:

- Stress management: avoid excessive stocking density, feed restriction and aggressive bird handling;
- Environmental management: remove sharp equipment or objects, encourage exercise, minimize slat height, manage litter and water quality,

availability of sufficient feed and water space (prevention of stress) are key factors;

- Monitoring vaccination and health programmes: monitor and control coccidiosis, utilize appropriate vaccination programmes, ensure proper biosecurity in the hatchery and on farm, ensure vaccine equipment hygiene is practised.

Because *S. aureus* is ubiquitously present there must be co-factors that cause the difference in prevalence between severe, mild and no problems with synovitis. Also intramuscular and subcutaneous injections can introduce *S. aureus* bacteria, which can be subclinical present inside the body (possibly intracellular) and express at older age because of an immune compromising problem like stress. These stress factors are wide-ranged such as bird handling procedures like catching, injecting, chasing and catching, transport and delivery. Severe problems in breeder flocks with hock joint inflammation (synovitis) have been observed at all ages, but especially during the second half of the rearing period, after transport and at the start of production which is possibly because of the oestrogenic and immunosuppressive effect.

Especially during the first weeks of the rearing period, *Enterococcus* spp. infections can cause losses from severe lameness due to arthritis and bacterial chondronecrosis with osteomyelitis (femur head necrosis). Sometimes in this period, a sudden increase in mortality due to (valvular) endocarditis occurs, from which predominantly *E. faecalis* or *E. hirae* can be cultured (personal communication).

Dissection, bacteriological examination and sensitivity testing will show whether there is a choice to treat with a specific antibacterial drug. Because this drug must penetrate joints, it needs to be administered in an adequate high dosage and for a long period (at least 5 consecutive days), but the pieces of advice and leaflets of pharmaceutical companies need to be followed. Sometimes the problems re-occur, even after a first treatment with antibiotics.

2.4 Spinal abscesses (Vertebral osteoarthritis)

Problems with spinal abscesses caused by *Enterococcus cecorum* in young broiler breeder males normally have an onset around 6–12 weeks of age (WOA) and usually last for 4–6 weeks. The growing abscess compresses the spinal cord and therefore an affected bird shows very typical clinical signs (sitting on hocks, walking backward). At the moment, there is little understanding about the cause and occurrence of this condition (Jones, 2009). Only early treatment with specific antibiotics for which *E. cecorum* shows extreme sensitivity (penicillin, virginiamycin, lincomycin-spectinomycin) seems to have a positive effect (A. Steentjes, 2021, personal communication).

2.5 Final conclusion

A rearing period without disease problems or failures in feeding, water supply, climate and ventilation problems is of huge importance for flock uniformity, development of the birds (BW), feather package, reproduction, production persistency and mortality rate during the laying period.

3 Diseases during the laying period

Females are very susceptible to diseases at the start of production: even a typical 'young bird disease' like coccidiosis can occur in a young laying flock at this age. Also, in females between 23 and 30 WOA, *E. coli*-infections are often taking place, sometimes as a primary pathogen but more often *E. coli* arises as a consequence of other problems such as poor climate conditions, inadequate ventilation (high air speed, too high CO_2 and NH_3 concentration, ventilation of cold air flow under the slight floor during the night), respiratory infections like IB, avian metapneumovirus (AMPV, also known as Turkey rhinotracheitis virus) and red mite infestations (Nolan et al., 2020). These *E. coli* problems can be categorized as 'secondary' *E. coli* infections.

During the laying period, females are most sensitive in the period from onset of lay until peak production, sometimes resulting in high mortality. This is probably because of hormonal changes (oestrogens) and the stress of chasing males which is influenced by a high male-to-female ratio (van Emous, 2010). The most prevalent lesion in female birds due to *E. coli* is usually peritonitis (*E. coli* peritonitis syndrome, or EPS). Because of high-energy feeding and feeding for optimal production, metabolic disorders like haemorrhagic fatty liver syndrome and leg problems can occur during production. Furthermore, several viral and bacterial infections like IB, AMPV, ND virus (NDV), avian influenza, infectious laryngotracheitis virus (ILTV), MG, MS and *Avibacterium paragallinarum* can damage production and eggshell quality and also can be harmful for fertility, hatchability and chick quality. Infections with parasites, both internal parasites like worms and *Histomonas meleagridis*, and external parasites like *Dermanyssus gallinae* ('red mites') are very common, even on farms with high biosecurity standards.

3.1 E. coli *peritonitis syndrome*

It is worldwide accepted that many mortality cases due to *E. coli* infections during the laying phase are secondary as the consequence of another primary cause (Jones, 2009). Primary causes are often related to stress, such as high bird density, high percentage of males, aggressiveness of males, insufficient feed or feed space, house and climate environmental problems (i.e. temperature,

NH_3 concentration, CO_2 concentration, fine dust concentration, etc.) and other priming infections, especially respiratory diseases like IB, AMPV, and MG (Fragoso et al., 2021). Nevertheless, there is a possibility that specific *E. coli* strains with certain pathogenic capacities can cause disease and problems on their own. EPS may be the main clinical outcome in laying birds; other clinical outcomes are septicaemia or more localized infections such as airsacculitis, cellulitis or salpingitis. (Nolan et al., 2020).

3.2 Infectious bronchitis

Infection of broiler breeders with IBV can take place at every age and is currently endemic all around the world (Collett, 2009; Jackwood and de Wit, 2020). Depending on factors such as age, co-infections, immune status at the time of the infection and pathogenicity of the IB strain, the consequences of an IBV infection will differ.

Infections with pathogenic strains at very young age (before 14–21 days) in birds without relevant levels of maternally derived antibodies and/or protection by early vaccination may result in permanent damage of the oviduct what will result in so-called false layers (de Wit et al., 2011) These false layers have all the hormones to lay eggs, but they never will, due to the lack of an active oviduct. Risk factors for this phenomenon are multi-age rearing and import of hatching eggs from an area where the IBV strain involved is not present. Early infection with nephropathogenic strain in birds without proper protection can also result in nephritis, gout and mortality. Infections during the laying period can cause major drops in egg production, loss of egg quality and hatchability. In the prevention of IB, biosecurity, an effective vaccination programme, stress prevention and a well-controlled climate inside the house by mechanical ventilation are important.

In vaccination programmes of breeder birds, successive IB vaccinations are prominently implemented, preferably starting with live vaccines including a classical (Massachusetts) and variant vaccine at the hatchery to achieve early cross-protection to reduce the risk of early infections with new strains that might cause permanent damage. These first vaccinations in the hatchery and subsequent live vaccinations in the rearing period must be followed by a high-quality inactivated IBV vaccine with at least one IBV antigen (preferably more) to achieve high levels of systemic antibodies against the relevant field strains at the end of the rearing period. High antibody levels against IBV are indicative of high protection against infection or at least protective enough to prevent clinical problems. Nevertheless, infection with a new IB strain for which the vaccination programme was not effective can still cause respiratory signs and a severe drop in production with eggshell alterations. Furthermore, health problems because of IB in the PS and subsequent eggshell abnormalities may have consequences for the offspring. Poor eggshell quality may cause easier penetration of bacteria, and malfunction of the reproductive system possibly alters the transmission of nutrients to the egg. To keep the local protection of the respiratory tract at a high level during the laying period, especially in regions with an increased level of field challenge, live IB vaccines can be applied periodically, e.g. with intervals of 4–6 weeks.

3.3 Other respiratory infections during the laying period: infectious laryngotracheitis, Avian pneumovirus, Avibacterium paragallinarum *and* Pasteurella multocida

Although because of adequate, well-thought-out vaccination programmes and strict biosecurity measures field infections with e.g. ILTV, Avian pneumovirus (TRT-virus), or bacterial infections with e.g. *Avibacterium paragallinarum*, *Pasteurella multocida* are rare, they still take place.

Even when flocks are properly vaccinated, still field infections can take place. In some flocks of broiler breeders where already at the hatchery (*in ovo* injection or by injection in the 1-day-old chicks) ILT vaccination was carried out by a ILT-vector vaccine still ILT-field virus (or alterrated ILT vaccine virus) can cause clinical problems.

For ILT protection (especially in poultry-dense areas where the infection pressure of ILTV occasionally is higher), eyedrop vaccination with live ILT vaccine even after a vector-ILT vaccination is something that should be considered.

For coryza (*A. paragallinarum*) and fowl cholera (*P. multocida*), commercial vaccines are available in many countries. Also the production of autogenous vaccines against isolates of these bacteria is used if commercial vaccines are not registered, available or do not show the expected effect against the strains that are involved.

Rodent control should be checked via a regular monitoring programme for especially *P. multocida*, which is also beneficial to prevent other infectious diseases, like salmonellosis and last but not least: avian influenza.

To diagnose the cause of respiratory problems (to exclude as soon as possible major disease problems like Highly Pathogenic Avian Influenza, ND and MG), multiplex PCR is available in modern labs. Also coryza, TRT, ILT and MS can be well diagnosed with these diagnostics.

3.4 Fowlpox

Infection with fowlpox virus can still take place if a flock is not uniformly protected from the vaccination during the rearing period. In exceptional cases, during the laying period, an emergency vaccination has to be carried out.

To prevent this kind of threat of disease and a lot of extra work because of the lack of protection against fowlpox virus, rearing organizations (rearing farms) should always check the pox reactions (skin lesions) 7–10 days after pox vaccination via the wingweb injection. A score less than 90% should not be accepted: a re-vaccination should then be considered.

3.5 *Neoplastic diseases such as Marek's disease, Avian Leucosis and* Reticulo Endotheliosis Virus

General clinical signs like loss of appetite, slimming, and subsequent increased mortality can be caused by tumours in visceral organs (liver, spleen, proventriculus, ovary, kidneys, muscle tissue, including the heart). In almost all cases, special laboratory testing (especially histopathology, but also PCR) is needed to diagnose which disease is involved.

Because of strict biosecurity during the early rearing period in combination with proper vaccination at the hatchery, Marek's disease is rarely seen in the modern broiler breeder husbandry. *Marek* virus is very resistant and can easily spread by dust with feather and skin particles. After infection (especially at a rearing farm where neurological signs and tumours have been seen), a thorough cleaning and strict disinfection are necessary. At rearing farms with Marek's disease in a previous flock, only older birds (4–6 WOA) that are well protected should be placed for a subsequent round.

Tumours as a result of infection with *Avian leucosis* virus are in the modern broiler breeder husbandry exceptional because of screening and selection by the breeding companies all over the world. Still, there are Leucosis virus strains (e.g. *ALV-J*) that have shown horizontal spreading. Because there is no vaccination possible, strict biosecurity is necessary to prevent infection, illness, mortality and last but not least: to prevent vertical transmission of ALV to the offspring. In many parts of the world, this approach has proven to be successful.

Other virus infections (e.g. *Reticulo Endotheliosis Virus, REV*) can also cause tumorous alterations. Because there are no vaccines available against this kind of neoplastic disease, biosecurity measures are of greatest importance.

3.6 *Worm infections*

Worm infections are rarely observed at breeder-rearing farms with good biosecurity (McDougald, 2020). However, on almost every GPS and PS production farm, even farms with very high biosecurity standards, worm infections are discovered someday. The roundworm species *Heterakis gallinarum* and *Ascaridia galli* and the tapeworm species *Raillietina cesticillus* are common worm species present at broiler breeder farms. Also other *Raillietina* spp. and *Davainea proglottina* are worth mentioning.

Capillaria spp. (the most pathogenic is *C. obsignata*) can cause morbidity and mortality in heavy infections.

Tetrameres americana is seen in the Middle East but not for example not in Europe.

And for free-range broiler breeders, *Syngamus trachea* could be of importance.

H. gallinarum and *A. galli* are worms with a direct cycle, but some others like *Raillietina* spp. need insects, like dark beetles, as a vector and therefore control of these intermediate hosts is necessary. To monitor possible worm infections, regular analyses (at least once per 3–4 weeks) at the farm level by dissections and/or faecal sampling (for egg per gram faeces counting) should be performed. Special attention should be paid to *H. gallinarum* because this nematode serves as a transport host (vector) for the protozoan parasite *Histomonas meleagridis*, the causative agent of blackhead disease.

Worm infections can be treated via traditional anthelmintics like benzimidazoles (e.g. fenbendazole, flubendazole). Especially on breeder farms with a history of *H. gallinarum* and *Histomonas meleagridis* infections, it is important to remove carriers by deworming early and frequent with a benzimidazole type dewormer, capable of killing the larvae of *H. gallinarum*.

3.7 Histomonosis (blackhead disease)

For many years, blackhead disease, caused by *Histomonas meleagridis,* seemed to be of minor importance, despite it causing in some cases up to 10% mortality, increased culling and nonuniformity in breeder flocks (GD, 2018; Beckstead, 2019). Since the ban of nitroimidazole antihistomonads (e.g. dimetridazole) in Europe, North America and other parts of the world, no specific treatment is available to treat this disease. This resulted in the re-emergence of the disease in recent years (Liebhart and Hess, 2020).

Outbreaks may be observed at 8-WOA, but outbreaks have been reported earlier than 4 WOA (Chadwick et al., 2020). The cecal roundworm *H. gallinarum* is the intermediate host and earth worms play a role as an accessory host. Infection after the onset of lay can result in morbidity and mortality of up to 10% of the flock; however normally only a few percent of the birds show clinical signs. Usually, an infection with *H. meleagridis* in that sensitive period causes a subsequent increase in mortality due to *E. coli* (peritonitis). Whether intestinal *E. coli* causes colibacillosis by bacterial dissemination to systemic organs due to intestinal destruction caused by *H. meleagridis* or by direct penetration of the cecal wall is not yet understood (Abdelhamid et al., 2020). Negative effects on production performance can include poor flock uniformity, delayed onset of lay, and a decrease in the quality and quantity of eggs produced. Conversely, some *H. meleagridis* outbreaks in breeder facilities go unnoticed because of the absence of clinical signs or gross pathologic lesions (Chadwick et al., 2020). To prevent the introduction of *H. meleagridis*, biosecurity measures (control of possible 'mechanical vectors' like earthworms, arthropods such as flies, grasshoppers, sowbugs, crickets and snails, although poorly documented) are very important. Worm monitoring programmes and strategic deworming are necessary as Heterakis seems to be the most important vector of *H. meleagridis* (McDougald, 2020).

3.8 Ruptured tendon

Especially in meat-type breeders, spontaneous rupture of the gastrocnemius tendon is a recognized cause of lameness and is primarily influenced by a rapid BW growth (Duff and Randall, 1986). In some birds, there might be a predisposition towards a spontaneous rupture of the gastrocnemius tendon (thus a congenital nature of the problem), confirmed by the lack of simultaneous tenosynovitis or arthritis or even by the lack of infectious aetiology. This condition has been primarily attributed to tenosynovitis caused by reovirus or staphylococcus infections; however, rupture by other causes can also occur. In a case report by Crespo and Shivaprasad (2011), the cause of the rupture was attributed to aggressive males, due to an increase in male to female ratio from 7–8 males per 100 females at 22–23 WOA to 9–10 males per 100 females at 31 WOA. To prevent ruptured tendons management, advice from the breeder companies (Aviagen, 2015a) about feed (diet composition, feed allowance) feeder space per bird, growth curve, stocking density should be followed. Predisposing factors can also be the height of slats, feed equipment and drinking water lines which can cause trauma as a consequence of jumping from too high elements. Very high (>40 cm) slats should be avoided while this kind of traumatic and sudden move could result in acute rupture of the hock tendon.

3.9 Amyloidosis

Amyloid arthropathy is observed occasionally in adult broiler breeders manifested by lame birds with severely swollen hock joints (Landman et al., 1998; Ovelgönne et al., 2001; Steentjes et al., 2002). At postmortem, there is an accumulation of amyloid in the lumen of the tibiotarsal joint. Sometimes other joints are also involved (knee, shoulder, foot). Occasionally, there are no signals during production, but the abnormalities are observed at slaughter, increasing rejections. Amyloid is a result of chronic inflammation from which *E. faecalis*, *E. coli*, *S. aureus*, other bacteria and *M. synoviae* (MS) have been cultured. (Landman et al., 1998). Steentjes et al. (2002) described field cases of unilateral amyloid arthropathy in rearing broiler breeders after an intramuscular Marek vaccination being a possible route of infection of *E. faecalis*. They advised special attention to hatchery hygiene, especially in relation to Marek vaccination with a half automatic device for intramuscular injection. In more recent cases, the precise aetiology of amyloid arthropathy is unknown; however, preventing stress and chronic infections are considered beneficial tools.

4 Vertically transmitted diseases

Because GPS and PS produce hatching eggs from which PS pullets and broilers are hatched, respectively, they not only transmit immunity (by means of antibodies against specific diseases) but possibly also transmit infectious agents (bacteria and viruses).

4.1 Mycoplasma gallisepticum

One of the most important vertically transmitted diseases is MG, which can have enormous economic impact on production efficiency in the poultry meat chain (Collett, 2009). A frequent monitoring of breeding flocks is necessary to prevent MG-infected hatching eggs in the hatchery and to keep the birds free from MG without vaccination in the reproductive flocks. Ter Veen et al. (2021) described a decrease in MG seroprevalence and the introduction of new genotypes in Dutch commercial poultry during the years 2001–2018. They concluded that an MG monitoring and control programme contributed to a significant decrease in the seroprevalence of MG in the Netherlands. Because MG infections in breeding poultry are rare because of the biosecurity measures that have been improved during the years and these MG-positive PS flocks are culled immediately, the contribution of vertical transmission to the introduction of MG in commercial poultry in the Netherlands has minimized over the years.

Nevertheless, if MG infection is detected on a breeder farm (by positive PCR and/or positive serology), there might already have been a horizontal spread to other houses on the same farm, to neighbouring farms and possibly also to the hatchery. For countries and/or integrations that need to have an MG-free status, the hatching eggs from MG-positive flocks should be removed from setters and hatchers as soon as possible because newly hatched chicks can spread MG horizontally inside the hatchery and the equipment of the hatchery can be contaminated. Moreover, morbidity and mortality in young chicks, hatched from MG-positive hatching eggs can be very disadvantageous. (Ferguson-Noel et al., 2020). On the other hand, in countries (and/or integrations) where MG is endemic, vaccination of rearing birds is a possible approach to prevent clinical signs and vertical spreading in the broiler breeders.

4.2 Salmonella *spp.*

Compared to *Mycoplasma* spp., *Salmonella spp.* are much more resilient and can easily survive in a poultry house between production cycles (Collett, 2009). By implementing high biosecurity standards in combination with intensified standards for Salmonella reduction in feed, intensive sampling on farms and a top-down approach, the poultry sector was able to reduce the prevalence of Salmonella infections in broiler breeders and broilers to a minimum (Gast and Porter, 2020). Although vaccination in breeders by repeated live vaccine administration via the drinking water, injecting killed commercial vaccines (Feberwee et al., 2000) and/or autogenous vaccines has shown positive results, vaccination alone will not be sufficient: it has to be combined with improving biosecurity measures to be successful on the long term. Intensive cleaning and disinfection after removal of a flock and before new flocks are delivered (tested negative before transport), in combination with strict restrictions for visitors and new equipment are very important. Also rodent and insect control measures should be incorporated (Gast and Porter, 2020).

Especially on the level of the breeding companies (e.g. at GGPS and at GPS-level) the strict implementation of biosecurity only, without the use of Salmonella vaccines, has resulted in almost zero incidents of Salmonella-introductions at this farm level. One of the most effective measures is heat treatment of the feed (6 min at > 80°C). Furthermore, when there is a need to have a minimum risk to introduce Salmonella to the hatchery, breeder flocks should be sampled for Salmonella presence every week (A. Steentjes, 2021, personal communication). In breeding organizations, this is organized on all kinds of levels with very strict procedures, e.g. with sterile sampling material and strict protocols and instructions for sampling.

4.3 Chicken infectious anaemia

Chicken infectious anaemia (CIA) is caused by the CIA virus which has been confirmed as the cause of disease in flocks between 2 and 4 WOA with syndromes suggestive of infectious anemia: growth was retarded, mortality was generally between 10% and 20%, but occasionally it reached 60% (Schat and van Santen, 2020). The only specific sign of CIAV infection is anemia, with a peak at 14–16 days post-infectious (PI). Affected birds may die between 12 and 28 days PI. If mortality occurs, it generally does not exceed 30%. Surviving chicks completely recover from depression and anemia by 20–28 days PI, although retarded recovery and increased mortality may be associated with secondary bacterial and viral infections. Secondary infections, causing more severe clinical signs are frequently seen in field cases.

Serological negative breeder flocks are sensitive to the CIA virus and can easily transmit the virus vertically via the hatching egg to the offspring. In the vertically infected offspring the virus causes (often in combination with *S. aureus*) typical signs of blue wing disease (BWD), a gangrenous dermatitis of the wing. To prevent infection with CIA, a successful vaccination during the rearing period is necessary. A check for the presence of antibodies against CIA at least four to six weeks before transport is highly recommended to be able to revaccinate in time when needed. Engström (1999) described a total of ten outbreaks of blue wing disease (BWD) in PS of broilers and layers, all of which were offspring of three late seroconverting GPS-flocks. Outbreaks of BWD were reported in offspring flocks from all broiler breeder flocks that had not developed antibodies to CIA before coming into lay, except for those that had been vaccinated (Engström, 1999). Therefore, the best advice is to check always at the end of the rearing period the presence of antibodies against CIA (and AE).

4.4 Avian Encephalomyelitis

Avian Encephalomyelitis virus (AEV) affects the central nervous system of very young chickens without sufficient maternal antibodies against AEV, often with high mortality. These outbreaks are mostly caused by vertical transmission from acutely infected breeders that had not been successfully vaccinated against AEV. These breeders will show a transient drop in egg production and reduced hatchability (Saif et al., 2020). In a case report (van Veen et al., 1998), an outbreak of AE in seven broiler flocks is described, with signs of ataxia and paralysis. The breeding flock had a temporary drop in egg production and reduced hatchability caused by late embryonic mortality. The breeding flock had not been vaccinated against AE during the rearing period and the problems were caused by vertical transmission. Moreover, rearing birds need to be properly vaccinated between 8 and 10 WOA (Calnek et al., 1961) in order to prevent a temporary drop in egg production and dissemination of the virus by the egg-borne route (vertical transmission). Vaccination against AE is possible in a combined vaccination (AE and Fowlpox) with wing web injection and/or via drinking water with a live vaccine. Not earlier than between 8 and 10 WOA pullets should be vaccinated because of the pathogenicity of the vaccine for younger birds. Breeder pullet flocks should be tested at least two weeks before the transport to the laying facility for the presence of antibodies for AE. If AE titres are too low or not present, revaccination should be carried out at last 4 weeks before the onset of lay.

4.5 Avian reovirus (strains that cause viral arthritis and tenosynovitis)

A lot of reovirus strains are known, but only some of these strains are recognized as strains that cause viral arthritis and tenosynovitis. Birds are most susceptible to avian reovirus (ARV) at a young age.

Reovirus can be vertically transmitted. The control of viral arthritis is facilitated mainly by vaccination of breeders to induce high levels of neutralizing antireovirus antibodies. Those antibodies are expected to prevent infection of the breeder flock that might be followed by vertical transmission of the virus to offspring via the yolk sac, and to deliver maternal derived antibodies to the progeny. Appropriate vaccination of broiler breeders can contribute to achieve high effective antibody levels at hatch and prolong the period of protection. (Pitcovski and Goyal, 2020)

Commercially produced live attenuated (e.g. based on S1133 strain, or derivates like strains 2177, 1733 and 2408 of ARV) and inactivated vaccines are available for broiler breeders. Furthermore, some integrations use autogenous vaccines, following the emergence of new antigenically different variants. To adequately vaccinate against reotenosynovitis depending on circumstances and field pressure different vaccination programmes exist:

- Priming with live vaccine followed by one time inactivated vaccine;
- Priming with live vaccine followed by two times inactivated vaccine;
- Two times inactivated, without priming with live vaccine;
- Serological testing for Reo-antibodies, and if not present: priming with live vaccine, followed by one time killed.

Immunity against reovirus must consist of high and uniform levels of antibodies within a flock. The vaccination is expected to protect against tenosynovitis as well as the malabsorption syndrome.

4.6 Adenovirus-infections

Many Adenovirusses have been found in poultry, only some of them are pathogenic, depending on serotype and strain and can cause problems in young birds. Especially vertically transmitted pathogenic aviadenoviruses can cause serious disease in the young birds in the form of Inclusion Body Hepatitis (IBH), Hepatitis-hydropericardium syndrome (HHS), and adenoviral gizzard erosion (AGE) (Fitzgerald et al., 2020). The consequence of proper biosecurity is that reproductive poultry flocks (as GGPS and GPS) preferably are reared in separate locations where they will not be infected or immunized by Adenoviruses during early life. If introduction with Adenovirus occurs in a seronegative breeder flock, there is always a possibility of vertical transmission to the offspring. In some regions in the world (e.g. South America, Pakistan) commercial and autogenous vaccines against Adenovirus are used. In Western Europe a specific autogenous vaccine against IBH seems to be beneficial.

5 Vaccination programme in broiler breeders

To produce disease-free, viable chicks by breeder operations (integrations, hatcheries) strict biosecurity measures are absolute necessary. Collett (2009) described that managing disease challenges in breeders should include an orchestrated mix of immune modulation (e.g. immunization by vaccination) and disease prevention and control. The primary reason to vaccinate breeder birds is to induce protection against pathogens minimizing the risk of becoming infected and if it did anyway, the detrimental effect on their health would be minimal. The second reason is to hyperimmunize females to maximize the amount of maternally derived antibodies (MDAs, 'passive immunity') that are passed through the egg to the offspring. Depending on the pathogen and level of MDA, MDA can protect progeny in the first weeks of life, after which its own immune system becomes fully functional (Davison et al., 2008). The required vaccinations of broiler breeders depend on the local disease situation and varies all over the world because of:

- Local disease conditions: prevalence (disease pressure);
- Legislation, e.g. compulsory vaccination against ND or Salmonella;
- Availability and registration (permission) of vaccines;
- Costs of vaccines.

Examples of vaccination programmes in Europe (the Netherlands), the USA and Asia (India) for broiler breeders are presented in Tables 1–3.

At the end of the rearing period, when revaccination is still possible, antibody titres should be checked in every house to find out whether vaccination failures can be discovered before flocks are transferred to the broiler breeder farms, especially before the birds start to produce hatching eggs. This is particularly necessary to prevent vertical transmission of viral diseases like Avian Encephalomyelitis virus (AE), CIA and reovirus.

5.1 Maternal immunity (Maternally derived antibodies)

The higher the level of circulating antibody in the hen's blood, the higher the amount of Maternally derived antibodies (MDA) provided for the newly hatched chick. MDA is not long-lived and most of the MDAs have been catabolized by

Table 1 Vaccination programme for broiler breeders in Europe (the Netherlands) (A. Steentjes, 2021, personal communication)[a]

Age	Vaccine	Route
1 day	Marek's disease[b]+ Infectious laryngotracheitis	Subcutaneous
	Infectious bronchitis (Massachusetts + IB variant)	Spray
3-18 days	Newcastle disease	Spray
7 days	(Possible) reovirus-live	Injection s.c.
9 days	Coccidiosis	Drinking water
12-14 days	Infectious bronchitis (variant, e.g. IB 4-91/793B)	Spray
27/28 days	Infectious bursal disease	Drinking water
35 days	Avian metapneumovirus	Spray
6 weeks	Newcastle disease	Spray/aerosol
7 weeks	Chicken infectious anaemia	Drinking water (or injection)
8 weeks	Fowlpox + Avian encephalomyelitis (AE)	Injection
	Salmonellosis (SE + ST)	Injection
	(possible) Reo-live	Injection sc.
10 weeks	Infectious bronchitis (variant vaccine QX)	Spray
12 weeks	Newcastle disease	Spray/aerosol
16 weeks	Multivalent inactivated vaccine containing various combinations of NDV, IBV, AMPV, EDS or IBDV antigens[c]	Injection
	Reovirus	Injection
	Salmonellosis (S.e. + S.t. + S. C-group)[d]	Injection
18 weeks	Infectious bronchitis (QX)	Spray

[a] A lot of variations are possible, mostly depending on the choice of the integration or hatchery and also because of broiler breeder farm history: e.g.: (1) vaccination with live and/or killed Salmonella vaccines, (2) vaccination with live *E. coli* (O78-serotype) one or more spray vaccinations, (3)vaccination with *O. rhinotracheale* by two injections, (4) vaccinations with specific strains of *E. coli* and/or *P. multocida* (often autogenous vaccines).

[b] Marek's disease is a bivalent vaccine consisting of serotype 1 Rispens vaccine with a recombinant herpesvirus of turkeys (serotype 3 MDV) with two inserts of genes of infectious laryngotracheitis virus.

[c] NDV = Newcastle disease virus, IBV = Infectious bronchitis virus, TRTV = Turkey rhinotracheitis virus, EDS = Egg drop syndrome, IBDV = Infectious bursal disease virus.

[d] S.e. = Salmonella enterica serovar Enteritis, S. t. = Salmonella enterica serovar Typhimurium, S. C-group = Salmonella C ('exotic' Salmonellae spp.). Instead of two oil emulsion vaccinations or combined with these two oil emulsion vaccinations also, Salmonella is vaccinated three times with live Salmonella (S.e. and S.t. or only S.e.) by drinking water.

3 to 4 WOA. Chicks with a high level of the correct MDA can protect against challenges with IBDV and CIA. MDA can also minimize the effect of very early infections with NDV, IBV and reovirus (van der Heide et al., 1976; Darbyshire and Peters, 1985), although the duration of this type of immunity is relatively short. MDAs may have a considerable effect on replication of some live vaccines. MDA can neutralize IBDV vaccines that are applied too early. The decline of IBDV

Table 2 Vaccination programme for broiler breeders in the United States (R. Koopman, 2021, personal communication)[a]

Age	Vaccine	Route
1 day	Marek's disease[b]	Subcutaneous
	Coccidiosis	Spray
	Reovirus-live	Subcutaneous
	Infectious bursal disease (89/03)	Subcutaneous
	Infectious bronchitis (Massachusetts)	Spray
14 days	Infectious bronchitis (Massachusetts + variant)	Spray
21 days	Reovirus-live	Drinking water
28 days	Infectious bursal disease	Drinking water
6 weeks	Infectious bronchitis (Ma5) + Newcastle disease (Clone 30)	Spray
10 weeks	Infectious bronchitis (Mass, Del072)	Spray
12 weeks	Fowlpox + Avian encephalomyelitis (AE) + chicken infectious anaemia	Wingweb
15 weeks	Infectious bronchitis (Mass) + Newcastle disease (Clone 30)	Spray
18 weeks	NDV, IBV, IBDV, Reovirus-dead	Injection
	Salmonellosis (S.e.)	Injection
	Fowl cholera	Wingweb

[a] A lot of variations are possible, mostly depending on the choice of the integration or hatchery.
[b] Marek's disease is vaccinated with Rispens and HVT.
NDV = Newcastle disease virus, IBV = Infectious bronchitis virus, IBDV = Infectious bursal disease virus.

MDA, for example, is the most important factor determining the optimal timing for vaccination in the field against IBDV. It creates a considerable challenge to time the first vaccination correctly, before field challenge can infect the flock. PS flocks that have been boosted with an IBDV inactivated vaccine at the end of the rearing period will produce DOCs with increased levels of MDA against IBDV. As a result of this, this progeny can be vaccinated at an older age compared to progeny of only with live IBDV vaccine vaccinated PS flocks. Various formulae are used to predict the optimal age for IBDV vaccination, such as the Deventer formula (de Wit, 2004).

6 Management effects on health breeders

6.1 Breeder house

Compared to broilers, broiler breeders have a relatively long production cycle that averages between 60 and 65 WOA which implicates that there are more possibilities to develop health problems (Zavala and Barnes, 2012). Breeders are given numerous vaccines to protect against mortality, egg production

Table 3 Vaccination programme for broiler breeders in Asia (India) (R. Koopman, 2021, personal communication)[a]

Age	Vaccine	Route
1 day	IBH (120)	Spray
	Marek's disease (Rispens+HVT)	Subcutaneous
	CocciVac D	Spray
5 days	VVND	Subcutaneous
	ND Clone 30	Eye drop
10 days	IBH (ST 4)	Injection
14 days	Infectious bursal disease D-78	Drinking water
21 days	ND+IB Mass	Eye drop
	IBD MB strain	Drinking water
	ND+IB multi	Injection
35 days	Fowlpox	Wingweb
	MGTS 11	Eye drop
	Vaxsafe MS	Eye drop
6 weeks	Salmonella SG9R	Subcutaneous
7 weeks	ND killed	Injection
	ND Clone 30	Eye drop
9 weeks	IC-Coryza	Injection
	FC	Injection
10 weeks	AE-Pox	Wingweb
	Tri Reo	Subcutaneous
	JPI Thymovac	Drinking water
12 weeks	NB+IB Mass	Eye drop
	Salmonella SG9R	Subcutaneous
	VVND	Injection
14 weeks	ND+IBD+IB multi	Injection
16 weeks	ND Clone 30	Eye drop
	FC+IC combined	Injection
18 weeks	Tri Reo	Subcutaneous
20 weeks	ND+IBD+IB multi	Injection
	ND+IB Mass	Eye drop
22 weeks	VVND	Injection
	IBH-ST4	Injection
26 weeks	ND+IB multi	Injection
36 weeks	ND+IBD+IB multi	Injection
	VVND	Injection
40 weeks	IBH-ST4	Injection
	Tri Reo	Injection
46 weeks	VVND	Injection
	ND killed	Injection
55 weeks	VVND	Injection
	ND killed	Injection

[a] A lot of variations are possible, mostly depending on the choice of the integration or hatchery.

drops, and provide high maternal immunity to chicks, but if the birds are not handled and vaccinated properly, serious losses can occur. Trauma can result from handling, entrapment in slats or equipment, or high slat height. Mortality in males due to accidents, foraging/scratching in litter or in carcasses as a result of feed restriction can introduce pathogens to the birds (Roza et al., 2006). On the other hand, over-conditioned breeder females become too heavy, which results in a number of disease problems including early ovarian regression due to super-ovulation, salpingoperitonitis, prolapse of the cloaca and vagina, and decreased egg production (Chen et al., 2006). It is noticed that poor air quality in the pullet or breeder house leads to chronic, obstructive pulmonary disease from inhaled foreign material that is characterized by e.g. multifocal granulomatous pneumonia (Zavala and Barnes, 2012).

In a study of van Emous et al. (2015) a high prevalence of gastrocnemius tendons between 25 and 40 WOA was found when breeders were fed high compared to medium and low energy diets. The exact cause for ruptures of the tendons is unknown, however, in the literature several possible factors are mentioned. It could be caused by genetic selection for fast growth and increased BW (Duff and Randall, 1986), aggressiveness of males resulting in chasing females and jumping from the slatted floor, an increased height between the slatted floor and the ground or seasonal influences (Crespo and Shivaprasad, 2011). In the study of van Emous et al. (2015) it was concluded that a lower feed intake (due to the higher dietary level) resulting in more competition and (hyper)activity like running and jumping during feeding, induced a higher risk of damaged tendons.

The layout of the house can cause different physical injuries in breeders. Standard perches can cause more breast blisters (Von Wachenfelt and Berndtson, 2014) and keel bone fractures (Gebhardt-Henrich et al., 2018) and thereby impairing their health and welfare. The combination of increased BW and less mobility of breeders while ageing is mentioned as the most important factors of these abnormalities (Von Wachenfelt and Berndtson, 2014; Gebhardt-Henrich et al., 2018). A study was carried out between 40 and 60 WOA to evaluate the preference for different elevated roosting places (slats and perches) during the laying period in broiler breeders (van Emous and Mens, in press). More and severe breast blisters were found in birds roosting on perches and drinking lines compared to the slat. The findings in this study are in line with the study of von Wachenfelt and Berndtson (2014) who recorded 90% and 30% breast blisters in 58-week-old breeders roosting on perches and slats, respectively.

Contact dermatitis is painful for broilers and breeders and is associated with health, welfare and performance problems (de Jong and Guémené, 2011). Compared to broilers, only few experiments have been carried out on contact dermatitis (footpad lesions and hock burn) in broiler breeders (Wolanski et al., 2004; Kaukonen et al., 2016; van den Over et al., 2021). Prevalence of severe foot pad dermatitis in the second phase of the laying period ranging between 25% and 83% and normally increased when ageing (Kaukonen et al., 2016). Contact dermatitis is correlated with decreasing litter quality, caused by increased amount of litter and increasing litter moisture and thereby increased ammonia release (Martland, 1985; Lien et al., 1998; Maguire et al., 2006).

6.2 Breeder age (egg size) on offspring

The first week mortality in PS pullets and broilers is affected by egg storage, season, breed, feed company, year and hatchery (Yassin et al., 2009). Besides these factors, breeder age is a very important factor on broiler health and mortality. Broiler chicks from young (<30 WOA) and old broiler breeders (>50 WOA) showed often a higher first week mortality (Wilson, 1991; Peebles et al., 2004; Pedroso et al., 2005). Yassin et al. (2009) found a first week mortality offspring in breeders at 25, between 38, 44 and 60 weeks of age of 1.82%, 1.02% and 1.20%, respectively.

In general, smaller eggs contain a larger proportion of albumen dry matter, a smaller proportion of yolk dry matter (DM), a thicker shell, smaller yolk sac and a lower chick weight (Vieira and Moran, 1998). In a study of Vieira and Moran (1999) a direct relationship between the nutrients provided by the yolk sac and the production performance of the chicks was observed. The yolk sac is a direct source of energy, however young breeders (before 30 WOA) have a reduced yolk lipid mobilization and a reduced lipoprotein transfer to mobilize the energy (Yassin et al., 2009). These impaired metabolic processes are normally associated with reduced viability during the first days in the offspring (Latour et al., 1998). Another reason for an increased mortality in offspring of young breeders is the lower BW and feed intake of offspring during the first week which causes impaired viability of the offspring (Maiorka et al., 2004). Especially nutritional interventions as higher sulphur amino acids and linoleic acid can prevent lower egg size during the initial laying period (van Emous, 2015).

First week mortality of offspring from old breeders can be caused by bad navel quality and navel-yolk sac infections which are more often found in these chicks (Yassin et al., 2009). Besides this, eggs from older breeder flocks hatch earlier with an increased risk of dehydration of the offspring (Suarez et al., 1997). Adjustment of hatchery management in case of earlier collecting of hatching eggs, or shorter hatching period, can prevent this phenomenon (Yassin et al., 2009).

6.3 Floor eggs

First-class eggs are clean, not broken, do not contain cracks and are layed in the nest boxes and not on the floor (de Reu, 2006; Khabisi et al., 2012; van den

Brand et al., 2016). In general, floor eggs are mostly dirty (Berrang et al., 1997), contain more bacteria on the eggshell (Berrang et al., 1997; de Reu et al., 2011; van den Brand et al., 2016), often broken, and/or contain more cracks than clean nest eggs (de Reu, 2006). Compared to clean nest eggs, floor eggs with more cracks and bacteria on the eggshell are more receptible for bacterial penetration before and during incubation (Smeltzer et al., 1979). The increased microbial infection rate of the yolk sac in floor eggs causes a higher embryonic and offspring mortality (Deeming et al., 2002). Especially during initial lay, the percentage of floor eggs can go up to 20% (Cooper and Appleby, 1996; Sheppard and Duncan, 2011). Moreover, hatching a relative low percentage floor eggs can result in higher antibiotic use in the offspring since all broilers must be treated when pathogens become manifest (Lourens and Steentjes, 2008). Many management factors during the rearing and laying period are identified to prevent floor eggs at the breeder farm (Aviagen, 2015b). The three key management factors to prevent breeders laying eggs on the floor are:

1 Training: to get the birds to use the nests starting already the rearing period;
2 Environment: keep the birds in the nests;
3 Egg laying: removing alternative nesting sites.

6.4 Feed restriction

Ad libitum feeding of breeder pullets results in excessive BW with negative effects on reproductive performance (Savory et al., 1993; Heck et al., 2004). Pullets fed ad libitum showed a five times higher mortality, 34% lower egg production and a 74% higher feed conversion ratio between 32 and 40 WOA in comparison to restricted fed birds (Heck et al., 2004). To prevent these negative effects, feed restriction during the pullet period is necessary (D'Eath et al., 2009). Already in the early 1990s, researchers (e.g. Hocking et al., 1993; Savory et al., 1993, 1996) described the effects of the restricted feeding programme in broiler breeders on behavioural signs of stress and frustration (de Jong and van Emous, 2017). Besides this, it has also been reported that physiological indicators of (chronic) stress were observed, such as increased plasma corticosterone levels and increased heterophil: lymphocyte ratios in the blood (Hocking et al., 1996; Savory and Mann, 1997; de Jong et al., 2003; Ekmay et al., 2010). Research suggests that adrenocorticotropin (ACTH) and glucocorticoids, which are commonly elevated in feed restricted birds, can have a negative effect on the immune system (Montiel, 2016). Higher levels of ACTH and corticosterone can have a suppressive effect on innate and induced immunity (e.g. Shini and Kaiser, 2009; Moe et al., 2010). The harmful effect of stress on the innate immune system can decrease the efficacy of pathogens control programmes in birds (Latshaw, 1991; Montiel, 2016).

In a recent study the impact of nutrient restriction of the breeder females on growth rate and immune responses in the offspring was investigated (Bowling et al., 2018). Broiler breeder hens were feed restricted from 24 WOA and maintained at three different BW; 3.4, 3.6, and 4.0 kg. Chicks were hatched and vaccinated at day 7 with IBV vaccine and at day 16, 18, and 20 they were given an immune challenge of lipopolysaccharide. It was concluded that offspring from heavy breeders had increased antibodies to the vaccination of IBV at day 35 compared with offspring of low BW hens. They postulated that this is also an indication of an improved immune response in these birds. Restricted breeders on the lowest BW showed behaviours indicative of increased stress (object pecking) and an increased heterophil: lymphocyte ratio resulted in offspring that had reduced growth rate and immune response.

6.5 Feeding programmes

Breeder pullets receive a limited amount of feed, once a day in the morning, and birds feed clean-up time is mostly less than 1 h per day (de Jong and van Emous, 2017). Therefore, optimal nutrient provision to all individual birds for optimal BW with maximizing flock BW uniformity is a challenge (Zuidhof et al., 2015). Besides qualitative feed restriction (e.g. lower energy levels), quantitative feed restriction programmes as skip-a-day, skip-another-day, etc. have been developed to improve BW uniformity in rearing (Leeson and Summers, 2005). At a skip-a-day programme, the amount of feed every second day is doubled and feed clean-up time is increased which leads to a more equal distribution of feed and a more uniform flock (de Beer and Coon, 2007). Feeding programmes as skip-a-day results in higher stress levels than everyday feeding, evidenced by higher corticosterone and insulin-like factor plasma levels (Ekmay et al., 2010). As previously mentioned, elevated levels of ACTH and corticosterone had an impaired effect on innate or induced immunity. There seems to be a possible decreasing effect of increased stress levels on the efficacy of control programmes on important pathogens for both avian and human health (Montiel, 2016). This is more pronounced when effect and success of these programmes are based on cell-mediated or innate immunity. It is postulated that transient stress situations in layers affects the transmission of SE from birds to consumption eggs, which is the main source of infection to human paratyphoid outbreaks in the past 20 years (Humphrey, 2006).

An additional advantage of daily feeding is the 10% higher feed efficiency to obtain target weight at the end of rear compared to feeding programmes (de Beer and Coon, 2007; Montiel, 2016). The reason for the difference in efficiency is that birds fed feeding programmes store nutrients (fat and protein) on feed days which must be utilized for maintenance and growth on the feed less day (de Beer and Coon, 2007; Leeson and Summers, 2009). This physiological process of deposition and re-utilization of nutrients is not 100% efficient resulting in a decreased efficiency and higher feed requirement.

6.6 Feeding system

During the rearing period feed is provided via pan feeders, chain feeders or spin feeders (spreading pellet feed in the litter) (de Jong and van Emous, 2017). In a rearing study, breeder pullets were divided into three different feeding programmes: skip-a-day, every day in feeders, every day on the litter (Montiel, 2016). Birds were challenged with intravenous inoculation of *E. coli* at 17, 24 and 35 WOA and oral administration of SE at 22, 35 and 65 WOA. The innate immune response was measured by determining the SE colonization rate in spleens and ceca, and the time of clearance of *E. coli* after intravenous inoculation. The study showed that Salmonella colonization was higher at 22 WOA in the birds fed every day in the feeders. No statistical differences in Salmonella colonization were found between the different groups at 35 WOA. At 65 WOA colonization rate was lower in the birds fed every day on the litter. They also found a lower *E. coli* load (cfu/ml of blood) in the birds fed every day on the litter measured 10 minutes after intravenous inoculation at 17 WOA, compared to the birds fed skip-a-day and every day fed in the feeders. However, no differences between treatments were observed at 20 and 60 minutes after inoculation or at any time at 24 or 35 WOA. They observed lower plasma corticosterone levels between 31 and 48 hours after feeding for the birds fed every day on the litter compared to skip-a-day fed birds. Higher corticosterone levels have been associated with impaired cellular and humoral immune responses (e.g. Shini and Kaiser, 2009; Moe et al., 2010). Montiel (2016) hypothesized that the lower corticosterone levels of the daily fed birds on the litter was caused by the increased feeding space and minimized stress derived from feed competition and more stable pecking order. It seems that feeding pullets on the litter improved immune system which can have a positive effect on health and performance.

7 Nutritional effects on health breeders

7.1 Diluted diets

Due to the severe feed restriction during the rearing period, most welfare research on breeder pullets are focusing on various feeding methods such as using alternative feed composition or alternative feeding methods. Effects of low-density broiler breeder diets during the rearing and laying period on breeder and offspring performance and mortality were studied (Enting et al., 2007). The treatments involved four groups with diluted diets and a control group with a normal density diet (Treatment 1: 2,600 kcal of Apparent Metabolizable Energy (AME)/kg during rear and 2,800 kcal of AME/kg during lay). In the other treatments nutrient densities were decreased through inclusion of palm kernel meal, wheat bran, wheat gluten feed, and sunflower seed meal by 11% (treatment 2) and 21% (treatment 3). Treatment 4 included diets with the same nutrient dilution (11%) as treatment 2 but including oat hulls and sugar beet pulp. No effect on mortality in breeders was found during the rearing and laying period, however mortality of offspring was affected. Enting et al. (2007) found that mortality of offspring from 60-week-old breeders fed the low-density diets was lower.

7.2 Crude protein/amino acids

Broiler breeder nutrition plays a major role on BW development, reproduction, egg size, shell quality and mortality in both breeders and offspring. Hocking et al. (2002) indicated that diets with a low crude protein (CP) (10%) content between 15 weeks and 18 weeks of age result in a doubling of mortality during lay. This was also found in a study of van Emous et al. (2015) who fed breeder pullets between 2 and 22 WOA, on average, 3.7% lower daily protein (and amino acids) intake. The latter authors found a tendency to an increased total mortality over the entire laying period (8.1% vs. 6.3%) when breeder pullets received a low compared to high daily protein intake. Van Emous et al. (2015) hypothesized that feeding low protein levels during the rearing period may negatively affect the immune system due to the indispensable need for certain amino acids (arginine, glutamine, and cysteine) for the development of the immune system (Kidd, 2004).

Feeding low-protein diets during the rearing period decreased embryonic mortality of the offspring between d 10 and 21 of incubation with 0.9% (4.2% vs. 5.1%) during the first phase of lay (22–45 weeks of age) (van Emous et al., 2015). This resulted in a 1.3% and 1.1% increased hatchability of set and fertile eggs, respectively. Van Emous et al. (2015) found that the abdominal fat pad content and not breast muscle content at the end of the rearing period was an important factor to affect embryonic mortality during the first phase of lay. Thus, a higher abdominal fat pad content at the end of rearing resulted in a decreased embryonic mortality. A sound explanation for this phenomenon might be that egg composition was affected by the differences in abdominal fat pad content of the body.

In a recent study, breeder pullets were fed an experimental diet with standard Lysine level and elevated (+15%) levels of specific amino acids (methionine/cystine, threonine, arginine, valine, proline, glycine and serine) (J. Lesuisse, 2020, personal communication). This experimental diet resulted in a 1.6% lower prevalence of ruptured tendons compared to the control diet during the laying period. They hypothesised that this was caused by the higher intake of the non-essential amino acids' proline and hydroxyproline who play a role in collagen content of the muscle and tendon repair (Koh et al., 2017).

Feeding breeders, a low daily protein intake between 45 and 60 WOA improved hatchability of fertilized eggs caused by a lower embryonic mortality (van Emous et al., 2015). These results agree with Pearson and Herron (1982), Whitehead et al. (1985) and Lopez and Leeson (1995). The effect of low daily protein intake on higher hatchability was also observed in a study of Pearson and Herron (1982) who found that lowering daily protein intake (21.3 vs. 27.0 g/bird) resulted in a decreased mortality and malformation of embryos. The decreased embryonic mortality in birds fed the low compared to the high daily protein intake in the study of van Emous et al. (2015) might be explained by the lower egg weight (68.7 g vs. 69.1 g). Larger eggs have a higher eggshell conductance (EC) due to an increased pore density or pore size (Shafey, 2002). A higher EC increased vital gas exchange and water loss which causes, respectively, an increased early and late embryonic mortality (Peebles et al., 1987). To improve hatchability and chick quality, it is advised to provide low protein diets during the rearing and laying phase.

7.3 Vitamins and trace minerals

Studies on the direct effects of vitamins and trace minerals in breeder nutrition on the breeders themselves are very limited. Most of the studies focussed on the effects of vitamins and trace minerals on the offspring. There is substantial evidence that confirms the negative effects of deficiencies of vitamins and trace minerals in breeder diets on chick quality and offspring performance (Kidd, 2003; Calini and Sirri, 2007; Hocking, 2007). Vitamins and trace minerals with a major impact on chick quality, chick viability and skeletal development are the fat-soluble vitamins D and E and the trace elements selenium, zinc and manganese (Chang et al., 2016).

7.3.1 Vitamins

Sufficient daily vitamin D3 intake in breeders is very important for the health of the breeders but also for optimal offspring development, viability and carcass development (Sunde et al., 1978). Incidence of tibial dyschondroplasia in offspring was lower in young (30 and 37 WOA), however not in older (45 and 57 WOA) broiler breeders receiving between 2,000 and 4,000 2,000 IU vitamin D3/kg (Atencio et al., 2005a). Moreover, feeding breeders a better biological value of vitamin D3 (25-OH-D3) resulted in lower embryonic mortality and higher bone ash in the offspring (Atencio et al., 2005b). Vitamin E can enhance immunity in the offspring by an improved adaptive antibody transfer from the breeder to the offspring (Boa-Amponsem et al., 2001). An improved immune system of the offspring was also found in a study of de Beer (2009) who supplemented vitamins B, E, and K 20% above Aviagen breeder recommendations resulting in a 2.2% lower mortality at slaughter age in the offspring.

Studies of Haq et al. (1996) and Hossain et al. (1998) clearly showed that the NDV titre of the offspring is influenced by maternal immunity by the vitamin E level in the breeder diet. Breeders were fed experimental diets containing four levels of supplementary vitamin E above the basal diet (25, 50, 75 and 100 mg/kg) between 24 and 54 WOA (Hossain et al., 1998). They found that broiler antibody titres to inactivated (oil emulsion) ND vaccine at day 14 were increased with increasing vitamin E level supplements in the diets. They concluded that higher vitamin E level in broiler breeder diet results in an increased chick's ability to improve their immune response. This was confirmed by Johnson (2013), who found that replacement of synthetic vitamin E with a natural vitamin E in breeder diets increased *ex vivo E. coli* bactericidal capacity of chicks from 58 WOA breeders, and oxidative burst in chicks from 46- to 58-week-old breeders. Adding 100 IU/kg vitamin E in breeder diets is recommended by the breeder companies to support optimal chick health and performance (Chang et al., 2016).

In a study of Urso et al. (2015) breeders were fed, in a 2 × 2 factorial design, 2 vitamin E levels (30 and 120 mg/kg) and two selenium sources (sodium selenite and zinc-L- selenomethionine). They concluded that dietary supplementation of zinc-L- selenomethionine and vitamin E (120 mg/kg feed) could be used to improve chick quality expressed in lower embryonic mortality and higher chick weight. The combination of increased vitamin E and selenium levels was also studied in an experiment of Surai (2000). Broiler breeder diets were supplemented with vitamin E (40–200 mg/kg α-tocopherol) and selenium (200 and 400 μg/kg) and they found that the liver antioxidant system in 1- and 5-day-old chicks was improved with potential benefits for chick health.

In 2007 in the Netherlands, due to a mistake by a premix producer several feed companies delivered feed without Vitamin D (cholecalciferol) resulting in drop in egg production, soft-shelled eggs, eating floor eggs, reduced hatchability (Steentjes, 2021, personal communication). In one broiler breeder farm during several weeks feed with this premix was delivered resulting in dramatically poor hatchability and DOCs were not able to open the egg with their (too soft) egg-tooth. After supplementation of Vitamin D via drinking water there was an exponential recovery of the hatchability within a few days.

7.3.2 Trace minerals

Selenium (Se) plays an important role in good semen quality and as an antioxidant defence of the embryo for optimal flock fertility and hatchability (Surai and Fisinin, 2014). The organic form of Se is identified as a potential important antioxidant and the studies showed that supplementing Se to breeders diets increased Se concentrations in the eggs and offspring (Surai, 2000; Kemp et al., 2001; Pappas et al., 2005, 2006; Jlali et al., 2013; Couloigner et al., 2015). Feeding breeders 500 μg/kg of organic Se resulted in an increased antioxidant status of eggs and a 4% higher hatchability, due to a lower embryonic mortality (Rajashree et al., 2014).

It is concluded by Kidd (2003) that higher zinc concentrations improved cellular immune function and early survival of the offspring. Supplementation of a combination of organic zinc and manganese has been observed to improve offspring liveability and some immune parameters and measures of cardiac function from day old to d 34 (Virden et al., 2003, 2004). Moreover, there is also evidence that substantially higher maternal dietary concentrations of Se, zinc and manganese than recommended may improve immune function and liveability of offspring (Hocking, 2007).

8 Future trends in research

In general, a major part of broiler breeder research concerning health is about the consequences for the offspring. This is logical because the health of a relatively small number of breeder birds affects an enormous number of offspring. Interventions (vaccinations, biosecurity, management and nutrition) affect directly the breeders and indirectly the offspring with major advantages for the total poultry meat chain. More research is necessary to prevent the direct negative effects of severe feed restriction on the breeder's immunity, health and welfare. In this field, also the offspring of breeders needs attention as they are a result of feed-restricted parents with possible detrimental effects on health and immunity of the broilers. Till now, e.g. most of the effects of nutrition on offspring performance, metabolism and immune response are unknown or, moreover, the underlying mechanisms are unclear. Therefore, future research should focus on the physiological effects of nutrition in broiler breeders during the rearing and laying period on offspring growth performance, endocrine status and lipid metabolism in the offspring.

More insight into the optimal housing and management conditions for breeders is needed to prevent *E. coli* infections. Primarily *E. coli* comes to expression under stress factors such as bird density, feed restriction, percentage of males present, aggressiveness of males, too less feed or feed space, house and climate environment like temperature, NH_3 concentration, CO_2 concentration, fine dust concentration, etc. Also, more research is necessary to understand the factors affecting rupture tendons at breeders during initial lay. Till now, no obvious causes are identified; however, it is postulated that a combination of several factors such as fast growth, feed restriction, aggressiveness of males, height of slatted floor, etc. plays a role in the prevalence of rupture tendons.

Because reduction of antibiotic use and antibiotic resistance is of big interest for public health, it is a challenge to limit treatments with antimicrobials in the poultry reproduction sector including at hatcheries. It is helpful to discover what are immunocompromising factors during the rearing and production period that cause mortality from bacterial infections by *E. coli*, *S. aureus* and Enterococcae. And still there is need to find out how in general immunity against these diseases can be improved, especially in PS chicks, during the first days of life.

There is also a lack of reliable, epidemiological data about diseases of breeder flocks around the world. Collection and comparative analysis are needed to improve the management of current disease challenges in breeders.

9 Where to look for further information

The best source of information on health in poultry and thus the most complete and definitive reference to all aspects of poultry diseases, Diseases of Poultry, the 14th edition with the editor-in chief David E. Swayne.

Centres of expertise in the Netherlands include:

- Utrecht University (UU) https://www.uu.nl/en/organisation/faculty-of-veterinary-medicine.
- Wageningen Bioveterinary Research (WBVR) https://www.wur.nl/nl/onderzoek-resultaten/onderzoeksinstituten/bioveterinary-research.htm.
- The Animal Health Service (Royal GD) https://www.gdanimalhealth.com/?sc_lang=en.
- Veterinary Centre Someren (VCS) https://www.vc-someren.nl/english/ Wageningen Livestock Research (WLR) https://www.wur.nl/en/research-results/research-institutes/livestock-research.htm.

10 References

Abdelhamid, M. K., Quijada, N. M., Dzieciol, M., Hatfaludi, T., Bilic, I., Selberherr, E., Liebhart, D., Hess, C., Hess, M. and Paudel, S. (2020). Co-infection of chicken layers with Histomonas meleagridis and avian pathogenic Escherichia coli is associated with dysbiosis, cecal colonization and translocation of the bacteria from the gut lumen, *Front. Microbiol.* 11: 586437.

Atencio, A., Edwards, H. M. Jr. and Pesti, G. M. (2005a). Effects of vitamin D3 dietary supplementation of broiler breeder hens on the performance and bone abnormalities of the offspring, *Poult. Sci.* 84(7): 1058–1068.

Atencio, A., Pesti, G. M. and Edwards, H. M. Jr. (2005b). Twenty-five hydroxycholecalciferol as a cholecalciferol substitute in broiler breeder hen diets and its effect on the performance and general health of the offspring, *Poult. Sci.* 84(8): 1277–1285.

Aviagen (2015a), *Management of Broiler Breeders to Improve Leg Health during Production, Aviagen Brief,* Available at: http://en.aviagen.com/tech-center/download/952/AviagenBriefLegHealth2015-EN.pdf.

Aviagen (2015b). *Best Practice in the Breeder House*. Preventing Floor Eggs, Aviagen, UK.

Beckstead, R. B. (2019). Histomoniasis in Poultry (Blackhead, Infectious enterohepatitis), Available at: https://www.msdvetmanual.com/poultry/histomoniasis/histomoniasis-in-poultry.

Berrang, M. E., Frank, J. F., Buhr, R. J., Bailey, J. S., Cox, N. A. and Mauldin, J. M. (1997). Microbiology of sanitized broiler hatching eggs through the egg production period, *J. Appl. Poult. Res.* 6(3): 298–305.

Boa-Amponsem, K., Price, S. E. H., Dunnington, E. A. and Siegel, P. B. (2001). Effect of route of inoculation on humoral immune response of White Leghorn chickens selected for high or low antibody response to sheep red blood cells, *Poult. Sci.* 80(8): 1073–1078.

Bowling, M., Forder, R., Hughes, R. J., Weaver, S. and Hynd, P. I. (2018). Effect of restricted feed intake in broiler breeder hens on their stress levels and thegrowth and immunology of their offspring, *Transl. Anim. Sci.* 2018: 263–271.

Calini, F. and Sirri, F. (2007). Breeder nutrition and offspring performance, *Rev. Bras. Cienc Avic* 9(2): 77–83.

Calnek, B. W., Taylor, P. J. and Sevoian, M. (1961). Studies on avian encephalomyelitis. V. development and application of an oral vaccine, *Avian Dis.* 5(3): 297–312.

Chadwick, E., Malheiros, R., Oviedo, E., Cordova Noboa, H. A., Quintana Ospina, G. A., Wisaquillo, M. C. A., Sigmon, C. and Beckstead, R. (2020). Early infection with Histomonas meleagridis has limited effects on broiler breeder hens' growth and egg production and quality, *Poult. Sci.* 99(9): 4242–4248.

Chang, A., Halley, J. and Silva, M. (2016). Can feeding the broiler breeder improve chick quality and offspring performance?, *Anim. Prod. Sci.* 56(8): 1254–1262.

Chen, S. E., McMurty, J. P. and Walzem, R. L. (2006). Overfeeding-induced ovarian dysfunction in broiler breeder hens is associated with lipotoxicity, *Poult. Sci.* 85(1): 70–81.

Collett, S. R. (2009). Managing current disease challenges in breeders. In: Hocking, P. (Ed.), *Biology of Breeding Poultry*, 414–434, Carfax Publishing Company, Abingdon, Oxfordshire.

Cooper, J. J. and Appleby, M. C. (1996). Individual variation in prelaying behaviour and the incidence of floor eggs, *Br. Poult. Sci.* 37(2): 245–253.

Couloigner, F., Rouffineau, F., Geraert, P. A. and Mercier, Y. (2015). Dietary selenium supplementation of breeders influences growth and selenium status of offspring, *Aust. Poul. Sci. Symp.* 26: 80–83.

Crespo, R. and Shivaprasad, H. L. (2011). Rupture of gastrocnemius tendon in broiler breeder hens, *Avian Dis.* 55(3): 495–498.

Darbyshire, J. H. and Peters, R. W. (1985). Humoral antibody response and assessment of protection following primary vaccination of chicks with maternally derived antibody against avian infectious bronchitis virus, *Res. Vet. Sci.* 38(1): 14–21.

Davison, F., Kaspers, B. and Schat, K. A. (2008). *Avian Immunology*, Elsevier, London.

D'Eath, R. B., Tolkamp, B. J., Kyriazakis, I. and Lawrence, A. B. (2009). 'Freedom from hunger' and preventing obesity: the animal welfare implications of reducing food quantity or quality, *Anim. Behav.* 77(2): 275–288.

de Beer, M. and Coon, C. N. (2007). The effect of different feed restriction programs on reproductive performance, efficiency, frame size, and uniformity in broiler breeder hens, *Poult. Sci.* 86(9): 1927–1939.

de Beer, M. (2009). Current approaches to feeding broiler breeders. In: *Proceedings of the 17th European Symposium of Poultry Nutrition, Edinburgh, Scotland, World's Poult. Sci. Assoc.*, Attleborough, 104–114.

de Wit, J. J. (2004). Gumboro disease: estimation of optimal time of vaccination by the Deventer formula, Available at: http://www.gumboro.com/control/Vaccination/vaccination-timing/asp.

de Wit, J. J., Nieuwenhuisen-van Wilgen, J., Hoogkamer, A., Van de Sande, H., Zuidam, G. J. and Fabri, T. H. F. (2011). Induction of cystic oviducts and protection against early challenge with infectious bronchitis virus serotype D388 (genotype QX) by maternally derived antibodies and by early vaccination, *Avian Pathol.* 40(5): 463–471.

Deeming, D. C., Clyburn, V., Williams, K. and Dixon, R. A. (2002). In ovo microbial contamination of the yolk sac of unhatched broiler and pheasant embryos, *Avian Poult. Biol. Rev.* 13: 240–241.

de Jong, I. C., van Voorst, A. S. and Blokhuis, H. J. (2003). Parameters for quantification of hunger in broiler breeders, *Physiol. Behav.* 78(4–5): 773–783.

de Jong, I. C. and Guémené, D. (2011). Major welfare issues in broiler breeders, *Worlds Poult. Sci. J.* 67(1): 73–82.

de Jong, I. C. and van Emous, R. A. (2017). Broiler breeding flocks: management and animal welfare. In: Applegate, T. (Ed.), *Achieving Sustainable Production of Poultry Meat*, 211–230, Burleigh Dodds Science Publishing Limited, Sawston, Cambridge.

de Reu, K. (2006). *Bacteriological Contamination and Infection of Shell Eggs in the Production Chain*, PhD thesis, University of Ghent, Belgium.

de Reu, K., Grijspeerdt, K., Heyndrickx, M., Uyttendaele, M., Debevere, J. and Herman, L. (2011). Bacterial shell contamination in the egg collection chains of different housing systems for laying hens, *Br. Poult. Sci.* 47(2): 163–172.

Dierikx, C. M., van der Goot, J. A., Smith, H. E., Kant, A. and Mevius, D. J. (2013). Presence of ESBL/AmpC-producing Escherichia coli in the broiler production pyramid: a descriptive study, *PLoS ONE* 8(11): e79005/ https://doi.org/10.1371/journal.pone.0079005.

Duff, S. R. and Randall, C. J. (1986). Tendon lesions in broiler fowls, *Res. Vet. Sci.* 40(3): 333–338.

Ekmay, R. D., de Beer, M., Rosebrough, R. W., Richards, M. P., McMurtry, J. P. and Coon, C. N. (2010). The role of feeding regimens in regulating metabolism of sexually mature broiler breeders, *Poult. Sci.* 89(6): 1171–1181.

Engström, B. E. (1999). Prevalence of antibody to chicken anaemia virus (CAV) in Swedish chicken breeding flocks correlated to outbreaks of blue wing disease (BWD) in their offspring, *Acta Vet. Scand.* 40(2): 97–107.

Enting, H., Boersma, W. J. A., Cornelissen, J. B., van Winden, S. C. L., Verstegen, M. W. A. and van der Aar, P. J. (2007). The effect of low-density broiler breeder diets on performance and immune status of their offspring, *Poult. Sci.* 86(2): 282–290.

Feberwee, A., de Vries, T. S., Elbers, A. R. W. and de Jong, W. A. (2000). Results of a *Salmonella enteritidis* vaccination field trial in broiler-breeder flocks in the Netherlands, *Avian Dis.* 44(2): 249–255.

Ferguson-Noel, N., Armour, N. K., Noormohammadi, A. H., El-Gazzar, M. and Bradbury, J. M. (2020). Mycoplasmosis. In: Swayne, D. E., Boulianne, M., Logue, C. M., McDougald, L. R., Nair, V., Suarez, D. L., de Wit, S., Grimes, T., Johnson, D., Kromm, M., Prajitno, T. Y., Rubinoff, I. and Zavala, G. (Eds), *Diseases of Poultry* (14th edn.), 907–965, Wiley-Blackwell, Hoboken, NJ.

Fitzgerald, S. D., Rautenschlein, S., Mahsoub, H. M., Pierson, F. W., Reed, W. M. and Jack, S. W. (2020). Adenovirus infections. In: Swayne, D. E., Boulianne, M., Logue, C. M., McDougald, L. R., Nair, V., Suarez, D. L., de Wit, S., Grimes, T., Johnson, D., Kromm, M., Prajitno, T. Y., Rubinoff, I. and Zavala, G. (Eds), *Diseases of Poultry* (14th edn.), 321–363, Wiley-Blackwell, Hoboken, NJ.

Fragoso, J. S., Moreno, J. C. A., Infiesta, P. C., Benevides, W. S., Lucio, E. T. and Pizarro, M. (2021). Principal lesions and patterns of mortality observed in a broiler breeder flock during the laying period, Braz, *J. Poult. Sci.* 23(2): 1–6.

Gast, R. K. and Porter, R. E. Jr. (2020). Salmonella infections. In: Swayne, D. E., Boulianne, M., Logue, C. M., McDougald, L. R., Nair, V., Suarez, D. L., de Wit, S., Grimes, T., Johnson, D., Kromm, M., et al. (Eds), *Diseases of Poultry* (14th edn.), 717–753, Wiley-Blackwell, Hoboken, NJ.

GD (2018). *Poultry Monitoring*, Poultry (In Dutch), General Dynamics, Deventer, the Netherlands.

Gebhardt-Henrich, S. G., Toscano, M. J. and Wurbel, H. (2018). Use of aerial perches and perches on aviary tiers by broiler breeders, *Appl. Anim. Behav. Sci.* 203: 24–33.

Haq, A. U., Bailey, C. A. and Chinnah, A. (1996). Effect of β-carotene, canthaxanthin, lutein, and vitamin E on neonatal immunity of chicks when supplemented in the broiler breeder diets, *Poult. Sci.* 75(9): 1092–1097.

Havenstein, G. B., Ferket, P. R. and Qureshi, M. A. (2003a). Growth, livability, and feed conversion of 1957 versus 2001 broilers when fed representative 1957 and 2001 broiler diets, *Poult. Sci.* 82(10): 1500–1508.

Havenstein, G. B., Ferket, P. R. and Qureshi, M. A. (2003b). Carcass composition and yield of 1957 versus 2001 broilers when fed representative 1957 and 2001 broiler diets, *Poult. Sci.* 82(10): 1509–1518.

Heck, A., Onagbesan, O., Tona, K., Metayer, S., Puterflam, J., Jego, Y., Trevidy, J. J., Decuypere, E., Williams, J., Picard, M. and Bruggeman, V. (2004). Effects of ad libitum feeding on performance of different strains of broiler breeders, *Br. Poult. Sci.* 45(5): 695–703.

Hocking, P. M., Maxwell, M. H. and Mitchell, M. A. (1993). Welfare assessment of broiler breeder and layer females subjected to food restriction and limited access to water during rearing, *Br. Poult. Sci.* 34(3): 443–458.

Hocking, P. M., Maxwell, M. H. and Mitchell, M. A. (1996). Relationships between the degree of food restriction and welfare indices in broiler breeder females, *Br. Poult. Sci.* 37(2): 263–278.

Hocking, P. M., Bernard, R. and Robertson, G. W. (2002). Effects of low dietary protein and different allocations of food during rearing and restricted feeding after peak rate of

lay on egg production, fertility and hatchability in female broiler breeders, *Br. Poult. Sci.* 43(1): 94–103.
Hocking, P. M. (2007). Optimum feed composition of broiler breeder diets to maximise offspring performance. In: *Proceedings of the 16th European Symposium of Poultry Nutrition, World Poultry Science Association*, Strasbourg, 101–108.
Hossain, S. M., Barreto, S. L., Bertechini, A. G., Rios, A. M. and Silva, C. G. (1998). Influence of dietary vitamin E level on egg production of broiler breeders, and on the growth and immune response of offspring in comparison with the offspring from eggs injected with vitamin E, *Anim. Feed Sci. Technol.* 73(3–4): 307–317.
Humphrey, T. (2006). Are happy chickens safer chickens? Poultry welfare and disease susceptibility, *Br. Poult. Sci.* 47(4): 379–391.
Jackwood, M. and de Wit, J. J. (2020). Infectious bronchitis. In: Swayne, D. E., Boulianne, M., Logue, C. M., McDougald, L. R., Nair, V., Suarez, D. L., de Wit, S., Grimes, T., Johnson, D., Kromm, M., Prajitno, T. Y., Rubinoff, I. and Zavala, G. (Eds), *Diseases of Poultry* (14th edn.), 167–188, Wiley-Blackwell, Hoboken, NJ.
Jlali, M., Briens, M., Rouffineau, F., Mercerand, F., Geraert, P. A. and Mercier, Y. (2013). Effect of 2-hydroxy-4-methylselenobutanoic acid as a dietary selenium supplement to improve the selenium concentration of table eggs, *J. Anim. Sci.* 91(4): 1745–1752.
Johnson, M. L. (2013). *The Effect of Maternal Antioxidant Nutrient Supplementation and Age on chick Post-Hatch Innate Immune Function*, PhD Thesis, University of Alberta, Canada.
Jones, K. H. (2009). *Health Management of the Modern Broiler Breeder Male*, Avaigen Brief, June 2009, Aviagen, USA.
Kaukonen, E., Norring, M. and Valros, A. (2016). Effect of litter quality on foot pad dermatitis, hock burns and breast blisters in broiler breeders during the production period, *Avian Pathol.* 45(6): 667–673.
Kemp, C., Wylie, L. and Fisher, C. (2001). Broiler breeder nutrition, nutrient transfer and broiler performance. In: *Blankenberghe 13th Eur. Symposium on Poultry Nutrition*, World Poultry Science Association, 61–67.
Khabisi, M. M., Salahi, A. and Mousavi, S. N. (2012). The influence of egg shell crack types on hatchability and chick quality, *Turk. J. Vet. Anim. Sci.* 36: 289–295.
Kidd, M. T. (2003). A treatise on chicken dam nutrition that impacts on offspring, *Worlds Poult. Sci. J.* 59(4): 475–494.
Kidd, M. T. (2004). Nutritional modulation of immune function in broilers, *Poult. Sci.* 83(4): 650–657.
Koh, I. H., Kang, H. J., Oh, W. T., Hong, J. J. and Choi, Y. R. (2017). Correlation between change in muscle excursion and collagen content after tendon rupture and delayed repair, *J. Orthop. Surg. Res.* 12(1): 15.
Landman, W. J. M., van den Boogaard, A. E. J. M., Doornenbal, P., Tooten, P. C. J., Elbers, A. R. W. and Gruys, E. (1998). The role of various agents in chicken AA, *Amyloid* 5(4): 266–278.
Latshaw, J. D. (1991). Nutrition - mechanisms of immunosuppression, *Vet. Immunol. Immunopathol.* 30(1): 111–120.
Latour, M. A., Peebles, E. D., Doyle, S. M., Pansky, T., Smith, T. W. and Boyle, C. R. (1998). Broiler breeder age and dietary fat influence the yolk fatty acid profiles of fresh egg and newly hatched chicks, *Poult. Sci.* 77(1): 47–53.
Leeson, S. and Summers, J. D. (2005). *Commercial Poultry Nutrition* (3rd edn.), University books, Guelph, Ontario, Canada.

Leeson, S. and Summers, J. D. (2009). *Broiler Breeder Production*, University Books, Guelph, Canada.

Liebhart, D. and Hess, M. (2020). Spotlight on histomonosis (blackhead disease): a re-emerging disease in turkeys and chickens, *Avian Pathol.* 49(1): 1–4.

Lien, R. J., Hess, J. B., Conner, D. E., Wood, C. W. and Shelby, R. A. (1998). Peanut hulls as a litter source for broiler breeder replacement pullets, *Poult. Sci.* 77(1): 41–46.

Lopez, G. and Leeson, S. (1995). Response of broiler breeders to low-protein diets. 1. Adult breeder performance, *Poult. Sci.* 74(4): 685–695.

Lourens, A. and Steentjes, A. (2008). *Zootechnical and Veterinary Factors at the Broiler Breeder Farm That Affect Broiler Mortality Are Described* (in Dutch), Report 194, Animal Sciences Group of Wageningen university and research center, Lelystad, The Netherlands.

Maguire, R. O., Plumstead, P. W. and Brake, J. (2006). Impact of diet, moisture, location, and storage on soluble phosphorus in broiler breeder manure, *J. Environ. Qual.* 35(3): 858–865.

Maiorka, A., Fischer da Silva, A. V. and Santin, E. (2004). Broiler breeder age and dietery energy level on performance and pancreas lipase and trypsin activities of 7-days old chicks, *Int. J. Poult. Sci.* 3(3): 234–237.

Martland, M. F. (1985). Ulcerative dermatitis in broiler chickens: the effects of wet litter, *Avian Pathol.* 14(3): 353–364.

McDougald, L. R. (2020). Internal parasites. In: Swayne, D. E., Boulianne, M., Logue, C. M., McDougald, L. R., Nair, V., Suarez, D. L., de Wit, S., Grimes, T., Johnson, D., Kromm, M., Prajitno, T. Y., Rubinoff, I. and Zavala, G. (Eds), *Diseases of Poultry* (14th edn.), 1157–1191, Wiley-Blackwell, Hoboken, NJ.

Moe, R. O., Guémené, D., Bakken, M., Larsen, H. J., Shini, S., Lervik, S., Skjerve, E., Michel, V. and Tauson, R. (2010). Effects of housing conditions during the rearing and laying period on adrenal reactivity, immune response and heterophil to lymphocyte (H/L) ratios in laying hens, *Animal* 4(10): 1709–1715.

Montiel, E. R. (2016). *Influence of Feeding Programs on Innate and Adaptive Immunity in Broiler Breeders*, PhD Diss, The Unitversity of Georgia, Athens.

Nolan, L. K., Vaillancourt, J. P., Barbieri, N. L. and Logue, C. M. (2020). Colibacillosis. In: Swayne, D. E., Boulianne, M., Logue, C. M., McDougald, L. R., Nair, V., Suarez, D. L., de Wit, S., Grimes, T., Johnson, D., Kromm, M., et al. (Eds), *Diseases of Poultry* (14th edn.), 770–830, Wiley-Blackwell, Hoboken, NJ.

Ovelgönne, J. H., Landman, W. J. M., Gruys, E., Gielkens, A. L. J. and Peeters, B. P. H. (2001). Identical amyloid precursor proteins in two breeds of chickens which differ in susceptibility to develop AA, *Amyloid* 8(1): 41–51.

Pappas, A. C., Karadas, F., Surai, P. F. and Speake, B. K. (2005). The selenium intake of the female chicken influences the selenium status of her offspring, *Comp. Biochem. Physiol., B Biochem. Mol. Biol.* 142(4): 465–474.

Pappas, A. C., Acamovic, T., Surai, P. F. and McDevitt, R. M. (2006). Maternal organo-selenium compounds and polyunsaturated fatty acids affect offspring performance and levels of selenium and docosahexaenoic acid in the chick tissues, *Poult. Sci.* 85(9): 1610–1620.

Pearson, R. A. and Herron, K. M. (1982). Effects of maternal energy and protein intakes on the incidence of malformations and malpositions of the embryo and time of death during incubation, *Br. Poult. Sci.* 23(1): 71–77.

Pedroso, A. A., Andrade, M. A., Cafe, M. B., Leandro, N. S. M., Menten, J. F. M. and Stringhini, J. H. (2005). Fertility and hatchability of eggs laid in the pullet-to-breeder transition period and in the initial production period, *Anim. Reprod. Sci.* 90(3-4): 355-364.

Peebles, E. D., Brake, J. D. and Gildersleeve, R. P. (1987). Effects of cuticle removal and incubation humidity on embryonic development and hatchability of broilers, *Poult. Sci.* 66(5): 834-840.

Peebles, E. D., Keirs, R. W., Bennett, L. W., Cummings, T. S., Whitmarsh, S. K. and Gerard, P. D. (2004). Relationship among posthatch physiological parameters in broiler chicks hatched from young breeder hens and subjected to delayed brooding placement, *Int. J. Poult. Sci.* 3(9): 578-585.

Pitcovski, J. and Goyal, S. M. (2020). Avian reovirus infections. In: Swayne, D. E., Boulianne, M., Logue, C. M., McDougald, L. R., Nair, V., Suarez, D. L., de Wit, S., Grimes, T., Johnson, D., Kromm, M., et al. (Eds), *Diseases of Poultry* (14th edn.), 382-400, Wiley-Blackwell, Hoboken, NJ.

Rajashree, K., Muthukumar, T. and Karthikeyan, N. (2014). Comparative study of the effects of organic selenium on hen performance and productivity of broiler breeders, *Br. Poult. Sci.* 55(3): 367-374.

Renema, R. A., Rustad, M. E. and Robinson, F. E. (2007). Implications of changes to commercial broiler and broiler breeder body weight targets over the past 30 years, *Worlds Poult. Sci. J.* 63(3): 457-472.

Roza, K., Martin, M. and Barnes, H. J. (2006). Litter impaction of the lower intestinal tract in male broiler breeders, *Avian Dis.* 50(3): 460-462.

Saif, Y. M., Marlier, D., Smyth, V. J., Noormohammadi, A. H., Guy, J. S., Suarez, D. L., Meng, X. J. and Shivaprasad, H. L. (2020). Other viral infections. In: Swayne, D. E., Boulianne, M., Logue, C. M., McDougald, L. R., Nair, V., Suarez, D. L., de Wit, S., Grimes, T., Johnson, D., Kromm, M., Prajitno, T. Y., Rubinoff, I. and Zavala, G. (Eds), *Diseases of Poultry* (14th edn.), 498-547, Wiley-Blackwell, Hoboken, NJ.

Savory, C. J., Maros, K. and Rutter, S. M. (1993). Assessment of hunger in growing broiler breeders in relation to a commercial restricted feeding programme, *Anim. Welf.* 2: 131-152.

Savory, C. J., Hocking, P. M., Mann, J. S. and Maxwell, M. H. (1996). Is broiler breeder welfare improved by using qualitative rather than quantitative food restriction to limit growth rate?, *Anim. Welf.* 5: 105-127.

Savory, C. J. and Mann, J. S. (1997). Is there a role for corticosterone in expression of abnormal behaviour in restricted-fed fowls?, *Physiol. Behav.* 62(1): 7-13.

Schat, K. A. and van Santen, V. L. (2020). Chicken infectious anemia and circovirus infections in commercial flocks. In: Swayne, D. E., Boulianne, M., Logue, C. M., McDougald, L. R., Nair, V., Suarez, D. L., de Wit, S., Grimes, T., Johnson, D., Kromm, M., Prajitno, T. Y., Rubinoff, I. and Zavala, G. (Eds), *Diseases of Poultry* (14th edn.), 284-320, Wiley-Blackwell, Hoboken, NJ.

Shafey, T. M. (2002). Effects of egg size and eggshell conductance on hatchability traits of meat and layer breeder flocks, *Asian. Australas J. Anim. Sci.* 15(1): 1-6.

Sheppard, K. C. and Duncan, I. J. H. (2011). Feeding motivation on the incidence of floor eggs and extraneously calcified eggs laid by broiler breeder hens, *Br. Poult. Sci.* 52(1): 20-29.

Shini, S. and Kaiser, P. (2009). Effects of stress, mimicked by administration of corticosterone in drinking water, on the expression of chicken cytokine and chemokine genes in lymphocytes, *Stress* 12(5): 388–399.

Smeltzer, T. I., Orange, K., Peel, B. and Runge, G. (1979). Bacterial penetration in floor and nest box eggs from mean and layer birds, *Aust. Vet. J.* 55(12): 592–593.

Steentjes, A. H. A., Veldman, K. T., Mevius, D. J. and Landman, W. J. M. (2002). Molecular epidemiology of unilateral AA in broiler breeders associated with Enterococcus faecalis, *Avian Pathol.* 31(1): 31–39.

Suarez, M. E., Wilson, H. R., Mather, F. B., Wilcox, C. J. and Mcpherson, B. N. (1997). Effects of strain and age of the broiler breeder female on incubation time and chick weight, *Poult. Sci.* 76(7): 1029–1036.

Sunde, M., Turk, C. M. and DeLuca, H. F. (1978). The essentiality of vitamin D metabolites for chick development, *Science*. 200: 1067–1069.

Surai, P. F. (2000). Effect of selenium and vitamin E content of the maternal diet on the antioxidant system of the yolk and the developing chick, *Br. Poult. Sci.* 41(2): 235–243.

Surai, P. F. and Fisinin, V. I. (2014). Selenium in poultry breeder nutrition: an update, *Anim. Feed Sci. Technol.* 191: 1–15.

ter Veen, C., Dijkman, R., de Wit, J. J., Gyuranecz, M. and Feberwee, A. (2021). Decrease of Mycoplasma gallisepticum seroprevalence and introduction of new genotypes in Dutch commercial poultry during the years 2001–2018, *Avian Pathol.* 50(1): 52–60.

Urso, U. R. A., Dahlke, F., Mairoka, A., Bueno, I. J. M., Schneider, A. F., Surek, D. and Rocha, C. (2015). Vitamin E and selenium in broiler breeder diets: effect on live performance, hatching process, and chick quality, *Poult. Sci.* 94(5): 976–983.

USDA (2021). Livestock and poultry: world markets and trade, United States Department of Agriculture Foreign Agricultural Service. Available at: https://apps.fas.usda.gov/psdonline/circulars/livestock_poultry.pdf

van den Brand, H., Sosef, M. P., Lourens, A. and van Harn, J. (2016). Effects of floor eggs on hatchability and later life performance in broiler chickens, *Poult. Sci.* 95(5): 1025–1032.

van den Oever, A. C. M., Candelotto, L., Kemp, B., Rodenburg, T. B., Bolhuis, J. E., Graat, E. A. M., van de Ven, L. J. F., Guggisberg, D. and Toscano, M. J. (2021). Influence of a raised slatted area in front of the nest on leg health, mating behaviour and floor eggs in broiler breeders, *Animal* 15(2): 100109.

van der Heide, L., Kalbac, M. and Hall, W. C. (1976). Infectious tenosynovitis (viral arthritis): influence of maternal antibodies on the development of tenosynovitis lesions after experimental infection of day old chickens with tenosynovitis virus, *Avian Dis.* 20(4): 641–648.

van Emous, R. A. (2010). Innovative housing in broiler breeder production. In: *Proceedings of the 2nd International Symposium: Highlights in Nutrition and Welfare in Poultry Production, Wageningen, Therapeutic Netherlands*, 37–44, Animal Science Group, Lelystad, The Netherlands.

van Emous, R. A. (2015). *Body Composition and Reproduction in Broiler Breeders: Impact of Feeding Strategies*, PhD Diss.,Wageningen University, Wageningen, the Netherlands.

van Emous, R. A., Kwakkel, R. P., van Krimpen, M. M. and Hendriks, W. H. (2015). Effects of dietary protein levels during rearing and dietary energy levels during lay on body composition and reproduction in broiler breeder females, *Poult. Sci.* 94(5): 1030–1042.

van Veen, L., Dwars, M. and Fabri, T. (1998). Ataxia and paralysis in broilers, as a result of an infection with Vibrational Disease virus on the breeding farm (in Dutch), *Tijdschr. Diergeneeskd.* 123(11): 344–348.

Verrette, L., Fairbrother, J. M. and Boulianne, M. (2019). Effect of cessation of ceftiofur and substitution with lincomycin-spectinomycin on extended-spectrum-beta-lactamase/AmpC genes and multidrug resistance in Escherichia coli from a Canadian broiler production pyramid, *Appl. Environ. Microbiol.* 85(13): e00037-19. https://doi.org/10.1128/AEM.00037-19.

Vieira, S. L. and Moran, E. T. Jr. (1998). Eggs and chicks from broiler breeders of extremely different ages, *J. Appl. Poult. Res.* 7(4): 372–376.

Vieira, S. L. and Moran, E. T. Jr. (1999). Effects of egg of origin and chick post-hatch nutrition on broiler live performance and meat yields, *Worlds Poult. Sci. J.* 55(2): 125–142.

Virden, W. S., Yeatman, J. B., Barber, S. J., Zumwalt, C. D., Ward, T. L., Johnson, A. B. and Kidd, M. T. (2003). Hen mineral nutrition impacts offspring livability, *J. Appl. Poult. Res.* 12(4): 411–416.

Virden, W. S., Yeatman, J. B., Barber, S. J., Willeford, K. O., Ward, T. L., Fakle, T. M., Wideman, R. F. Jr. and Kidd, M. T. (2004). Immune system and cardiac functions of offspring chicks from dams fed diets differing in zinc and manganese level and source, *Poult. Sci.* 83(3): 344–351.

Von Wachenfelt, E. and Berndtson, E. (2014). Usage of perches in meat fowls, Sveriges Lantbruksuniversitet Fakulteten för landskapsarkitektur, trädgårds- och äxtproduktionsvetenskap, Report 2014, 3.

Whitehead, C. C., Pearson, R. A. and Herron, K. M. (1985). Biotin requirements of broiler breeders fed diets of different protein content and effect of insufficient biotin on the viability of offspring, *Br. Poult. Sci.* 26(1): 73–82.

Wilson, H. R. (1991). Interrelationships of egg size, chick size, post hatching growth and hatchability, *Worlds Poult. Sci. J.* 47(1): 5–20.

Wolanski, N. J., Renema, R. A., Robinson, F. E. and Wilson, J. L. (2004). End-of-season carcass and reproductive traits in original and replacement male broiler breeders, *J. Appl. Poult. Res.* 13(3): 451–460.

Yassin, H., Velthuis, A. G. J., Boerjan, M. and van Riel, J. W. (2009). Field study on broilers first-week mortality, *Poult. Sci.* 88(4): 798–804.

Zavala, G. and Barnes, H. J. (2012). Broiler breeder diseases - a review, Available at: https://en.engormix.com/poultry-industry/articles/broiler-breeder-diseases-t35368.htm.

Zuidhof, M. J., Schneider, B. L., Carney, V. L., Korver, D. R. and Robinson, F. E. (2014). Growth, efficiency, and yield of commercial broilers from 1957, 1978, and 2005, *Poult. Sci.* 93(12): 2970–2982.

Zuidhof, M. J., Holm, D. E., Renema, R. A., Jalal, M. A. and Robinson, F. E. (2015). Effects of broiler breeder management on pullet body weight and carcass uniformity, *Poult. Sci.* 94(6): 1389–1397.

Index